SCHAUM'S OUTLINE OF

THEORY AND PROBLEMS

OF

FEEDBACK and CONTROL SYSTEMS

SI (METRIC) EDITION

•

BY

JOSEPH J. DiSTEFANO, III, Ph.D.

Department of Engineering
University of California, Los Angeles

ALLEN R. STUBBERUD, Ph.D.

Department of Engineering
University of California, Los Angeles

IVAN J. WILLIAMS, Ph.D.

Nortronics Division, Northrop Corp.

SCHAUM'S OUTLINE SERIES
McGraw-Hill Book Company

New York · St Louis · San Francisco · Auckland
Bogotá · Guatemala · Hamburg · Lisbon · London
Madrid · Mexico · Montreal · New Delhi · Panama · Paris
San Juan · São Paulo · Singapore · Sydney · Tokyo · Toronto

Preface

Feedback, one of the most fundamental processes existing in nature, is present in almost all dynamic systems, including those within man, among men, and between men and machines. However, feedback concepts have been utilized almost exclusively by engineers. As a result, the theory of feedback control systems has been developed as an engineering discipline for analyzing and designing practical control systems and other technological devices. Recognition that this theory is directly applicable to formulating and solving problems in many other fields is becoming widespread, but its use has been limited because of its heavy orientation toward technological applications.

The objective of this book is to present a comprehensive, concise and modern treatment of the fundamentals of feedback and linear control systems theory. It is intended not only for engineers, but also for physical, biological and behavioral scientists, economists, mathematicians and students of these disciplines. Knowledge of basic physics and calculus are the only requirements for understanding the material. All of the mathematical tools required beyond calculus, including topics in differential equations, Laplace transforms and complex variables, are developed within the book as they are needed. Special control systems nomenclature is presented in a rational manner and each new term is defined before it is used.

The book is designed for use as a textbook for a formal course in the fundamentals of feedback and control systems or as a supplement to all modern standard texts. It should also be of considerable value as a reference book.

The material is presented in a form that is applicable to any kind of control system and the examples generally provide specific illustrations of the theory. Each new topic is introduced either by section or by chapter. At the end of each chapter, there are solved problems consisting of mathematical proofs, theoretical and practical extensions of the theory, and many applications from various fields illustrating the ideas presented in the chapter.

The book is divided into two parts. The first half, chapters 1 through 9, considers the conceptual framework and terminology of feedback and control systems, modern techniques for formulating and solving linear constant coefficient ordinary differential equations, the mathematical basis of linear systems, the Laplace transform, methods for determining the stability of linear systems, transfer functions, block diagrams and signal flow graphs, error constants, sensitivity analysis, and the classification of control systems. Chapter 10 introduces the second half of the book with a discussion of the objectives and the methods of the analysis and design of linear feedback control systems. Then, in chapters 11 through 18, the four classical analysis and design techniques, i.e., the Nyquist, Bode, Root-locus, and Nichols Chart methods, are individually treated, with each analysis and each design method considered in a separate chapter. The final chapter of this book is an introduction to advanced topics in control systems theory.

The authors wish to express their sincere appreciation to all who helped in the preparation of the manuscript and especially to the staff of the Schaum Publishing Company for their unfailing cooperation.

JOSEPH J. DiSTEFANO, III
ALLEN R. STUBBERUD
IVAN J. WILLIAMS

Los Angeles, Calif.
May, 1967

CONTENTS

CONTENTS

Contents

Chapter 1

Introduction

1.1 CONTROL SYSTEMS: WHAT THEY ARE

In modern usage the meaning of the word *system* has become nebulous. So let us begin by defining it, first abstractly then slightly more specifically in relation to scientific literature.

Definition 1.1a: A system is an arrangement, set, or collection of things connected or related in such a manner as to form an entirety or whole.

Definition 1.1b: A system is an arrangement of physical components connected or related in such a manner as to form and/or act as an entire unit.

The word **control** is usually taken to mean *regulate, direct,* or *command*. Combining the above definitions, we have

Definition 1.2: A control system is an arrangement of physical components connected or related in such a manner as to command, direct, or regulate itself or another system.

In the most abstract sense it is possible to consider every physical object a control system. Everything alters its environment in some manner, if not actively then passively — like a mirror *directing* a beam of light shining on it at some acute angle. The mirror, Fig. 1-1, may be considered an elementary control system, controlling the beam of light according to the simple equation "the angle of reflection α equals the angle of incidence α".

In engineering and science we usually restrict the meaning of control systems to apply to those systems whose major function is to *dynamically* or *actively* command, direct, or regulate. The system shown in Fig. 1-2, consisting of a mirror pivoted at one end and adjusted up and down with a screw at the other end, is properly termed a *control system*. The angle of reflected light is regulated by means of the screw.

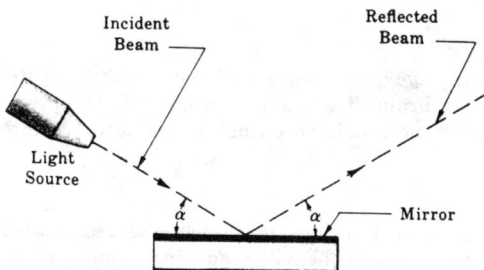

Fig. 1-1

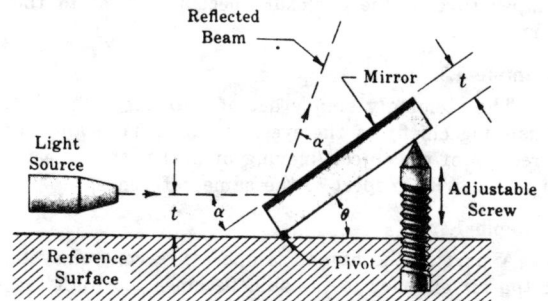

Fig. 1-2

1.2 EXAMPLES OF CONTROL SYSTEMS

Control systems abound in man's environment. But before exemplifying this, we define two terms: *input* and *output*, which help in identifying or defining the control system.

Definition 1.3: The **input** is the stimulus or excitation applied *to* a control system from an external energy source, usually in order to produce a specified response *from* the control system.

Definition 1.4: The **output** is the actual response obtained from a control system. It may or may not be equal to the specified response implied by the input.

The purpose of the control system usually identifies or defines the output and input. If the output and input are given, it is possible to identify or define the nature of the system's components.

Control systems may have more than one input or output. Often all inputs and outputs are well defined by the system description. But sometimes they are not. For example, an atmospheric electrical storm may intermittently interfere with radio reception, producing an unwanted output from a loudspeaker in the form of static. This "noise" output is not usually specified for the simple identification of a radio receiving system, but is part of the total output as defined above. For the purpose of simply identifying a system, spurious inputs producing undesirable outputs are not normally considered as inputs and outputs in the system description. But it is usually necessary to carefully consider these extra inputs and outputs when the system is examined in detail.

There are three basic types of control systems:

1. Man-made control systems
2. Natural, including biological control systems
3. Control systems whose components are both man-made and natural.

Example 1.1.

An *electric switch* is a man-made control system, controlling the flow of electricity. By definition, the apparatus or person flipping the switch is not a part of this control system.

Flipping the switch on or off may be considered as the input. That is, the input can be in one of two states — on or off. The output is the flow or nonflow (two states) of electricity.

The electric switch is probably one of the most rudimentary control systems.

Example 1.2.

A *thermostatically controlled heater or furnace automatically regulating the temperature of a room or enclosure* is a control system. The input to this system is a. reference temperature, which is usually specified by appropriately setting a thermostat. The output is the actual temperature of the enclosure.

When the thermostat detects that the output is less than the input, the furnace provides heat until the temperature of the enclosure becomes equal to the reference input. Then the furnace is automatically turned off.

Example 1.3.

The seemingly simple act of *pointing at an object with a finger* requires a biological control system consisting chiefly of the eyes, the arm, hand and finger, and the brain of a man. The input is the precise direction of the object (moving or not) with respect to some reference, and the output is the actual pointed direction with respect to the same reference.

Example 1.4.

A part of the human temperature control system is the *perspiration system.* When the temperature of the air exterior to the skin becomes too high the sweat glands secrete heavily, inducing cooling of the skin by evaporation. Secretions are reduced when the desired cooling effect is achieved, or when the air temperature falls sufficiently.

The input to this system is "normal" or comfortable skin temperature. The output is the actual skin temperature.

Example 1.5.

The control system consisting of *a man driving an automobile* has components which are clearly both man-made and biological. The driver wants to keep the automobile in the appropriate lane of the roadway. He accomplishes this by constantly watching the direction of the automobile with respect to the direction of the road. In this case, the direction or heading of the road, represented by the painted guide line or lines on either side of his lane may be considered as the input. The heading of the automobile is the output of the system. The driver controls this output by constantly measuring it with his eyes and brain, and correcting it with his hands on the steering wheel. The major components of this control system are the driver's hands, eyes and brain, and the vehicle.

1.3 CLASSIFICATION OF CONTROL SYSTEMS

Control systems are classified into two general categories: *open-loop* and *closed-loop* systems. The distinction is determined by the **control action**, which is that quantity responsible for activating the system to produce the output.

Definition 1.5: An **open-loop** control system is one in which the control action is independent of the output.

Definition 1.6: A **closed-loop** control system is one in which the control action is somehow dependent on the output.

Two outstanding features of open-loop control systems are:

1. Their ability to perform accurately is determined by their calibration. To **calibrate** means to establish or reestablish the input-output relation to obtain a desired system accuracy.

2. They are not generally troubled with problems of *instability*, a concept to be subsequently discussed in detail.

Closed-loop control systems are more commonly called *feedback* control systems, and are considered in more detail beginning in the next section.

In order to classify a control system as open-loop or closed-loop, the components of the system must be clearly distinguished from components that interact with, but are not part of the system. For example, a human operator may or may not be a component of a system.

Example 1.6.

An *automatic toaster* is an open-loop control system because it is controlled by a timer. The time required to make "good toast" must be estimated by the user, who is not a part of the system. Control over the quality of toast (the output) is removed once the time, which is both the input and the control action, has been set.

Example 1.7.

An *autopilot mechanism and the airplane it controls* is a closed-loop (feedback) control system. Its purpose is to maintain a specified airplane heading, despite atmospheric changes. . It performs this task by continuously measuring the actual airplane heading, and automatically adjusting the airplane control surfaces (rudder, flaps, etc.) so as to bring the actual airplane heading into correspondence with the specified heading. The human pilot or operator who presets the autopilot is not part of the control system.

1.4 FEEDBACK

Feedback is that characteristic of closed-loop control systems which distinguishes them from open-loop systems. The study of feedback control systems is the major objective of this book.

Definition 1.7: **Feedback** is that property of a closed-loop system which permits the output (or some other controlled variable of the system) to be compared with the input to the system (or an input to some other internally situated component or subsystem of the system) so that the appropriate control action may be formed as some function of the output and input.

More generally, feedback is said to exist in a system when a closed sequence of cause-and-effect relations exists between system variables.

In essence, every passive system (one containing no energy sources) may be viewed as a feedback system. We shall consider only those closed-loop control systems where the existence and purpose of feedback are readily identified.

Example 1.8.

The concept of feedback is clearly illustrated by the autopilot mechanism of Example 1.7.

The input is the specified heading, which may be set on a dial of the airplane control panel, and the output is the actual heading, as determined by automatic navigation instruments. A comparison device continuously monitors the input and output. When the two are in correspondence, control action is not required. When a difference exists between the input and output, the comparison device delivers a control action signal to the controller, the autopilot mechanism. The controller provides the appropriate signals to the control surfaces of the airplane to reduce the input-output difference.

Feedback may be effected by a mechanical or electrical connection from the navigation instruments, measuring the heading, to the comparison device.

1.5 CHARACTERISTICS OF FEEDBACK

The most important features the presence of feedback imparts to a system are the following:

1. Increased accuracy. For example, the ability to faithfully reproduce the input.

2. Reduced sensitivity of the ratio of output to input to variations in system characteristics (Chapter 9).

3. Reduced effects of nonlinearities (Chapter 3) and distortion.

4. Increased bandwidth. The **bandwidth** of a system is that range of frequencies (of the input) over which the system will respond satisfactorily.

5. Tendency toward oscillation or instability. This characteristic is considered in detail in Chapter 5.

1.6 THE CONTROL SYSTEMS ENGINEERING PROBLEM

The nature of control systems engineering is the consideration of two problems: the *analysis* and the *design* of a control system configuration.

Analysis is the investigation of the properties of an existing system. The **design** problem is the choice and arrangement of control systems components to perform a specific task.

Two methods exist for design:

1. Design by analysis

2. Design by synthesis

Design by analysis is accomplished by modifying the characteristics of an existing or standard system configuration, and **design by synthesis** by defining the form of the system directly from its specifications. The former method is employed in the design portions of this book.

1.7 REPRESENTATION OF THE PROBLEM: THE MODEL

In order to solve a systems problem, the specifications or description of the system configuration and its components must be put into a form amenable to analysis, design, and evaluation.

Three basic representations (models) of physical components and systems are extensively employed in the study of control systems:

1. Differential equations and other mathematical relations

2. Block diagrams

3. Signal flow graphs

Block diagrams and **signal flow graphs** are shorthand, graphical representations of either the schematic diagram of a physical system, or the set of mathematical equations characterizing its parts. Block diagrams are considered in detail in Chapters 2 and 7, and signal flow graphs in Chapter 8.

Mathematical models, in the form of system equations, are employed when detailed relationships are required. Every control system may theoretically be characterized by mathematical equations. The solution of these equations represents the system's behavior. Often this solution is difficult if not impossible to find. In these cases, certain simplifying assumptions must be made in the mathematical description. For a large number of control systems these approximations and simplifications lead to systems describable by linear ordinary differential equations. Moreover, techniques for solving these equations are well documented in the literature of mathematics and engineering. Hence the major part of this book is restricted to the consideration of control systems which can be described by linear ordinary differential equations. For those readers having a knowledge of calculus and elementary complex numbers, Chapters 3 and 4 provide a treatment of linear ordinary differential equations and their solution. The material is presented from the point of view of its application to feedback systems and modern control system theory.

1.8 CONTROL SYSTEMS SCIENCE

A major emphasis in modern control technology is the development of mathematical models for physical situations. Common mathematical and physical principles are also utilized in order to understand the characteristics of feedback systems as they relate to the transmission or processing of the abstract quantity *information*. Thus control systems engineering spans not only the entire breadth of the engineering sciences, but the biological and social sciences as well.

These added dimensions have created so many new problems that systems analysis and design has virtually become a science.

In order to communicate with as many readers as possible, the emphasis in this text is on mathematical and physical principles, the basic languages of the sciences. Specific applications of these principles, taken chiefly from the engineering and biological sciences, are found in the solved problems at the end of each chapter.

Solved Problems

INPUT AND OUTPUT

1.1. Identify those quantities which are the input and output for the pivoted, adjustable mirror of Fig. 1-2.

> The input is the angle of inclination of the mirror θ, which is varied by turning the screw. The output is the angular position of the reflected beam $\theta + \alpha$ from the reference surface.

1.2.   Identify a possible input and a possible output for a rotational generator of electricity.

The input may be the rotational speed of the prime mover (e.g. a steam turbine), in revolutions per minute. Assuming the generator has no load attached to its output terminals, the output may be the induced voltage at the output terminals.

Alternatively, the input can be expressed as angular momentum of the prime mover shaft, and the output in units of electrical power (watts) with a load attached to the generator.

1.3.   Identify the input and output for an automatic washing machine.

Most (but not all) washing machines are operated in the following manner. After the clothes to be washed have been put into the machine, the soap or detergent, bleach, and water are entered in the proper amounts. The washing and wringing cycle-time is then set on a timer and the washer is energized. When the cycle is completed, the machine shuts itself off.

If the proper amounts of detergent, bleach, and water, and the appropriate temperature of the water are predetermined or specified by the machine manufacturer, or automatically entered by the machine itself, then the input is the time (in minutes) for the wash and wring cycle. The timer is usually set by a human operator.

The output of a washing machine is more difficult to identify. Let us define *clean* as the absence of all foreign substances from the items to be washed. Then we can identify the output as the percentage of cleanliness. Therefore, at the start of a cycle the output is less than 100 percent, and at the end of a cycle the ideal output is equal to 100 percent (*clean* clothes are not always obtained).

For most coin-operated machines the cycle-time is preset, and the machine begins operating when the coin is entered. In this case, the percentage of cleanliness can be controlled by adjusting the amount of detergent, bleach, water, and the temperature of the water. We may consider all of these quantities as inputs.

Other combinations of inputs and outputs are also possible.

1.4.   Identify the components, input, and output, and describe the operation of the biological control system consisting of a human being reaching for an object.

The basic components of this control system are the brain, arm and hand, and eyes.

The brain sends the required nervous system signal to the arm and hand to reach for the object. This signal is amplified in the muscles of the arm and hand, which serve as power actuators for the system. The eyes are employed as a sensing device, continuously "feeding back" the position of the hand to the brain.

Hand position is the output for the system. The input is object position.

The objective of the control system is to reduce the distance between hand position and object position to zero. Fig. 1-3 is a schematic diagram. The dashed lines and arrows represent the direction of information flow.

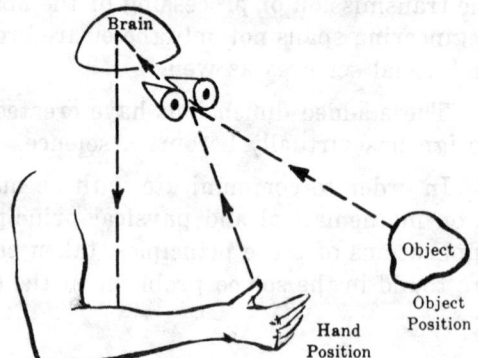

Fig. 1-3

OPEN-LOOP AND CLOSED-LOOP SYSTEMS

1.5.   Explain how a closed-loop automatic washing machine might operate.

Assume that all quantities described as possible inputs in Problem 1.3, namely cycle-time, water volume, water temperature, amount of detergent, and amount of bleach, can be adjusted by devices such as valves and heaters.

A closed-loop automatic washer would continuously or periodically measure the percentage of cleanliness (output) of the items being washed, adjust the input quantities accordingly, and turn itself off when 100 percent cleanliness has been achieved.

1.6. How are the following open-loop systems calibrated: (a) automatic washing machine, (b) automatic toaster, (c) voltmeter?

(a) Automatic washing machines are calibrated by estimating any combination of the following input quantities: (1) amount of detergent, (2) amount of bleach, (3) amount of water, (4) temperature of the water, (5) cycle-time.

On some washing machines one or more of these inputs is (are) predetermined by the manufacturer. The remaining quantities must be estimated by the user and depend upon factors such as degree of hardness of the water, type of detergent, and type or strength of the bleach. Once this calibration has been determined for a specific type of wash (e.g. all white clothes, very dirty clothes) it does not normally have to be redetermined during the lifetime of the machine. If the machine breaks down and replacement parts are installed, recalibration is probably necessary.

(b) Although the timer dial for most automatic toasters is calibrated by the manufacturer (e.g. light-medium-dark), the amount of heat produced by the heating element may vary over a wide range. In addition, the efficiency of the heating element normally deteriorates with age. Hence the amount of time required for "good toast" must be estimated by the user, and this setting must be periodically readjusted. At first, the toast is usually too light or too dark. After several successively different estimates, the required toasting time for a desired quality of toast is obtained.

(c) In general, a voltmeter is calibrated by comparing it with a known-voltage standard source, and appropriately marking the reading scale at specified intervals.

1.7. Identify the control action in the systems of Problems 1.1, 1.2, and 1.4.

For the mirror system of Problem 1.1 the control action is equal to the input, that is, the angle of inclination of the mirror θ. For the generator of Problem 1.2 the control action is equal to the input, the rotational speed or angular momentum of the prime mover shaft. The control action of the human reaching system of Problem 1.4 is equal to the distance between hand and object position.

1.8. Which of the control systems in Problems 1.1, 1.2, and 1.4 are open-loop? Closed-loop?

Since the control action is equal to the input for the systems of Problems 1.1 and 1.2, no feedback exists and the systems are open-loop. The human reaching system of Problem 1.4 is closed-loop because the control action is dependent upon the output, hand position.

1.9. Identify the control action in Examples 1.1 through 1.5.

The control action for the electric switch of Example 1.1 is equal to the input, the on or off command. The control action for the heating system of Example 1.2 is equal to the difference between the reference and actual room temperatures. For the finger pointing system of Example 1.3, the control action is equal to the difference between the actual and pointed direction of the object. The perspiration system of Example 1.4 has its control action equal to the difference between the "normal" and actual skin surface temperature. The difference between the direction of the road and the heading of the automobile is the control action for the human driver and automobile of Example 1.5.

1.10. Which of the control systems in Examples 1.1 through 1.5 are open-loop? Closed-loop?

The electric switch of Example 1.1 is open-loop because the control action is equal to the input, and therefore independent of the output. For the remaining Examples 1.2-1.5 the control action is clearly a function of the output. Hence they are closed-loop systems.

FEEDBACK

1.11. Consider the voltage divider network of Fig. 1-4 below. The output is v_2 and the input is v_1.

(a) Write an equation for v_2 as a function of v_1, R_1, and R_2. That is, write an equation for v_2 which yields an open-loop system.

(b) Write the equation for v_2 in closed-loop form, that is, v_2 as a function of v_1, v_2, R_1, and R_2.

This problem illustrates how a passive network can be characterized as either an open-loop or a closed-loop system.

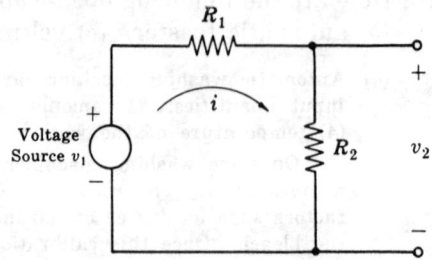

Fig. 1-4

(a) From Kirchhoff's voltage and current laws we have

$$v_2 = R_2 i, \qquad i = \frac{v_1}{R_1 + R_2}$$

Therefore

$$v_2 = \left(\frac{R_2}{R_1 + R_2}\right) v_1 = f(v_1, R_1, R_2)$$

(b) Writing the current i in a slightly different form, we have $i = \dfrac{v_1 - v_2}{R_1}$. Hence

$$v_2 = R_2\left(\frac{v_1 - v_2}{R_1}\right) = \left(\frac{R_2}{R_1}\right) v_1 - \left(\frac{R_2}{R_1}\right) v_2 = f(v_1, v_2, R_1, R_2)$$

1.12. Explain how the classical economic concept known as the Law of Supply and Demand can be interpreted as a feedback control system. Choose the market price (selling price) of a particular item as the output of the system, and assume the objective of the system is to maintain price stability.

The Law can be stated in the following manner: The market *demand* for the item decreases as its price increases. The market *supply* usually increases as its price increases. The Law of Supply and Demand says that a stable market price is achieved if and only if the supply is equal to the demand.

The manner in which the price is regulated by the supply and the demand can be described with feedback control concepts. Let us choose the following four basic elements for our system: the Supplier, the Demander, the Pricer, and the Market where the item is bought and sold. These elements generally represent very complicated processes.

The input to our economic system is price stability. A more convenient way to describe this input is *zero price fluctuation*. The output is the actual market price.

The system operates as follows: The Pricer receives a command (zero) for price stability. He (it) estimates a price for the Market transaction with the help of information from his (its) memory or records of past transactions. This price causes the Supplier to produce or supply a certain number of items, and the Demander to demand a number of items. The difference between the supply and the demand is the control action for this system. If the control action is nonzero, that is, if the supply is not equal to the demand, the Pricer initiates a change in the market price in a direction which makes the supply eventually equal to the demand. Hence both the Supplier and the Demander may be considered the feedback, since they determine the control action.

MISCELLANEOUS PROBLEMS

1.13. (a) Explain the operation of ordinary traffic signals which control automobile traffic at roadway intersections. (b) Why are they open-loop control systems? (c) How can traffic be controlled more efficiently? (d) Why is the system of (c) closed-loop?

(a) Traffic lights control the flow of traffic by successively confronting the traffic in a particular direction (e.g. north-south) with a red (stop) and then a green (go) light. When one direction has the green signal, the cross traffic in the other direction (east-west) has the red. Most traffic signal red and green light intervals are predetermined by a calibrated timing mechanism.

(b) All control systems operated by preset timing mechanisms are open-loop. The control action is equal to the input, the red and green times.

(c) Besides preventing collisions, it is a function of traffic signals to generally control the *volume* of traffic. For the open-loop system described above, the volume of traffic does not influence the preset red and green timing intervals. In order to make traffic flow more smoothly, the green light timing interval must be made longer than the red in the direction containing the greater traffic volume. Often a traffic policeman performs this task.

The ideal system would measure the volume of traffic in all directions, compare them, and use the difference to control the red and green time intervals.

(d) The system of (c) is closed-loop because the control action (the difference between the volume of traffic in each direction) is a function of the output (actual traffic volume flowing past the intersection in each direction).

1.14. (a) Describe the components and variables of the biological control apparatus involved in walking in a prescribed direction. (b) Why is walking a closed-loop operation? (c) Under what conditions would the human walking apparatus become an open-loop system?

(a) The major components involved in walking are the brain, eyes, and legs and feet. The input may be chosen as the desired walk direction, and the output the actual walk direction. The control action is determined by the eyes, which detect the difference between the input and output and send this information to the brain. The brain commands the legs and feet to walk in the prescribed direction.

(b) Walking is a closed-loop operation because the control action is a function of the output.

(c) If the eyes are closed, the feedback loop is broken and the system becomes open-loop.

1.15. Devise a control system to fill a container with water after it is emptied through a stopcock at the bottom. The system must automatically shut off the water when the container is filled.

The simplified schematic diagram, Fig. 1-5, illustrates the principle of the ordinary water closet tank filling system.

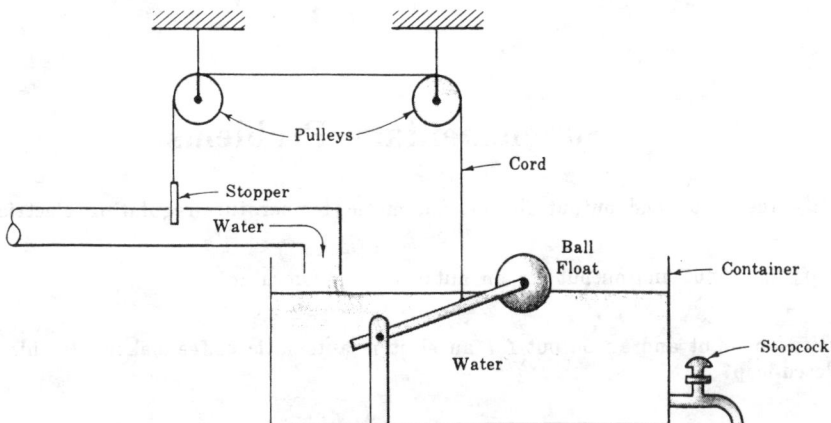

Fig. 1-5

The ball floats on the water. As the ball gets closer to the top of the container, the stopper decreases the flow of water. When the container becomes full, the stopper shuts off the flow of water.

1.16. Devise a simple control system which automatically turns on a room lamp at dusk, and turns it off in daylight. Show a schematic of your system.

A simple system that accomplishes this task is shown in the schematic diagram, Fig. 1-6.

At dusk, the photocell, which functions as a light-sensitive switch, closes the lamp circuit, thereby lighting the room. The lamp stays lighted until daylight, at which time the photocell detects the bright outdoor light and opens the lamp circuit.

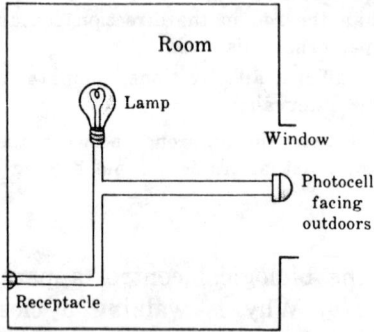

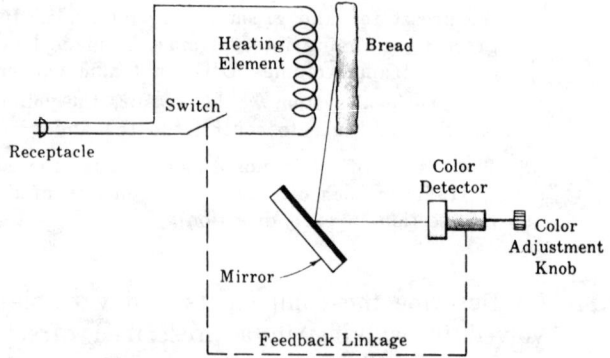

Fig. 1-6   Fig. 1-7

1.17. Devise a closed-loop automatic toaster.

Assume each heating element supplies the same amount of heat to both sides of the bread, and toast quality can be determined by its color. A simplified schematic diagram of one possible way to apply the feedback principle to a toaster is shown in Fig. 1-7. Only one side of the toaster is illustrated.

The toaster is initially calibrated for a desired toast quality by means of the color adjustment knob. This setting never needs readjustment unless the toast quality criterion changes. When the switch is closed, the bread is toasted until the color detector "sees" the desired color. Then the switch is automatically opened by means of the feedback linkage, which may be electrical or mechanical.

Supplementary Problems

1.18. Identify the input and output for an automatic temperature-regulating electric oven.

1.19. Identify the input and output for an automatic refrigerator.

1.20. Identify an input and an output for an electric automatic coffee-maker. Is this system open-loop or closed-loop?

1.21. Devise a control system to automatically raise and lower a lift-bridge to permit ships to pass. No continuous human operator is permissible. The system must function entirely automatically.

1.22. Explain the operation and identify the pertinent quantities and components of an automatic, radar-controlled anti-aircraft gun. Assume that no operator is required except to initially put the system into an operational mode.

1.23.   Can the electrical network of Fig. 1-8 be deemed a feedback control system?

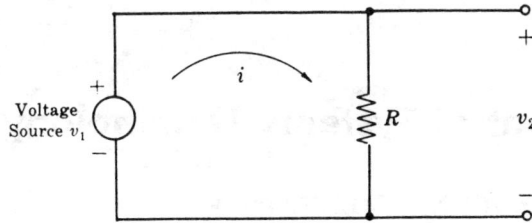

Fig. 1-8

1.24.   Devise a control system for positioning the rudder of a ship from a control room located far from the rudder. The objective of the control system is to steer the ship in a desired heading.

1.25.   What inputs in addition to the command for a desired heading would you expect to find acting on the system of Problem 1.24?

1.26.   Can the application of "laissez faire capitalism" to an economic system be interpreted as a feedback control system? Why?

1.27.   Does the operation of a stock exchange such as the New York Stock Exchange fit the model of the Law of Supply and Demand described in Problem 1.12? How?

1.28.   Does a purely socialistic economic system fit the model of the Law of Supply and Demand described in Problem 1.12? Why (or why not)?

ANSWERS TO SUPPLEMENTARY PROBLEMS

1.18.   The input is the reference temperature. The output is actual oven temperature.

1.19.   The input is the reference temperature. The output is the actual refrigerator temperature.

1.20.   One possible input for the automatic electric coffeemaker is the amount of coffee used. In addition, most coffeemakers have a dial which can be set for weak, medium, or strong coffee. This setting usually regulates a timing mechanism. The brewing-time is therefore another possible input. The output of any coffeemaker can be chosen as coffee strength. The coffeemakers described above are open-loop.

Chapter 2

Control Systems Terminology

2.1 BLOCK DIAGRAMS: FUNDAMENTALS

A **block diagram** is a shorthand, pictorial representation of the cause and effect relationship between the input and output of a physical system. It provides a convenient and useful method for characterizing the functional relationships among the various components of a control system. System *components* are alternatively called *elements* of the system. The simplest form of the block diagram is the single *block*, with one input and one output:

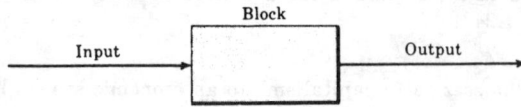

The interior of the rectangle representing the block usually contains a description of or the name of the element, or the symbol for the mathematical operation to be performed on the input to 'yield the output. The *arrows* represent the direction of unilateral information or signal flow.

Example 2.1.

The operations of addition and subtraction have a special representation. The block becomes a small circle, called a **summing point**, with the appropriate plus or minus sign associated with the arrows entering the circle. The output is the algebraic sum of the inputs. Any number of inputs may enter a summing point.

Example 2.2.

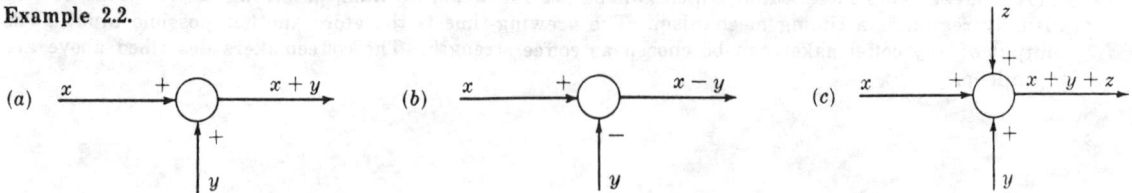

Some authors put a cross in the circle:

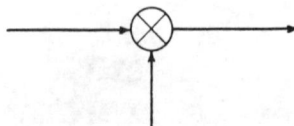

This notation is avoided here because it is sometimes confused with the multiplication operation.

In order to employ the same signal or variable as an input to more than one block or summing point, a **takeoff** **point** is used. This permits the signal to proceed unaltered along several different paths to several destinations.

Example 2.3.

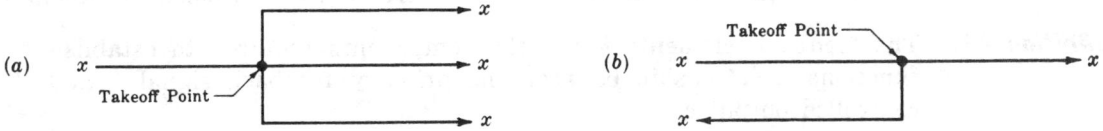

(a) ... Takeoff Point ...

(b) ... Takeoff Point ...

2.2 BLOCK DIAGRAM OF A FEEDBACK CONTROL SYSTEM

The blocks representing the various components of a control system are connected in a fashion which characterizes their functional relationship within the system. The basic configuration of a simple closed-loop (feedback) control system is illustrated in the block diagram below, Fig. 2-1. It is emphasized that the arrows of the closed loop, connecting one block with another, represent the direction of flow of *control* energy or information, and not the main source of energy for the system. For example, the major source of energy for the thermostatically controlled furnace of Example 1.2, Page 2, may be chemical, from burning fuel oil or coal. But this energy source would not appear in the closed control loop of the system.

The generalized feedback control system is given by Fig. 2-1.

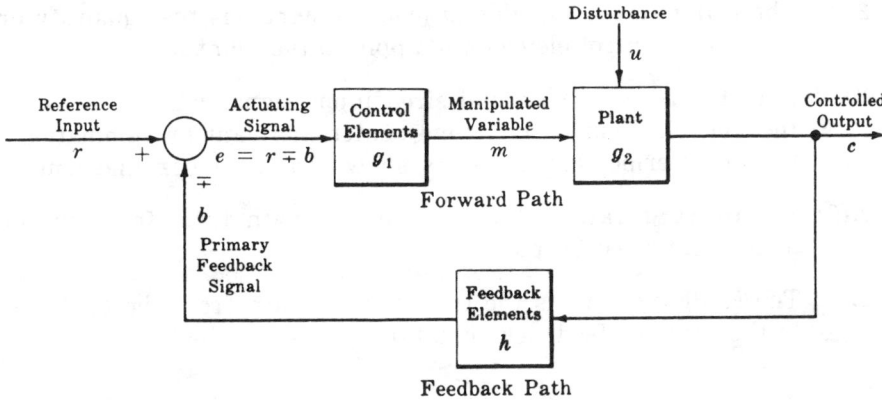

Fig. 2-1

2.3 TERMINOLOGY OF THE CLOSED-LOOP BLOCK DIAGRAM

It is important that the terms used in the closed-loop block diagram be clearly understood and remembered.

Lower case letters are used to represent the input and output variables of each element as well as the symbols for the blocks g_1, g_2, and h. These quantities represent functions of time, unless otherwise specified.

Example 2.4. $r = r(t)$

Capital letters denote Laplace transformed quantities as functions of the complex variable s, or Fourier transformed quantities (frequency functions) as functions of the pure imaginary variable $j\omega$. Functions of s are usually abbreviated to the capital letter appearing alone. Frequency functions are never abbreviated.

Example 2.5. $R(s)$ is abbreviated to R. $R(j\omega)$ is never abbreviated.

The letters r, c, m, etc. were chosen to preserve the generic nature of the block diagram.

Definition 2.1: The **plant** g_2, also called the **controlled system,** is the body, process, or machine, of which a particular quantity or condition is to be controlled.

Definition 2.2: The **control elements** g_1, also called the **controller,** are the components required to generate the appropriate control signal m applied to the plant.

Definition 2.3: The **feedback elements** h are the components required to establish the functional relationship between the primary feedback signal b and the controlled output c.

Definition 2.4: The **reference input** r is an external signal applied to a feedback control system in order to command a specified action of the plant. It often represents ideal plant output behavior.

Definition 2.5: The **controlled output** c is that quantity or condition of the plant which is controlled.

Definition 2.6: The **primary feedback signal** b is a signal which is a function of the controlled output c, and which is algebraically summed with the reference input r to obtain the actuating signal e.

Definition 2.7: The **actuating signal** e, also called the **error** or **control action,** is the algebraic sum consisting of the reference input r plus or minus (usually minus) the primary feedback b.

Definition 2.8: The **manipulated variable** m (control signal) is that quantity or condition which the control elements g_1 apply to the plant g_2.

Definition 2.9: A **disturbance** u is an undesired input signal which affects the value of the controlled output c. It may enter the plant by summation with m, or via an intermediate point, as shown in the block diagram of Fig. 2-1.

Definition 2.10: The **forward path** is the transmission path from the actuating signal e to the controlled output c.

Definition 2.11: The **feedback path** is the transmission path from the controlled output c to the primary feedback signal b.

2.4 SUPPLEMENTARY TERMINOLOGY

Several more terms require definition and illustration at this time. Others will be presented in subsequent chapters as required.

Definition 2.12: A **transducer** is a device which converts one energy form into another.

For example, one of the most common transducers in control systems applications is the *potentiometer*, which converts mechanical position into an electrical voltage:

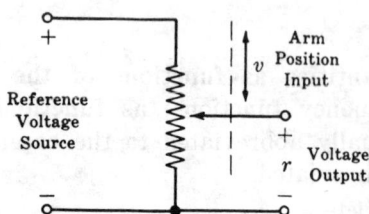

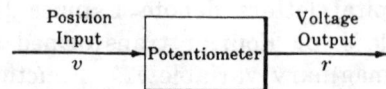

Schematic   Block Diagram

Definition 2.13:   The **command** v is an input signal, usually equal to the **reference input** r. But when the energy form of the command v is not the same as that of the primary feedback b, a transducer is required between the command v and the reference input r as shown in Fig. (a).

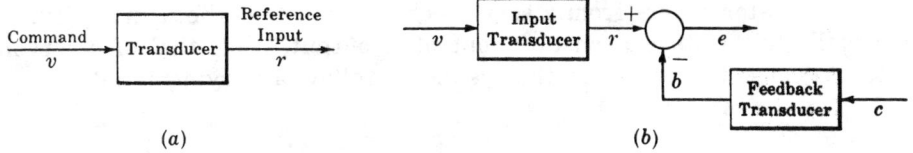

(a)   (b)

Definition 2.14:   When the feedback element consists of a transducer, and a transducer is required at the input, that part of the control system illustrated in the above block diagram (b) is called the **error detector**.

Definition 2.15:   **Negative feedback** means that the summing point is a subtracter; or

$$e = r - b$$

Definition 2.16:   **Positive feedback** means that the summing point is an adder; or

$$e = r + b$$

Definition 2.17:   A **stimulus** is any externally introduced input signal affecting the controlled output c.

For example, the reference input r and a disturbance u are stimuli.

Definition 2.18:   The **time response** of a system or element is the output as a function of time, following the application of a prescribed input under specified operating conditions.

2.5 SERVOMECHANISMS

The specialized feedback control system called a *servomechanism* deserves special attention due to its prevalence in industrial applications and control systems literature.

Definition 2.19:   A **servomechanism** is a power-amplifying feedback control system in which the controlled variable c is mechanical position, or a time derivative of position such as velocity or acceleration.

Example 2.6.

An *automobile power-steering apparatus* is a servomechanism. The command input is the angular position of the steering wheel. A small rotational torque applied to the steering wheel is amplified hydraulically, resulting in a force adequate to modify the output, the angular position of the front wheels. The block diagram of such a system may be represented by Fig. 2-2 below. Negative feedback is necessary in order to return the control valve to the neutral position, reducing the torque from the hydraulic amplifier to zero when the desired wheel position has been achieved.

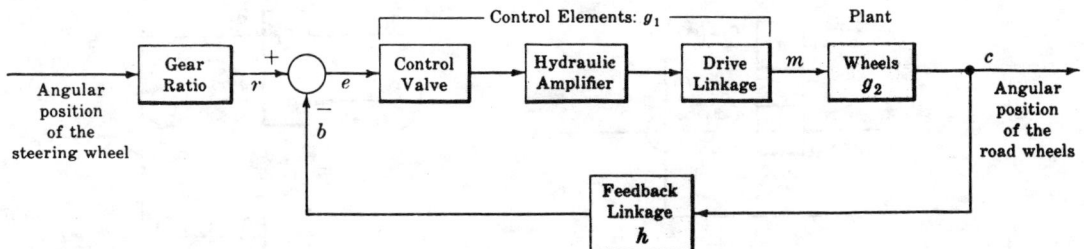

Fig. 2-2

2.6 REGULATORS

Definition 2.20:   A **regulator** or **regulating system** is a feedback control system in which
the reference input or command is constant for long periods of time,
often for the entire time interval during which the system is operational.

A regulator differs from a servomechanism in that the primary function of a regulator
is usually to maintain a constant controlled output, while that of a servomechanism is most
often to cause the output of the system to follow a varying input.

Solved Problems

BLOCK DIAGRAMS

2.1.   Consider the following equations in which $x_1, x_2, \ldots, x_n$ are variables, and $a_1, a_2, \ldots, a_n$
are general coefficients or mathematical operators:

$$(a) \quad x_3 = a_1 x_1 + a_2 x_2 - 5$$

$$(b) \quad x_n = a_1 x_1 + a_2 x_2 + \cdots + a_{n-1} x_{n-1}$$

Draw a block diagram for each equation, identifying all blocks, inputs, and outputs.

(a) In the form the equation is written, x_3 is the output. The terms on the right-hand side of the
equation are combined at a summing point as shown in Fig. 2-3.

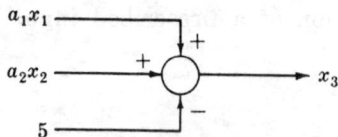
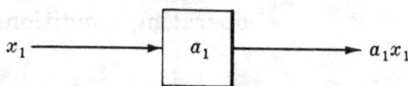

Fig. 2-3   **Fig. 2-4**

The $a_1 x_1$ term is represented by a single block, with x_1 as its input and $a_1 x_1$ as its output.
Therefore the coefficient a_1 is put inside the block as shown in Fig. 2-4. a_1 may represent
any mathematical operation. For example, if a_1 were a constant, the block operation would
be "multiply the input x_1 by the constant a_1." It is usually clear from the description or
context of a problem what is meant by the symbol, operator, or description inside the block.

The $a_2 x_2$ term is represented in the same manner.

The block diagram for the entire equation is therefore shown in Fig. 2-5.

(b) Following the same line of reasoning as in part (a), the block diagram for

$$x_n = a_1 x_1 + a_2 x_2 + \cdots + a_{n-1} x_{n-1}$$

is shown in Fig. 2-6.

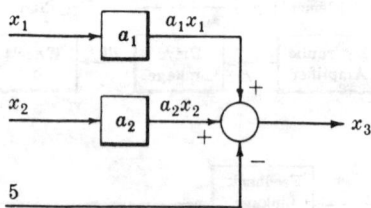
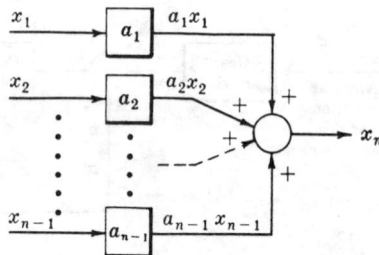

Fig. 2-5   **Fig. 2-6**

2.2. Draw block diagrams for each of the following equations:

$$(a) \quad x_2 = a_1\left(\frac{dx_1}{dt}\right), \qquad (b) \quad x_3 = \frac{d^2x_2}{dt^2} + \frac{dx_1}{dt} - x_1, \qquad (c) \quad x_4 = \int x_3\, dt$$

(*a*) Two operations are specified by this equation, a_1 and differentiation d/dt. Therefore the block diagram contains two blocks as shown in Fig. 2-7.

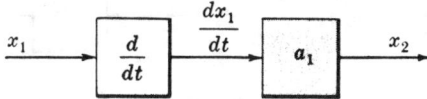

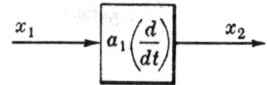

Fig. 2-7   Fig. 2-8

If a_1 is a constant, the a_1 block may be combined with the d/dt block since no confusion results as shown in Fig. 2-8. Note that if a_1 were an unknown operator, the reversal of blocks d/dt and a_1 would not necessarily result in an output equal to x_2:

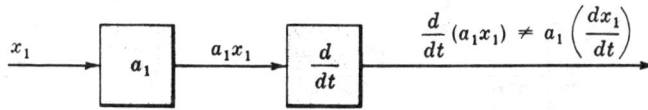

(*b*) By the same line of reasoning as in part (*a*), there are two possible block diagrams for the equation for x_3:

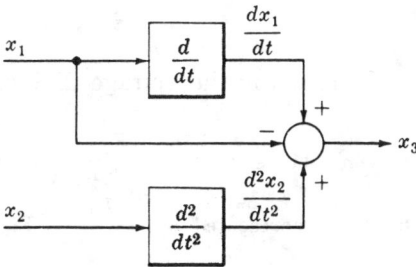
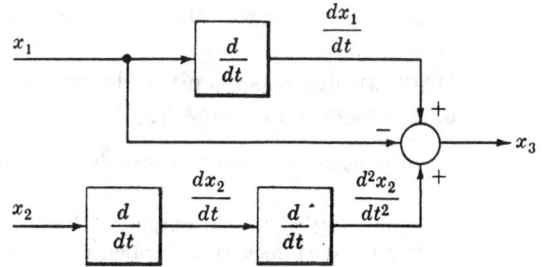

(*c*) The integration operation can be represented in block diagram form as

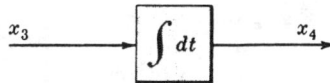

2.3. Draw a block diagram for the pivoted, adjustable mirror mechanism of Section 1.1, Page 1, with the output identified as in Problem 1.1, Page 5. Assume that each 360 degree rotation of the screw raises or lowers the mirror k degrees. Identify all the signals and components of the control system in the diagram.

 The schematic diagram of the system is repeated in Fig. 2-9 for convenience. Note that the distance of the light beam from the reference surface is equal to the thickness t of the mirror.

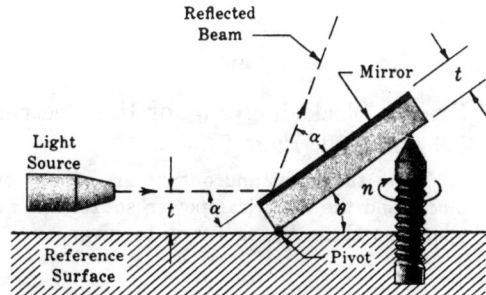

Fig. 2-9

Whereas the input was defined as θ in Problem 1.1, the specifications for this problem imply an input equal to the number of rotations of the screw. Let n be the number of rotations of the screw such that $n = 0$ when $\theta = 0$ degrees. Therefore, n and θ can be related by a block described by the constant k, since $\theta = kn$ as shown in Fig. 2-10.

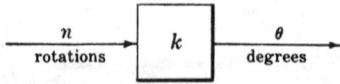

Fig. 2-10

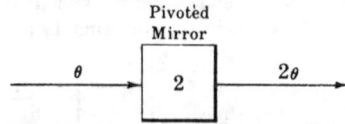

Fig. 2-11

The output of the system was determined in Problem 1.1 as $\theta + \alpha$. But since the light source is directed parallel to the reference surface, then $\alpha = \theta$. Therefore the output is equal to 2θ, and the mirror can be represented by a constant equal to 2 in a block as shown in Fig. 2-11.

The complete open-loop block diagram is given by Fig. 2-12.

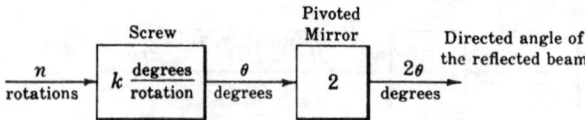

Fig. 2-12

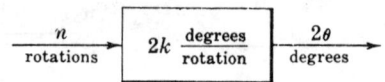

Fig. 2-13

For this simple example we also note that the output 2θ is equal to $2kn$ rotations of the screw. This yields the simpler block diagram of Fig. 2-13.

2.4. Draw an open-loop and a closed-loop block diagram for the voltage divider network of Problem 1.11, Page 7.

The open-loop equation was determined in Problem 1.11 as $v_2 = \left(\dfrac{R_2}{R_1 + R_2} \right) v_1$, where v_1 is the input and v_2 is the output. Therefore the block is represented by $\dfrac{R_2}{R_1 + R_2}$, Fig. 2-14, and clearly the operation is multiplication.

The closed-loop equation is $v_2 = \left(\dfrac{R_2}{R_1} \right) v_1 - \left(\dfrac{R_2}{R_1} \right) v_2$. The actuating signal is $v_1 - v_2$. The closed-loop, negative feedback block diagram is easily constructed with the only block represented by $\dfrac{R_2}{R_1}$ as shown in Fig. 2-15.

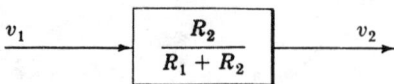

Fig. 2-14

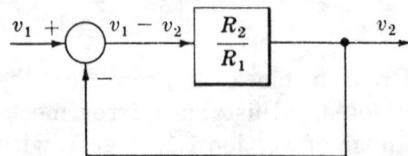

Fig. 2-15

2.5. Draw a block diagram for the electric switch of Example 1.1, Page 2 (see Problems 1.9 and 1.10, Page 7).

Both the input and output are binary or two-state variables. The switch is represented by a block, and the electrical power source the switch controls is not part of the control system. The open-loop block diagram may be given by

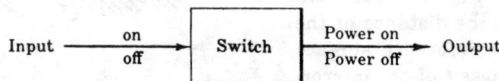

2.6. Draw simple block diagrams for the control systems in Examples 1.2-1.5, Page 2.

From Problem 1.10, Page 7, we note that these systems are closed-loop, and from Problem 1.9 the actuating signal (control action) for the system in each example is equal to the input minus the output. Therefore negative feedback exists in each system.

For the thermostatically controlled furnace of Example 1.2, the thermostat may be chosen as the summing point, since this is the device which determines whether or not the furnace shall be turned on.

The eyes may be represented by a summing point in both the human pointing system of Example 1.3 and the man-automobile system of Example 1.5. The eyes perform the function of monitoring the input and output.

For the perspiration system of Example 1.4, the summing point is not so easily defined. For the sake of simplicity let us call it the Nervous System.

The block diagrams are easily constructed as shown below from the information given above and the list of components, inputs, and outputs given in the Examples.

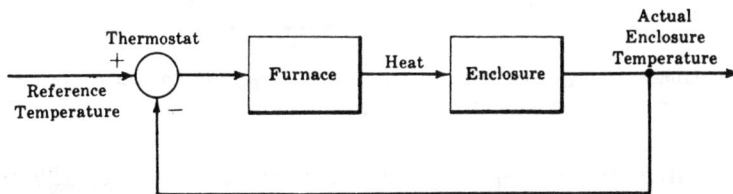

Example 1.2

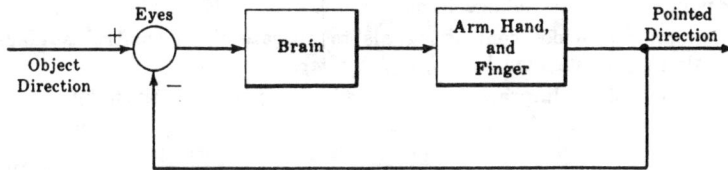

Example 1.3

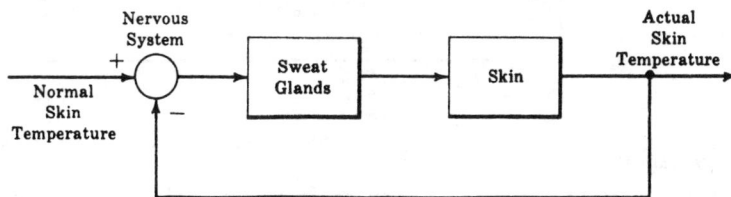

Example 1.4

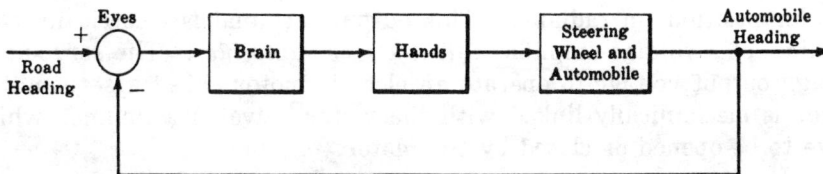

Example 1.5

The arrows drawn from one component to another for the block diagrams of the biological systems in Examples 1.3-1.5 may represent electrical, chemical, or mechanical signals controlled by the central nervous system.

BLOCK DIAGRAMS OF FEEDBACK CONTROL SYSTEMS

2.7. Draw a block diagram for the water-filling system described in Problem 1.15, Page 9. Which component or components comprise the plant? The controller? The feedback?

The container is the plant because the water level *of the container* is being controlled (see Definition 2.1). The stopper may be chosen as the control element or controller, and the ball-float, cord, and associated linkage as the feedback elements. The block diagram is given by

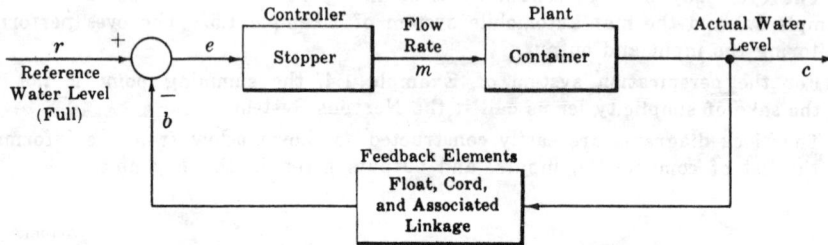

The feedback is negative because the water flow rate to the container must decrease as the water level rises in the container.

2.8. Draw a simple block diagram for the feedback control system of Examples 1.7, Page 3, and 1.8, Page 4, the airplane with an autopilot.

The plant for this system is the airplane, including its control surfaces and navigational instruments. The controller is the autopilot mechanism, and the summing point is the comparison device. The feedback linkage may be simply represented by an arrow from the output to the summing point, since this linkage is not well-defined in Example 1.8.

The autopilot provides control signals to operate the n control surfaces (rudder, flaps, etc.). These signals may be denoted by $m_1, m_2, \ldots, m_n$.

The simplest block diagram for this feedback system is given by

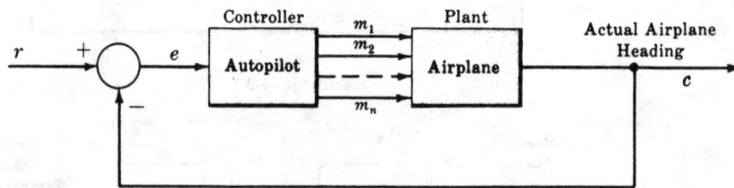

SERVOMECHANISMS

2.9. Draw a schematic and a block diagram from the following description of a position servomechanism whose function is to open and close a water valve.

At the input of the system there is a rotating-type potentiometer connected across a battery voltage source. Its movable (third) terminal is calibrated in terms of angular position (in radians). This output terminal is electrically connected to one terminal of a voltage amplifier called a *servo-amplifier*. The servo-amplifier supplies enough output voltage to operate an electric motor called a *servomotor*. The servo-motor is mechanically linked with the water valve in a manner which permits the valve to be opened or closed by the motor.

Assume the loading effect of the valve on the motor is negligible. A 360 degree rotation of the motor shaft completely opens the valve. In addition, the movable terminal of a second potentiometer connected in parallel at its fixed terminals with the input potentiometer is mechanically connected to the motor shaft. It is electrically connected to the remaining input terminal of the servo-amplifier. The potentiometer ratios are set so that they are equal when the valve is closed.

When a command is given to open the valve, the servomotor rotates in the appropriate direction. As the valve opens, the second potentiometer, which is called the feedback potentiometer, rotates in the same direction as the input potentiometer. It stops when the potentiometer ratios are again equal.

The schematic diagram, Fig. 2-16, is easily drawn from the above description. The mechanical connections are shown as dashed lines.

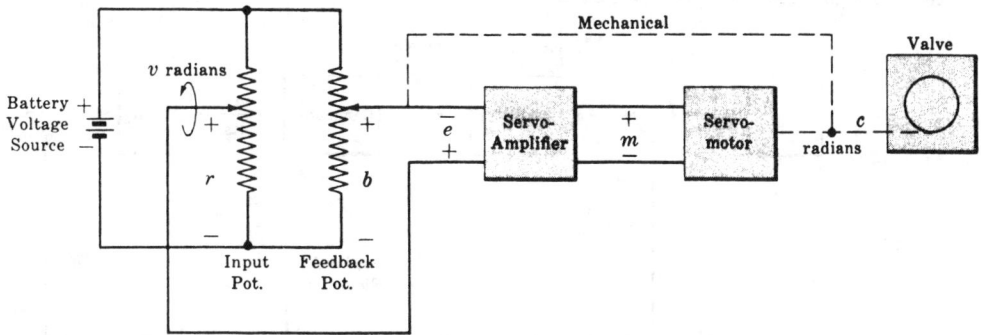

Fig. 2-16

The block diagram for this system, Fig. 2-17, is easily drawn from the schematic diagram.

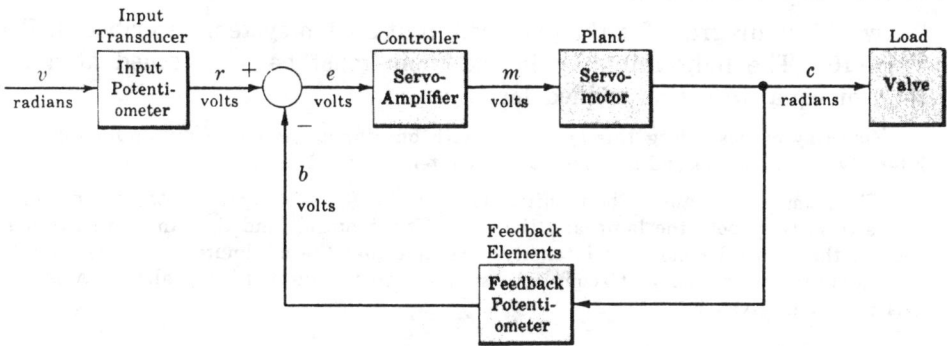

Fig. 2-17

2.10. Draw a block diagram for the elementary speed control system (velocity servomechanism) given by the following schematic diagram:

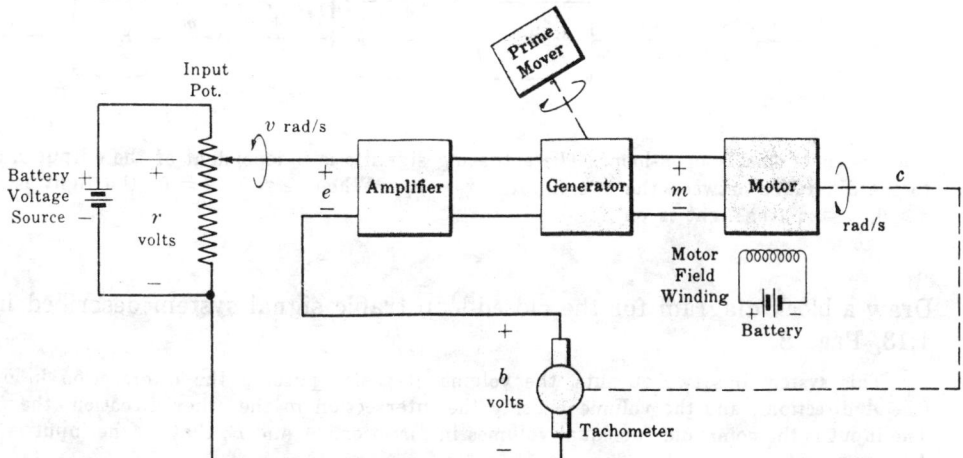

The potentiometer is a rotating-type, calibrated in rad/s, and the prime-mover speed, motor field winding and input potentiometer currents are constant functions of time. No load is attached to the motor shaft.

The battery voltage sources for both the input potentiometer and motor field winding, and the prime-mover source for the generator are not part of the control loop of this servomechanism. The output of each of these sources is a constant function of time, and can be accounted for in the mathematical description of the input potentiometer, generator, and motor, respectively. Therefore the block diagram for this system is

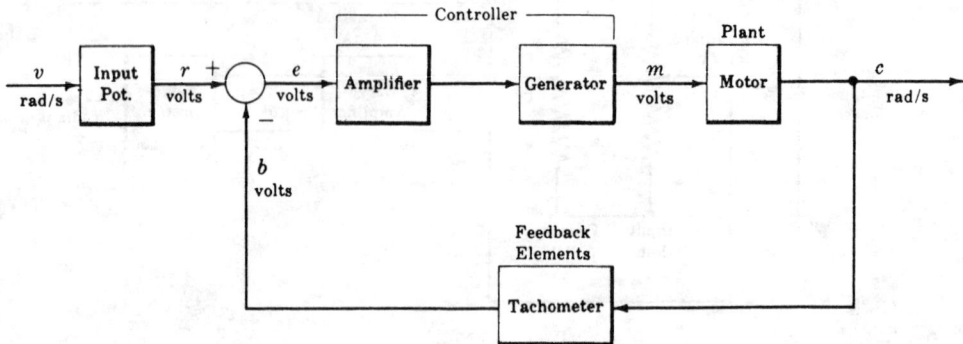

MISCELLANEOUS PROBLEMS

2.11. Draw a block diagram for the photocell light switch system described in Problem 1.16, Page 10. The light intensity in the room must be maintained at a level greater than or equal to a prespecified level.

One way of describing this system is with one input chosen as minimum reference room-light intensity r_1, and the second as room sunlight intensity r_2. The output c is actual room-light intensity.

The room is the plant. The manipulated variable (control signal) is the amount of light supplied to the room from both the lamp and the sun. The photocell and the lamp are the control elements because they control room-light intensity. Assume that the minimum reference room-light intensity r_1 is equal to the intensity of room-light supplied by the lighted lamp alone. A block diagram for this system is given by

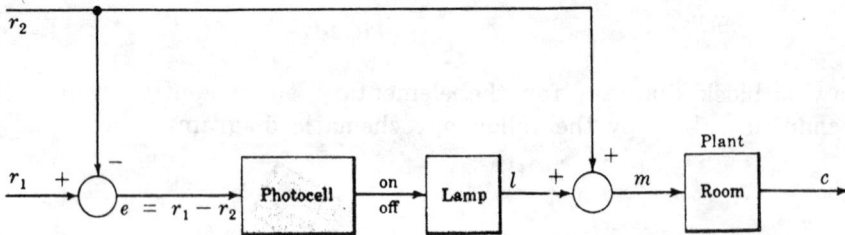

The system is clearly open-loop. The actuating signal e is independent of the output c, and is equal to the difference between the two inputs $r_1 - r_2$. When $e \leqq 0$, $l = 0$ (the light is off). When $e > 0$, $l = r_1$ (the light is on).

2.12. Draw a block diagram for the closed-loop traffic signal system described in Problem 1.13, Page 8.

This system has two outputs, the volume of traffic passing the intersection in one direction (the A direction), and the volume passing the intersection in the other direction (the B direction). The input is the command for equal volumes in directions A and B; that is, the input is zero volume difference.

Let us call the mechanism which computes the appropriate red and green timing intervals the Red-Green Time Interval Computer. This device, in addition to the traffic signal, make up the control elements. The plants are the roadway in direction A and the roadway in direction B. The block diagram is given by

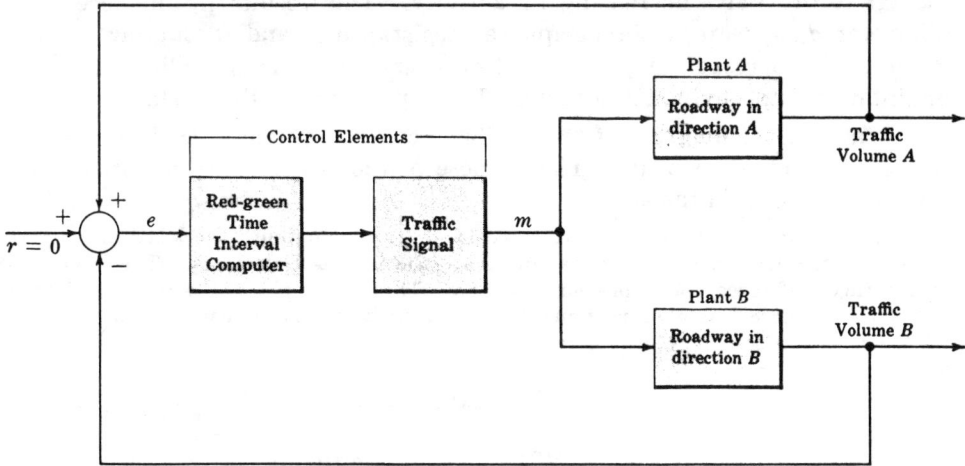

2.13. Draw a block diagram illustrating the economic Law of Supply and Demand, as described in Problem 1.12, Page 8.

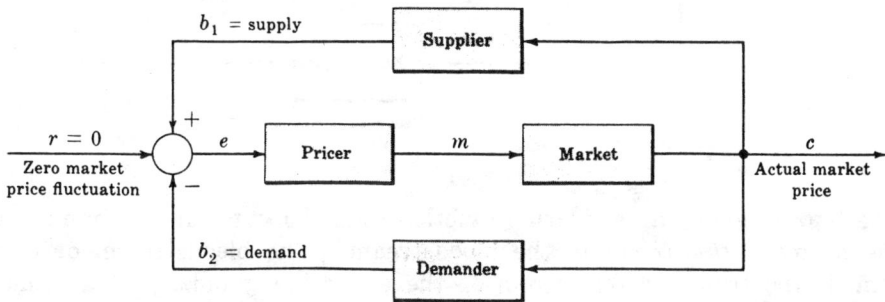

2.14. The following simplified version of a biological mechanism regulating human arterial blood pressure is an example of a feedback control system.

An adequate, relatively constant pressure must be maintained in the blood vessels (arteries, arterioles, and capillaries) supplying the tissues. This pressure is usually measured in the aorta (artery), and is called the *blood pressure p*. It is typically equal to 10-16 kPa (Pascal, N/m²). Let us assume that p is equal to 13 kPa (on the average) in a normal individual.

One of the fundamental equations of circulatory physiology is the general equation for arterial blood pressure:
$$p = Q\rho$$
where Q is the cardiac output, or the volume flow rate of blood from the heart to the aorta, and ρ is the peripheral resistance offered to blood flow by the arterioles. ρ is inversely proportional to the fourth power of the diameter d of the vessels (arterioles).

Now d is controlled by the *vasomotor center* (VMC), situated in the medulla at the base of the brain. Increased activity of the VMC decreases d, and vice-versa.

Although several factors affect VMC activity, the *baroreceptor cells*, located in a region of the arterial tree known as the arterial sinus, are the most important. Baroreceptor activity *inhibits* the VMC, and therefore functions in a feedback mode. If p increases, the baroreceptors send signals along the vagus and glossopharyngeal nerves to the VMC, decreasing its activity. This results in an increase in arteriole diameter d, a decrease in peripheral resistance ρ, and (assuming constant cardiac output Q) a corresponding drop in blood pressure p. This feedback network serves to maintain an approximately constant blood pressure in the aorta.

Draw a block diagram of the feedback control system described above, identifying all signals and components. The purpose of the system is to maintain constant blood pressure (13 kPa) in the aorta.

Let the aorta be the plant, represented by Q (cardiac output); the VMC and arterioles may be chosen as the controller; the baroreceptors are the feedback elements. The input p_0 is the average normal (reference) blood pressure, 13 kPa. The output p is the actual blood pressure. Since $\rho = k(1/d)^4$ where k is a proportionality constant, the arterioles can be represented in the block by $k(\cdot)^4$. The block diagram is given by

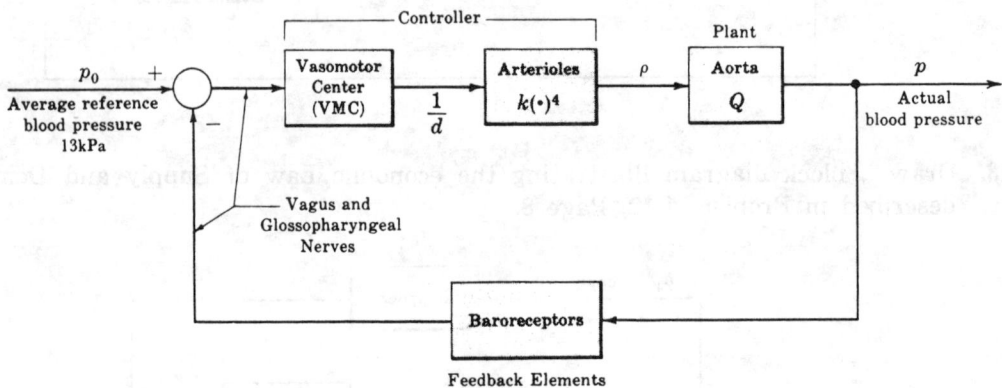

2.15. The *thyroid* gland, an endocrine (ductless) gland located in the human neck, secretes the hormone *thyroxin* into the bloodstream. The bloodstream, or circulatory system, is the transmission system of the endocrine glands, just as conductive wires are the transmission system for the flow of electrons that produce electrical currents, or pipes and tubes may be the transmission system for hydrodynamic fluid flow. Thyroxin acts in human body tissues by generally stimulating metabolism, causing a rise in oxygen consumption and an increase in body temperature. Like most human physiological processes, the production of this hormone by the thyroid gland is very carefully and automatically controlled. The amount of thyroxin in the bloodstream is regulated by the secretion of a hormone from the *pituitary* gland, an endocrine gland suspended from the base of the brain. This "control" hormone is appropriately called *thyroid stimulating hormone* (TSH). When the level of thyroxin in the circulatory system is higher than that required by the organism, TSH secretion is inhibited (reduced), causing a reduction in the activity of the thyroid. Hence less thyroxin is released by the thyroid. Malfunction of this feedback control system, causing either underactivity or overactivity of the thyroid gland, may result in swelling of this gland in the neck, a disease known as *goiter*.

Draw a simple block diagram of this system, identifying all components and signals.

Let the plant be the thyroid gland, the controlled variable the level of thyroxin in the bloodstream. The pituitary gland is the controller, and the manipulated variable is the amount of TSH it secretes. The block diagram is given by

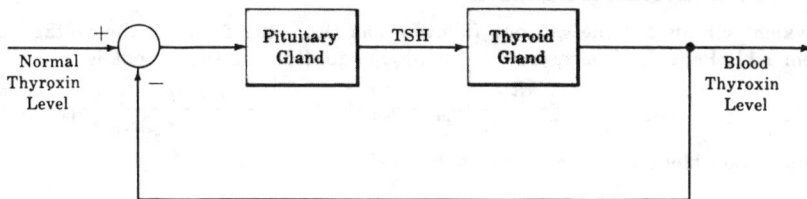

Supplementary Problems

2.16. The schematic diagram of a triode voltage amplifier called a *cathode follower* is given by Fig. 2-18.

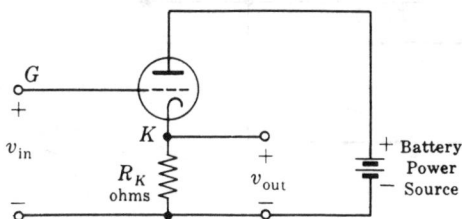

Fig. 2-18

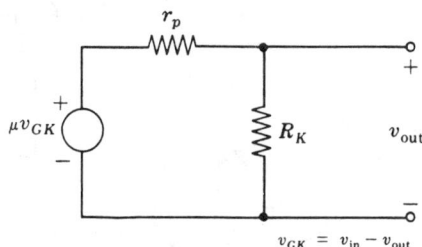

Fig. 2-19

An equivalent circuit for this amplifier is shown in Fig. 2-19 where r_p is the internal plate resistance of, and μ is a parameter of the particular triode. Draw both an open-loop and a closed-loop block diagram for this circuit with an input v_{in} and an output v_{out}.

2.17. Draw a block diagram for the human walking system of Problem 1.14, Page 9.

2.18. Draw a block diagram for the human reaching system described in Problem 1.4, Page 6.

2.19. Draw a block diagram for the automatic temperature-regulated oven of Problem 1.18, Page 10.

2.20. Draw a block diagram for the closed-loop automatic toaster of Problem 1.17, Page 10.

2.21. State the common dimensional units for the input and output of the following transducers: (*a*) accelerometer, (*b*) generator of electricity, (*c*) thermistor (temperature-sensitive resistor), (*d*) thermocouple.

2.22. Which systems in Problems 2.1 to 2.8 and 2.11 to 2.18 are servomechanisms?

2.23. The endocrine gland known as the *adrenal cortex* is located on top of each kidney. It secretes several hormones, one of which is *cortisol* (hydrocortisone). Cortisol plays an important part in regulating the metabolism of carbohydrates, proteins, and fats, particularly in times of mental and physical stress. An excess of cortisol in the bloodstream leads to a high blood sugar (glucose) level and the disease known as *diabetes mellitus*. Cortisol production is controlled by the adrenocorticotrophic hormone (ACTH) from the pituitary gland. High blood cortisol inhibits ACTH production. Draw a block diagram of this feedback control system.

2.24. Draw block diagrams for each of the following elements, first with voltage v as input and current i as output, and then vice-versa: (*a*) resistance R, (*b*) capacitance C, (*c*) inductance L.

2.25. Draw block diagrams for each of the following mechanical systems, where force is the input and position the output: (*a*) a dashpot, (*b*) a spring, (*c*) a mass, (*d*) a mass, spring, and dashpot connected in series and fastened at one end (mass position is the output).

2.26. Draw a block diagram of a (*a*) parallel, (*b*) series R-L-C network.

2.27. Which systems described in the problems of this chapter are regulators?

ANSWERS TO SUPPLEMENTARY PROBLEMS

2.16. The equivalent circuit for the cathode follower has the same form as the voltage divider network of Problem 1.11, Page 7. Therefore the open-loop equation for the output is

$$v_{\text{out}} = \frac{\mu R_K}{r_p + R_K}(v_{\text{in}} - v_{\text{out}}) = \left(\frac{\mu R_K}{r_p + (1 + \mu)R_K}\right)v_{\text{in}}$$

and the open-loop block diagram is given by

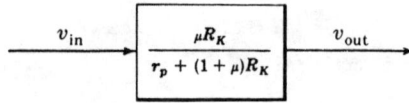

The closed-loop output equation is simply $v_{\text{out}} = \dfrac{\mu R_K}{r_p + R_K}(v_{\text{in}} - v_{\text{out}})$, and the closed-loop block diagram is given by

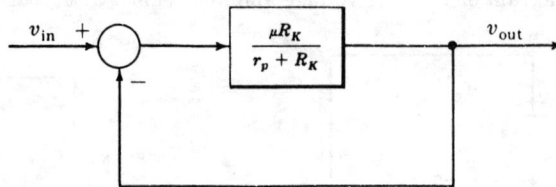

2.17.

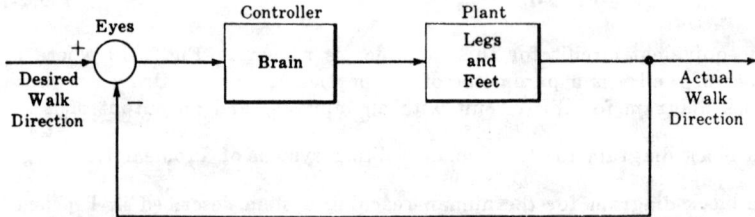

2.18.

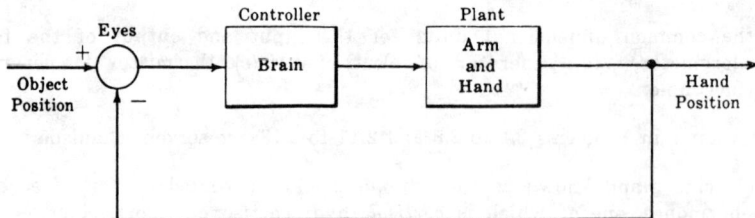

2.19.

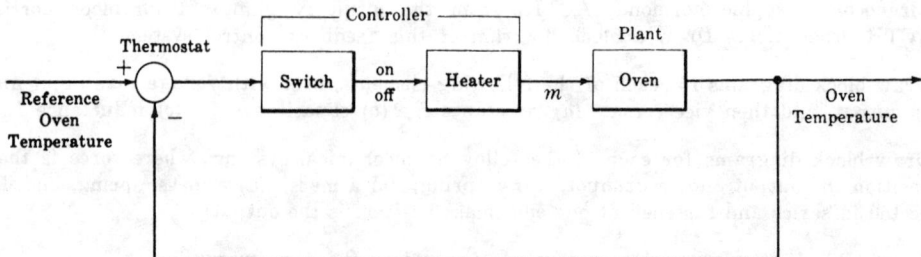

When $e > 0$ $(r > b)$, the switch turns the heater on. When $e \leqq 0$, the heater is turned off.

2.20.

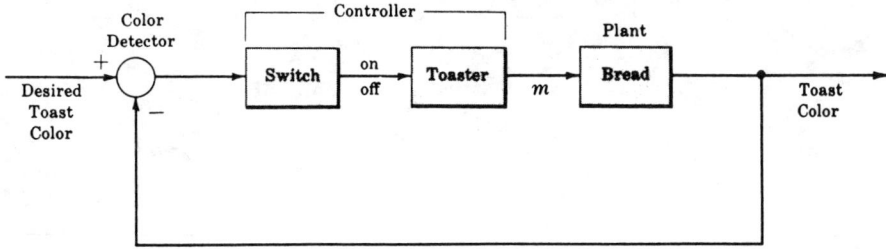

2.21. (a) The input to an accelerometer is acceleration. The output is displacement of a mass, voltage, or another quantity proportional to acceleration.

(b) See Problem 1.2, Page 6.

(c) The input to a thermistor is temperature. The output is an electrical quantity measured in ohms, volts, or amperes.

(d) The input to a thermocouple is a temperature difference. The output is a voltage.

2.22. The following problems describe systems which are servomechanisms: Examples 1.3 and 1.5 in Problem 2.6, Problems 2.7, 2.8, 2.17 and 2.18.

2.23.

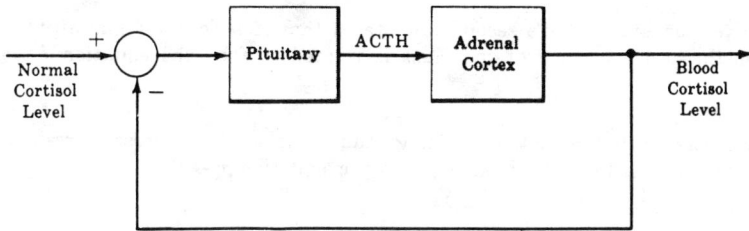

2.27. The systems of Examples 1.2 and 1.4 in Problem 2.6, and the systems of Problems 2.7, 8, 12, 13, 14, 15, 19, 20, and 23 are regulators.

Chapter 3

Linear Systems and Differential Equations

3.1 EQUATIONS OF PHYSICAL SYSTEMS

A property common to all basic laws of physics is that certain fundamental quantities can be defined by numerical values. The physical laws define relationships between these fundamental quantities and are usually represented by equations.

Example 3.1.

The scalar version of Newton's second law states that if a force of magnitude f is applied to a mass of M units then the acceleration a of the mass is related to f by the equation $f = Ma$.

Example 3.2.

Ohm's law states that if a voltage of magnitude v is applied across a resistor of R units then the current i through the resistor is related to v by the equation $v = Ri$.

3.2 DIFFERENTIAL EQUATIONS

One class of equations which has broad application in the description of physical laws is differential equations.

Definition 3.1: A **differential equation** is any algebraic or transcendental equality which involves either differentials or derivatives.

Differential equations are useful for relating rates of change of variables and other parameters.

Example 3.3.

Newton's second law (Example 3.1) can be written alternatively as a relationship between force f, mass M, and the rate of change of the velocity v of the mass with respect to time t; that is, $f = M\dfrac{dv}{dt}$.

Example 3.4.

Ohm's law (Example 3.2) can be written alternatively as a relationship between voltage v, resistance R, and the time rate of passage of charge through the resistor; that is, $v = R\dfrac{dq}{dt}$.

Example 3.5.

Newton's second law (Example 3.3) can be written in the form $f\,dt = M\,dv$.

Example 3.6.

The diffusion equation in one dimension describes the relationship between the time rate of change of a quantity T in a body (e.g., heat concentration in an iron bar) and the positional rate of change of T: $\dfrac{\partial T}{\partial x} = k\dfrac{\partial T}{\partial t}$ where k is a proportionality constant, x is a position variable, and t is time.

3.3 PARTIAL AND ORDINARY DIFFERENTIAL EQUATIONS

Definition 3.2:   A **partial differential equation** is an equality involving one or more dependent and two or more independent variables, together with partial derivatives of the dependent with respect to the independent variables.

Definition 3.3:   An **ordinary (total) differential equation** is an equality involving one or more dependent variables, one independent variable, and one or more derivatives of the dependent variables with respect to the independent variable.

Example 3.7.

The diffusion equation $\frac{\partial T}{\partial x} = k\frac{\partial T}{\partial t}$ is a partial differential equation. $T = T(x, t)$ is the dependent variable which represents the concentration of some quantity at some position and some time in the body. The independent variable x defines the position in the body, and the independent variable t defines the time.

Example 3.8.

Newton's second law (Example 3.3) is an ordinary differential equation: $f = M\frac{dv}{dt}$. The velocity $v = v(t)$ and the force $f = f(t)$ are dependent variables and the time t is the independent variable.

Example 3.9.

Ohm's law (Example 3.4) is an ordinary differential equation: $v = R\frac{dq}{dt}$. The charge $q = q(t)$ and the voltage $v = v(t)$ are dependent variables and the time t is the independent variable.

Example 3.10.

A differential equation of the form

$$a_n\frac{d^ny}{dt^n} + a_{n-1}\frac{d^{n-1}y}{dt^{n-1}} + \cdots + a_1\frac{dy}{dt} + a_0y = x(t)$$

or, more compactly,
$$\sum_{i=0}^{n} a_i\frac{d^iy(t)}{dt^i} = x(t) \tag{3.1}$$

where a_i, $i = 0, 1, \ldots, n$, are constants, is an ordinary differential equation. $y(t)$ *and* $x(t)$ **are dependent variables, and** t **is the independent variable.**

3.4 TIME-VARIABLE AND TIME-INVARIANT DIFFERENTIAL EQUATIONS

Definition 3.4:   A **time-variable differential equation** is a differential equation in which one or more terms depend *explicitly* on the independent variable time t.

Definition 3.5:   A **time-invariant differential equation** is a differential equation in which none of the terms depends *explicitly* on the independent variable time t.

Definition 3.5 implies that the process defined by a time-invariant differential equation does not vary with time.

Example 3.11.

The differential equation $t^2\frac{d^2y}{dt^2} + y = x$ where x and y are dependent variables is *time-variable* since the term $t^2\frac{d^2y}{dt^2}$ depends *explicitly* on t through the coefficient t^2.

Example 3.12.

Any differential equation of the form

$$\sum_{i=0}^{n} a_i\frac{d^iy}{dt^i} = \sum_{i=0}^{m} b_i\frac{d^ix}{dt^i} \tag{3.2}$$

where the coefficients $a_0, a_1, \ldots, a_n, b_0, b_1, \ldots, b_m$ are constants is *time-invariant* since the equation depends only *implicitly* on t through the dependent variables x and y and their derivatives.

3.5 LINEAR AND NONLINEAR DIFFERENTIAL EQUATIONS

A **term** of an ordinary differential equation consists of products and quotients of explicit functions of the independent variable t, and functions of the dependent variables and their derivatives. For example, $(5/\cos t)(d^2y/dt^2)$ is a term of first degree in the dependent variable y, and $2xy^3(dy/dt)$ is a term of fifth degree in the dependent variables x and y.

Definition 3.6:   A **linear term** is one which is first degree in the dependent variables and their derivatives.

Definition 3.7:   A **linear differential equation** is a differential equation consisting of a sum of linear terms. All others are **nonlinear differential equations**.

If a differential equation contains terms which are higher powers, products, or transcendental functions of the dependent variables, it is nonlinear. Such terms include $\left(\dfrac{dy}{dt}\right)^3$, $x\dfrac{dy}{dt}$, and $\sin x$, respectively.

Example 3.13.

The diffusion equation $\dfrac{\partial T}{\partial x} = k\dfrac{\partial T}{\partial t}$ is a linear (partial) differential equation. The terms $\dfrac{\partial T}{\partial x}$ and $k\dfrac{\partial T}{\partial t}$ are first degree.

Example 3.14.

The ordinary differential equations $\left(\dfrac{dy}{dt}\right)^2 + y = 0$ and $\dfrac{d^2y}{dt^2} + \cos y = 0$ are nonlinear because $(dy/dt)^2$ is second degree in the first equation, and $\cos y$ in the second equation is *not* first degree, which is true of all transcendental functions.

Example 3.15.

The ordinary differential equation $\dfrac{d^2y}{dt^2} + x\dfrac{dy}{dt} + y = x$ where x and y are dependent variables is a nonlinear differential equation since $x(dy/dt)$ is second degree in the dependent variables x and y.

Example 3.16.

Any ordinary differential equation

$$\sum_{i=0}^{n} a_i(t)\frac{d^i y}{dt^i} = \sum_{i=0}^{m} b_i(t)\frac{d^i x}{dt^i} \tag{3.3}$$

where the coefficients $a_i(t)$ and $b_i(t)$ depend only upon the independent variable t, is a linear differential equation.

3.6 LINEARITY AND SUPERPOSITION

The concept of linearity has been presented in Definition 3.7 as a property of a class of differential equations. In this section, linearity is discussed as a property of general systems in which there is one independent variable, time t. In Chapters 1 and 2, the concepts of system, input, and output were defined. The following definition of linearity is based on these earlier definitions.

Definition 3.8:   A **linear system** is a system which has the property that if:

   (a) an input $x_1(t)$ produces an output $y_1(t)$, and

   (b) an input $x_2(t)$ produces an output $y_2(t)$, then

   (c) an input $c_1x_1(t) + c_2x_2(t)$ produces an output $c_1y_1(t) + c_2y_2(t)$ for all pairs of inputs $x_1(t)$ and $x_2(t)$ and all pairs of constants c_1 and c_2.

Linear systems can often be represented by linear differential equations.

Example 3.17.

Any system is *linear* if its input-output relationship can be described by the linear differential equation

$$\sum_{i=0}^{n} a_i(t) \frac{d^i y}{dt^i} = \sum_{i=0}^{m} b_i(t) \frac{d^i x}{dt^i} \qquad (3.4)$$

where $y = y(t)$ is the system output, and $x = x(t)$ is the system input.

Example 3.18.

Any system is linear if its input-output relationship can be described by the integral

$$y(t) = \int_{-\infty}^{\infty} w(t, \tau)\, x(\tau)\, d\tau \qquad (3.5)$$

where $w(t, \tau)$ is a function which embodies the internal physical properties of the system, $y(t)$ is the output, and $x(t)$ is the input.

The relationship between the systems of Examples 3.17 and 3.18 is discussed in Section 3.13.

The concept of linearity can be represented by the

Principle of Superposition: The response $y(t)$ of a linear system due to several inputs $x_1(t), x_2(t), \ldots, x_n(t)$ acting simultaneously is equal to the sum of the responses of each input acting alone. That is, if $y_i(t)$ is the response due to the input $x_i(t)$, then

$$y(t) = \sum_{i=1}^{n} y_i(t)$$

Example 3.19.

A linear system is described by the linear algebraic equation

$$y(t) = 2\, x_1(t) + x_2(t)$$

where $x_1(t) = t$ and $x_2(t) = t^2$ are inputs, and $y(t)$ is the output. When $x_1(t) = t$ and $x_2(t) = 0$, then $y(t) = y_1(t) = 2t$. When $x_1(t) = 0$ and $x_2(t) = t^2$, then $y(t) = y_2(t) = t^2$. The total output resulting from $x_1(t) = t$ and $x_2(t) = t^2$ is then equal to

$$y(t) = y_1(t) + y_2(t) = 2t + t^2$$

The principle of superposition apparently follows directly from the definition of linearity (Definition 3.8). Also, it should be noted that any system which satisfies the principle of superposition is linear. The concepts of *linearity* and *superposition* are equivalent.

Definition 3.9: The input-output relationship of a linear system can often be described by the integral

$$y(t) = \int_{-\infty}^{\infty} w(t, \tau)\, x(\tau)\, d\tau$$

where $x(t)$ is the input and $y(t)$ the resulting output. The function of two variables $w(t, \tau)$ which embodies the physical properties of the system is called the **weighting function** of the system.

Definition 3.10: The integral of the form

$$y(t) = \int_{-\infty}^{\infty} w(t, \tau)\, x(\tau)\, d\tau$$

is called a **convolution integral**.

If a linear system is described by a convolution integral, then the output $y(t)$ at a given time t is a *weighted* sum (an integral in the limit) of the values of the input over the interval from negative to positive infinity. That is, the contribution to the output $y(t)$ from the input $x(\tau)$ is a weighted value of $x(\tau)$, where the weighting is provided by the weighting function $w(t, \tau)$.

3.7 CAUSALITY AND PHYSICALLY REALIZABLE SYSTEMS

The properties of a physical system restrict the form of its output. This restriction is embodied in the concept of *causality*.

Definition 3.11: A system in which time is the independent variable is called **causal** if the output depends only on the present and past values of the input. That is, if $y(t)$ is the output, then $y(t)$ depends only on the input $x(\tau)$ for values of $\tau \leq t$.

The implication of Definition 3.11 is that a *causal* system is one which cannot anticipate what its future input will be. Accordingly, causal systems are sometimes called **physically realizable** systems. An important consequence of causality (physical realizability) is that the weighting function $w(t, \tau)$ of a causal system is identically zero for $\tau > t$, that is, future values of the input are weighted zero.

3.8 LINEARIZED AND PIECEWISE-LINEAR SYSTEMS

In reality, no physical system can be described *exactly* by a linear constant-coefficient differential equation; however, many systems can be represented over a *limited operating range*, or *approximated* by such equations.

Example 3.20.

Consider the spring-mass system of Fig. 3-1 where the spring force $f_s(x)$ is a nonlinear function of the displacement x measured from the rest position as shown in Fig. 3-2.

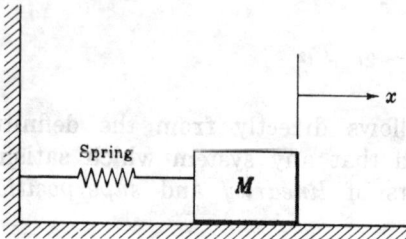

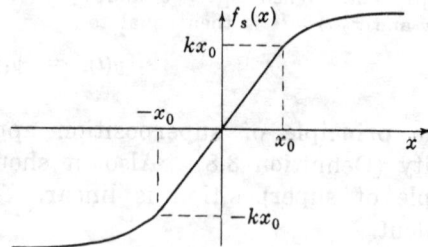

Fig. 3-1   Fig. 3-2

The equation of motion of the mass is $M\dfrac{d^2x}{dt^2} + f_s(x) = 0$. However, if the absolute magnitude of the displacement does not exceed x_0, then $f_s(x) = kx$ where k is a constant. In this case the equation of motion is a constant coefficient linear equation given by $M\dfrac{d^2x}{dt^2} + kx = 0$ and is valid for $|x| \leq x_0$.

Example 3.21.

Consider again the system of Example 3.20. Suppose this time that the displacement x exceeds x_0. To treat this problem let the spring force curve be approximated by three straight lines as shown in Fig. 3-3 below.

The system is then approximated by a piecewise linear system; that is, the system is described by the linear equation $M\dfrac{d^2x}{dt^2} + kx = 0$ where $|x| \leq x_1$, and by the equations $M\dfrac{d^2x}{dt^2} \pm F_1 = 0$ when $|x| > x_1$. The + sign is used if $x > x_1$ and the − sign if $x < -x_1$.

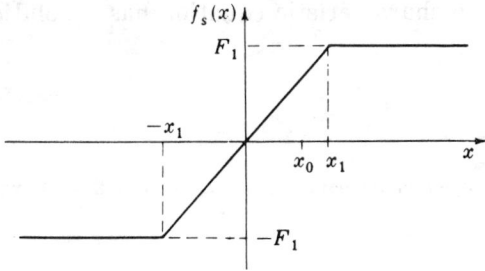

Fig. 3-3

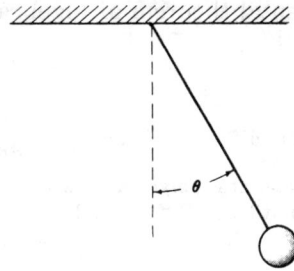

Fig. 3-4

Example 3.22.

Consider the nonlinear equation describing the motion of a pendulum (see **Fig. 3-4**)

$$\frac{d^2\theta}{dt^2} + (g/l)\sin\theta = 0$$

where l is the length of the pendulum bob and g is the acceleration of gravity. If small motions of the pendulum about the "operating point" $\theta = 0$ are of interest, then the equation of motion can be linearized about this operating point. This is done by forming a Taylor series expansion of the nonlinear term $(g/l)\sin\theta$ about the point $\theta = 0$ and retaining only the first degree terms. The nonlinear equation is

$$\frac{d^2\theta}{dt^2} + (g/l)\sin\theta = \frac{d^2\theta}{dt^2} + (g/l)\sum_{n=0}^{\infty}\frac{\theta^n}{n!}\left(\frac{d^n}{d\theta^n}(\sin\theta)\bigg|_{\theta=0}\right)$$

$$= \frac{d^2\theta}{dt^2} + (g/l)\left[\theta - \frac{\theta^3}{3!} + \cdots\right] = 0$$

The linear equation is $\frac{d^2\theta}{dt^2} + (g/l)\theta = 0$, valid for small variations in θ.

3.9 THE DIFFERENTIAL OPERATOR D AND THE CHARACTERISTIC EQUATION

Consider the nth order linear constant coefficient differential equation

$$\frac{d^n y}{dt^n} + a_{n-1}\frac{d^{n-1}y}{dt^{n-1}} + \cdots + a_1\frac{dy}{dt} + a_0 y = x \qquad (3.6)$$

It is convenient to define a **differential operator**

$$D \equiv \frac{d}{dt}$$

and more generally an **nth order differential operator**

$$D^n \equiv \frac{d^n}{dt^n}$$

The differential equation can now be written

$$D^n y + a_{n-1}D^{n-1}y + \cdots + a_1 Dy + a_0 y = x$$

or

$$(D^n + a_{n-1}D^{n-1} + \cdots + a_1 D + a_0)y = x$$

Definition 3.12: The polynomial in D

$$D^n + a_{n-1}D^{n-1} + \cdots + a_1 D + a_0 \qquad (3.7)$$

is called the **characteristic polynomial.**

Definition 3.13: The equation

$$D^n + a_{n-1}D^{n-1} + \cdots + a_1 D + a_0 = 0 \qquad (3.8)$$

is called the **characteristic equation.**

The fundamental theorem of algebra states that the characteristic equation has n solutions $D = D_1$, $D = D_2$, ..., $D = D_n$.

Example 3.23.

Consider the differential equation $\dfrac{d^2y}{dt^2} + 3\dfrac{dy}{dt} + 2y = x$.

The characteristic polynomial is $D^2 + 3D + 2$. The characteristic equation is $D^2 + 3D + 2 = 0$ which has the two solutions: $D = -1$ and $D = -2$.

3.10 LINEAR INDEPENDENCE AND FUNDAMENTAL SETS

Definition 3.14:   A set of n functions of time $f_1(t), f_2(t), \ldots, f_n(t)$ is called **linearly independent** if the only set of constants $c_1, c_2, \ldots, c_n$ for which

$$c_1 f_1(t) + c_2 f_2(t) + \cdots + c_n f_n(t) = 0$$

for all t are the constants $c_1 = c_2 = \cdots = c_n = 0$.

Example 3.24.

The functions t and t^2 are linearly independent functions since

$$c_1 t + c_2 t^2 = t(c_1 + c_2 t) = 0$$

implies that $c_1/c_2 = -t$. There are *no constants* which satisfy this relationship.

A *homogeneous* nth order linear differential equation of the form

$$\sum_{i=0}^{n} a_i \frac{d^i y}{dt^i} = 0$$

has at least one set of n linearly independent solutions.

Definition 3.15:   Any set of n *linearly independent* solutions of a homogeneous nth order linear differential equation is called a **fundamental set.**

There is no unique fundamental set. From a given fundamental set other fundamental sets can be generated by the following technique. Suppose that $y_1(t), y_2(t), \ldots, y_n(t)$ is a fundamental set for an nth order linear differential equation. Then a set of n functions $z_1(t), z_2(t), \ldots, z_n(t)$ can be formed:

$$z_1(t) = \sum_{i=1}^{n} a_{1i} y_i(t), \quad z_2(t) = \sum_{i=1}^{n} a_{2i} y_i(t), \quad \ldots, \quad z_n(t) = \sum_{i=1}^{n} a_{ni} y_i(t) \tag{3.9}$$

where the a_{ji} are a set of n^2 constants. Each $z_i(t)$ is a solution of the differential equation. This set of n solutions is a *fundamental set* if the determinant

$$\begin{vmatrix} a_{11} & a_{12} & \cdots & a_{1n} \\ a_{21} & a_{22} & \cdots & a_{2n} \\ \cdots\cdots\cdots\cdots\cdots\cdots \\ a_{n1} & a_{n2} & \cdots & a_{nn} \end{vmatrix} \neq 0$$

Example 3.25.

The equation for simple harmonic motion, $\dfrac{d^2y}{dt^2} + \omega^2 y = 0$, has as a fundamental set

$$y_1 = \sin \omega t \qquad y_2 = \cos \omega t$$

A second fundamental set is*

$$z_1 = \cos \omega t + j \sin \omega t = e^{j\omega t} \qquad z_2 = \cos \omega t - j \sin \omega t = e^{-j\omega t}$$

---

* The *complex exponential function* e^w, where $w = u + jv$ for real u and v, and $j = \sqrt{-1}$, is defined in complex variable theory by $e^w \equiv e^u(\cos v + j \sin v)$. Therefore, $e^{\pm j\omega t} = \cos \omega t \pm j \sin \omega t$.

Distinct Roots

In general, if the characteristic equation

$$\sum_{i=0}^{n} a_i D^i = 0$$

has distinct roots $D_1, D_2, \ldots, D_n$, then a fundamental set for the homogeneous equation

$$\sum_{i=0}^{n} a_i \frac{d^i y}{dt^i} = 0$$

is the set of functions $y_1 = e^{D_1 t}$, $y_2 = e^{D_2 t}$, $\ldots$, $y_n = e^{D_n t}$.

Example 3.26.

The differential equation $\dfrac{d^2 y}{dt^2} + 3 \dfrac{dy}{dt} + 2y = 0$ has the characteristic equation $D^2 + 3D + 2 = 0$ whose roots are $D = D_1 = -1$ and $D = D_2 = -2$. A fundamental set for this equation is $y_1 = e^{-t}$ and $y_2 = e^{-2t}$.

Repeated Roots

If the characteristic equation has repeated roots, then for each root D_i of multiple n_i there are n_i elements of the fundamental set $e^{D_i t}, te^{D_i t}, \ldots, t^{n_i - 1} e^{D_i t}$.

Example 3.27.

The equation $\dfrac{d^2 y}{dt^2} + 2 \dfrac{dy}{dt} + y = 0$ whose characteristic equation is $D^2 + 2D + 1 = 0$ with the repeated root $D = -1$, has a fundamental set consisting of e^{-t} and te^{-t}.

3.11 SOLUTION OF LINEAR CONSTANT COEFFICIENT ORDINARY DIFFERENTIAL EQUATIONS

Consider the class of differential equations of the form

$$\sum_{i=0}^{n} a_i \frac{d^i y}{dt^i} = \sum_{i=0}^{m} b_i \frac{d^i x}{dt^i} \tag{3.10}$$

where t is time, the coefficients a_i and b_i are constant, $x = x(t)$ *(the input) is a known time function*, and $y = y(t)$ (the output) is the unknown solution of the equation. If this equation describes a physical system, then generally $m \le n$, and n is called the **order** of the differential equation. To completely specify the problem so that *a unique solution* $y(t)$ can be obtained, two additional items must be specified: (1) the interval of time over which a solution is desired and (2) a set of *n initial conditions* for $y(t)$ and its first $n-1$ derivatives. The time interval for the class of problems considered is defined by $0 \le t < +\infty$. This interval is used in the remainder of this book unless otherwise specified. The set of initial conditions is

$$y(0), \ \left. \frac{dy}{dt}\right|_{t=0}, \ \ldots, \ \left. \frac{d^{n-1} y}{dt^{n-1}}\right|_{t=0} \tag{3.11}$$

A problem defined over this interval and with these initial conditions is called an **initial value problem.**

The solution of a differential equation of this class can be divided into two parts, a *free response*, and a *forced response*. The sum of these two responses constitutes the *total response*, or solution $y(t)$, of the equation.

3.12 THE FREE RESPONSE

The **free response** of a differential equation is the solution of the differential equation when the input $x(t)$ is identically zero.

If the input $x(t)$ is identically zero, then the differential equation assumes the form

$$\sum_{i=0}^{n} a_i \frac{d^i y}{dt^i} = 0 \qquad (3.12)$$

The solution $y(t)$ of such an equation depends only on the n initial conditions in (3.11).

Example 3.28.

The solution of the homogeneous first order differential equation $dy/dt + y = 0$ with the initial condition $y(0) = c$, is $y(t) = ce^{-t}$. This can be verified by direct substitution. ce^{-t} is the free response of any differential equation of the form $dy/dt + y = x$ with the initial condition $y(0) = c$.

The *free response* of a differential equation can always be written as a linear combination of the elements of a *fundamental set*. That is, if $y_1(t), y_2(t), \ldots, y_n(t)$ is a fundamental set, then *any* free response $y_a(t)$ of the differential equation can be represented in the form

$$y_a(t) = \sum_{i=1}^{n} c_i \, y_i(t) \qquad (3.13)$$

where the *constants* c_i are determined in terms of the initial conditions

$$y(0), \quad \frac{dy}{dt}\Big|_{t=0}, \quad \ldots, \quad \frac{d^{n-1}y}{dt^{n-1}}\Big|_{t=0}$$

from the set of n algebraic equations

$$y(0) = \sum_{i=1}^{n} c_i y_i(0), \quad \frac{dy}{dt}\Big|_{t=0} = \sum_{i=1}^{n} c_i \frac{dy_i}{dt}\Big|_{t=0}, \quad \ldots, \quad \frac{d^{n-1}y}{dt^{n-1}}\Big|_{t=0} = \sum_{i=1}^{n} c_i \frac{d^{n-1}y_i}{dt^{n-1}}\Big|_{t=0} \qquad (3.14)$$

The linear independence of the $y_i(t)$ guarantees that a solution to these equations can be obtained for c_i, $i = 1, 2, \ldots, n$.

Example 3.29.

The free response $y_a(t)$ of the differential equation

$$\frac{d^2 y}{dt^2} + 3 \frac{dy}{dt} + 2y = x$$

with initial conditions $y(0) = 0$, $\dfrac{dy}{dt}\Big|_{t=0} = 1$ is determined by letting

$$y_a(t) = c_1 e^{-t} + c_2 e^{-2t}$$

where c_1 and c_2 are unknown coefficients and e^{-t} and e^{-2t} are a fundamental set for the equation (Example 3.26). Since $y_a(t)$ must satisfy the initial conditions, that is,

$$y_a(0) = y(0) = 0 = c_1 + c_2, \qquad \frac{dy_a(t)}{dt}\Big|_{t=0} = \frac{dy}{dt}\Big|_{t=0} = 1 = -c_1 - 2c_2$$

then $c_1 = 1$ and $c_2 = -1$. The free response is therefore given by $y_a(t) = e^{-t} - e^{-2t}$.

3.13 THE FORCED RESPONSE

The **forced response** $y_b(t)$ of a differential equation is the solution of the differential equation when all the initial conditions

$$y(0), \quad \frac{dy}{dt}\Big|_{t=0}, \quad \ldots, \quad \frac{d^{n-1}y}{dt^{n-1}}\Big|_{t=0}$$

are identically zero.

The implication of this definition is that the forced response depends only on the input $x(t)$. The *forced response* for a linear constant coefficient ordinary differential equation can be written in terms of a convolution integral

$$y_b(t) = \int_0^t w(t-\tau) \left[\sum_{i=0}^{m} b_i \frac{d^i x(\tau)}{d\tau^i} \right] d\tau \qquad (3.15)$$

where $w(t - \tau)$ is the *weighting function of the differential equation*. This form of the convolution integral assumes that the weighting function describes a *causal* system. This assumption is maintained below.

The weighting function of a linear constant coefficient ordinary differential equation can be written as

$$w(t) \;=\; \sum_{i=1}^{n} c_i\, y_i(t) \qquad t \geqq 0$$
$$\;=\; 0 \qquad\qquad\qquad t < 0 \tag{3.16}$$

where $c_1, \ldots, c_n$ are constants and the set of functions $y_1(t), y_2(t), \ldots, y_n(t)$ is a fundamental set of the differential equation. It should be noted that $w(t)$ *is a free response of the differential equation* and therefore requires n initial conditions for complete specification. These conditions fix the values of the constants $c_1, c_2, \ldots, c_n$. The initial conditions which all weighting functions of linear differential equations must satisfy are

$$w(0) \;=\; 0, \quad \left.\frac{dw}{dt}\right|_{t=0} \;=\; 0, \quad \ldots, \quad \left.\frac{d^{n-2} w}{dt^{n-2}}\right|_{t=0} \;=\; 0, \quad \left.\frac{d^{n-1} w}{dt^{n-1}}\right|_{t=0} \;=\; 1 \tag{3.17}$$

Example 3.30.

The weighting function of the differential equation $\dfrac{d^2y}{dt^2} + 3\dfrac{dy}{dt} + 2y = x$ is a linear combination of e^{-t} and e^{-2t} (a fundamental set of the equation). That is,

$$w(t) \;=\; c_1 e^{-t} + c_2 e^{-2t}$$

c_1 and c_2 are determined from the two algebraic equations

$$w(0) \;=\; 0 \;=\; c_1 + c_2, \qquad \left.\frac{dw}{dt}\right|_{t=0} \;=\; 1 \;=\; -c_1 - 2c_2$$

The solution is $c_1 = 1$, $c_2 = -1$ and the weighting function is $w(t) = e^{-t} - e^{-2t}$.

Example 3.31.

For the differential equation of Example 3.30, if $x(t) = 1$, then the forced response $y_b(t)$ of the equation is

$$y_b(t) \;=\; \int_0^t w(t-\tau)\, x(\tau)\, d\tau \;=\; \int_0^t \left[e^{-(t-\tau)} - e^{-2(t-\tau)} \right] d\tau$$
$$\;=\; e^{-t} \int_0^t e^{\tau}\, d\tau \;-\; e^{-2t} \int_0^t e^{2\tau}\, d\tau \;=\; \tfrac{1}{2}(1 - 2e^{-t} + e^{-2t})$$

3.14   THE TOTAL RESPONSE

The **total response** of a linear constant coefficient differential equation is the sum of the *free response* and the *forced response*.

Example 3.32.

The total response $y(t)$ of the differential equation $\dfrac{d^2y}{dt^2} + 3\dfrac{dy}{dt} + 2y = 1$ with initial conditions $y(0) = 0$ and $\left.\dfrac{dy}{dt}\right|_{t=0} = 1$ is the sum of the free response $y_a(t)$ determined in Example 3.29 and the forced response $y_b(t)$ determined in Example 3.31. Thus

$$y(t) \;=\; y_a(t) + y_b(t) \;=\; (e^{-t} - e^{-2t}) + \tfrac{1}{2}(1 - 2e^{-t} + e^{-2t}) \;=\; \tfrac{1}{2}(1 - e^{-2t})$$

3.15   THE STEADY-STATE AND TRANSIENT RESPONSES

The *steady-state response* and *transient response* are another pair of quantities whose sum is equal to the total response. These terms are often used for specifying control system performance. They are defined as follows:

Definition 3.16:   The **steady-state response** is that part of the total response which *does not* approach zero as time approaches infinity.

Definition 3.17:   The **transient response** is that part of the total response which approaches zero as time approaches infinity.

Example 3.33.

The total response for the differential equation in Example 3.32 was determined as $y = \frac{1}{2} - \frac{1}{2}e^{-t}$. Clearly, the steady state response is given by $y_{ss} = \frac{1}{2}$. Since $\lim\limits_{t \to \infty} [-\frac{1}{2}e^{-t}] = 0$, the transient response is $y_T = -\frac{1}{2}e^{-t}$.

3.16 SINGULARITY FUNCTIONS

In the study of control systems and the differential equations which describe them, a particular family of functions, the *singularity functions*, is extensively used. Each member of this family is related to the others by one or more integrations or differentiations. Of this family, the three most widely used functions are the unit step function, the unit impulse function, and the unit ramp function.

Definition 3.18:   A **unit step function** is a function of time denoted by $u(t - t_0)$ and defined by

$$u(t - t_0) = \begin{cases} 1 & t > t_0 \\ 0 & t \leq t_0 \end{cases} \qquad (3.18)$$

A graphical plot of a unit step function is given in Fig. 3-5.

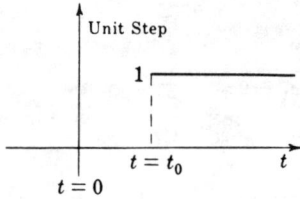
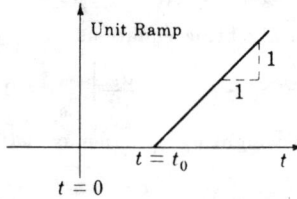
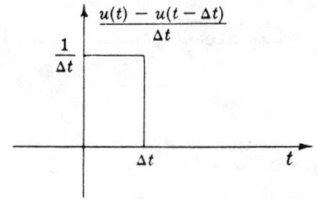

Fig. 3-5   **Fig. 3-6**   **Fig. 3-7**

Definition 3.19:   A **unit ramp function** is a function of time which is the integral of a unit step function given by

$$\int_{-\infty}^{t} u(\tau - t_0)\, d\tau = \begin{cases} t - t_0 & t > t_0 \\ 0 & t \leq t_0 \end{cases} \qquad (3.19)$$

A graphical plot of a unit ramp function is given in Fig. 3-6.

Definition 3.20:   A **unit impulse function** $\delta(t)$ may be defined by

$$\delta(t) = \lim_{\substack{\Delta t \to 0 \\ \Delta t > 0}} \left[\frac{u(t) - u(t - \Delta t)}{\Delta t} \right] \qquad (3.20)^*$$

where $u(t)$ is the unit step function.

The pair $\begin{cases} \Delta t \to 0 \\ \Delta t > 0 \end{cases}$ may be abbreviated by $\Delta t \to 0^+$, meaning that Δt approaches zero *from the right*. The quotient in brackets represents a rectangle of height $1/\Delta t$ and width Δt as shown in Fig. 3-7. The limiting process produces a function whose height approaches infinity and width approaches zero. The area under the curve is equal to 1 for all values of Δt. That is,

$$\int_{-\infty}^{\infty} \delta(t)\, dt = 1$$

---

* In a formal sense, Equation *(3.20)* defines the *one-sided derivative* of the unit step function. But neither the limit nor the derivative exist in the ordinary mathematical sense. A more detailed discussion of these points can be found in reference [1]. However, Definition 3.20 is **satisfactory for the purposes** of this book.

The unit impulse function has the following very important property:

Screening Property: The integral of the product of a unit impulse function $\delta(t - t_0)$ and a function $f(t)$ which is continuous at $t = t_0$ over an interval which includes t_0, is equal to the function $f(t)$ evaluated at t_0; that is,

$$\int_{-\infty}^{\infty} f(t)\, \delta(t - t_0)\, dt \;=\; f(t_0) \tag{3.21}$$

Definition 3.21: The **unit impulse response** of a linear system is the output $y(t)$ of the system when the input $x(t) = \delta(t)$ and all initial conditions are zero.

Example 3.34.

If the input-output relationship of a linear system is given by the convolution integral

$$y(t) \;=\; \int_0^t w(t - \tau)\, x(\tau)\, d\tau$$

then the unit impulse response $y_\delta(t)$ of the system is

$$y_\delta(t) \;=\; \int_0^t w(t - \tau)\, \delta(\tau)\, d\tau \;=\; \int_{-\infty}^{\infty} w(t - \tau)\, \delta(\tau)\, d\tau \;=\; w(t) \tag{3.22}$$

since $w(t - \tau) = 0$ for $\tau > t$, $\delta(\tau) = 0$ for $\tau < 0$, and the screening property of the unit impulse has been used to evaluate the integral.

Definition 3.22: The **unit step response** is the output $y(t)$ when the input $x(t) = u(t)$ and all initial conditions are zero.

Definition 3.23: The **unit ramp response** is the output $y(t)$ when the input $x(t) = t$ for $t > 0$, $x(t) = 0$ for $t \leq 0$, and all initial conditions are zero.

3.17 SECOND ORDER SYSTEMS

In the study of control systems, linear constant coefficient second order differential equations of the form

$$\frac{d^2 y}{dt^2} + 2\zeta \omega_n \frac{dy}{dt} + \omega_n^2 y \;=\; \omega_n^2 x \tag{3.23}$$

are very important because higher order systems can often be approximated by second order systems. The constant ζ is called the **damping ratio**, and the constant ω_n is called the **undamped natural frequency** of the system. The forced response of this equation for inputs x belonging to the class of singularity functions is of particular interest. That is, the *forced response* to a unit impulse, unit step, or unit ramp is the same as the *unit impulse response, unit step response,* or *unit ramp response* of a system represented by this equation.

Assuming that $0 \leq \zeta \leq 1$, the characteristic equation for Equation (*3.23*) is

$$D^2 + 2\zeta\omega_n D + \omega_n^2 \;=\; (D + \zeta\omega_n - j\omega_n\sqrt{1 - \zeta^2})(D + \zeta\omega_n + j\omega_n\sqrt{1 - \zeta^2}) \;=\; 0$$

Hence the roots are

$$D_1 \;=\; -\zeta\omega_n + j\omega_n\sqrt{1 - \zeta^2} \;\equiv\; -\alpha + j\omega_d \qquad D_2 \;=\; -\zeta\omega_n - j\omega_n\sqrt{1 - \zeta^2} \;\equiv\; -\alpha - j\omega_d$$

where $\alpha \equiv \zeta\omega_n$ is called the **damping coefficient**, and $\omega_d \equiv \omega_n\sqrt{1 - \zeta^2}$ is called the **damped natural frequency**. α is the inverse of the **time constant** τ of the system; that is, $\tau = 1/\alpha$.

The weighting function of Equation (3.23) is $w(t) = \frac{1}{\omega_d} e^{-\alpha t} \sin \omega_d t$. The unit step response is given by

$$y_u(t) = \int_0^t w(t-\tau)\,\omega_n^2\,d\tau = 1 - \frac{\omega_n e^{-\alpha t}}{\omega_d} \sin(\omega_d t + \phi) \qquad (3.24)$$

where $\phi \equiv \tan^{-1}(\omega_d/\alpha)$.

The following figure is a parametric representation of the unit step response. Note that the abscissa of this family of curves is normalized time $\omega_n t$, and the parameter defining each curve is the damping ratio ζ.

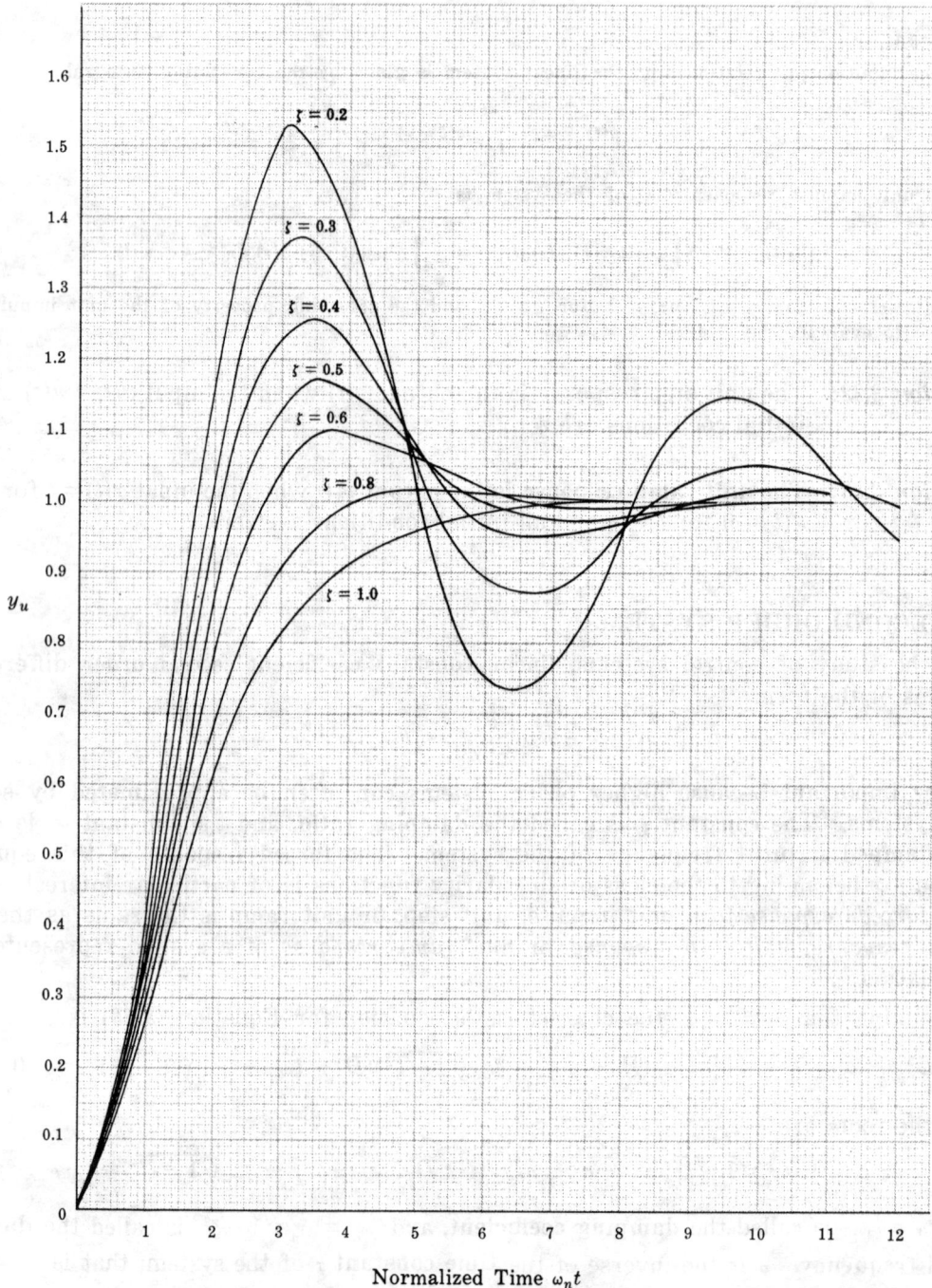

Normalized Time $\omega_n t$

Fig. 3-8

Solved Problems

EQUATIONS OF PHYSICAL SYSTEMS

3.1. Faraday's law states that the voltage v induced between the terminals of an inductor is equal to the time rate of change of flux linkages. (A flux linkage is defined as one line of magnetic flux linking one turn of the winding of the inductor.) Suppose it is experimentally determined that the number of flux linkages λ is related to the current i in the inductor as shown in the adjacent graph. The curve is approximately a straight line for $-I_0 \le i \le I_0$. Determine a differential equation, valid for $-I_0 \le i \le I_0$, which relates the induced voltage v and current i.

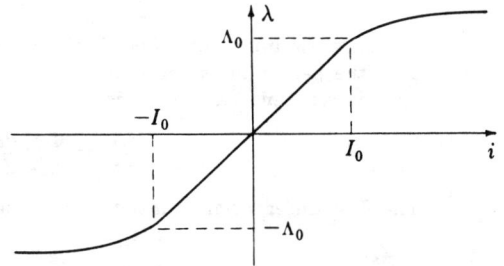

Faraday's law can be written as $v = d\lambda/dt$. It is seen from the graph that

$$\lambda = (\Lambda_0/I_0)i = Li \qquad -I_0 \le i \le I_0$$

where $L \equiv \Lambda_0/I_0$ is called the *inductance* of the inductor. The equation relating v and i is obtained by substituting Li for λ:

$$v = \frac{d\lambda}{dt} = \frac{d}{dt}(Li) = L\frac{di}{dt} \qquad \text{where} \quad -I_0 \le i \le I_0$$

3.2. Determine a differential equation relating the voltage $v(t)$ and the current $i(t)$ for $t \ge 0$ for the electrical network given in the adjoining figure. Assume the capacitor is uncharged at $t = 0$, the current i is zero at $t = 0$, and the switch S closes at $t = 0$.

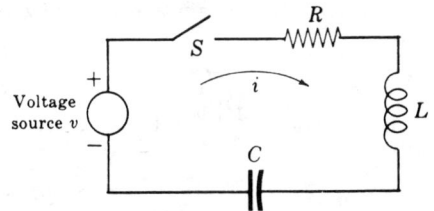

By Kirchhoff's voltage law, the applied voltage $v(t)$ is equal to the sum of the voltage drops v_R, v_L and v_C across the resistor R, the inductor L and the capacitor C, respectively. Thus

$$v = v_R + v_L + v_C = Ri + L\frac{di}{dt} + \frac{1}{C}\int_0^t i(\tau)\,d\tau$$

To eliminate the integral, both sides of the equation are differentiated with respect to time, resulting in the desired differential equation:

$$L\frac{d^2i}{dt^2} + R\frac{di}{dt} + \frac{i}{C} = \frac{dv}{dt}$$

3.3. Kepler's first two laws of planetary motion state that:

1. The orbit of a planet is an ellipse with the sun at a focus of the ellipse.

2. The radius vector drawn from the sun to a planet sweeps over equal areas in equal times.

Find a pair of differential equations which describes the motion of a planet about the sun, using Kepler's first two laws.

From Kepler's first law, the motion of a planet satisfies the equation of an ellipse:

$$r = \frac{p}{1 + e\cos\theta}$$

where r and θ are defined in the figure below, and $p \equiv b^2/a = a(1 - e^2)$.

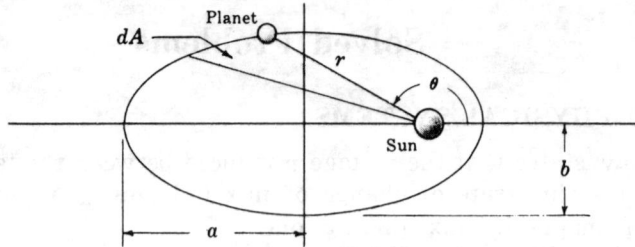

In an infinitesimal time dt the angle θ increases by an amount $d\theta$. The area swept out by r over the period dt is therefore equal to $dA = \frac{1}{2}r^2\,d\theta$. The rate at which the area is swept out by r is constant (Kepler's second law). Hence

$$\frac{dA}{dt} = \frac{1}{2}r^2\frac{d\theta}{dt} = \text{constant} \quad \text{or} \quad r^2\frac{d\theta}{dt} = k$$

The first differential equation is obtained by differentiating this result with respect to time:

$$2r\frac{dr}{dt}\frac{d\theta}{dt} + r^2\frac{d^2\theta}{dt^2} = 0 \quad \text{or} \quad 2\frac{dr}{dt}\frac{d\theta}{dt} + r\frac{d^2\theta}{dt^2} = 0$$

The second equation is obtained by differentiating the equation of the ellipse:

$$\frac{dr}{dt} = \left[\frac{pe\sin\theta}{(1 + e\cos\theta)^2}\right]\frac{d\theta}{dt}$$

Using the results that $\dfrac{d\theta}{dt} = \dfrac{k}{r^2}$ and $(1 + e\cos\theta) = \dfrac{p}{r}$, $\dfrac{dr}{dt}$ can be rewritten as

$$\frac{dr}{dt} = \frac{ek}{p}\sin\theta$$

Differentiating again and replacing $r^2(d\theta/dt)$ with k yields

$$\frac{d^2r}{dt^2} = \left(\frac{e}{p}\right)\left(\frac{k^2}{r^2}\right)\cos\theta$$

But $\cos\theta = \dfrac{1}{e}\left[\dfrac{p}{r} - 1\right]$. Hence

$$\frac{d^2r}{dt^2} = \frac{k^2}{pr^2}\left[\frac{p}{r} - 1\right] = \frac{k^2}{r^3} - \frac{k^2}{pr^2}$$

Substituting $r(d\theta/dt)^2$ for k^2/r^3, we obtain the required second differential equation:

$$\frac{d^2r}{dt^2} - r\left(\frac{d\theta}{dt}\right)^2 + \frac{k^2}{pr^2} = 0 \quad \text{or} \quad \frac{d^2r}{dt^2} - r\left(\frac{d\theta}{dt}\right)^2 = -\frac{k^2}{pr^2}$$

3.4. A mathematical model for a feature of nervous system organization called *lateral inhibition* has been produced as a result of the work of several authors [2, 3, 4]. Lateral inhibitory phenomena can be simply described as inhibitory electrical inter- action among laterally spaced, neighboring neurons (nerve cells). Each neuron in this model has a response c, measured by the frequency of discharge of pulses in its axon (the connection "cable" or "wire"). The response is determined by an excitation r supplied by an external stimulus, and is diminished by whatever inhibitory influences are acting on the neurons as a result of the activity of neigh- boring neurons. In a system of n neurons, the steady-state response of the kth neuron is given by

$$c_k = r_k - \sum_{i=1}^{n} a_{k-i}c_i$$

where the constant a_{k-i} is the inhibitory coefficient of the action of neuron i on k. It depends only on the separation of the kth and ith neurons, and can be interpreted as a *spatial weighting function*. In addition, $a_m = a_{-m}$ (symmetrical spatial interaction).

(a) If the effect of neuron i on k is not immediately felt, but exhibits a small time lag Δt, how should this model be modified?

(b) If the input $r_k(t)$ is determined only by the output c_k, Δt seconds prior to t $(r_k(t) = c_k(t - \Delta t))$, determine an approximate differential equation for the system of part (a).

(a) The equation becomes
$$c_k(t) = r_k(t) - \sum_{i=1}^{n} a_{k-i} c_i(t - \Delta t)$$

(b) Substituting $c_k(t - \Delta t)$ for $r_k(t)$,
$$c_k(t) - c_k(t - \Delta t) = -\sum_{i=1}^{n} a_{k-i} c_i(t - \Delta t)$$

Dividing both sides by Δt,
$$\frac{c_k(t) - c_k(t - \Delta t)}{\Delta t} = -\sum_{i=1}^{n} \left(\frac{a_{k-i}}{\Delta t}\right) c_i(t - \Delta t)$$

The left-hand side is approximately equal to dc_k/dt for small Δt. If we additionally assume that $c_i(t - \Delta t) \cong c_i(t)$ for small Δt, then we get the approximate differential equation
$$\frac{dc_k}{dt} + \sum_{i=1}^{n} \left(\frac{a_{k-i}}{\Delta t}\right) c_i(t) = 0$$

This type of equation is called a *differential-difference equation*.

CLASSIFICATIONS OF DIFFERENTIAL EQUATIONS

3.5. Classify the following differential equations according to whether they are ordinary or partial. Indicate the dependent and independent variables.

(a) $\dfrac{dx}{dt} + \dfrac{dy}{dt} + x + y = 0 \qquad x = x(t), \quad y = y(t)$

(b) $\dfrac{\partial f}{\partial x} + \dfrac{\partial f}{\partial y} + x + y = 0 \qquad f = f(x, y)$

(c) $\dfrac{d}{dt}\left[\dfrac{\partial f}{\partial x}\right] = 0 \qquad f = x^2 + \dfrac{dx}{dt}$

(d) $\dfrac{df}{dx} = x \qquad f = y^2(x) + \dfrac{dy}{dx}$

(a) Ordinary; independent variable t; dependent variables x and y.

(b) Partial; independent variables x and y; dependent variable f.

(c) Since $\dfrac{\partial f}{\partial x} = 2x$, then $\dfrac{d}{dt}\left[\dfrac{\partial f}{\partial x}\right] = 2\dfrac{dx}{dt} = 0$ which is an ordinary differential equation; independent variable t; dependent variable x.

(d) $\dfrac{df}{dx} = 2y\dfrac{dy}{dx} + \dfrac{d^2y}{dx^2} = x$, which is an ordinary differential equation; independent variable x; dependent variable y.

3.6. Classify the following linear differential equations according to whether they are time-variable or time-invariant. Indicate any time-variable terms.

(a) $\dfrac{d^2y}{dt^2} + 2y = 0 \qquad$ (c) $\left(\dfrac{1}{t+1}\right)\dfrac{d^2y}{dt^2} + \left(\dfrac{1}{t+1}\right)y = 0$

(b) $\dfrac{d}{dt}(t^2y) = 0 \qquad$ (d) $\dfrac{d^2y}{dt^2} + (\cos t)y = 0$

(a) Time-invariant.

(b) $\dfrac{d}{dt}(t^2 y) = 2ty + t^2 \dfrac{dy}{dt} = 0$. Dividing through by t, $t\dfrac{dy}{dt} + 2y = 0$ which is time-variable. The time-variable term is $t(dy/dt)$.

(c) Multiplying through by $t+1$, we obtain $\dfrac{d^2 y}{dt^2} + y = 0$ which is time-invariant.

(d) Time-variable. The time-variable term is $(\cos t)y$.

3.7. Classify the following differential equations according to whether they are linear or nonlinear. Indicate the dependent and independent variables and any nonlinear terms.

(a) $t\dfrac{dy}{dt} + y = 0$ $y = y(t)$ (d) $(\cos t)\dfrac{d^2 y}{dt^2} + (\sin 2t)y = 0$ $y = y(t)$

(b) $y\dfrac{dy}{dt} + y = 0$ $y = y(t)$ (e) $(\cos y)\dfrac{d^2 y}{dt^2} + \sin 2y = 0$ $y = y(t)$

(c) $\dfrac{dy}{dt} + y^2 = 0$ $y = y(t)$ (f) $(\cos x)\dfrac{d^2 y}{dt^2} + \sin 2x = 0$ $y = y(t),$
$x = x(t)$

(a) Linear; independent variable t; dependent variable y.

(b) Nonlinear; independent variable t; dependent variable y; nonlinear term $y(dy/dt)$.

(c) Nonlinear; independent variable t; dependent variable y; nonlinear term y^2.

(d) Linear; independent variable t; dependent variable y.

(c) Nonlinear; independent variable t; dependent variable y; nonlinear terms $(\cos y)\, d^2 y/dt^2$ and $\sin 2y$.

(f) Nonlinear; independent variable t; dependent variables x and y; nonlinear terms $(\cos x)\, d^2 y/dt^2$ and $\sin 2x$.

3.8. Why are all transcendental functions *not* of first degree?

Transcendental functions, such as the logarithmic, trigonometric, and hyperbolic functions and their corresponding inverses, are not of first degree because they are either defined by or can be written as infinite series. Hence their degree is in general equal to *infinity*. For example,

$$\sin x = \sum_{n=1}^{\infty} (-1)^{n-1} \frac{x^{2n-1}}{(2n-1)!} = x - \frac{x^3}{3!} + \frac{x^5}{5!} - \cdots$$

where the first term is first degree, the second is third degree, and so on.

LINEARITY AND SUPERPOSITION

3.9. Using the definition of linearity, Definition 3.8, show that any differential equation of the form

$$\sum_{i=0}^{n} a_i(t) \frac{d^i y}{dt^i} = x$$

where y is the output and x is the input, is linear.

Let x_1 and x_2 be two arbitrary inputs, and let y_1 and y_2 be the corresponding outputs. Then

$$\sum_{i=0}^{n} a_i(t) \frac{d^i y_1}{dt^i} = x_1 \quad \text{and} \quad \sum_{i=0}^{n} a_i(t) \frac{d^i y_2}{dt^i} = x_2$$

Now form
$$c_1 x_1 + c_2 x_2 = c_1 \left[\sum_{i=0}^{n} a_i(t) \frac{d^i y_1}{dt^i} \right] + c_2 \left[\sum_{i=0}^{n} a_i(t) \frac{d^i y_2}{dt^i} \right]$$

$$= \sum_{i=0}^{n} a_i(t) \frac{d^i(c_1 y_1)}{dt^i} + \sum_{i=0}^{n} a_i(t) \frac{d^i(c_2 y_2)}{dt^i}$$

$$= \sum_{i=0}^{n} a_i(t) \frac{d^i}{dt^i}(c_1 y_1 + c_2 y)$$

Since this equation holds for all c_1 and c_2, the equation is linear.

3.10. Show that a system described by the convolution integral

$$y(t) = \int_{-\infty}^{\infty} w(t, \tau)\, x(\tau)\, d\tau$$

is linear. y is the output and x the input.

Let x_1 and x_2 be two arbitrary inputs and let

$$y_1 = \int_{-\infty}^{\infty} w(t, \tau)\, x_1(\tau)\, d\tau, \qquad y_2 = \int_{-\infty}^{\infty} w(t, \tau)\, x_2(\tau)\, d\tau$$

Now let $c_1 x_1 + c_2 x_2$ be a third input and form

$$\int_{-\infty}^{\infty} w(t, \tau)\, [c_1 x_1(\tau) + c_2 x_2(\tau)]\, d\tau = c_1 \int_{-\infty}^{\infty} w(t, \tau)\, x_1(\tau)\, d\tau + c_2 \int_{-\infty}^{\infty} w(t, \tau)\, x_2(\tau)\, d\tau$$

$$= c_1 y_1 + c_2 y_2$$

Since this relationship holds for all c_1 and c_2, the convolution integral is a linear operation (or transformation).

3.11. Use the Principle of Superposition to determine the output y of the following system:

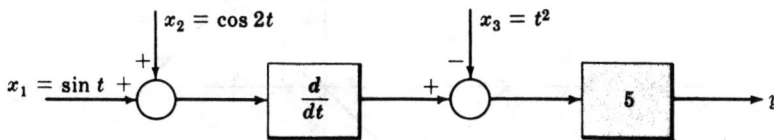

For $x_2 = x_3 = 0$, $y_1 = 5\frac{d}{dt}(\sin t) = 5\cos t$. For $x_1 = x_3 = 0$, $y_2 = 5\frac{d}{dt}(\cos 2t) = -10 \sin 2t$. For $x_1 = x_2 = 0$, $y_3 = -5t^2$. Therefore

$$y = y_1 + y_2 + y_3 = 5(\cos t - 2 \sin 2t - t^2)$$

3.12. A linear system is described by the weighting function

$$w(t, \tau) = e^{-|t-\tau|} \qquad \text{for all } t, \tau$$

Suppose the system is excited by an input

$$x(t) = t \qquad \text{for all } t$$

Find the output $y(t)$.

The output is given by the convolution integral (Definition 3.10)

$$y(t) = \int_{-\infty}^{\infty} e^{-|t-\tau|}\,\tau\,d\tau = \int_{-\infty}^{t} e^{-(t-\tau)}\,\tau\,d\tau + \int_{t}^{\infty} e^{(t-\tau)}\,\tau\,d\tau$$

$$= e^{-t}\int_{-\infty}^{t} e^{\tau}\,\tau\,d\tau + e^{t}\int_{t}^{\infty} e^{-\tau}\,\tau\,d\tau$$

$$= e^{-t}\left[e^{\tau}(\tau-1)\Big|_{-\infty}^{t}\right] + e^{t}\left[e^{-\tau}(-\tau-1)\Big|_{t}^{\infty}\right] = 2t$$

CAUSALITY

3.13. Two systems are defined by the relationships between their inputs and outputs as follows:

    System 1: The input is $x(t)$, and at the same instant of time the output is $y(t) = x(t+T)$, $T > 0$.

    System 2: The input is $x(t)$ and at the same instant of time the output is $y(t) = x(t-T)$, $T > 0$.

Are either of these systems causal?

    In System 1, the output depends only on the input T seconds in the future. Thus it is not causal. An operation of this type is called **prediction**.

    In System 2, the output depends only on the input T seconds in the past. Thus it is causal. An operation of this type is called a **time delay**.

LINEARIZED AND PIECEWISE-LINEAR SYSTEMS

3.14. The differential equation of a certain physical system is given by

$$\frac{d^3y}{dt^3} + 4\,\frac{d^2y}{dt^2} + f(y) = 0$$

where $f(y)$ has the graph

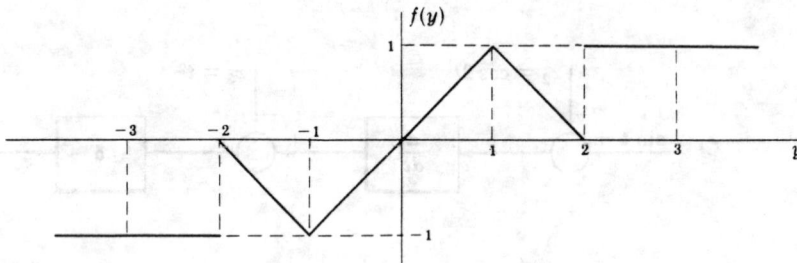

Write this nonlinear system as a piecewise-linear system.

    The system can be represented by the set of equations:

$$\frac{d^3y}{dt^3} + 4\,\frac{d^2y}{dt^2} - 1 = 0 \qquad y < -2$$

$$\frac{d^3y}{dt^3} + 4\,\frac{d^2y}{dt^2} - y - 2 = 0 \qquad -2 \le y < -1$$

$$\frac{d^3y}{dt^3} + 4\,\frac{d^2y}{dt^2} + y = 0 \qquad -1 \le y \le 1$$

$$\frac{d^3y}{dt^3} + 4\,\frac{d^2y}{dt^2} - y + 2 = 0 \qquad 1 < y \le 2$$

$$\frac{d^3y}{dt^3} + 4\,\frac{d^2y}{dt^2} + 1 = 0 \qquad 2 < y$$

3.15. A solution of the nonlinear differential equation

$$\frac{d^2y}{dt^2} + y\cos y \;=\; x$$

when the input $x = 0$ is $y = 0$. Linearize the differential equation about this input and output using a Taylor series expansion of the function $\frac{d^2y}{dt^2} + y\cos y - x$ about the point $x = y = 0$.

The Taylor series expansion of $\cos y$ about $y = 0$ is

$$\cos y \;=\; \sum_{n=0}^{\infty} \frac{y^n}{n!}\left[\frac{d^n}{dy^n}(\cos y)\Big|_{y=0}\right] \;=\; 1 - \frac{1}{2!}y^2 + \cdots$$

Therefore

$$\frac{d^2y}{dt^2} + y\cos y - x \;=\; \frac{d^2y}{dt^2} + y\left(1 - \frac{y^2}{2!} + \cdots\right) - x$$

Keeping only first degree terms, the linearized equation is $\frac{d^2y}{dt^2} + y = x$. This equation is valid only for small deviations about the operating point $x = y = 0$.

THE CHARACTERISTIC EQUATION

3.16. Find the characteristic polynomial and characteristic equation for each system:

(a) $\dfrac{d^4y}{dt^4} + 9\dfrac{d^2y}{dt^2} + 7y \;=\; x$ (b) $\dfrac{d^4y}{dt^4} + 9\dfrac{d^2y}{dt^2} + 7y \;=\; \sin t$

(a) Putting $D^n \equiv \dfrac{d^n}{dt^n}$ for $n = 2$ and $n = 4$, the characteristic polynomial is $D^4 + 9D^2 + 7$; and the characteristic equation is $D^4 + 9D^2 + 7 = 0$.

(b) Although the equation given in (b) is nonlinear by Definition 3.7 (the term $\sin t$ is not first degree in y), we can treat it as a linear equation if we arbitrarily put $\sin t \equiv x$, and treat x as a second dependent variable representing the input. In this case, part (b) has the same answer as part (a).

3.17. Determine the solution of the characteristic equation of the preceding problem.

Let $D^2 \equiv E$. Then $D^4 = E^2$, and the characteristic equation becomes quadratic:

$$E^2 + 9E + 7 \;=\; 0, \quad E \;=\; -\frac{9 \pm \sqrt{53}}{2}, \quad \text{and} \quad D \;=\; \pm\sqrt{\frac{-9 \pm \sqrt{53}}{2}}$$

LINEAR INDEPENDENCE AND FUNDAMENTAL SETS

3.18. Show that a sufficient condition for a set of n functions $f_1, f_2, \ldots, f_n$ to be linearly independent is that the determinant

$$\begin{vmatrix} f_1 & f_2 & \cdots & f_n \\[6pt] \dfrac{df_1}{dt} & \dfrac{df_2}{dt} & \cdots & \dfrac{df_n}{dt} \\[6pt] \cdots\cdots\cdots\cdots\cdots\cdots\cdots\cdots \\[6pt] \dfrac{d^{n-1}f_1}{dt^{n-1}} & \dfrac{d^{n-1}f_2}{dt^{n-1}} & \cdots & \dfrac{d^{n-1}f_n}{dt^{n-1}} \end{vmatrix}$$

be nonzero. This determinant is called the **Wronskian** of the functions $f_1, f_2, \ldots, f_n$.

Assuming the f_i are differentiable at least $n-1$ times, let $n-1$ derivatives of

$$c_1 f_1 + c_2 f_2 + \cdots + c_n f_n = 0$$

be formed as follows, where the c_i are unknown constants:

$$c_1 \frac{df_1}{dt} + c_2 \frac{df_2}{dt} + \cdots + c_n \frac{df_n}{dt} = 0$$

$$\cdots\cdots\cdots\cdots\cdots\cdots\cdots\cdots\cdots\cdots\cdots\cdots\cdots$$

$$c_1 \frac{d^{n-1}f_1}{dt^{n-1}} + c_2 \frac{d^{n-1}f_2}{dt^{n-1}} + \cdots + c_n \frac{d^{n-1}f_n}{dt^{n-1}} = 0$$

These equations may be considered as n simultaneous linear homogeneous equations in the n unknown constants $c_1, c_2, \ldots, c_n$, with coefficients given by the elements of the Wronskian. It is well known that these equations have a nonzero solution for $c_1, c_2, \ldots, c_n$ (that is, not all c_i equal to zero) if and only if the determinant of the coefficients (the Wronskian) is equal to zero. Hence if the Wronskian is nonzero, then the only solution for $c_1, c_2, \ldots, c_n$ is the degenerate one, $c_1 = c_2 = \cdots = c_n = 0$. Clearly, this is equivalent to saying that if the Wronskian is nonzero the functions $f_1, f_2, \ldots, f_n$ are linearly independent, since the only solution to $c_1 f_1 + c_2 f_2 + \cdots + c_n f_n = 0$ is then $c_1 = c_2 = c_3 = \cdots = c_n = 0$. Hence a sufficient condition for the linear independence of $f_1, f_2, \ldots, f_n$ is that the Wronskian be nonzero. This condition is not *necessary*; that is, there exist sets of linearly independent functions for which the Wronskian *is* zero.

3.19. Show that the functions $1, t, t^2$ are linearly independent.

The Wronskian of these three functions (see Problem 3.18) is

$$\begin{vmatrix} 1 & t & t^2 \\ 0 & 1 & 2t \\ 0 & 0 & 2 \end{vmatrix} = 2$$

Since the Wronskian is nonzero, the functions are linearly independent.

3.20. Determine a fundamental set for the differential equations:

(a) $\dfrac{d^3y}{dt^3} + 5\dfrac{d^2y}{dt^2} + 8\dfrac{dy}{dt} + 4y = x$ (b) $\dfrac{d^3y}{dt^3} + 4\dfrac{d^2y}{dt^2} + 6\dfrac{dy}{dt} + 4y = x$

(a) The characteristic polynomial is $D^3 + 5D^2 + 8D + 4$ which can be written in factored form: $(D+2)(D+2)(D+1)$. Corresponding to the root $D_1 = -1$ there is a solution e^{-t}, and corresponding to the repeated root $D_2 = D_3 = -2$ are the two solutions e^{-2t} and te^{-2t}. The three solutions constitute a fundamental set.

(b) The characteristic polynomial is $D^3 + 4D^2 + 6D + 4$ which can be written in factored form: $(D+1+j)(D+1-j)(D+2)$.

A fundamental set is then $e^{(-1-j)t}$, $e^{(-1+j)t}$ and e^{-2t}.

3.21. For the differential equations of Problem 3.20, find fundamental sets different from those found in Problem 3.20.

(a) Choose any 3×3 nonzero determinant, say

$$\begin{vmatrix} 1 & 2 & -1 \\ -3 & 2 & 0 \\ 1 & 3 & -2 \end{vmatrix} = -5$$

Using the elements of the first row as coefficients a_{1i} for the fundamental set $e^{-t}, e^{-2t}, te^{-2t}$ found in Problem 3.20, form

$$z_1 = e^{-t} + 2e^{-2t} - te^{-2t}$$

Using the second row, form $z_2 = -3e^{-t} + 2e^{-2t}$

From the third row, form $z_3 = e^{-t} + 3e^{-2t} - 2te^{-2t}$

The functions z_1, z_2 and z_3 constitute a fundamental set.

(b) For this equation generate the second fundamental set by letting

$$z_1 = e^{-2t}$$

$$z_2 = \tfrac{1}{2}e^{(-1+j)t} + \tfrac{1}{2}e^{(-1-j)t} = e^{-t}\left(\frac{e^{-jt} + e^{jt}}{2}\right)$$

$$= e^{-t}\left(\frac{\cos t - j\sin t + \cos t + j\sin t}{2}\right) = e^{-t}\cos t$$

$$z_3 = \frac{1}{2j}e^{(-1+j)t} - \frac{1}{2j}e^{(-1-j)t} = e^{-t}\left(\frac{e^{-jt} - e^{jt}}{2j}\right)$$

$$= e^{-t}\left(\frac{\cos t + j\sin t - \cos t + j\sin t}{2j}\right) = e^{-t}\sin t$$

The coefficient determinant in this case is

$$\begin{vmatrix} 1 & 0 & 0 \\ 0 & \frac{1}{2} & \frac{1}{2} \\ 0 & \frac{1}{2j} & -\frac{1}{2j} \end{vmatrix} = -\frac{1}{2j}$$

SOLUTION OF LINEAR CONSTANT COEFFICIENT ORDINARY DIFFERENTIAL EQUATIONS

3.22. Show that any free response $y_a(t) = \sum_{k=1}^{n} c_k y_k(t)$ satisfies $\sum_{i=0}^{n} a_i \dfrac{d^i y}{dt^i} = 0$.

By the definition of a fundamental set, $y_k(t)$, $k = 1, 2, \ldots, n$ satisfies $\sum_{i=0}^{n} a_i \dfrac{d^i y_k}{dt^i} = 0$.

Substituting $\sum_{k=1}^{n} c_k y_k(t)$ into this differential equation yields

$$\sum_{i=0}^{n} a_i \frac{d^i}{dt^i}\left[\sum_{k=1}^{n} c_k y_k(t)\right] = \sum_{i=0}^{n}\sum_{k=1}^{n} a_i \frac{d^i}{dt^i}(c_k y_k(t)) = \sum_{k=1}^{n} c_k\left[\sum_{i=0}^{n} a_i \frac{d^i y_k(t)}{dt^i}\right] = 0$$

The last equality is obtained because the term in the brackets is zero for all k.

3.23. Show that the forced response given by Equation 3.15,

$$y_b(t) = \int_0^t w(t-\tau)\left[\sum_{i=0}^{m} b_i \frac{d^i x(\tau)}{d\tau^i}\right] d\tau$$

satisfies the differential equation

$$\sum_{i=0}^{n} a_i \frac{d^i y}{dt^i} = \sum_{i=0}^{m} b_i \frac{d^i x}{dt^i}$$

For simplification, let $r(t) \equiv \sum_{i=0}^{m} b_i \dfrac{d^i x}{dt^i}$. Then $y_b(t) = \displaystyle\int_0^t w(t-\tau)\, r(\tau)\, d\tau$ and

$$\frac{dy_b}{dt} = \int_0^t \frac{\partial w(t-\tau)}{\partial t}\, r(\tau)\, d\tau + w(t-\tau)\, r(t)\Big|_{\tau=t} = \int_0^t \frac{\partial w(t-\tau)}{\partial t}\, r(\tau)\, d\tau + 0 \cdot r(t)$$

Similarly,

$$\frac{d^2 y_b}{dt^2} = \int_0^t \frac{\partial^2 w(t-\tau)}{\partial t^2}\, r(\tau)\, d\tau, \qquad \ldots, \qquad \frac{d^{n-1} y_b}{dt^{n-1}} = \int_0^t \frac{\partial^{n-1} w(t-\tau)}{\partial t^{n-1}}\, r(\tau)\, d\tau$$

since, by Equation (3.17), $\left. \dfrac{\partial^i w(t-\tau)}{\partial t^i}\right|_{\tau=t} = \left. \dfrac{d^i w(t)}{dt^i}\right|_{t=0} = 0$ for $i = 0, 1, 2, \ldots, n-2$.

The nth derivative is

$$\frac{d^n y_b}{dt^n} = \int_0^t \frac{\partial^n w(t-\tau)}{\partial t^n}\, r(\tau)\, d\tau + \frac{\partial^{n-1} w(t-\tau)}{\partial t^{n-1}}\Big|_{t=\tau} \cdot r(t) = \int_0^t \frac{\partial^n w(t-\tau)}{\partial t^n}\, r(\tau)\, d\tau + r(t)$$

since, by Equation (3.17), $\left. \dfrac{\partial^{n-1} w(t-\tau)}{\partial t^{n-1}}\right|_{t=\tau} = \left. \dfrac{d^{n-1} w(t)}{dt^{n-1}}\right|_{t=0} = 1$. The summation of the n derivatives is

$$\sum_{i=0}^{n} a_i \frac{d^i y_b}{dt^i} = \int_0^t \left[\sum_{i=0}^{n} a_i \frac{\partial^i w(t-\tau)}{\partial t^i}\right] r(\tau)\, d\tau + r(t)$$

Finally, making the change of variables $t - \tau = \theta$ in the bracketed term yields

$$\sum_{i=0}^{n} a_i \frac{\partial^i w(\theta)}{\partial \theta^i} = \sum_{i=0}^{n} a_i \frac{d^i w(\theta)}{d\theta^i} = 0$$

because $w(\theta)$ is a free response (see Section 3.13 and Problem 3.22). Hence

$$\sum_{i=0}^{n} a_i \frac{d^i y_b}{dt^i} = r(t) \equiv \sum_{i=0}^{m} b_i \frac{d^i x}{dt^i}$$

3.24. Find the free response of the differential equation

$$\frac{d^3 y}{dt^3} + 4 \frac{d^2 y}{dt^2} + 6 \frac{dy}{dt} + 4y = x$$

with initial conditions $y(0) = 1$, $\left. \dfrac{dy}{dt}\right|_{t=0} = 0$, and $\left. \dfrac{d^2 y}{dt^2}\right|_{t=0} = -1$.

From the results of Problems 3.20 and 3.21, a fundamental set for this equation is: e^{-2t}, $e^{-t} \cos t$, $e^{-t} \sin t$. Hence the free response can be written as

$$y_a(t) = c_1 e^{-2t} + c_2 e^{-t} \cos t + c_3 e^{-t} \sin t$$

The initial conditions provide the following set of algebraic equations for c_1, c_2, c_3:

$$y_a(0) = c_1 + c_2 = 1, \qquad \left. \frac{dy_a}{dt}\right|_{t=0} = -2c_1 - c_2 + c_3 = 0, \qquad \left. \frac{d^2 y_a}{dt^2}\right|_{t=0} = 4c_1 - 2c_3 = -1$$

from which $c_1 = \tfrac{1}{2}$, $c_2 = \tfrac{1}{2}$, $c_3 = \tfrac{3}{2}$. Therefore the free response is

$$y_a(t) = \tfrac{1}{2} e^{-2t} + \tfrac{1}{2} e^{-t} \cos t + \tfrac{3}{2} e^{-t} \sin t$$

3.25. Find the weighting function of the differential equation

$$\frac{d^2 y}{dt^2} + 4 \frac{dy}{dt} + 4y = 3 \frac{dx}{dt} + 2x$$

The characteristic equation is $D^2 + 4D + 4 = (D+2)^2 = 0$ with the repeated root $D = -2$. A fundamental set is therefore given by e^{-2t}, te^{-2t}, and the weighting function has the form

$$w(t) = c_1 e^{-2t} + c_2 t e^{-2t}$$

with the initial conditions

$$w(0) = [c_1 e^{-2t} + c_2 t e^{-2t}]\Big|_{t=0} = c_1 = 0, \quad \frac{dw}{dt}\Big|_{t=0} = [-2c_1 e^{-2t} + c_2 e^{-2t} - 2c_2 t e^{-2t}]\Big|_{t=0} = c_2 = 1$$

Thus $w(t) = t e^{-2t}$.

3.26. Find the forced response of the differential equation (Problem 3.25)

$$\frac{d^2 y}{dt^2} + 4 \frac{dy}{dt} + 4y = 3 \frac{dx}{dt} + 2x$$

where $x(t) = e^{-3t}, \ t \geq 0$.

The forced response is given by Equation (*3.15*) as

$$y_b(t) = \int_0^t w(t-\tau)\left[3\frac{dx}{d\tau} + 2x\right] d\tau = 3 \int_0^t w(t-\tau) \frac{dx}{d\tau} d\tau + 2 \int_0^t w(t-\tau)\, x \, d\tau$$

Integrating the first integral by parts,

$$\int_0^t w(t-\tau)\frac{dx}{d\tau} d\tau = w(t-\tau)\,x(\tau)\Big|_0^t - \int_0^t \frac{\partial w(t-\tau)}{\partial \tau} x \, d\tau$$

$$= w(0)\,x(t) - w(t)\,x(0) - \int_0^t \frac{\partial w(t-\tau)}{\partial \tau} x \, d\tau$$

But $w(0) = 0$; hence the forced response can be written as

$$y_b(t) = \int_0^t \left[-3\frac{\partial w(t-\tau)}{\partial \tau} + 2w(t-\tau)\right] x(\tau)\, d\tau - 3w(t)\,x(0)$$

From Problem 3.25, $w(t-\tau) = (t-\tau)e^{-2(t-\tau)}$; hence

$$\left[-3\frac{\partial w(t-\tau)}{\partial \tau} + 2w(t-\tau)\right] = 3e^{-2(t-\tau)} - 4(t-\tau)\,e^{-2(t-\tau)}$$

and the forced response is

$$y_b(t) = 3e^{-2t}\int_0^t e^{2\tau} e^{-3\tau} d\tau - 4t e^{-2t}\int_0^t e^{2\tau} e^{-3\tau} d\tau + 4e^{-2t}\int_0^t \tau e^{2\tau} e^{-3\tau} d\tau - 3t e^{-2t}$$

$$= 7[e^{-2t} - e^{-3t} - t e^{-2t}]$$

3.27. Find the output y of a system described by the differential equation

$$\frac{d^2 y}{dt^2} + 3 \frac{dy}{dt} + 2y = 1 + t$$

with initial conditions $y(0) = 0$ and $\dfrac{dy}{dt}\Big|_{t=0} = 1$.

Let $x_1 \equiv 1$, $x_2 \equiv t$. The response y due to x_1 alone was determined in Example 3.32 as $y_1 = \frac{1}{2}(1 - e^{-2t})$. The free response y_i for the differential equation was found in Example 3.29 to be $y_a = e^{-t} - e^{-2t}$. The forced response due to x_2 is given by Equation (*3.15*). Using the weighting function determined in Example 3.30, the forced response due to x_2 is

$$y_2 = \int_0^t w(t-\tau)\,x_2(\tau)\,d\tau = \int_0^t [e^{-(t-\tau)} - e^{-2(t-\tau)}]\tau \, d\tau$$

$$= e^{-t}\int_0^t \tau e^{\tau} d\tau - e^{-2t}\int_0^t \tau e^{2\tau} d\tau = \frac{1}{4}[4e^{-t} - e^{-2t} + 2t - 3]$$

Thus the forced response is

$$y_b = y_1 + y_2 = \tfrac{1}{4}[4e^{-t} - 3e^{-2t} + 2t - 1]$$

and the total response is

$$y = y_a + y_b = \tfrac{1}{4}[8e^{-t} - 7e^{-2t} + 2t - 1]$$

3.28. Find the transient and steady state responses of a system described by the differential equation

$$\frac{d^2y}{dt^2} + 3\frac{dy}{dt} + 2y = 1 + t$$

with the initial conditions $y(0) = 0$ and $\dfrac{dy}{dt}\bigg|_{t=0} = 1$.

The total response for this equation was determined in Problem 3.27 as

$$y = \tfrac{1}{4}[8e^{-t} - 7e^{-2t} + 2t - 1]$$

Since $\lim\limits_{t\to\infty}[\tfrac{1}{4}(8e^{-t} - 7e^{-2t})] = 0$, the transient response is $y_T = \tfrac{1}{4}(8e^{-t} - 7e^{-2t})$. The steady state response is $y_{ss} = \tfrac{1}{4}(2t - 1)$.

SINGULARITY FUNCTIONS

3.29. Evaluate: (a) $\displaystyle\int_5^8 t^2\,\delta(t-6)\,dt$,   (b) $\displaystyle\int_0^4 \sin t\,\delta(t-7)\,dt$.

(a) Using the screening property of the unit impulse function, $\displaystyle\int_5^8 t^2\,\delta(t-6)\,dt = t^2\bigg|_{t=6} = 36$.

(b) Since the interval of integration $0 \leq t \leq 4$ does not include the position of the unit impulse function $t = 7$, then $\displaystyle\int_0^4 \sin t\,\delta(t-7)\,dt = 0$.

3.30. Show that the unit step response $y_u(t)$ of a causal linear system described by the convolution integral

$$y(t) = \int_0^t w(t-\tau)\,x(\tau)\,d\tau$$

is related to the unit impulse response $y_\delta(t)$ by the equation $y_u(t) = \displaystyle\int_0^t y_\delta(\tau)\,d\tau$.

The unit step response is given by $y_u(t) = \displaystyle\int_0^t w(t-\tau)\,u(\tau)\,d\tau$ where $u(t)$ is a unit step function. In Example 3.34 it was shown that $y_\delta(t) = w(t)$. Hence

$$y_u(t) = \int_0^t y_\delta(t-\tau)\,u(\tau)\,d\tau = \int_0^t y_\delta(t-\tau)\,d\tau$$

Now make the change of variable $\theta = t - \tau$. Then $d\tau = -d\theta$, $\tau = 0$ implies $\theta = t$, $\tau = t$ implies $\theta = 0$, and the integral becomes

$$y_u(t) = -\int_t^0 y_\delta(\theta)\,d\theta = \int_0^t y_\delta(\theta)\,d\theta$$

3.31. Show that the unit ramp response $y_r(t)$ of a causal linear system described by the convolution integral (see Problem 3.30) is related to the unit impulse response $y_\delta(t)$ and the unit step response $y_u(t)$ by the equation

$$y_r(t) = \int_0^t y_u(\tau')\, d\tau' = \int_0^t \int_0^{\tau'} y_\delta(\theta)\, d\theta\, d\tau'$$

Proceeding as in Problem 3.30 with $w(t-\tau) = y_\delta(t-\tau)$ and τ changed to $t - \tau'$, we get

$$y_r(t) = \int_0^t y_\delta(t-\tau)\, \tau\, d\tau = \int_0^t (t-\tau')\, y_\delta(\tau')\, d\tau' = \int_0^t t\, y_\delta(\tau')\, d\tau' - \int_0^t \tau'\, y_\delta(\tau')\, d\tau'$$

From Problem 3.30, the first term can be written as $t \int_0^t y_\delta(\tau')\, d\tau' = t\, y_u(t)$. The second term can be integrated by parts, yielding $\int_0^t \tau'\, y_\delta(\tau')\, d\tau' = \tau'\, y_u(\tau') \Big|_0^t - \int_0^t y_u(\tau')\, d\tau'$ where $dy_u(\tau') = y_\delta(\tau')\, d\tau'$. Therefore

$$y_r(t) = t\, y_u(t) - t\, y_u(t) + \int_0^t y_u(\tau')\, d\tau' = \int_0^t y_u(\tau')\, d\tau'$$

Again using the result of Problem 3.30 we obtain the required equation.

SECOND ORDER SYSTEMS

3.32. Show that the weighting function of the second order differential equation

$$\frac{d^2y}{dt^2} + 2\zeta\omega_n \frac{dy}{dt} + \omega_n^2 y = \omega_n^2 x$$

is given by $w(t) = \dfrac{1}{\omega_d} e^{-\alpha t} \sin \omega_d t$ where $\alpha \equiv \zeta\omega_n$, $\omega_d \equiv \omega_n \sqrt{1-\zeta^2}$, $0 \leqq \zeta \leqq 1$.

The characteristic equation $\qquad D^2 + 2\zeta\omega_n D + \omega_n^2 = 0$

has the roots
$$D_1 = -\zeta\omega_n + j\omega_n \sqrt{1-\zeta^2} = -\alpha + j\omega_d$$
$$D_2 = -\zeta\omega_n - j\omega_n \sqrt{1-\zeta^2} = -\alpha - j\omega_d$$

One fundamental set is $y_1 = e^{-\alpha t} e^{j\omega_d t}$, $y_2 = e^{-\alpha t} e^{-j\omega_d t}$; and the weighting function can be written as

$$w(t) = c_1 e^{-\alpha t} e^{-j\omega_d t} + c_2 e^{-\alpha t} e^{j\omega_d t}$$

where c_1 and c_2 are, as yet, unknown coefficients. $w(t)$ can be rewritten as

$$w(t) = e^{-\alpha t}[c_1 \cos \omega_d t - jc_1 \sin \omega_d t + c_2 \cos \omega_d t + jc_2 \sin \omega_d t]$$
$$= (c_1 + c_2)e^{-\alpha t} \cos \omega_d t + j(c_2 - c_1)e^{-\alpha t} \sin \omega_d t$$
$$= A e^{-\alpha t} \cos \omega_d t + B e^{-\alpha t} \sin \omega_d t$$

where $A \equiv c_1 + c_2$ and $B \equiv j(c_2 - c_1)$ are unknown coefficients determined from the initial conditions given by Equation (*3.17*), Page 37. That is,

$$w(0) = [A e^{-\alpha t} \cos \omega_d t + B e^{-\alpha t} \sin \omega_d t]\Big|_{t=0} = A = 0$$

and
$$\frac{dw}{dt}\Big|_{t=0} = B e^{-\alpha t}[\omega_d \cos \omega_d t - \alpha \sin \omega_d t]\Big|_{t=0} = B\omega_d = 1$$

Hence
$$w(t) = \frac{1}{\omega_d} e^{-\alpha t} \sin \omega_d t$$

3.33. Determine the damping ratio ζ, undamped natural frequency ω_n, damped natural frequency ω_d, damping coefficient α, and time constant τ for the following second order system:

$$2\frac{d^2y}{dt^2} + 4\frac{dy}{dt} + 8y = 8x$$

Dividing both sides of the equation by 2, $\frac{d^2y}{dt^2} + 2\frac{dy}{dt} + 4y = 4x$. Comparing the coefficients of this equation with those of Equation (*3.23*), we obtain $2\zeta\omega_n = 2$ and $\omega_n^2 = 4$ with the solutions $\omega_n = 2$ and $\zeta = \frac{1}{2} = 0.5$. Now $\omega_d = \omega_n\sqrt{1-\zeta^2} = \sqrt{3}$, $\alpha = \zeta\omega_n = 1$, and $\tau = 1/\alpha = 1$.

3.34. The **overshoot** of a second order system in response to a unit step input is the difference between the maximum value attained by the output and the steady-state solution. Determine the overshoot for the system of Problem 3.33 using the normalized family of curves given in Section 3.17.

Since the damping ratio of this system is $\zeta = 0.5$, the normalized curve corresponding to $\zeta = 0.5$ is used. This curve has its maximum value (peak) at $\omega_n t = 3.4$. From Problem 3.33, $\omega_n = 2$; hence the time t_p at which the peak occurs is $t_p = 3.4/\omega_n = 3.4/2 = 1.7$ s. The value attained at this time is 1.17, and the overshoot is $1.17 - 1.00 = 0.17$.

Supplementary Problems

3.35. Which of the following terms are first degree in the dependent variable $y = y(t)$?
(a) t^2y, (b) $\tan y$, (c) $\cos t$, (d) e^{-y}, (e) te^{-t}.

3.36. Show that a system defined by the equation $y = mx + b$, where y is the output, x is the input, and m and b are nonzero constants, is nonlinear according to Definition 3.8.

3.37. Show that any differential equation of the form $\displaystyle\sum_{i=0}^{n} a_i(t)\frac{d^iy}{dt^i} = \sum_{i=0}^{m} b_i(t)\frac{d^ix}{dt^i}$ satisfies Definition 3.8. (See Example 3.17 and Problem 3.9.)

3.38. Find a linear differential equation which describes small variations Δy about a *known solution* $y_0(t)$ of the nonlinear differential equation $\frac{d^2y}{dt^2} + y\frac{dy}{dt} + y^2 = 0$.

3.39. Show that the functions $\cos t$ and $\sin t$ are linearly independent.

3.40. Show that the functions $\sin nt$ and $\sin kt$ where n and k are integers are linearly independent if $n \neq k$.

3.41. Show that the functions t and t^2 constitute a fundamental set for the differential equation

$$t^2\frac{d^2y}{dt^2} - 2t\frac{dy}{dt} + 2y = 0$$

3.42. Find a fundamental set for $\frac{d^3y}{dt^3} + 6\frac{d^2y}{dt^2} + 21\frac{dy}{dt} + 26y = x$.

3.43. Find the free response of the linear differential equation

$$t^2 \frac{d^2y}{dt^2} - 2t \frac{dy}{dt} + 2y = 0$$

with the initial conditions $y(0) = 0$, $\left.\dfrac{dy}{dt}\right|_{t=0} = 1$. A fundamental set for this equation consists of the functions t and t^2.

3.44. Find the free response of $\dfrac{d^3y}{dt^3} + 6\dfrac{d^2y}{dt^2} + 11\dfrac{dy}{dt} + 6y = 0$ with the initial conditions $y(0) = 1$, $\left.\dfrac{dy}{dt}\right|_{t=0} = \left.\dfrac{d^2y}{dt^2}\right|_{t=0} = 0$.

3.45. Find the weighting function of the differential equation $\dfrac{d^2y}{dt^2} - \dfrac{dy}{dt} - 2y = x$.

3.46. Find the forced response of $\dfrac{d^2y}{dt^2} + 3\dfrac{dy}{dt} + 2y = x$ when $x = \sin t$.

3.47. Evaluate $\displaystyle\int_{-\infty}^{\infty} e^{-t} \cos 5t \ \delta(t-1) \ dt$.

3.48. Find the undamped natural frequency, the damping ratio, the damped natural frequency, and the time constant of the second order differential equation $\dfrac{d^2y}{dt^2} + 5\dfrac{dy}{dt} + 7y = 7x$.

3.49. Show that the peak value of the unit step response of a second order system occurs at time $t_p = \pi/(\omega_n \sqrt{1 - \zeta^2})$.

Answers to Supplementary Problems

3.35. $t^2 y$

3.38. $\dfrac{d^2\Delta y}{dt^2} + y_0 \dfrac{d\Delta y}{dt} + \left(\dfrac{dy_0}{dt} + 2y_0\right)\Delta y = 0$

3.42. $e^{-2t}, \ e^{-2t}\sin 3t, \ e^{-2t}\cos 3t$

3.43. $y_a(t) = t$

3.44. $y_a(t) = 3e^{-t} - 3e^{-2t} + e^{-3t}$

3.45. $w(t) = -\frac{1}{3}e^{-t} + \frac{1}{3}e^{2t}$

3.46. $y_b(t) = \frac{1}{10}[5e^{-t} - 2e^{-2t} + \sin t - 3\cos t]$

3.47. 0.1042

3.48. $\omega_n = \sqrt{7}$ rad/s, $\quad \zeta = \dfrac{5}{2\sqrt{7}} = 0.943$, $\quad \omega_d = 0.866$ rad/s, $\quad \tau = 0.4$ s

Chapter 4

The Laplace Transform

4.1 INTRODUCTION

Several techniques used in solving engineering problems are based on the replacement of functions of a real variable (usually time or distance) by certain frequency dependent representations, or by functions of a complex variable dependent upon frequency. A typical example is the use of Fourier series to solve certain electrical problems. One such problem consists of finding the current in some part of a linear electrical network in which the input voltage is a periodic or repeating waveform. The periodic voltage may be replaced by its Fourier series representation, and the current produced by each term of the series can then be determined. The total current is the sum of the individual currents (superposition). This technique often results in a substantial savings in computational effort.

A transformation technique relating time functions to frequency dependent functions of a complex variable is presented in the next few sections of this chapter. It is called the *Laplace transform*. The application of this mathematical transformation to solving linear constant coefficient differential equations is discussed in the remaining sections, and provides the basis for the analysis and design techniques developed in subsequent chapters.

4.2 THE LAPLACE TRANSFORM

The Laplace transform is defined in the following manner:

Definition 4.1: Let $f(t)$ be a real function of a real variable t defined for $t > 0$. Then

$$\mathcal{L}\left[f(t)\right] \equiv F(s) \equiv \lim_{\substack{T \to \infty \\ \epsilon \to 0}} \int_\epsilon^T f(t)e^{-st}\,dt = \int_{0^+}^\infty f(t)e^{-st}\,dt, \qquad 0 < \epsilon < T$$

is called the **Laplace transform** of $f(t)$. s is a complex variable defined by $s \equiv \sigma + j\omega$, where σ and ω are real variables* and $j = \sqrt{-1}$.

Note that the lower limit on the integral is $t = \epsilon > 0$. This definition of the lower limit is sometimes useful in dealing with functions which are discontinuous at $t = 0$. When *explicit* use is made of this limit, it will be abbreviated $t = \lim_{\epsilon \to 0} \epsilon \equiv 0^+$, as shown above in the integral on the right.

The real variable t always denotes *time*.

---

* The real part σ of a complex variable s is often written as Re(s) (the real part of s) and the imaginary part ω as Im(s) (the imaginary part of s). Parentheses are placed around s only when there is a possibility of confusion.

Definition 4.2: If $f(t)$ is defined and single-valued for $t > 0$ and $F(\sigma)$ is absolutely convergent for some real number σ_0, that is,

$$\int_{0^+}^{\infty} |f(t)|\, e^{-\sigma_0 t}\, dt \;=\; \lim_{\substack{T \to \infty \\ \epsilon \to 0}} \int_{\epsilon}^{T} |f(t)|\, e^{-\sigma_0 t}\, dt \;<\; +\infty, \qquad 0 < \epsilon < T$$

then $f(t)$ is **Laplace transformable** for $\operatorname{Re}(s) > \sigma_0$.

Example 4.1.

The function e^{-t} is Laplace transformable since

$$\int_{0^+}^{\infty} |e^{-t}|\, e^{-\sigma_0 t}\, dt \;=\; \int_{0^+}^{\infty} e^{-(1+\sigma_0)t}\, dt \;=\; \left. \frac{1}{-(1+\sigma_0)}\, e^{-(1+\sigma_0)t} \right|_{0^+}^{\infty} \;=\; \frac{1}{1+\sigma_0} \;<\; +\infty$$

if $1 + \sigma_0 > 0$ or $\sigma_0 > -1$.

Example 4.2.

The Laplace transform of e^{-t} is

$$\mathcal{L}\,[e^{-t}] \;=\; \int_{0^+}^{\infty} e^{-t}\, e^{-st}\, dt \;=\; \left. \frac{-1}{(s+1)}\, e^{-(s+1)t} \right|_{0^+}^{\infty} \;=\; \frac{1}{s+1} \qquad \text{for } \operatorname{Re}(s) > -1$$

4.3 THE INVERSE LAPLACE TRANSFORM

The Laplace transform transforms a problem from the real variable time domain into the complex variable s domain. After a solution of the transformed problem has been obtained in terms of s, it is necessary to "invert" this transform in order to obtain the time domain solution. The transformation from the s domain into the t domain is called the *inverse Laplace transform*.

Definition 4.3: Let $F(s)$ be the Laplace transform of a function $f(t)$, $t > 0$. The contour integral

$$\mathcal{L}^{-1}\,[F(s)] \;\equiv\; f(t) \;=\; \frac{1}{2\pi j} \int_{c-j\infty}^{c+j\infty} F(s)\, e^{st}\, ds$$

where $j = \sqrt{-1}$ and $c > \sigma_0$ (σ_0 as given in Definition 4.2) is called the **inverse Laplace transform** of $F(s)$.

It is seldom necessary in practice to perform the contour integration defined in Definition 4.3. For applications of the Laplace transform in this book, it is never necessary. A simple technique for evaluating the inverse transform for most control system problems is presented in Section 4.8.

4.4 SOME PROPERTIES OF THE LAPLACE TRANSFORM AND ITS INVERSE

The Laplace transform and its inverse have several important properties which can be used advantageously in the solution of linear constant coefficient differential equations. They are:

1. The Laplace transform is a *linear transformation* between functions defined in the t domain and functions defined in the s domain. That is, if $F_1(s)$ and $F_2(s)$ are the Laplace transforms of $f_1(t)$ and $f_2(t)$, respectively, then $a_1 F_1(s) + a_2 F_2(s)$ is the Laplace transform of $a_1 f_1(t) + a_2 f_2(t)$, where a_1 and a_2 are arbitrary constants.

2. The inverse Laplace transform is a *linear transformation* between functions defined in the s domain and functions defined in the t domain. That is, if $f_1(t)$ and $f_2(t)$ are the inverse Laplace transforms of $F_1(s)$ and $F_2(s)$, respectively, then $b_1 f_1(t) + b_2 f_2(t)$ is the inverse Laplace transform of $b_1 F_1(s) + b_2 F_2(s)$, where b_1 and b_2 are arbitrary constants.

3. The Laplace transform of the *derivative df/dt* of a function $f(t)$ whose Laplace transform is $F(s)$ is

$$\mathcal{L}\left[df/dt\right] = sF(s) - f(0^+)$$

where $f(0^+)$ is the initial value of $f(t)$, evaluated as the one-sided limit of $f(t)$ as t approaches zero from positive values.

4. The Laplace transform of the *integral* $\int_0^t f(\tau)\, d\tau$ of a function $f(t)$ whose Laplace transform is $F(s)$ is

$$\mathcal{L}\left[\int_0^t f(\tau)\, d\tau\right] = \frac{F(s)}{s}$$

5. The initial value $f(0^+)$ of the function $f(t)$ whose Laplace transform is $F(s)$ is

$$f(0^+) = \lim_{t \to 0} f(t) = \lim_{s \to \infty} sF(s) \qquad t > 0$$

This relation is called the *Initial Value Theorem*.

6. The final value $f(\infty)$ of the function $f(t)$ whose Laplace transform is $F(s)$ is

$$f(\infty) = \lim_{t \to \infty} f(t) = \lim_{s \to 0} sF(s)$$

if $\lim_{t \to \infty} f(t)$ exists. This relation is called the *Final Value Theorem*.

7. The Laplace transform of a function $f(t/a)$ (*Time Scaling*) is

$$\mathcal{L}\left[f(t/a)\right] = aF(as)$$

where $F(s) = \mathcal{L}\left[f(t)\right]$.

8. The inverse Laplace transform of the function $F(s/a)$ (*Frequency Scaling*) is

$$\mathcal{L}^{-1}\left[F(s/a)\right] = af(at)$$

where $\mathcal{L}^{-1}\left[F(s)\right] = f(t)$.

9. The Laplace transform of the function $f(t-T)$ (*Time Delay*) where $T > 0$ and $f(t-T) = 0$ for $t \leq T$, is

$$\mathcal{L}\left[f(t-T)\right] = e^{-sT}F(s)$$

where $F(s) = \mathcal{L}\left[f(t)\right]$.

10. The Laplace transform of the function $e^{-at}f(t)$ is given by

$$\mathcal{L}\left[e^{-at}f(t)\right] = F(s+a)$$

where $F(s) = \mathcal{L}\left[f(t)\right]$. (*Complex Translation*)

11. The Laplace transform of the *product of two functions* $f_1(t)$ and $f_2(t)$ is given by the *complex convolution integral*

$$\mathcal{L}\left[f_1(t) \cdot f_2(t)\right] = \frac{1}{2\pi j} \int_{c-j\infty}^{c+j\infty} F_1(\omega)\, F_2(s-\omega)\, d\omega$$

where $F_1(s) = \mathcal{L}\left[f_1(t)\right]$, $F_2(s) = \mathcal{L}\left[f_2(t)\right]$.

12. The inverse Laplace transform of the *product of the two transforms* $F_1(s)$ and $F_2(s)$ is given by the *convolution integrals*

$$\mathcal{L}^{-1}\left[F_1(s) \cdot F_2(s)\right] = \int_{0^+}^t f_1(\tau)\, f_2(t-\tau)\, d\tau = \int_{0^+}^t f_2(\tau)\, f_1(t-\tau)\, d\tau$$

where $\mathcal{L}^{-1}[F_1(s)] = f_1(t)$, $\mathcal{L}^{-1}[F_2(s)] = f_2(t)$.

Example 4.3.

The Laplace transforms of the functions e^{-t} and e^{-2t} are $\mathcal{L}[e^{-t}] = \dfrac{1}{s+1}$, $\mathcal{L}[e^{-2t}] = \dfrac{1}{s+2}$. Then by Property 1,

$$\mathcal{L}[3e^{-t} - e^{-2t}] = 3\mathcal{L}[e^{-t}] - \mathcal{L}[e^{-2t}] = \frac{3}{s+1} - \frac{1}{s+2} = \frac{2s+5}{s^2+3s+2}$$

Example 4.4.

The inverse Laplace transforms of the functions $\dfrac{1}{s+1}$ and $\dfrac{1}{s+3}$ are

$$\mathcal{L}^{-1}\left[\frac{1}{s+1}\right] = e^{-t}, \qquad \mathcal{L}^{-1}\left[\frac{1}{s+3}\right] = e^{-3t}$$

Then by Property 2,

$$\mathcal{L}^{-1}\left[\frac{2}{s+1} - \frac{4}{s+3}\right] = 2\mathcal{L}^{-1}\left[\frac{1}{s+1}\right] - 4\mathcal{L}^{-1}\left[\frac{1}{s+3}\right] = 2e^{-t} - 4e^{-3t}$$

Example 4.5.

The Laplace transform of $\dfrac{d}{dt}(e^{-t})$ can be determined by application of Property 3. Since $\mathcal{L}[e^{-t}] = \dfrac{1}{s+1}$ and $\lim\limits_{t \to 0} e^{-t} = 1$, then

$$\mathcal{L}\left[\frac{d}{dt}(e^{-t})\right] = s\left(\frac{1}{s+1}\right) - 1 = \frac{-1}{s+1}$$

Example 4.6.

The Laplace transform of $\displaystyle\int_0^t e^{-\tau}\, d\tau$ can be determined by application of Property 4. Since $\mathcal{L}[e^{-t}] = \dfrac{1}{s+1}$, then

$$\mathcal{L}\left[\int_0^t e^{-\tau}\, d\tau\right] = \frac{1}{s}\left(\frac{1}{s+1}\right) = \frac{1}{s(s+1)}$$

Example 4.7.

The Laplace transform of e^{-3t} is $\mathcal{L}[e^{-3t}] = \dfrac{1}{s+3}$. The initial value of e^{-3t} can be determined by the Initial Value Theorem as $\lim\limits_{t \to 0} e^{-3t} = \lim\limits_{s \to \infty} s\left(\dfrac{1}{s+3}\right) = 1$.

Example 4.8.

The Laplace transform of the function $(1 - e^{-t})$ is $\dfrac{1}{s(s+1)}$. The final value of this function can be determined from the Final Value Theorem as $\lim\limits_{t \to \infty} (1 - e^{-t}) = \lim\limits_{s \to 0} \dfrac{s}{s(s+1)} = 1$.

Example 4.9.

The Laplace transform of e^{-t} is $\dfrac{1}{s+1}$. The Laplace transform of e^{-3t} can be determined by application of Property 7 (Time Scaling), where $a = \frac{1}{3}$: $\mathcal{L}[e^{-3t}] = \frac{1}{3}\left[\dfrac{1}{(\frac{1}{3}s+1)}\right] = \dfrac{1}{s+3}$.

Example 4.10.

The inverse transform of $\dfrac{1}{s+1}$ is e^{-t}. The inverse transform of $\dfrac{1}{\frac{1}{3}s+1}$ can be determined by application of Property 8 (Frequency Scaling): $\mathcal{L}^{-1}\left[\dfrac{1}{\frac{1}{3}s+1}\right] = 3e^{-3t}$.

Example 4.11.

The Laplace transform of the function e^{-t} is $\dfrac{1}{s+1}$. The Laplace transform of the function defined as

$$f(t) = \begin{cases} e^{-(t-2)} & t > 2 \\ 0 & t \le 2 \end{cases}$$

can be determined by Property 9, with $T = 2$: $\mathcal{L}[f(t)] = e^{-2s} \cdot \mathcal{L}[e^{-t}] = \dfrac{e^{-2s}}{s+1}$.

Example 4.12.

The Laplace transform of $\cos t$ is $\frac{s}{s^2+1}$. The Laplace transform of $e^{-2t}\cos t$ can be determined from Property 10 with $a=2$: $\quad \mathcal{L}[e^{-2t}\cos t] = \frac{s+2}{(s+2)^2+1} = \frac{s+2}{s^2+4s+5}$.

Example 4.13.

The Laplace transform of the product $e^{-2t}\cos t$ can be determined by application of Property 11 (Complex Convolution). That is, since $\mathcal{L}[e^{-2t}] = \frac{1}{s+2}$ and $\mathcal{L}[\cos t] = \frac{s}{s^2+1}$, then

$$\mathcal{L}[e^{-2t}\cos t] = \frac{1}{2\pi j}\int_{c-j\infty}^{c+j\infty}\left(\frac{\omega}{\omega^2+1}\right)\left(\frac{1}{s-\omega+2}\right)d\omega = \frac{s+2}{s^2+4s+5}$$

The details of this contour integration are not carried out here because they are too complicated (see, for example, Reference [5]) and unnecessary. The Laplace transform of $e^{-2t}\cos t$ was very simply determined in Example 4.12 using Property 10. There are, however, many instances in more advanced treatments of automatic control in which complex convolution can be used effectively.

Example 4.14.

The inverse Laplace transform of the function $F(s) = \frac{s}{(s+1)(s^2+1)}$ can be determined by application of Property 12. Since $\mathcal{L}^{-1}\left[\frac{1}{s+1}\right] = e^{-t}$ and $\mathcal{L}^{-1}\left[\frac{s}{s^2+1}\right] = \cos t$, then

$$\mathcal{L}^{-1}\left[\left(\frac{1}{s+1}\right)\left(\frac{s}{s^2+1}\right)\right] = \int_{0^+}^{t}e^{-(t-\tau)}\cos\tau\,d\tau = e^{-t}\int_{0^+}^{t}e^{\tau}\cos\tau\,d\tau = \tfrac{1}{2}(\cos t + \sin t - e^{-t})$$

4.5 SHORT TABLE OF LAPLACE TRANSFORMS

The following is a short table of Laplace transforms. It is not complete, but when used in conjunction with the properties of the Laplace transform described in Section 4.4 and the partial fraction expansion techniques described in Section 4.7, it is adequate to handle all of the problems in this book. A more complete table of Laplace transform pairs is found in the Appendix.

TABLE 4.1

Time Function		Laplace Transform
Unit Impulse	$\delta(t)$	1
Unit Step	$u(t)$	$1/s$
Unit Ramp	t	$1/s^2$
Polynomial	t^n	$n!/s^{n+1}$
Exponential	e^{-at}	$\frac{1}{s+a}$
Sine Wave	$\sin\omega t$	$\frac{\omega}{s^2+\omega^2}$
Cosine Wave	$\cos\omega t$	$\frac{s}{s^2+\omega^2}$
Damped Sine Wave	$e^{-at}\sin\omega t$	$\frac{\omega}{(s+a)^2+\omega^2}$
Damped Cosine Wave	$e^{-at}\cos\omega t$	$\frac{s+a}{(s+a)^2+\omega^2}$

Table 4.1 can be used to find both Laplace transforms and inverse Laplace transforms. To find the Laplace transform of a time function which can be represented by some combination of the elementary functions given in Table 4.1, the appropriate transforms are chosen from the table and are combined using the Properties in Section 4.4.

Example 4.15.

The Laplace transform of the function $f(t) = e^{-4t} + \sin(t-2) + t^2 e^{-2t}$ is determined as follows. The Laplace transforms of e^{-4t}, $\sin t$, and t^2 are given in the table as

$$\mathcal{L}\,[e^{-4t}] \;=\; \frac{1}{s+4}, \quad \mathcal{L}\,[\sin t] \;=\; \frac{1}{s^2+1}, \quad \mathcal{L}\,[t^2] \;=\; \frac{2}{s^3}$$

Application of Properties 9 and 10, respectively, yields

$$\mathcal{L}\,[\sin(t-2)] \;=\; \frac{e^{-2s}}{s^2+1}, \quad \mathcal{L}\,[t^2 e^{-2t}] \;=\; \frac{2}{(s+2)^3}$$

Then Property 1 (linearity) gives

$$\mathcal{L}\,[f(t)] \;=\; \frac{1}{s+4} \;+\; \frac{e^{-2s}}{s^2+1} \;+\; \frac{2}{(s+2)^3}$$

To find the inverse of the transform of a combination of those in Table 4.1, the corresponding time functions (inverse transforms) are determined from the table and combined appropriately using the Properties in Section 4.4.

Example 4.16.

The inverse Laplace transform of $F(s) = \left(\dfrac{s+2}{s^2+4}\right) \cdot e^{-s}$ can be determined as follows. $F(s)$ is first rewritten as $F(s) \;=\; \dfrac{se^{-s}}{s^2+4} + \dfrac{2e^{-s}}{s^2+4}$. Now

$$\mathcal{L}^{-1}\left[\frac{s}{s^2+4}\right] \;=\; \cos 2t, \quad \mathcal{L}^{-1}\left[\frac{2}{s^2+4}\right] \;=\; \sin 2t$$

Application of Property 9 for $t>1$ yields

$$\mathcal{L}^{-1}\left[\frac{se^{-s}}{s^2+4}\right] \;=\; \cos 2(t-1), \quad \mathcal{L}^{-1}\left[\frac{2e^{-s}}{s^2+4}\right] \;=\; \sin 2(t-1)$$

Then Property 2 (Linearity) gives

$$\begin{aligned} \mathcal{L}^{-1}\,[F(s)] \;&=\; \cos 2(t-1) + \sin 2(t-1) \qquad t>1 \\ &=\; 0 \qquad\qquad\qquad\qquad\qquad\quad\ t \le 1 \end{aligned}$$

4.6 APPLICATION OF LAPLACE TRANSFORMS TO THE SOLUTION OF LINEAR CONSTANT COEFFICIENT DIFFERENTIAL EQUATIONS

The application of Laplace transforms to the solution of linear constant coefficient differential equations is of major importance in linear control system problems. Two classes of equations of general interest are treated in this section. The first of these has the form

$$\sum_{i=0}^{n} a_i \frac{d^i y}{dt^i} \;=\; x \tag{4.1}$$

where y is the output, x is the input, the coefficients a_i, $i = 0, 1, \ldots, n-1$, are constants, and $a_n = 1$. The initial conditions for this equation are written as

$$\frac{d^k y}{dt^k}\bigg|_{t=0^+} \;\equiv\; y_0^k, \qquad k = 0, 1, \ldots, n-1$$

where y_0^k are constants. The Laplace transform of Equation (4.1) is given by

$$\sum_{i=0}^{n}\left[\,a_i\left(s^i\,Y(s)\;-\;\sum_{k=0}^{i-1}s^{i-1-k}\,y_0^k\right)\right]\;=\;X(s) \qquad (4.2)$$

and the transform of the output is

$$Y(s)\;=\;\frac{X(s)}{\displaystyle\sum_{i=0}^{n}a_i s^i}\;+\;\frac{\displaystyle\sum_{i=0}^{n}\sum_{k=0}^{i-1}a_i s^{i-1-k}\,y_0^k}{\displaystyle\sum_{i=0}^{n}a_i s^i} \qquad (4.3)$$

Note that the right side of Equation (4.3) is the sum of two terms: a term dependent only on the input transform, and a term dependent only on the initial conditions. In addition, note that the denominator of both terms in Equation (4.3), that is,

$$\sum_{i=0}^{n}a_i s^i\;=\;s^n\;+\;a_{n-1}s^{n-1}\;+\;\cdots\;+\;a_i s\;+\;a_0$$

is the $characteristic\ polynomial$ of Equation (4.1) (see Section 3.9).

The time solution $y(t)$ of Equation (4.1) is the inverse Laplace transform of $Y(s)$, that is,

$$y(t)\;=\;\mathcal{L}^{-1}\left[\frac{X(s)}{\displaystyle\sum_{i=0}^{n}a_i s^i}\right]\;+\;\mathcal{L}^{-1}\left[\frac{\displaystyle\sum_{i=0}^{n}\sum_{k=0}^{i-1}a_i s^{i-1-k}\,y_0^k}{\displaystyle\sum_{i=0}^{n}a_i s^i}\right] \qquad (4.4)$$

The first term on the right is the $forced\ response$ and the second term is the $free\ response$ of the system represented by Equation (4.1).

Direct substitution into Equations (4.2), (4.3) and (4.4) yields the transform of the differential equation, the solution transform $Y(s)$, or the time solution $y(t)$, respectively. But it is often easier to apply directly the properties of Section 4.4 to determine these quantities, especially when the order of the differential equation is low.

Example 4.17.

The Laplace transform of the differential equation

$$\frac{d^2 y}{dt^2}\;+\;3\frac{dy}{dt}\;+\;2y\;=\;u(t)\;=\;\text{unit step}$$

with initial conditions $y(0^+) = -1$ and $\left.\dfrac{dy}{dt}\right|_{t=0^+} = 2$ can be written directly from Equation (4.2) by first identifying n, a_i, and y_0^k: $n = 2$, $y_0^0 = -1$, $y_0^1 = 2$, $a_0 = 2$, $a_1 = 3$, $a_2 = 1$. Substitution of these values into Equation (4.2) yields

$$2Y + 3(sY + 1) + 1(s^2 Y + s - 2)\;=\;\frac{1}{s}\quad\text{or}\quad (s^2 + 3s + 2)Y\;=\;\frac{-(s^2 + s - 1)}{s}$$

It should be noted that when $i = 0$ in (4.2), the summation interior to the brackets is, by definition,

$$\left.\sum_{k=0}^{i-1}\right|_{i=0}\;=\;\sum_{k=0}^{k=-1}\;=\;0$$

The Laplace transform of the differential equation can also be determined in the following manner. The transform of $d^2 y/dt^2$ is given by

$$\mathcal{L}\left[\frac{d^2 y}{dt^2}\right]\;=\;s^2\,Y(s)\;-\;s\,y(0^+)\;-\;\left.\frac{dy}{dt}\right|_{t=0^+}$$

This equation is a direct consequence of Property 3, Section 4.4 (see Problem 4.17, Page 76). With this information the transform of the differential equation can be determined by applying Property 1 (Linearity) of Section 4.4; that is,

$$\mathcal{L}\left[\frac{d^2y}{dt^2} + 3\frac{dy}{dt} + 2y\right] = \mathcal{L}\left[\frac{d^2y}{dt^2}\right] + \mathcal{L}\left[3\frac{dy}{dt}\right] + \mathcal{L}[2y] = (s^2 + 3s + 2)Y + s + 1 = \mathcal{L}[u(t)] = \frac{1}{s}$$

The output transform $Y(s)$ is determined by rearranging the above equation and is

$$Y(s) = \frac{-(s^2 + s - 1)}{s(s^2 + 3s + 2)}$$

The output time solution $y(t)$ is the inverse transform of $Y(s)$. A method for determining the inverse transform of functions like $Y(s)$ above is presented in Sections 4.7 and 4.8.

Now consider constant coefficient equations of the form

$$\sum_{i=0}^{n} a_i \frac{d^i y}{dt^i} = \sum_{i=0}^{m} b_i \frac{d^i x}{dt^i} \tag{4.5}$$

where y is the output, x is the input, $a_n = 1$, and $m \leqq n$. The Laplace transform of Equation (4.5) is given by

$$\sum_{i=0}^{n}\left[a_i\left(s^i Y(s) - \sum_{k=0}^{i-1} s^{i-1-k} y_0^k\right)\right] = \sum_{i=0}^{m}\left[b_i\left(s^i X(s) - \sum_{k=0}^{i-1} s^{i-1-k} x_0^k\right)\right] \tag{4.6}$$

where $x_0^k = \dfrac{d^k x}{dt^k}\bigg|_{t=0^+}$.   The output transform $Y(s)$ is

$$Y(s) = \left[\frac{\displaystyle\sum_{i=0}^{m} b_i s^i}{\displaystyle\sum_{i=0}^{n} a_i s^i}\right] X(s) - \frac{\displaystyle\sum_{i=0}^{m}\sum_{k=0}^{i-1} b_i s^{i-1-k} x_0^k}{\displaystyle\sum_{i=0}^{n} a_i s^i} + \frac{\displaystyle\sum_{i=0}^{n}\sum_{k=0}^{i-1} a_i s^{i-1-k} y_0^k}{\displaystyle\sum_{i=0}^{n} a_i s^i} \tag{4.7}$$

The time solution $y(t)$ is the inverse Laplace transform of $Y(s)$:

$$y(t) = \mathcal{L}^{-1}\left[\frac{\displaystyle\sum_{i=0}^{m} b_i s^i}{\displaystyle\sum_{i=0}^{n} a_i s^i} X(s) - \frac{\displaystyle\sum_{i=0}^{m}\sum_{k=0}^{i-1} b_i s^{i-1-k} x_0^k}{\displaystyle\sum_{i=0}^{n} a_i s^i}\right] + \mathcal{L}^{-1}\left[\frac{\displaystyle\sum_{i=0}^{n}\sum_{k=0}^{i-1} a_i s^{i-1-k} y_0^k}{\displaystyle\sum_{i=0}^{n} a_i s^i}\right] \tag{4.8}$$

The first term on the right is the *forced response*, and the second term is the *free response* of a system represented by Equation (4.5).

Note that the Laplace transform $Y(s)$ of the output $y(t)$ consists of ratios of polynomials in the complex variables s. Such ratios are generally called **rational (algebraic) functions.**

For problems in which initial conditions are not specified on $y(t)$ but on some other parameter of the system (such as the initial voltage across a capacitor not appearing at the output), y_0^k, $k = 0, 1, \ldots, n-1$ must be derived using the available information. For systems represented in the form of Equation (4.5), i.e. including derivative terms in x, computation of y_0^k will also depend on x_0^k. Problem 4.38, Page 84, illustrates these points.

The restriction $n \geqq m$ in Equation (4.5) is a practical constraint based on the fact that most systems of importance have a *smoothing* effect on their input. By a smoothing effect, it is meant that variations in the input are made less pronounced (at least no more pronounced) by the action of the system on the input. Since a differentiator generates the slope of a time function, it accentuates the variations of the function. An integrator, on the other hand, sums the area under the curve of a time function over an interval of time and thus averages (smooths) the variations of the function.

The action of a system described by Equation (4.5) on an input x can be described as follows. The output y is related to the input x by an operation which includes m differentiations and n integrations of the input. Hence in order that there be a smoothing effect (at least no accentuation of the variations) between the input and the output, there must be more (at least as many) integrations than differentiations; that is, $n \geq m$. This practical constraint should not be confused with the *physical* constraint of causality.

Example 4.18.

A certain system is described by the differential equation

$$\frac{d^2y}{dt^2} = \frac{dx}{dt}, \quad y(0^+) = \left.\frac{dy}{dt}\right|_{t=0^+} = 0$$

where the input x is graphed below. The corresponding functions dx/dt and

$$y(t) = \int_{0^+}^{t} \int_{0^+}^{\theta} \frac{dx}{d\alpha} d\alpha \, d\theta = \int_{0^+}^{t} x(\theta) \, d\theta$$

are also shown:

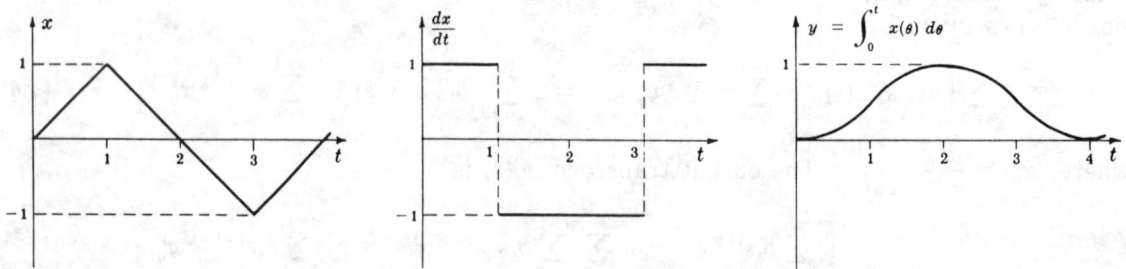

Note from these graphs that differentiation of x accentuates the variations in x while integration smooths them.

Example 4.19.

Consider a system described by the differential equation

$$\frac{d^2y}{dt^2} + 3\frac{dy}{dt} + 2y = \frac{dx}{dt} + 3x$$

with initial conditions $y_0^0 = 1$, $y_0^1 = 0$. If the input is given by $x(t) = e^{-4t}$, then the Laplace transform of the output $y(t)$ can be obtained by direct application of Equation (4.7) by first identifying m, n, a_i, b_i and x_0^0: $n = 2$, $a_0 = 2$, $a_1 = 3$, $a_2 = 1$, $m = 1$, $x_0^0 = \lim_{t \to 0} e^{-4t} = 1$, $b_0 = 3$, $b_1 = 1$. Substitution of these values into Equation (4.7) yields

$$Y(s) = \left(\frac{s+3}{s^2+3s+2}\right)\left(\frac{1}{s+4}\right) + \frac{s+3}{s^2+3s+2} + \frac{1}{s^2+3s+2}$$

This transform can also be obtained by direct application of Properties 1 and 3 of Section 4.4 to the differential equation, as was done in Example 4.17.

4.7 PARTIAL FRACTION EXPANSIONS

In Section 4.6 it was shown that the Laplace transforms encountered in the solution of linear constant coefficient differential equations are rational functions of s (i.e., ratios of polynomials in s). In this section an important representation of rational functions, the partial fraction expansion, is presented. It will be shown in the next section that this representation greatly simplifies the inversion of the Laplace transform of a rational function.

Consider the rational function

$$F(s) = \frac{\sum_{i=0}^{m} b_i s^i}{\sum_{i=0}^{n} a_i s^i} \tag{4.9}$$

where $a_n = 1$ and $n \geq m$. By the fundamental theorem of algebra, the denominator polynomial equation

$$\sum_{i=0}^{n} a_i s^i = 0$$

has n roots. Some of these roots may be repeated.

Example 4.20.

The polynomial $s^3 + 5s^2 + 8s + 4$ has three roots: $-2, -2, -1$. -2 is a repeated root.

Suppose the denominator polynomial equation above has n_1 roots equal to $-p_1$, n_2 roots equal to $-p_2$, ..., n_r roots equal to $-p_r$, where $\sum_{i=1}^{r} n_i = n$. Then

$$\sum_{i=0}^{n} a_i s^i = \prod_{i=1}^{r} (s + p_i)^{n_i}$$

The rational function $F(s)$ can then be written as

$$F(s) = \frac{\sum_{i=0}^{m} b_i s^i}{\prod_{i=1}^{r} (s + p_i)^{n_i}}$$

The **partial fraction expansion** representation of the rational function $F(s)$ is

$$F(s) = b_n + \sum_{i=1}^{r} \sum_{k=1}^{n_i} \frac{c_{ik}}{(s + p_i)^k} \tag{4.10 a}$$

where $b_n = 0$ unless $m = n$. The coefficients c_{ik} are given by

$$c_{ik} = \frac{1}{(n_i - k)!} \frac{d^{n_i - k}}{ds^{n_i - k}} \left[(s + p_i)^{n_i} F(s) \right] \Big|_{s = -p_i} \tag{4.10 b}$$

The particular coefficients c_{i1}, $i = 1, 2, \ldots, r$, are called the **residues** of $F(s)$ at $-p_i$, $i = 1, 2, \ldots, r$. If none of the roots are repeated, then

$$F(s) = b_n + \sum_{i=1}^{n} \frac{c_{i1}}{s + p_i} \tag{4.11 a}$$

where

$$c_{i1} = (s + p_i) F(s) \Big|_{s = -p_i} \tag{4.11 b}$$

Example 4.21.

Consider the rational function $F(s) = \dfrac{s^2 + 2s + 2}{s^2 + 3s + 2} = \dfrac{s^2 + 2s + 2}{(s + 1)(s + 2)}$. The partial fraction expansion of $F(s)$ is

$$F(s) = b_2 + \frac{c_{11}}{s + 1} + \frac{c_{21}}{s + 2}$$

The numerator coefficient of s^2 is $b_2 = 1$. The coefficients c_{11} and c_{21} are determined from Equation (4.11 b) as

$$c_{11} = (s+1) F(s) \Big|_{s=-1} = \frac{s^2 + 2s + 2}{s+2} \Big|_{s=-1} = 1$$

$$c_{21} = (s+2) F(s) \Big|_{s=-2} = \frac{s^2 + 2s + 2}{s+1} \Big|_{s=-2} = -2$$

Hence

$$F(s) = 1 + \frac{1}{s+1} - \frac{2}{s+2}$$

Example 4.22.

Consider the rational function $F(s) = \dfrac{1}{(s+1)^2 (s+2)}$. The partial fraction expansion of $F(s)$ is

$$F(s) = b_3 + \frac{c_{11}}{s+1} + \frac{c_{12}}{(s+1)^2} + \frac{c_{21}}{s+2}$$

The coefficients $b_3, c_{11}, c_{12}, c_{21}$ are given by

$$b_3 = 0$$

$$c_{11} = \frac{d}{ds} (s+1)^2 F(s) \Big|_{s=-1} = \frac{d}{ds} \frac{1}{s+2} \Big|_{s=-1} = -1$$

$$c_{12} = (s+1)^2 F(s) \Big|_{s=-1} = \frac{1}{s+2} \Big|_{s=-1} = 1$$

$$c_{21} = (s+2) F(s) \Big|_{s=-2} = 1$$

Thus

$$F(s) = -\frac{1}{s+1} + \frac{1}{(s+1)^2} + \frac{1}{s+2}$$

4.8 INVERSE TRANSFORMS USING PARTIAL FRACTION EXPANSIONS

In Section 4.6 it was shown that the solution to a linear constant coefficient ordinary differential equation can be determined by finding the inverse Laplace transform of a rational function. The general form of this operation can be written using Equation (*4.10*) as

$$\mathcal{L}^{-1} \left[\frac{\sum\limits_{i=0}^{m} b_i s^i}{\sum\limits_{i=0}^{n} a_i s^i} \right] = \mathcal{L}^{-1} \left[b_n + \sum_{i=0}^{r} \sum_{k=1}^{n_i} \frac{c_{ik}}{(s+p_i)^k} \right]$$

$$= b_n \delta(t) + \sum_{i=1}^{r} \sum_{k=1}^{n_i} \frac{c_{ik}}{(k-1)!} t^{k-1} e^{-p_i t}$$

where $\delta(t)$ is the unit impulse function and $b_n = 0$ unless $m = n$.

Example 4.23.

The inverse Laplace transform of the function $F(s) = \dfrac{s^2 + 2s + 2}{(s+1)(s+2)}$ is given by

$$\mathcal{L}^{-1} \left[\frac{s^2 + 2s + 2}{(s+1)(s+2)} \right] = \mathcal{L}^{-1} \left[1 + \frac{1}{s+1} - \frac{2}{s+2} \right]$$

$$= \mathcal{L}^{-1}[1] + \mathcal{L}^{-1} \left[\frac{1}{s+1} \right] - \mathcal{L}^{-1} \left[\frac{2}{s+2} \right] = \delta(t) + e^{-t} - 2e^{-2t}$$

Example 4.24.

The inverse Laplace transform of the function $F(s) = \dfrac{1}{(s+1)^2 (s+2)}$ is given by

$$\mathcal{L}^{-1}\left[\frac{1}{(s+1)^2(s+2)}\right] = \mathcal{L}^{-1}\left[-\frac{1}{s+1}+\frac{1}{(s+1)^2}+\frac{1}{s+2}\right]$$

$$= -\mathcal{L}^{-1}\left[\frac{1}{s+1}\right] + \mathcal{L}^{-1}\left[\frac{1}{(s+1)^2}\right] + \mathcal{L}^{-1}\left[\frac{1}{s+2}\right] = -e^{-t} + te^{-t} + e^{-2t}$$

4.9 DETERMINING ROOTS OF POLYNOMIALS

The results of Sections 4.7 and 4.8 indicate that finding the solution of a linear constant coefficient differential equation by Laplace transform techniques generally requires the determination of the roots of polynomial equations of the form

$$Q_n(s) = \sum_{i=0}^{n} a_i s^i = 0 \qquad (4.12)$$

where $a_n = 1$ and a_i, $i = 0, 1, \ldots, n-1$ are real constants.

The roots of a second order polynomial equation $s^2 + a_1 s + a_0 = 0$ can be obtained directly from the quadratic formula and are given by

$$s_1 = \frac{-a_1 + \sqrt{a_1^2 - 4a_0}}{2}, \quad s_2 = \frac{-a_1 - \sqrt{a_1^2 - 4a_0}}{2}$$

But for higher order polynomials such analytical expressions do not, in general, exist. The expressions that do exist are very complicated. Fortunately numerical techniques exist for determining these roots.

To aid in the use of these numerical techniques, the following general properties of $Q_n(s)$ are given:

1. If a repeated root of multiplicity n_i is counted as n_i roots, then $Q_n(s) = 0$ has exactly n roots. (Fundamental theorem of algebra.)

2. If $Q_n(s)$ is divided by the factor $s + p$ until a constant remainder is obtained, the remainder is $Q_n(-p)$.

3. $s + p$ is a factor of $Q_n(s)$ if and only if $Q_n(-p) = 0$ ($-p$ is a root of $Q_n(s) = 0$).

4. If $\sigma + j\omega$ (σ, ω real) is a root of $Q_n(s) = 0$, then $\sigma - j\omega$ is also a root of $Q_n(s) = 0$.

5. If n is odd, $Q_n(s) = 0$ has at least one real root.

6. The number of positive real roots of $Q_n(s) = 0$ cannot exceed the number of variations in sign of the coefficients in the polynomial $Q_n(s)$, and the number of negative roots cannot exceed the number of variations in sign of the coefficients of $Q_n(-s)$. (Descartes' rule of signs.)

Of the techniques available for iteratively determining the roots of a polynomial equation (or equivalently the factors of the polynomial), some can determine only real roots and others both real and complex roots. Both types are presented below.

Horner's Method

This method can determine the *real roots* of the polynomial equation $Q_n(s) = 0$. The steps to be followed are:

1. Evaluate $Q_n(s)$ for real integer values of s, $s = 0, \pm 1, \pm 2, \ldots$, until for two consecutive integer values such as k_0 and $k_0 + 1$, $Q_n(k_0)$ and $Q_n(k_0 + 1)$ have opposite signs. A real root then lies between k_0 and $k_0 + 1$. Assume this root is positive without loss of generality. A first approximation of the root is taken to be k_0. Corrections to this approximation are obtained in the remaining steps.

2. Determine a sequence of polynomials $Q_n^l(s)$ using the recursive relationship

$$Q_n^{l+1}(s) = Q_n^l(k_l/10^l + s) = \sum_{i=0}^{n} a_i^{l+1} s^i, \qquad l = 0, 1, 2, \ldots \qquad (4.13)$$

where $Q_n^0(s) = Q_n(s)$, and the values k_l, $l = 1, 2, \ldots$, are generated in Step 3.

3. Determine the integer k_l at each iteration by evaluating $Q_n^l(s)$ for real values of s given by $s = \dfrac{k}{10^l}$ for $k = 0, 1, 2, \ldots, 9$. For two consecutive values of k, say k_l and k_{l+1}, the values $Q_n(k_l/10^l)$ and $Q_n(k_{l+1}/10^l)$ have opposite signs.

4. Repeat until the desired accuracy of the root has been achieved. The approximation of the real root after the Nth iteration is given by

$$s_N = \sum_{l=0}^{N} \frac{k_l}{10^l} \qquad (4.14)$$

Each iteration increases the accuracy of the approximation by one decimal place.

Newton's Method

This method can determine *real roots* of the polynomial equation $Q_n(s) = 0$. The steps to be followed are:

1. Obtain a first approximation s_0 of a root by making an "educated" guess, or by a technique such as the one in Step 1 of Horner's method.

2. Generate a sequence of improved approximations until the desired accuracy is achieved by the recursive relationship

$$s_{l+1} = s_l - \frac{Q_n(s)}{\dfrac{d}{ds}[Q_n(s)]}\bigg|_{s=s_l}$$

which can be rewritten as

$$s_{l+1} = \frac{\sum_{i=0}^{n} (i-1) a_i s_l^i}{\sum_{i=1}^{n} i a_i s_l^{i-1}} \qquad (4.15)$$

where $l = 0, 1, 2, \ldots$.

This method does not provide a measure of the accuracy of the approximation. Indeed, there is no guarantee that the approximations converge to the correct value.

Lin-Bairstow Method

This method can determine both *real and complex roots* of the polynomial equation $Q_n(s) = 0$. More exactly, this method determines quadratic factors of $Q_n(s)$ from which two roots can be determined by the quadratic formula. The roots can, of course, be either real or complex. The steps to be followed are:

1. Obtain a first approximation of a quadratic factor

$$s^2 + \alpha_1 s + \alpha_0$$

of $Q_n(s) = \sum_{i=0}^{n} a_i s^i$ by some method, perhaps an "educated" guess. Corrections to this approximation are obtained in the remaining steps.

2. Generate a set of constants $b_{n-2}, b_{n-3}, \ldots, b_0, b_{-1}, b_{-2}$ from the recursive relationship

$$b_{i-2} = a_i - \alpha_1 b_{i-1} - \alpha_0 b_i$$

where $b_n = b_{n-1} = 0$, and $i = n, n-1, \ldots, 1, 0$.

3. Generate a set of constants $c_{n-2}, c_{n-3}, \ldots, c_1, c_0$ from the recursive relationship

$$c_{i-1} = b_{i-1} - \alpha_1 c_i - \alpha_0 c_{i+1}$$

where $c_n = c_{n-1} = 0$, and $i = n, n-1, \ldots, 1$.

4. Solve the two simultaneous equations

$$c_0 \Delta\alpha_1 + c_1 \Delta\alpha_0 = b_{-1}$$

$$(-\alpha_1 c_0 - \alpha_0 c_1)\Delta\alpha_1 + c_0 \Delta\alpha_0 = b_{-2}$$

for $\Delta\alpha_1$ and $\Delta\alpha_0$. The new approximation of the quadratic factor is

$$s^2 + (\alpha_1 + \Delta\alpha_1)s + (\alpha_0 + \Delta\alpha_0)$$

5. Repeat Steps 1-4 for the quadratic factor obtained in Step 4, until successive approximations are sufficiently close.

This method does not provide a measure of the accuracy of the approximation. Indeed, there is no guarantee that the approximations converge to the correct value.

Root Locus Method

This method can be used to determine both real and complex roots of the polynomial equation $Q_n(s) = 0$. The technique is discussed in Chapter 13.

4.10 COMPLEX PLANE: POLE-ZERO MAPS

The rational function $F(s)$ of the previous sections can be rewritten as

$$F(s) = \frac{b_m \sum_{i=0}^{m} \dfrac{b_i}{b_m} s^i}{\sum_{i=0}^{n} a_i s^i} = \frac{b_m \prod_{i=1}^{m} (s + z_i)}{\prod_{i=1}^{n} (s + p_i)}$$

where the terms $s + z_i$ are factors of the numerator polynomial and the terms $s + p_i$ are factors of the denominator polynomial.

Definition 4.4:   Those values of the complex variable s for which $|F(s)|$ (absolute value of $F(s)$) is zero are called the **zeros** of $F(s)$.

Definition 4.5:   Those values of the complex variable s for which $|F(s)|$ is infinite are called the **poles** of $F(s)$.

Example 4.25.

Let $F(s)$ be given by $F(s) = \dfrac{2s^2 - 2s - 4}{s^3 + 5s^2 + 8s + 6}$ which can be rewritten as

$$F(s) = \frac{2(s+1)(s-2)}{(s+3)(s+1+j)(s+1-j)}$$

$F(s)$ has *finite zeros* at $s = -1$ and $s = 2$, and a *zero* at $s = \infty$.
$F(s)$ has *finite poles* at $s = -3$, $s = -1-j$, and $s = -1+j$.

Poles and zeros are complex numbers determined by two real variables, one representing the real part and the other the imaginary part of the complex number. A pole or zero can therefore be represented as a point in rectangular coordinates. The *abscissa* of this point represents the real part and the *ordinate* the imaginary part. The abscissa is also called the σ-axis and the ordinate the $j\omega$-axis. The plane defined by this coordinate system is called the *complex plane* or the **s-plane**. That half of the plane in which $\sigma < 0$ is called the **left-half of the s-plane** (LHP), and that half in which $\sigma > 0$ is called the **right-half of the s-plane** (RHP).

The location of a pole in the s-plane is denoted symbolically by a cross ($\times$), and the location of a zero by a small circle ($\circ$). The s-plane including the locations of the finite poles and zeros of $F(s)$ is called the **pole-zero map** of $F(s)$.

Example 4.26.

The rational function $\qquad F(s) = \dfrac{(s+1)(s-2)}{(s+3)(s+1+j)(s+1-j)}$

has finite poles $s = -3$, $s = -1-j$ and $s = -1+j$, and finite zeros $s = -1$ and $s = 2$. The pole-zero map of $F(s)$ is shown in Fig. 4-1 below.

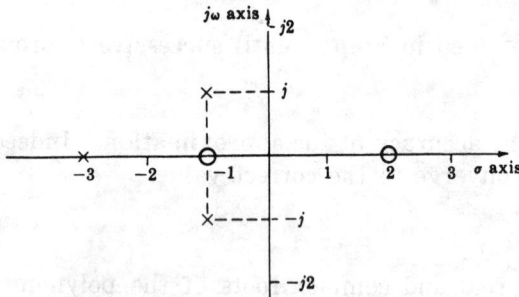

Fig. 4-1

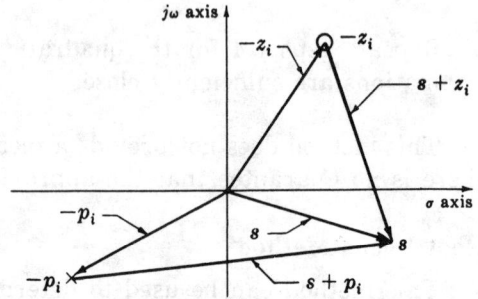

Fig. 4-2

4.11 GRAPHICAL EVALUATION OF RESIDUES

Let $F(s)$ be a rational function written in its factored form:

$$F(s) = \frac{b_m \displaystyle\prod_{i=1}^{m} (s+z_i)}{\displaystyle\prod_{i=1}^{n} (s+p_i)}$$

Since $F(s)$ is a complex function, it can be written in *polar form* as

$$F(s) = |F(s)|\, e^{j\phi} = |F(s)|\, \underline{/\phi}$$

where $|F(s)|$ is the absolute value of $F(s)$ and $\phi \equiv \arg F(s) = \tan^{-1}\left[\dfrac{\operatorname{Im} F(s)}{\operatorname{Re} F(s)}\right]$.

$F(s)$ can further be written in terms of the polar forms of the factors $s+z_i$ and $s+p_i$ as

$$F(s) = \frac{b_m \displaystyle\prod_{i=1}^{m} |s+z_i|}{\displaystyle\prod_{i=1}^{n} |s+p_i|} \quad \underline{\bigg/\left[\sum_{i=1}^{m} \phi_{iz} - \sum_{i=1}^{n} \phi_{ip}\right]}$$

where $s + z_i = |s+z_i|\,\underline{/\phi_{iz}}$ and $s + p_i = |s+p_i|\,\underline{/\phi_{ip}}$.

Each complex number s, z_i, p_i, $s+z_i$ and $s+p_i$ can be represented by a vector in the s-plane. If p is a general complex number, then the vector representing p has magnitude $|p|$ and direction defined by the angle

$$\phi = \tan^{-1}\left[\frac{\operatorname{Im} p}{\operatorname{Re} p}\right]$$

measured counterclockwise from the positive σ-axis.

A typical pole $-p_i$ and zero $-z_i$ are shown in Fig. 4-2, along with a general complex variable s. The *sum vectors* $s+z_i$ and $s+p_i$ are also shown. Note that the vector $s+z_i$ is a vector which starts at the zero $-z_i$ and terminates at s, and $s + p_i$ starts at the pole $-p_i$ and terminates at s.

For distinct poles of the rational function $F(s)$, the *residue* $c_{k1} \equiv c_k$ of the pole $-p_k$ is given by

$$c_k = (s + p_k) F(s) \Big|_{s = -p_k} = \frac{b_m(s + p_k) \prod_{i=1}^{m} (s + z_i)}{\prod_{i=1}^{n} (s + p_i)} \Bigg|_{s = -p_k}$$

These residues can be determined by the following graphical procedure:

1. Plot the pole-zero map of $(s + p_k) F(s)$.

2. Draw vectors on this map starting at the poles and zeros of $(s + p_k) F(s)$, and terminating at $-p_k$. Measure the magnitude (in the scale of the pole-zero map) of these vectors and the angles of the vectors measured from the positive real axis in the counterclockwise direction.

3. Obtain the magnitude $|c_k|$ of the residue c_k as the product of b_m and the magnitudes of the vectors from the zeros to $-p_k$, divided by the product of the magnitudes of the vectors from the poles to $-p_k$.

4. Determine the angle ϕ_k of the residue c_k as the sum of the angles of the vectors from the zeros to $-p_k$, minus the sum of the angles of the vectors from the poles to $-p_k$. This is true for positive b_m. If b_m is negative, then add $180°$ to this angle.

The residue c_k is given in polar form by

$$c_k = |c_k| e^{j\phi_k} = |c_k| \underline{/\phi_k}$$

or in rectangular form by $c_k = |c_k| \cos \phi_k + j |c_k| \sin \phi_k$

This graphical technique is not directly applicable for evaluating residues of multiple poles.

4.12 SECOND ORDER SYSTEMS

Many control systems can be described or approximated by the **general second order** differential equation

$$\frac{d^2y}{dt^2} + 2 \zeta \omega_n \frac{dy}{dt} + \omega_n^2 y = \omega_n^2 x$$

The positive coefficient ω_n is called the **undamped natural frequency** and the coefficient ζ is the **damping ratio** of the system.

The Laplace transform of $y(t)$, when the initial conditions are zero, is

$$Y(s) = \left[\frac{\omega_n^2}{s^2 + 2 \zeta \omega_n s + \omega_n^2} \right] X(s)$$

where $X(s) = \mathcal{L}[x(t)]$. The poles of the function $\dfrac{Y(s)}{X(s)} = \dfrac{\omega_n^2}{s^2 + 2 \zeta \omega_n s + \omega_n^2}$ are

$$s = -\zeta \omega_n \pm \omega_n \sqrt{\zeta^2 - 1}$$

Note that:

1. If $\zeta > 1$, both poles are negative and real.

2. If $\zeta = 1$, the poles are equal, negative, and real ($s = -\omega_n$).

3. If $0 < \zeta < 1$, the poles are complex conjugates with negative real parts ($s = -\zeta\omega_n \pm j\omega_n\sqrt{1-\zeta^2}$).

4. If $\zeta = 0$, the poles are imaginary and complex conjugate ($s = \pm j\omega_n$).

5. If $\zeta < 0$, the poles are in the right half of the s-plane (RHP).

Of particular interest in this book is Case 3, representing an **underdamped second order system.** The poles are complex conjugates with negative real parts and are located at

$$s = -\zeta\omega_n \pm j\omega_n\sqrt{1-\zeta^2}$$

or at

$$s = -\alpha \pm j\omega_d$$

where $\dfrac{1}{\alpha} \equiv \dfrac{1}{\zeta\omega_n}$ is called the **time constant** of the system and $\omega_d \equiv \omega_n\sqrt{1-\zeta^2}$ is called the **damped natural frequency** of the system. For fixed ω_n, the adjoining figure shows the locus of these poles as a function of ζ ($0 < \zeta < 1$). The locus is a semicircle of radius ω_n. The angle θ is related to the damping ratio by $\theta = \cos^{-1}\zeta$.

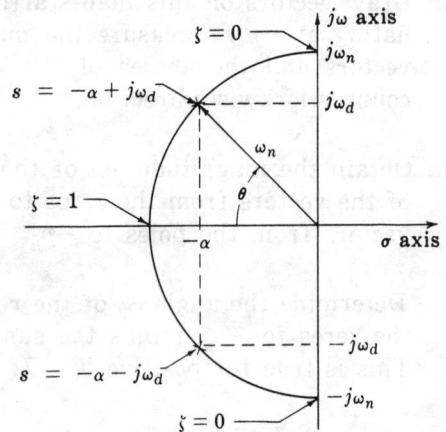

Solved Problems

LAPLACE TRANSFORMS FROM THE DEFINITION

4.1. Show that the unit step function $u(t)$ is Laplace transformable and determine its Laplace transform.

Direct substitution into the equation of Definition 4.2, Page 57, yields

$$\int_{0^+}^{\infty} |u(t)|\, e^{-\sigma_0 t}\, dt = \int_{0^+}^{\infty} e^{-\sigma_0 t}\, dt = -\frac{1}{\sigma_0} e^{-\sigma_0 t}\Big|_{0^+}^{\infty} = \frac{1}{\sigma_0} < +\infty$$

for $\sigma_0 > 0$. The Laplace transform is given by Definition 4.1, Page 56:

$$\mathcal{L}\left[u(t)\right] = \int_{0^+}^{\infty} u(t)\, e^{-st}\, dt = -\frac{1}{s} e^{-st}\Big|_{0^+}^{\infty} = \frac{1}{s} \qquad \text{for} \quad \operatorname{Re} s > 0$$

4.2. Show that the unit ramp function t is Laplace transformable and determine its Laplace transform.

Direct substitution into the equation of Definition 4.2 yields

$$\int_{0^+}^{\infty} |t|\, e^{-\sigma_0 t}\, dt = \frac{e^{-\sigma_0 t}}{\sigma_0^2}(-\sigma_0 t - 1)\Big|_{0^+}^{\infty} = \frac{1}{\sigma_0^2} < +\infty$$

for $\sigma_0 > 0$. The Laplace transform is given by Definition 4.1:

$$\mathcal{L}\,[t] \;=\; \int_{0+}^{\infty} t\,e^{-st}\,dt \;=\; \frac{e^{-st}}{s^2}(-st-1)\Big|_{0+}^{\infty} \;=\; \frac{1}{s^2} \qquad \text{for} \quad \text{Re } s > 0$$

4.3. Show that the sine function $\sin t$ is Laplace transformable and determine its Laplace transform.

The integral $\displaystyle\int_{0+}^{\infty} |\sin t|\,e^{-\sigma_0 t}\,dt$ can be evaluated by writing the integral over the positive half cycles of $\sin t$ as

$$\int_{n\pi}^{(n+1)\pi} \sin t\,e^{-\sigma_0 t}\,dt \;=\; \frac{e^{-\sigma_0 n\pi}}{\sigma_0^2+1}\,[e^{-\sigma_0\pi}+1]$$

for n even, and over negative half cycles of $\sin t$ as

$$-\int_{n\pi}^{(n+1)\pi} \sin t\,e^{-\sigma_0 t}\,dt \;=\; \frac{e^{-\sigma_0 n\pi}}{\sigma_0^2+1}\,[e^{-\sigma_0\pi}+1]$$

for n odd. Then

$$\int_{0+}^{\infty} |\sin t|\,e^{-\sigma_0 t}\,dt \;=\; \frac{e^{-\sigma_0\pi}+1}{\sigma_0^2+1}\sum_{n=0}^{\infty} e^{-\sigma_0 n\pi}$$

The summation converges for $e^{-\sigma_0\pi} < 1$ or $\sigma_0 > 0$ and can be written in closed form as

$$\sum_{n=0}^{\infty} e^{-\sigma_0 n\pi} \;=\; \frac{1}{1-e^{-\sigma_0\pi}}$$

Then

$$\int_{0+}^{\infty} |\sin t|\,e^{-\sigma_0 t}\,dt \;=\; \left[\frac{1+e^{-\sigma_0\pi}}{1-e^{-\sigma_0\pi}}\right]\!\left(\frac{1}{\sigma_0^2+1}\right) < +\infty \qquad \text{for} \quad \sigma_0 > 0$$

Finally,

$$\mathcal{L}\,[\sin t] \;=\; \int_{0+}^{\infty} \sin t\,e^{-st}\,dt \;=\; \frac{e^{-st}(-s\sin t - \cos t)}{s^2+1}\Big|_{0+}^{\infty} \;=\; \frac{1}{s^2+1} \qquad \text{for} \quad \text{Re } s > 0$$

4.4. Show that the Laplace transform of the unit impulse function is given by $\mathcal{L}\,[\delta(t)] = 1$.

Direct substitution of Equation (3.20), Page 38, into the equation of Definition 4.1, Page 56, yields

$$\int_{0+}^{\infty} \delta(t)\,e^{-st}\,dt \;=\; \int_{0+}^{\infty} \lim_{\Delta t\to 0}\left[\frac{u(t)-u(t-\Delta t)}{\Delta t}\right]e^{-st}\,dt$$

$$=\; \lim_{\Delta t\to 0}\left[\int_{0+}^{\infty}\frac{u(t)}{\Delta t}e^{-st}\,dt - \int_{0+}^{\infty}\frac{u(t-\Delta t)}{\Delta t}e^{-st}\,dt\right] \;=\; \lim_{\Delta t\to 0}\frac{1}{\Delta t}\left[\frac{1}{s}-\frac{e^{-\Delta t\,s}}{s}\right]$$

where the Laplace transform of $u(t)$ is $1/s$, as shown in Problem 4.1, and the second term is obtained using Property 9. Now $e^{-\Delta ts} = 1 - \Delta t\,s + \dfrac{(\Delta t\,s)^2}{2!} - \dfrac{(\Delta t\,s)^3}{3!} + \cdots$ (see reference [6]). Thus

$$\mathcal{L}\,[\delta(t)] \;=\; \lim_{\Delta t\to 0}\frac{1}{\Delta t}\left[\frac{1}{s}-\frac{e^{-\Delta t\,s}}{s}\right] \;=\; \lim_{\Delta t\to 0}\frac{1}{\Delta t}\left[\Delta t - \frac{(\Delta t)^2 s}{2!} + \frac{(\Delta t)^3 s^2}{3!} - \cdots\right] \;=\; 1$$

PROPERTIES OF THE LAPLACE TRANSFORM AND ITS INVERSE

4.5. Show that $\mathcal{L}\,[a_1 f_1(t) + a_2 f_2(t)] = a_1 F_1(s) + a_2 F_2(s)$ where $F_1(s) = \mathcal{L}\,[f_1(t)]$ and $F_2(s) = \mathcal{L}\,[f_2(t)]$ (Property 1, Page 57).

By definition,

$$\mathcal{L}\left[a_1 f_1(t) + a_2 f_2(t)\right] = \int_{0+}^{\infty} [a_1 f_1(t) + a_2 f_2(t)] e^{-st}\, dt$$

$$= \int_{0+}^{\infty} a_1 f_1(t) e^{-st}\, dt + \int_{0+}^{\infty} a_2 f_2(t) e^{-st}\, dt$$

$$= a_1 \int_{0+}^{\infty} f_1(t) e^{-st}\, dt + a_2 \int_{0+}^{\infty} f_2(t) e^{-st}\, dt$$

$$= a_1 \mathcal{L}\left[f_1(t)\right] + a_2 \mathcal{L}\left[f_2(t)\right] = a_1 F_1(s) + a_2 F_2(s)$$

4.6. Show that $\mathcal{L}^{-1}[a_1 F_1(s) + a_2 F_2(s)] = a_1 f_1(t) + a_2 f_2(t)$ where $\mathcal{L}^{-1}[F_1(s)] = f_1(t)$ and $\mathcal{L}^{-1}[F_2(s)] = f_2(t)$ (Property 2, Page 57).

By definition,

$$\mathcal{L}^{-1}[a_1 F_1(s) + a_2 F_2(s)] = \frac{1}{2\pi j} \int_{c-j\infty}^{c+j\infty} [a_1 F_1(s) + a_2 F_2(s)] e^{st}\, ds$$

$$= \frac{1}{2\pi j} \int_{c-j\infty}^{c+j\infty} a_1 F_1(s) e^{st}\, ds + \frac{1}{2\pi j} \int_{c-j\infty}^{c+j\infty} a_2 F_2(s) e^{st}\, ds$$

$$= a_1 \left[\frac{1}{2\pi j} \int_{c-j\infty}^{c+j\infty} F_1(s) e^{st}\, ds\right] + a_2 \left[\frac{1}{2\pi j} \int_{c-j\infty}^{c+j\infty} F_2(s) e^{st}\, ds\right]$$

$$= a_1 \mathcal{L}^{-1}[F_1(s)] + a_2 \mathcal{L}^{-1}[F_2(s)] = a_1 f_1(t) + a_2 f_2(t)$$

4.7. Show that the Laplace transform of the derivative df/dt of a function $f(t)$ is given by $\mathcal{L}[df/dt] = s F(s) - f(0^+)$ where $F(s) = \mathcal{L}[f(t)]$ (Property 3, Page 58).

By definition, $\mathcal{L}\left[\dfrac{df}{dt}\right] = \lim_{\substack{T \to \infty \\ \epsilon \to 0}} \int_{\epsilon}^{T} \dfrac{df}{dt} e^{-st}\, dt.$ Integrating by parts,

$$\lim_{\substack{T \to \infty \\ \epsilon \to 0}} \int_{\epsilon}^{T} \frac{df}{dt} e^{-st}\, dt = \lim_{\substack{T \to \infty \\ \epsilon \to 0}} \left[f(t) e^{-st} \Big|_{\epsilon}^{T} + s \int_{\epsilon}^{T} f(t) e^{-st}\, dt \right] = -f(0^+) + s F(s)$$

where $\lim_{\epsilon \to 0} f(\epsilon) = f(0^+)$.

4.8. Show that $\mathcal{L}\left[\displaystyle\int_{0}^{t} f(\tau)\, d\tau\right] = \dfrac{F(s)}{s}$ where $F(s) = \mathcal{L}[f(t)]$ (Property 4, Page 58).

By definition and a change in the order of integrations, we have

$$\mathcal{L}\left[\int_{0}^{t} f(\tau)\, d\tau\right] = \int_{0+}^{\infty} \int_{0}^{t} f(\tau)\, d\tau\, e^{-st}\, dt = \int_{0+}^{\infty} f(\tau) \int_{\tau}^{\infty} e^{-st}\, dt\, d\tau$$

$$= \int_{0+}^{\infty} f(\tau) \left[-\frac{1}{s} e^{-st} \Big|_{\tau}^{\infty} \right] d\tau = \int_{0+}^{\infty} f(\tau) \frac{e^{-s\tau}}{s}\, d\tau = \frac{F(s)}{s}$$

4.9. Show that $f(0^+) \equiv \lim_{t \to 0} f(t) = \lim_{s \to \infty} s F(s)$ where $F(s) = \mathcal{L}[f(t)]$ (Property 5, Page 58).

From Problem 4.7,

$$\mathcal{L}\left[\frac{df}{dt}\right] = s F(s) - f(0^+) = \lim_{\substack{T \to \infty \\ \epsilon \to 0}} \int_{\epsilon}^{T} \frac{df}{dt} e^{-st}\, dt$$

Now let $s \to \infty$, that is,

$$\lim_{s \to \infty} [s F(s) - f(0^+)] = \lim_{s \to \infty} \left[\lim_{\substack{T \to \infty \\ \epsilon \to 0}} \int_{\epsilon}^{T} \frac{df}{dt} e^{-st}\, dt \right]$$

Since the limiting processes can be interchanged, we have

$$\lim_{s \to \infty} \left[\lim_{\substack{T \to \infty \\ \epsilon \to 0}} \int_{\epsilon}^{T} \frac{df}{dt} e^{-st}\, dt \right] = \lim_{\substack{T \to \infty \\ \epsilon \to 0}} \int_{\epsilon}^{T} \frac{df}{dt} \left(\lim_{s \to \infty} e^{-st} \right) dt$$

But $\lim_{s \to \infty} e^{-st} = 0$. Hence the right side of the equation is zero and $\lim_{s \to \infty} s F(s) = f(0^+)$.

4.10. Show that if $\lim\limits_{t \to \infty} f(t)$ exists then $f(\infty) \equiv \lim\limits_{t \to \infty} f(t) = \lim\limits_{s \to 0} s\, F(s)$ where $F(s) = \mathcal{L}\,[f(t)]$ (Property 6).

From Problem 4.7,

$$\mathcal{L}\left[\frac{df}{dt}\right] = s\, F(s) - f(0^+) = \lim_{\substack{T \to \infty \\ \epsilon \to 0}} \int_\epsilon^T \frac{df}{dt} e^{-st}\, dt$$

Now let $s \to 0$, that is,

$$\lim_{s \to 0} [s\, F(s) - f(0^+)] = \lim_{s \to 0}\left[\lim_{\substack{T \to \infty \\ \epsilon \to 0}} \int_\epsilon^T \frac{df}{dt} e^{-st}\, dt\right]$$

Since the limiting processes can be interchanged, we have

$$\lim_{s \to 0}\left[\lim_{\substack{T \to \infty \\ \epsilon \to 0}} \int_\epsilon^T \frac{df}{dt} e^{-st}\, dt\right] = \lim_{\substack{T \to \infty \\ \epsilon \to 0}} \int_\epsilon^T \frac{df}{dt}\left(\lim_{s \to 0} e^{-st}\right) dt = \lim_{\substack{T \to \infty \\ \epsilon \to 0}} \int_\epsilon^T \frac{df}{dt}\, dt = f(\infty) - f(0^+)$$

Adding $f(0^+)$ to both sides of the last equation yields $\lim\limits_{s \to 0} s\, F(s) = f(\infty)$ if $f(\infty) = \lim\limits_{t \to \infty} f(t)$ exists.

4.11. Show that $\mathcal{L}\,[f(t/a)] = a\, F(as)$ where $F(s) = \mathcal{L}\,[f(t)]$ (Property 7).

By definition, $\mathcal{L}\,[f(t/a)] = \int_{0^+}^\infty f(t/a)\, e^{-st}\, dt$. Making the change of variable $\tau = t/a$,

$$\mathcal{L}\,[f(t/a)] = a \int_{0^+}^\infty f(\tau)\, e^{-(as)\tau}\, d\tau = a\, F(as)$$

4.12. Show that $\mathcal{L}^{-1}\,[F(s/a)] = a\, f(at)$ where $f(t) = \mathcal{L}^{-1}\,[F(s)]$ (Property 8).

By definition, $\mathcal{L}^{-1}\,[F(s/a)] = \dfrac{1}{2\pi j} \displaystyle\int_{c-j\infty}^{c+j\infty} F(s/a)\, e^{st}\, ds$. Making the change of variable $\omega = s/a$,

$$\mathcal{L}^{-1}\,[F(s/a)] = \frac{a}{2\pi j} \int_{c-j\infty}^{c+j\infty} F(\omega)\, e^{\omega(at)}\, d\omega = a\, f(at)$$

4.13. Show that $\mathcal{L}\,[f(t-T)] = e^{-sT}\, F(s)$ where $f(t-T) = 0$ for $t \leq T$ and $F(s) = \mathcal{L}\,[f(t)]$ (Property 9).

By definition,

$$\mathcal{L}\,[f(t-T)] = \int_{0^+}^\infty f(t-T)\, e^{-st}\, dt = \int_T^\infty f(t-T)\, e^{-st}\, dt$$

Making the change of variable $\theta = t - T$,

$$\mathcal{L}\,[f(t-T)] = \int_{0^+}^\infty f(\theta)\, e^{-s\theta} e^{-sT}\, d\theta = e^{-sT}\, F(s)$$

4.14. Show that $\mathcal{L}\,[e^{-at} f(t)] = F(s+a)$ where $F(s) = \mathcal{L}\,[f(t)]$ (Property 10).

By definition,

$$\mathcal{L}\,[e^{-at} f(t)] = \int_{0^+}^\infty e^{-at} f(t)\, e^{-st}\, dt = \int_{0^+}^\infty f(t)\, e^{-(s+a)t}\, dt = F(s+a)$$

4.15. Show that

$$\mathcal{L}\,[f_1(t) \cdot f_2(t)] = \frac{1}{2\pi j} \int_{c-j\infty}^{c+j\infty} F_1(\omega)\, F_2(s - \omega)\, d\omega$$

where $F_1(s) = \mathcal{L}\,[f_1(t)]$ and $F_2(s) = \mathcal{L}\,[f_2(t)]$ (Property 11, Page 58).

By definition,
$$\mathcal{L}\left[f_1(t)\cdot f_2(t)\right] = \int_{0+}^{\infty} f_1(t)\cdot f_2(t)\, e^{-st}\, dt$$

But $f_1(t) = \dfrac{1}{2\pi j}\displaystyle\int_{c-j\infty}^{c+j\infty} F_1(\omega)\, e^{\omega t}\, d\omega.$ Hence,

$$\mathcal{L}\left[f_1(t)\cdot f_2(t)\right] = \frac{1}{2\pi j}\int_{0+}^{\infty}\int_{c-j\infty}^{c+j\infty} F_1(\omega)\, e^{\omega t}\, d\omega\, f_2(t)\, e^{-st}\, dt$$

Interchanging the order of integrations yields

$$\mathcal{L}\left[f_1(t)\cdot f_2(t)\right] = \frac{1}{2\pi j}\int_{c-j\infty}^{c+j\infty} F_1(\omega)\int_{0+}^{\infty} f_2(t)\, e^{-(s-\omega)t}\, dt\, d\omega$$

Since $\displaystyle\int_{0+}^{\infty} f_2(t)\, e^{-(s-\omega)t}\, dt = F_2(s-\omega),$

$$\mathcal{L}\left[f_1(t)\cdot f_2(t)\right] = \frac{1}{2\pi j}\int_{c-j\infty}^{c+j\infty} F_1(\omega)\, F_2(s-\omega)\, d\omega$$

4.16. Show that $\qquad \mathcal{L}^{-1}\left[F_1(s)\cdot F_2(s)\right] = \displaystyle\int_{0+}^{t} f_1(\tau)\, f_2(t-\tau)\, d\tau$

where $f_1(t) = \mathcal{L}^{-1}\left[F_1(s)\right]$ and $f_2(t) = \mathcal{L}^{-1}\left[F_2(s)\right]$ **(Property 12, Page 58).**

By definition, $\quad \mathcal{L}^{-1}\left[F_1(s)\cdot F_2(s)\right] = \dfrac{1}{2\pi j}\displaystyle\int_{c-j\infty}^{c+j\infty} F_1(s)\, F_2(s)\, e^{st}\, ds$

But $F_1(s) = \displaystyle\int_{0+}^{\infty} f_1(\tau)\, e^{-s\tau}\, d\tau.$ Hence,

$$\mathcal{L}^{-1}\left[F_1(s)\, F_2(s)\right] = \frac{1}{2\pi j}\int_{c-j\infty}^{c+j\infty}\int_{0+}^{\infty} f_1(\tau)\, e^{-s\tau}\, d\tau\, F_2(s)\, e^{st}\, ds$$

Interchanging the order of integrations yields

$$\mathcal{L}^{-1}\left[F_1(s)\, F_2(s)\right] = \int_{0+}^{\infty}\frac{1}{2\pi j}\int_{c-j\infty}^{c+j\infty} F_2(s)\, e^{s(t-\tau)}\, ds\, f_1(\tau)\, d\tau$$

Since $\dfrac{1}{2\pi j}\displaystyle\int_{c-j\infty}^{c+j\infty} F_2(s)\, e^{s(t-\tau)}\, ds = f_2(t-\tau),$

$$\mathcal{L}^{-1}\left[F_1(s)\, F_2(s)\right] = \int_{0+}^{\infty} f_1(\tau)\, f_2(t-\tau)\, d\tau = \int_{0+}^{t} f_1(\tau)\, f_2(t-\tau)\, d\tau$$

where the second equality is true since $f_2(t-\tau) = 0$ for $\tau \geq t$.

4.17. Show that
$$\mathcal{L}\left[\frac{d^i y}{dt^i}\right] = s^i Y(s) - \sum_{k=0}^{i-1} s^{i-1-k}\, y_0^k$$

for $i > 0$ where $Y(s) = \mathcal{L}\left[y(t)\right]$ and $y_0^k = \dfrac{d^k y}{dt^k}\bigg|_{t=0+}$

This result can be shown by mathematical induction. For $i=1$,

$$\mathcal{L}\left[\frac{dy}{dt}\right] = s\, Y(s) - y(0^+) = s\, Y(s) - y_0^0$$

as shown in Problem 4.7. Now assume the result holds for $i = n-1$, that is,

$$\mathcal{L}\left[\frac{d^{n-1}y}{dt^{n-1}}\right] = s^{n-1} Y(s) - \sum_{k=0}^{n-2} s^{n-2-k}\, y_0^k$$

Then $\mathcal{L}\left[\dfrac{d^n y}{dt^n}\right]$ can be written as

$$\mathcal{L}\left[\frac{d^n y}{dt^n}\right] = \mathcal{L}\left[\frac{d}{dt}\left(\frac{d^{n-1}y}{dt^{n-1}}\right)\right] = s\,\mathcal{L}\left[\frac{d^{n-1}y}{dt^{n-1}}\right] - \frac{d^{n-1}y}{dt^{n-1}}\bigg|_{t=0^+}$$

$$= s\left(s^{n-1}Y(s) - \sum_{k=0}^{n-2} s^{n-2-k}y_0^k\right) - y_0^{n-1} = s^n Y(s) - \sum_{k=0}^{n-1} s^{n-1-k}y_0^k$$

For the special case $n = 2$, we have $\mathcal{L}\left[\dfrac{d^2 y}{dt^2}\right] = s^2 Y(s) - sy_0^0 - y_0^1$.

LAPLACE TRANSFORMS AND THEIR INVERSE FROM THE TABLE OF TRANSFORM PAIRS

4.18. Find the Laplace transform of $f(t) = 2e^{-t}\cos 10t - t^4 + 6e^{-(t-10)}$ for $t > 0$.

From the table of transform pairs,

$$\mathcal{L}\left[e^{-t}\cos 10t\right] = \frac{s+1}{(s+1)^2 + 10^2}, \quad \mathcal{L}\left[t^4\right] = \frac{4!}{s^5}, \quad \mathcal{L}\left[e^{-t}\right] = \frac{1}{s+1}$$

Using Property 9, $\mathcal{L}\left[e^{-(t-10)}\right] = \dfrac{e^{-10s}}{s+1}$. Using Property 1,

$$\mathcal{L}[f(t)] = 2\,\mathcal{L}\left[e^{-t}\cos 10t\right] - \mathcal{L}\left[t^4\right] + 6\,\mathcal{L}\left[e^{-(t-10)}\right] = \frac{2(s+1)}{s^2 + 2s + 101} - \frac{24}{s^5} + \frac{6e^{-10s}}{s+1}$$

4.19. Find the inverse Laplace transform of $F(s) = \dfrac{2e^{-0.5s}}{s^2 - 6s + 13} - \dfrac{s-1}{s^2 - 2s + 2}$ for $t > 0$.

$$\frac{2}{s^2 - 6s + 13} = \frac{2}{(s-3)^2 + 2^2}, \qquad \frac{s-1}{s^2 - 2s + 2} = \frac{s-1}{(s-1)^2 + 1}$$

The inverse transforms are determined directly from Table 4.1 as

$$\mathcal{L}^{-1}\left[\frac{2}{(s-3)^2 + 2^2}\right] = e^{3t}\sin 2t, \qquad \mathcal{L}^{-1}\left[\frac{s-1}{(s-1)^2 + 1}\right] = e^t \cos t$$

Using Property 9, *then* Property 2, results in

$$f(t) = \begin{cases} -e^t \cos t & 0 < t \leq 0.5 \\ e^{3(t-0.5)}\sin 2(t-0.5) - e^t \cos t & t > 0.5 \end{cases}$$

LAPLACE TRANSFORMS OF LINEAR CONSTANT COEFFICIENT DIFFERENTIAL EQUATIONS

4.20. Determine the output transform $Y(s)$ for the differential equation

$$\frac{d^3 y}{dt^3} + 3\frac{d^2 y}{dt^2} - \frac{dy}{dt} + 6y = \frac{d^2 x}{dt^2} - x$$

where y = output, x = input, and initial conditions are -

$$y(0^+) = \frac{dy}{dt}\bigg|_{t=0^+} = 0, \qquad \frac{d^2 y}{dt^2}\bigg|_{t=0^+} = 1$$

Using Property 3 or the result of Problem 4.17, the Laplace transforms of the terms of the equation are given as

$$\mathcal{L}\left[\frac{d^3 y}{dt^3}\right] = s^3 Y(s) - s^2 y(0^+) - s\frac{dy}{dt}\bigg|_{t=0^+} - \frac{d^2 y}{dt^2}\bigg|_{t=0^+} = s^3 Y(s) - 1$$

$$\mathcal{L}\left[\frac{d^2 y}{dt^2}\right] = s^2 Y(s) - s y(0^+) - \frac{dy}{dt}\bigg|_{t=0^+} = s^2 Y(s)$$

$$\mathcal{L}\left[\frac{dy}{dt}\right] = s Y(s) - y(0^+) = s Y(s) \qquad \mathcal{L}\left[\frac{d^2 x}{dt^2}\right] = s^2 X(s) - s x(0^+) - \frac{dx}{dt}\bigg|_{t=0^+}$$

where $Y(s) = \mathcal{L}[y(t)]$ and $X(s) = \mathcal{L}[x(t)]$. The Laplace transform of the given equation can now be written as

$$\mathcal{L}\left[\frac{d^3y}{dt^3}\right] + 3\mathcal{L}\left[\frac{d^2y}{dt^2}\right] - \mathcal{L}\left[\frac{dy}{dt}\right] + 6\mathcal{L}[y]$$

$$= s^3 Y(s) - 1 + 3s^2 Y(s) - s Y(s) + 6 Y(s)$$

$$= \mathcal{L}\left[\frac{d^2x}{dt^2}\right] - \mathcal{L}[x] = s^2 X(s) - s\,x(0^+) - \frac{dx}{dt}\bigg|_{t=0^+} - X(s)$$

Solving for $Y(s)$, we obtain

$$Y(s) = \frac{(s^2-1)X(s)}{s^3+3s^2-s+6} - \frac{s\,x(0^+) + \dfrac{dx}{dt}\bigg|_{t=0^+}}{s^3+3s^2-s+6} + \frac{1}{s^3+3s^2-s+6}$$

4.21. What part of the solution of Problem 4.20 is the transform of the free response? The forced response?

The transform of the free response $Y_a(s)$ is that part of the output transform $Y(s)$ which does not depend on the input $x(t)$, its derivatives or its transform; that is,

$$Y_a(s) = \frac{1}{s^3+3s^2-s+6}$$

The transform of the forced response $Y_b(s)$ is that part of $Y(s)$ which depends on $x(t)$, its derivative and its transform; that is,

$$Y_b(s) = \frac{(s^2-1)X(s)}{s^2+3s^2-s+6} - \frac{s\,x(0^+) + \dfrac{dx}{dt}\bigg|_{t=0^+}}{s^3+3s^2-s+6}$$

4.22. What is the characteristic polynomial for the differential equation of Problems 4.20 and 4.21?

The characteristic polynomial is the denominator polynomial which is common to the transforms of the free and forced responses (see Problem 4.21), that is, the polynomial $s^2 + 3s^2 - s + 6$.

4.23. Determine the output transform $Y(s)$ of the system of Problem 4.20 for an input $x(t) = 5\sin t$.

From the table on Page 60, $X(s) \equiv \mathcal{L}[x(t)] = \mathcal{L}[5\sin t] = \dfrac{5}{s^2+1}$.

The initial values of $x(t)$ and $\dfrac{dx}{dt}$ are: $x(0^+) = \lim_{t\to0} 5\sin t = 0$, $\dfrac{dx}{dt}\bigg|_{t=0^+} = \lim_{t\to0} 5\cos t = 5$.

Substituting these values into the output transform $Y(s)$ given in Problem 4.20,

$$Y(s) = \frac{s^2-9}{(s^3+3s^2-s+6)(s^2+1)}$$

PARTIAL FRACTION EXPANSIONS

4.24. A rational function $F(s)$ can be represented by

$$F(s) = \frac{\displaystyle\sum_{i=0}^{n} b_i s^i}{\displaystyle\prod_{i=1}^{r} (s+p_i)^{n_i}} = b_n + \sum_{i=1}^{r}\sum_{k=1}^{n_i} \frac{c_{ik}}{(s+p_i)^k} \qquad (4.10\,a)$$

where the second form is the partial fraction expansion of $F(s)$. Show that the constants c_{ik} are given by

$$c_{ik} = \frac{1}{(n_i-k)!} \frac{d^{n_i-k}}{ds^{n_i-k}}\left[(s+p_i)^{n_i} F(s)\right]\bigg|_{s=-p_i} \qquad (4.10\,b)$$

Let $(s + p_j)$ be the factor of interest and form

$$(s + p_j)^{n_j} F(s) \;=\; (s + p_j)^{n_j} b_n \;+\; \sum_{i=1}^{r} \sum_{k=1}^{n_i} \frac{(s + p_j)^{n_j} c_{ik}}{(s + p_i)^k}$$

This can be rewritten

$$(s + p_j)^{n_j} F(s) \;=\; (s + p_j)^{n_j} b_n \;+\; \sum_{i=1}^{j-1} \sum_{k=1}^{n_i} \frac{(s + p_j)^{n_j} c_{ik}}{(s + p_i)^k}$$

$$+\; \sum_{i=j+1}^{r} \sum_{k=1}^{n_i} \frac{(s + p_j)^{n_j} c_{ik}}{(s + p_i)^k} \;+\; \sum_{k=1}^{n_j} (s + p_j)^{n_j - k} c_{jk}$$

Now form

$$\left. \frac{d^{n_j - l}}{ds^{n_j - l}} [(s + p_j)^{n_j} F(s)] \right|_{s = -p_j}$$

Note that the first three terms on the right-hand side of $(s + p_j)^{n_j} F(s)$ will have a factor $s + p_j$ in the numerator even after being differentiated $n_j - l$ times ($l = 1, 2, \ldots, n_j$) and thus these three terms become zero when evaluated at $s = -p_j$. Therefore

$$\left. \frac{d^{n_j - l}}{ds^{n_j - l}} [(s + p_j)^{n_j} F(s)] \right|_{s = -p_j} \;=\; \left. \frac{d^{n_j - l}}{ds^{n_j - l}} \left[\sum_{k=1}^{n_j} (s + p_j)^{n_j - k} c_{jk} \right] \right|_{s = -p_j}$$

$$=\; \left. \sum_{k=1}^{l} (n_j - k)(n_j - k - 1) \cdots (l - k + 1)(s + p_j)^{(-k + l)} c_{jk} \right|_{s = -p_j}$$

Except for that term in the summation for which $k = l$, all the other terms are zero since they contain factors $s + p_j$. Then

$$\left. \frac{d^{n_j - l}}{ds^{n_j - l}} [(s + p_j)^{n_j} F(s)] \right|_{s = -p_j} \;=\; (n_j - l)(n_j - l - 1) \cdots (1) \, c_{jl}$$

or

$$c_{jl} \;=\; \left. \frac{1}{(n_j - l)!} \frac{d^{n_j - l}}{ds^{n_j - l}} [(s + p_j)^{n_j} F(s)] \right|_{s = -p_j}$$

4.25. Expand $Y(s)$ of Example 4.17 in a partial fraction expansion.

$Y(s)$ can be rewritten with the denominator polynomial in **factored form** as

$$Y(s) \;=\; \frac{-(s^2 + s - 1)}{s(s + 1)(s + 2)}$$

The partial fraction expansion of $Y(s)$ is (see Equation (*4.11*))

$$Y(s) \;=\; b_3 + \frac{c_{11}}{s} + \frac{c_{21}}{s + 1} + \frac{c_{31}}{s + 2}$$

where $b_3 = 0$, $\quad c_{11} = \left. \dfrac{-(s^2 + s - 1)}{(s + 1)(s + 2)} \right|_{s = 0} = \dfrac{1}{2}$, $\quad c_{21} = \left. \dfrac{-(s^2 + s - 1)}{s(s + 2)} \right|_{s = -1} = -1$,

$$c_{31} = \left. \frac{-(s^2 + s - 1)}{s(s + 1)} \right|_{s = -2} = -\frac{1}{2}$$

Thus

$$Y(s) \;=\; \frac{1}{2s} - \frac{1}{s + 1} - \frac{1}{2(s + 2)}$$

4.26. Expand $Y(s)$ of Example 4.19 in a partial fraction expansion.

$Y(s)$ can be rewritten with the denominator polynomial in **factored form** as

$$Y(s) \;=\; \frac{s^2 + 9s + 19}{(s + 1)(s + 2)(s + 4)}$$

The partial fraction expansion of $Y(s)$ is (see Equation (*4.11*)),

$$Y(s) \;=\; b_3 + \frac{c_{11}}{s + 1} + \frac{c_{21}}{s + 2} + \frac{c_{31}}{s + 4}$$

where $b_3 = 0$,

$$c_{11} = \left.\frac{s^2 + 9s + 19}{(s+2)(s+4)}\right|_{s=-1} = \frac{11}{3}, \qquad c_{21} = \left.\frac{s^2 + 9s + 19}{(s+1)(s+4)}\right|_{s=-2} = -\frac{5}{2},$$

$$c_{31} = \left.\frac{s^2 + 9s + 19}{(s+1)(s+2)}\right|_{s=-4} = -\frac{1}{6}$$

Thus
$$Y(s) = \frac{11}{3(s+1)} - \frac{5}{2(s+2)} - \frac{1}{6(s+4)}$$

INVERSE TRANSFORMS USING PARTIAL FRACTION EXPANSIONS

4.27. Determine $y(t)$ for the system of Example 4.17, Page 62.

From the result of Problem 4.25, the transform of $y(t)$ can be written as

$$\mathcal{L}\,[y(t)] \equiv Y(s) = \frac{1}{2s} - \frac{1}{s+1} - \frac{1}{2(s+2)}$$

Therefore
$$y(t) = \frac{1}{2}\mathcal{L}^{-1}\left[\frac{1}{s}\right] - \mathcal{L}^{-1}\left[\frac{1}{s+1}\right] - \frac{1}{2}\mathcal{L}^{-1}\left[\frac{1}{s+2}\right] = \frac{1}{2}\,[1 - 2e^{-t} - e^{-2t}] \qquad t > 0$$

4.28. Determine $y(t)$ for the system of Example 4.19, Page 64.

From the result of Problem 4.26, the transform of $y(t)$ can be written as

$$\mathcal{L}\,[y(t)] = Y(s) = \frac{11}{3(s+1)} - \frac{5}{2(s+2)} - \frac{1}{6(s+4)}$$

Therefore
$$y(t) = \frac{11}{3}\,e^{-t} - \frac{5}{2}\,e^{-2t} - \frac{1}{6}\,e^{-4t}$$

ROOTS OF POLYNOMIALS

4.29. Find an approximation of a real root of the polynomial equation

$$Q_3(s) = s^3 - 3s^2 + 4s - 5 = 0$$

to an accuracy of three significant figures using *Horner's method*.

By Descartes' rule of signs, $Q_3(s)$ has 3 variations in the signs of its coefficients (1 to -3, -3 to 4, and 4 to -5). Thus there may be three positive real roots. $Q_3(-s) = -s^3 - 3s^2 - 4s - 5$ has no sign changes; therefore $Q_3(s)$ has no negative real roots and only real values of s greater than zero need be considered.

Step 1 — We have $Q_3(0) = -5$, $Q_3(1) = -3$, $Q_3(2) = -1$, $Q_3(3) = 7$. Therefore $k_0 = 2$ and the first approximation is $s_0 = k_0 = 2$.

Step 2 — Determine $Q_3^1(s)$ as

$$Q_3^1(s) = Q_3^0(2+s) = (2+s)^3 - 3(2+s)^2 + 4(2+s) - 5 = s^3 + 3s^2 + 4s - 1$$

Step 3 — $Q_3^1(0) = -1$, $Q_3^1(1/10) = -0.569$, $Q_3^1(2/10) = -0.072$, $Q_3^1(3/10) = 0.497$. Hence $k_1 = 0.2$ and $s_1 = k_0 + k_1 = 2.2$.

Now repeat Step 2 to determine $Q_3^2(s)$:

$$Q_3^2(s) = Q_3^1(0.2+s) = (0.2+s)^3 + 3(0.2+s)^2 + 4(0.2+s) - 1 = s^3 + 3.6s^2 + 5.32s - 0.072$$

Repeating Step 3: $Q_3^2(0) = -0.072$, $Q_3^2(1/100) = -0.018$, $Q_3^2(2/100) = 0.036$. Hence $k_2 = 0.01$ and $s_2 = k_0 + k_1 + k_2 = 2.21$ which is an approximation of the root accurate to 3 significant figures.

4.30. Find an approximation of a real root of the polynomial equation given in Problem 4.29 using *Newton's method*. Perform 4 iterations and compare the result with the solution of Problem 4.29.

The sequence of approximations is defined by letting $n = 3$, $a_3 = 1$, $a_2 = -3$, $a_1 = 4$ and $a_0 = -5$ in the recursive relationship of Newton's method (Equation (4.15), Page 68). The result is

$$s_{l+1} = \frac{2s_l^3 - 3s_l^2 + 5}{3s_l^2 - 6s_l + 4} \qquad l = 0, 1, 2, \ldots$$

Let the first guess be $s_0 = 0$. Then

$$s_1 = \frac{5}{4} = 1.25 \qquad\qquad s_3 = \frac{2(3.55)^3 - 3(3.55)^2 + 5}{3(3.55)^2 - 6(3.55) + 4} = 2.76$$

$$s_2 = \frac{2(1.25)^3 - 3(1.25)^2 + 5}{3(1.25)^2 - 6(1.25) + 4} = 3.55 \qquad\qquad s_4 = \frac{2(2.76)^3 - 3(2.76)^2 + 5}{3(2.76)^2 - 6(2.76) + 4} = 2.35$$

The next iteration yields $s_5 = 2.22$ and the sequence is converging.

4.31. Find an approximation of a quadratic factor of the polynomial

$$Q_3(s) = s^3 - 3s^2 + 4s - 5$$

of Problems 4.29 and 4.30, using the *Lin-Bairstow method*. **Perform 2 iterations.**

Step 1 — Choose as a first approximation the factor $s^2 - s + 2$.

The constants needed in *Step 2* are $\alpha_1 = -1$, $\alpha_0 = 2$, $n = 3$, $a_3 = 1$, $a_2 = -3$, $a_1 = 4$, $a_0 = -5$.

Step 2 — From the recursive relationship

$$b_{i-2} = a_i - \alpha_1 b_{i-1} - \alpha_0 b_i$$

$i = n$, $n-1$, $\ldots$, 1, 0, the following constants are formed:

$$b_1 = a_3 = 1 \qquad\qquad b_0 = a_2 + b_1 = -2$$

$$b_{-1} = a_1 + b_0 - 2b_1 = 0 \qquad b_{-2} = a_0 + b_{-1} - 2b_0 = -1$$

Step 3 — From the recursive relationship

$$c_{i-1} = b_{i-1} - \alpha_1 c_i - \alpha_0 c_{i+1}$$

$i = n$, $n-1$, $\ldots$, 1, the following constants are determined:

$$c_1 = b_1 = 1 \qquad c_0 = b_0 + c_1 = -1$$

Step 4 — The simultaneous equations

$$c_0 \Delta\alpha_1 + c_1 \Delta\alpha_0 = b_{-1}$$

$$(-\alpha_1 c_0 - \alpha_0 c_1) \Delta\alpha_1 + c_0 \Delta\alpha_0 = b_{-2}$$

can now be written as

$$-\Delta\alpha_1 + \Delta\alpha_0 = 0$$

$$-3 \Delta\alpha_1 - \Delta\alpha_0 = -1$$

whose solution is $\Delta\alpha_1 = 1/4$, $\Delta\alpha_0 = 1/4$, and the new approximation of the quadratic **factor** is

$$s^2 - 0.75s + 2.25$$

If *Steps 1-4* are repeated for $\alpha_1 = -0.75$, $\alpha_0 = 2.25$, the second iteration produces

$$s^2 - 0.7861s + 2.2583$$

POLE-ZERO MAPS

4.32. Determine all of the poles and zeros of $F(s) = \dfrac{s^2 - 16}{s^5 - 7s^4 - 30s^3}$.

The finite poles of $F(s)$ are the roots of the denominator polynomial equation

$$s^5 - 7s^4 - 30s^3 = s^3(s + 3)(s - 10) = 0$$

Therefore $s = 0$, $s = -3$, and $s = 10$ are the finite poles of $F(s)$. $s = 0$ is a triple root of the equation and is called a **triple pole** of $F(s)$. These are the only values of s for which $|F(s)|$ is infinite and are all the poles of $F(s)$. The finite zeros of $F(s)$ are the roots of the numerator polynomial equation

$$s^2 - 16 = (s-4)(s+4) = 0$$

Therefore $s = 4$ and $s = -4$ are the *finite zeros* of $F(s)$. As $|s| \to \infty$, $F(s) \cong \frac{1}{s^3} \to 0$. Then $F(s)$ has a triple zero at $s = \infty$.

4.33. Draw a pole-zero map for the function of Problem 4.32.

From the solution of Problem 4.32, $F(s)$ has *finite zeros* at $s = 4$ and $s = -4$, and *finite poles* at $s = 0$ (a triple pole), $s = -3$ and $s = 10$. The pole-zero map is shown in the adjoining figure.

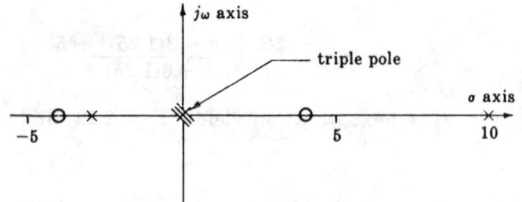

4.34. Using the graphical technique, evaluate the residues of the function

$$F(s) = \frac{20}{(s+10)(s+1+j)(s+1-j)}$$

The pole-zero map of $F(s)$ is

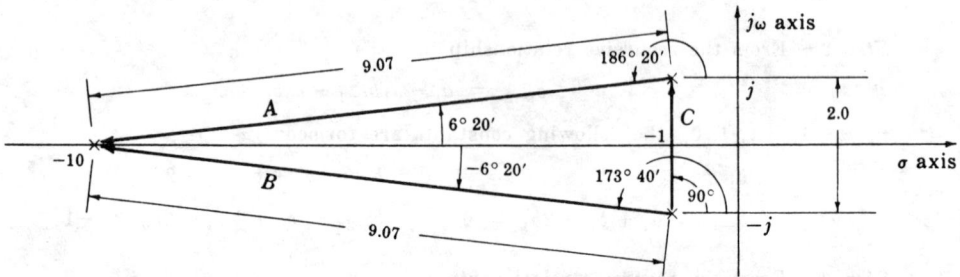

Included in this pole-zero map are the vector displacements between the poles. For example, A is the vector displacement of the pole $s = -10$ relative to the pole $s = -1+j$. Clearly then, $-A$ is the vector displacement of the pole $s = -1+j$ relative to the pole $s = -10$.

The magnitude of the residue at the pole $s = -10$ is $|c_1| = \frac{20}{|A||B|} = \frac{20}{(9.07)(9.07)} = 0.243$. The angle ϕ_1 of the residue at $s = -10$ is the negative of the sum of the angles of A and B, that is, $\phi_1 = -[186° 20' + 173° 40'] = -360°$. Hence $c_1 = 0.243$.

The magnitude of the residue at the pole $s = -1+j$ is $|c_2| = \frac{20}{|-A||C|} = \frac{20}{(9.07)(2)} = 1.102$. The angle ϕ_2 of the residue at the pole $s = -1+j$ is the negative of the sums of the angles of $-A$ and C: $\phi_2 = -[6° 20' + 90°] = -96° 20'$. Hence $c_2 = 1.102\underline{/-96° 20'} = -0.128 - j1.095$.

The magnitude of the residue at the pole $s = -1-j$ is $|c_3| = \frac{20}{|-B||-C|} = \frac{20}{(9.07)(2)} = 1.102$. The angle ϕ_3 of the residue at the pole $s = -1-j$ is the negative of the sum of the angles of $-B$ and $-C$: $\phi_3 = -[-90° - 6° 20'] = 96° 20'$. Hence $c_3 = 1.102\underline{/96° 20'} = -0.128 + j1.095$.

Note that the residues c_2 and c_3 of the complex conjugate poles are also complex conjugates. This is always true for the residues of complex conjugate poles.

4.35. Determine the *(a)* undamped natural frequency ω_n, *(b)* damping ratio ζ, *(c)* time constant τ, *(d)* damped natural frequency ω_d, and *(e)* characteristic equation for the second order system given by

$$\frac{d^2y}{dt^2} + 5\frac{dy}{dt} + 9y = 9x$$

Comparing this equation with the definitions of Section **4.12, Page 71,** we have

(a) $\omega_n^2 = 9$ or $\omega_n = 3$ rad/s   (c) $\tau = \dfrac{1}{\zeta\omega_n} = \dfrac{2}{5}$ s   (e) $s^2 + 5s + 9 = 0$

(b) $2\zeta\omega_n = 5$ or $\zeta = \dfrac{5}{2\omega_n} = \dfrac{5}{6}$ (d) $\omega_d = \omega_n \sqrt{1-\zeta^2} = 1.66$ rad/s

4.36. How and why can the following system be approximated by a second order system?

$$\frac{d^3y}{dt^3} + 12\frac{d^2y}{dt^2} + 22\frac{dy}{dt} + 20y = 20x$$

When the initial conditions on $y(t)$ and its derivatives are zero, the output transform is

$$\mathcal{L}\left[y(t)\right] \equiv Y(s) = \frac{20}{s^3 + 12s^2 + 22s + 20} X(s)$$

where $X(s) = \mathcal{L}\left[x(t)\right]$. This can be rewritten as

$$Y(s) = \frac{10}{41}\left(\frac{1}{s+10} - \frac{s}{s^2+2s+2}\right)X(s) + \frac{80}{41}\left(\frac{X(s)}{s^2+2s+2}\right)$$

The constant factor 80/41 of the second term is 8 times the constant factor 10/41 of the first term. The output $y(t)$ will then be dominated by the time function

$$\frac{80}{41}\mathcal{L}^{-1}\left[\frac{X(s)}{s^2+2s+2}\right]$$

The output transform $Y(s)$ can then be approximated by this second term; that is,

$$Y(s) \cong \frac{80}{41}\left(\frac{X(s)}{s^2+2s+2}\right) \cong \left(\frac{2}{s^2+2s+2}\right)X(s)$$

The second order approximation is $\dfrac{d^2y}{dt^2} + 2\dfrac{dy}{dt} + 2y = 2x.$

4.37. In Chapter 6 it will be shown that the output $y(t)$ of a time-invariant linear causal system with all initial conditions equal to zero is related to the input $x(t)$ in the Laplace transform domain by the equation $Y(s) = P(s)X(s)$ where $P(s)$ is called the *transfer function* of the system. Show that $p(t)$, the inverse Laplace transform of $P(s)$, is equal to the *weighting function* $w(t)$ of a system described by the constant coefficient differential equation $\displaystyle\sum_{i=0}^{n} a_i \frac{d^iy}{dt^i} = x.$

The forced response for a system described by the above equation is given by Equation 3.15, Page 36, with all $b_i = 0$ except $b_0 = 1$:

$$y(t) = \int_{0+}^{t} w(t-\tau)\, x(\tau)\, d\tau$$

and $w(t-\tau)$ is the weighting function of the differential equation.

The inverse Laplace transform of $Y(s) = P(s)X(s)$ is easily determined from the convolution integral of Property 12, Page 58, as

$$y(t) = \mathcal{L}^{-1}\left[Y(s)\right] = \mathcal{L}^{-1}\left[P(s)X(s)\right] = \int_{0+}^{t} p(t-\tau)\, x(\tau)\, d\tau$$

Hence $\displaystyle\int_{0+}^{t} w(t-\tau)\, x(\tau)\, d\tau = \int_{0+}^{t} p(t-\tau)\, x(\tau)\, d\tau$ or $w(t) = p(t)$

MISCELLANEOUS PROBLEM

4.38. For the RC network in the schematic given below

(a) Find a differential equation which relates the output voltage y and the input voltage x.

(b) Let the initial voltage across the capacitor C be $v_{co} = 1$ volt with the polarity shown, and let $x = 2e^{-t}$. Using the Laplace transform technique, find y.

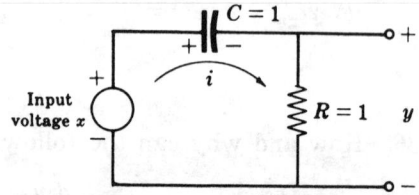

(a) From Kirchhoff's voltage law

$$x = v_{co} + (1/C) \int_0^t i\,dt + Ri = v_{co} + \int_0^t i\,dt + i$$

But $y = Ri = i$. Therefore $x = v_{co} + \int_0^t y\,dt + y$. Differentiating both sides of this integral equation yields the differential equation $\dot{y} + y = \dot{x}$.

(b) The Laplace transform of the differential equation found in Part (a) is

$$s\,Y(s) - y(0^+) + Y(s) = s\,X(s) - x(0^+)$$

where $X(s) = \mathcal{L}\,[2e^{-t}] = \dfrac{2}{s+1}$ and $x(0^+) = \lim_{t \to 0} 2e^{-t} = 2$. To find $y(0^+)$, limits are taken on both sides of the original voltage equation:

$$x(0^+) = \lim_{t \to 0} x(t) = \lim_{t \to 0} \left[v_{co} + \int_0^t y\,dt + y(t) \right] = v_{co} + y(0^+)$$

Hence $y(0^+) = x(0^+) - v_{co} = 2 - 1 = 1$. The transform of $y(t)$ is then

$$Y(s) = \frac{2s}{(s+1)^2} - \frac{1}{s+1} = -\frac{2}{(s+1)^2} + \frac{2}{s+1} - \frac{1}{s+1} = -\frac{2}{(s+1)^2} + \frac{1}{s+1}$$

Finally, $\quad y(t) = \mathcal{L}^{-1}\left[-\dfrac{2}{(s+1)^2} \right] + \mathcal{L}^{-1}\left[\dfrac{1}{s+1} \right] = -2te^{-t} + e^{-t}$

Supplementary Problems

4.39. Show that $\mathcal{L}\,[-t\,f(t)] = \dfrac{dF(s)}{ds}$ where $F(s) = \mathcal{L}\,[f(t)]$.

4.40. Using the convolution integral find the inverse transform of $\dfrac{1}{s(s+2)}$.

4.41. Determine the final value of the function $f(t)$ whose Laplace transform is

$$F(s) = \frac{2(s+1)}{s(s+3)(s+5)^2}$$

4.42. Determine the initial value of the function $f(t)$ whose Laplace transform is

$$F(s) = \frac{4s}{s^3 + 2s^2 + 9s + 6}$$

4.43.   Find the partial fraction expansion of the function $F(s) = \dfrac{10}{(s+4)(s+2)^3}$.

4.44.   Find the inverse Laplace transform $f(t)$ of the function $F(s) = \dfrac{10}{(s+4)(s+2)^3}$.

4.45.   Solve Problem 3.24, Page 50, using the Laplace transform technique.

4.46.   Using the Laplace transform technique, find the forced response of the differential equation
$$\frac{d^2y}{dt^2} + 4\frac{dy}{dt} + 4y = 3\frac{dx}{dt} + 2x$$
where $x(t) = e^{-3t}$, $t > 0$. Compare this solution with that obtained in Problem 3.26, **Page 51.**

4.47.   Using the Laplace transform technique, find the transient and steady-state responses of the system described by the differential equation $\dfrac{d^2y}{dt^2} + 3\dfrac{dy}{dt} + 2y = 1$ with initial conditions $y(0^+)$ and $\dfrac{dy}{dt}\bigg|_{t=0^+} = 1$.

4.48.   Using the Laplace transform technique, find the unit impulse response of the system described by the differential equation $\dfrac{d^3y}{dt^3} + \dfrac{dy}{dt} = x$.

Answers to Supplementary Problems

4.40. $\frac{1}{2}[1 - e^{-2t}]$

4.41.   2/75

4.42.   0

4.43. $F(s) = \dfrac{5}{(s+2)^3} - \dfrac{5}{2(s+2)^2} + \dfrac{5}{4(s+2)} - \dfrac{5}{4(s+4)}$

4.44. $f(t) = \dfrac{5t^2e^{-2t}}{2} - \dfrac{5te^{-2t}}{2} + \dfrac{5e^{-2t}}{4} - \dfrac{5e^{-4t}}{4}$

4.46. $y_b(t) = 7e^{-2t} - 7e^{-3t} - 7te^{-2t}$

4.47.   Transient response $= 2e^{-t} - \frac{3}{2}e^{-2t}$.   Steady state response $= \frac{1}{2}$.

4.48. $y_\delta(t) = 1 - \cos t$

Chapter 5

Stability

5.1 STABILITY DEFINITIONS

The stability of a system is determined by its response to inputs or disturbances. Intuitively, a stable system is one that will remain at rest unless excited by an external source and will return to rest if all excitations are removed. Stability can be precisely defined in terms of the impulse response of a system (see Chapters 3 and 4) as follows:

Definition 5.1a: A system is **stable** if its impulse response approaches zero as time approaches infinity.

Alternatively, the definition of a stable system can be based upon the response of the system to **bounded inputs**, that is, inputs whose magnitudes are less than some finite value for all time.

Definition 5.1b: A system is **stable** if every bounded input produces a bounded output.

Consideration of the *degree* of stability of a system often provides valuable information about its behavior. That is, if it is stable, how close is it to being unstable? This is the concept of **relative stability**. Usually, relative stability is expressed in terms of some allowable variation of a particular system parameter for which the system will remain stable. More precise definitions of relative stability indicators will be presented in later chapters.

5.2 CHARACTERISTIC ROOT LOCATIONS

It was shown in Chapter 4 that the impulse response of a linear time-invariant system is composed of a sum of exponential time functions whose exponents are the roots of the system characteristic equation. A necessary condition for the system to be stable is that the real parts of the roots of the characteristic equation have negative real parts. This insures that the impulse response will decay exponentially with time.

If the system has some roots with real parts equal to zero, but none with positive real parts, the system is said to be **marginally stable**. In this instance, the impulse response does not decay to zero although it is bounded. Additionally, certain inputs will produce unbounded outputs. Therefore marginally stable systems are unstable.

Example 5.1.

The system described by the Laplace transformed differential equation,

$$(s^2 + 1) \, Y(s) \;=\; X(s)$$

has the characteristic equation $\qquad s^2 + 1 \;=\; 0$

This equation has the two roots $\pm j$. Since these roots have zero real parts, the system is not stable. It is however marginally stable since the equation has no roots with positive real parts. In response to most inputs or disturbances, the system will oscillate with a bounded output. However, if the input is $x = \sin t$, the output will be

$$y \;=\; t \sin t$$

which is unbounded.

5.3 ROUTH STABILITY CRITERION

The Routh stability criterion is a method for determining system stability that can be applied to an nth order characteristic equation of the form

$$a_n s^n + a_{n-1} s^{n-1} + \cdots + a_1 s + a_0 = 0$$

The criterion is applied through the use of a **Routh table** defined as follows:

$$
\begin{array}{c|cccc}
s^n & a_n & a_{n-2} & a_{n-4} & \cdots \\
s^{n-1} & a_{n-1} & a_{n-3} & a_{n-5} & \cdots \\
\cdot & b_1 & b_2 & b_3 & \cdots \\
\cdot & c_1 & c_2 & c_3 & \cdots \\
\cdot & & & &
\end{array}
$$

where $a_n, a_{n-1}, \ldots, a_0$ are the coefficients of the characteristic equation and

$$b_1 \equiv \frac{a_{n-1} a_{n-2} - a_n a_{n-3}}{a_{n-1}}, \quad b_2 \equiv \frac{a_{n-1} a_{n-4} - a_n a_{n-5}}{a_{n-1}}, \quad \text{etc.}$$

$$c_1 \equiv \frac{b_1 a_{n-3} - a_{n-1} b_2}{b_1}, \quad c_2 \equiv \frac{b_1 a_{n-5} - a_{n-1} b_3}{b_1}, \quad \text{etc.}$$

The table is continued horizontally and vertically until only zeros are obtained. Any row can be multiplied by a constant before the next row is computed without disturbing the properties of the table. All the roots of this characteristic equation have negative real parts if and only if the elements of the first column of the Routh table have the same sign. Otherwise, the number of roots with positive real parts is equal to the number of changes of sign.

Example 5.2. $\qquad\qquad s^3 + 6s^2 + 12s + 8 = 0$

$$
\begin{array}{c|ccc}
s^3 & 1 & 12 & 0 \\
s^2 & 6 & 8 & 0 \\
s^1 & \dfrac{64}{6} & 0 & \\
s^0 & 8 & &
\end{array}
$$

Since there are no changes of sign in the first column of the table, all the roots of the equation have negative real parts.

Often it is desirable to determine a range of values of a particular system parameter for which the system is stable. This can be accomplished by writing the inequalities that will insure that there is no change of sign in the first column of the Routh table for the system. These inequalities then specify the range of allowable values of the parameter.

Example 5.3. $\qquad\qquad s^3 + 3s^2 + 3s + 1 + K = 0$

$$
\begin{array}{c|ccc}
s^3 & 1 & 3 & 0 \\
s^2 & 3 & 1+K & 0 \\
s^1 & \dfrac{8-K}{3} & 0 & \\
s^0 & 1+K & &
\end{array}
$$

In order that there be no sign changes in the first column, it is necessary that the conditions $8 - K > 0$, $1 + K > 0$ be satisfied. Thus the characteristic equation has roots with negative real parts if $-1 < K < 8$.

A row of zeros for the s^1 row of the Routh table indicates that the polynomial has a pair of roots which satisfy the **auxiliary equation** formed as follows:

$$As^2 + B = 0$$

where A and B are the first and second elements of the s^2 row.

Example 5.4.

In the previous example, the s^1 row is zero if $K = 8$. In this case, the auxiliary equation is $3s^2 + 9 = 0$. Therefore two of the roots of the characteristic equation are $s = \pm j\sqrt{3}$.

5.4 HURWITZ STABILITY CRITERION

The Hurwitz stability criterion is another method for determining whether or not all the roots of a characteristic equation have negative real parts. This criterion is applied through the use of determinants formed from the coefficients of the characteristic equation. It is assumed that the first coefficient, a_n, is positive. The determinants Δ_i for $i = 1, 2, \ldots, n-1$ are formed as the principal minor determinants of the determinant

$$\Delta_n = \begin{vmatrix} a_{n-1} & a_{n-3} & \cdots & \begin{bmatrix} a_0 \text{ if } n \text{ odd} \\ a_1 \text{ if } n \text{ even} \end{bmatrix} & 0 & \cdots & 0 \\ a_n & a_{n-2} & \cdots & \begin{bmatrix} a_1 \text{ if } n \text{ odd} \\ a_0 \text{ if } n \text{ even} \end{bmatrix} & 0 & \cdots & 0 \\ 0 & a_{n-1} & a_{n-3} & \multicolumn{3}{c}{\cdots\cdots\cdots\cdots\cdots} & 0 \\ 0 & a_n & a_{n-2} & \multicolumn{3}{c}{\cdots\cdots\cdots\cdots\cdots} & 0 \\ \multicolumn{7}{c}{\cdots\cdots\cdots\cdots\cdots\cdots\cdots\cdots\cdots} \\ 0 & \multicolumn{5}{c}{\cdots\cdots\cdots\cdots\cdots\cdots\cdots\cdots} & a_0 \end{vmatrix}$$

The determinants are thus formed as follows:

$$\Delta_1 = a_{n-1}$$

$$\Delta_2 = \begin{vmatrix} a_{n-1} & a_{n-3} \\ a_n & a_{n-2} \end{vmatrix} = a_{n-1}a_{n-2} - a_n a_{n-3}$$

$$\Delta_3 = \begin{vmatrix} a_{n-1} & a_{n-3} & a_{n-5} \\ a_n & a_{n-2} & a_{n-4} \\ 0 & a_{n-1} & a_{n-3} \end{vmatrix} = a_{n-1}a_{n-2}a_{n-3} + a_n a_{n-1}a_{n-5} - a_n a_{n-3}^2 - a_{n-4}a_{n-1}^2$$

and so on up to Δ_{n-1}.

All the roots of the characteristic equation have negative real parts if and only if $\Delta_i > 0$ for $i = 1, 2, \ldots, n$.

Example 5.5.

For $n = 3$,

$$\Delta_3 = \begin{vmatrix} a_2 & a_0 & 0 \\ a_3 & a_1 & 0 \\ 0 & a_2 & a_0 \end{vmatrix} = a_2 a_1 a_0 - a_0^2 a_3, \quad \Delta_2 = \begin{vmatrix} a_2 & a_0 \\ a_3 & a_1 \end{vmatrix} = a_2 a_1 - a_0 a_3, \quad \Delta_1 = a_2$$

Thus all the roots of the characteristic equation have negative real parts if

$$a_2 > 0, \quad a_2 a_1 - a_0 a_3 > 0, \quad a_2 a_1 a_0 - a_0^2 a_3 > 0$$

5.5 CONTINUED FRACTION STABILITY CRITERION

The continued fraction stability criterion can be applied to the characteristic equation by forming a continued fraction from the odd and even portions of the equation in the following manner. Let

$$Q(s) \equiv a_n s^n + a_{n-1} s^{n-1} + \cdots + a_1 s + a_0$$

$$Q_1(s) \equiv a_n s^n + a_{n-2} s^{n-2} + \cdots$$

$$Q_2(s) \equiv a_{n-1} s^{n-1} + a_{n-3} s^{n-3} + \cdots$$

Form the fraction Q_1/Q_2, and then divide the denominator into the numerator and invert the remainder to form a continued fraction as follows:

$$\frac{Q_1(s)}{Q_2(s)} = \frac{a_n s}{a_{n-1}} + \frac{\left(a_{n-2} - \dfrac{a_n a_{n-3}}{a_{n-1}}\right) s^{n-2} + \left(a_{n-4} - \dfrac{a_n a_{n-5}}{a_{n-1}}\right) s^{n-4} + \cdots}{Q_2}$$

$$= h_1 s + \cfrac{1}{h_2 s + \cfrac{1}{h_3 s + \cfrac{1}{h_4 s + \cfrac{1}{\ddots \cfrac{1}{h_n s}}}}}$$

If $h_1, h_2, \ldots, h_n$ are all positive, then all the roots of $Q(s)$ have negative real parts.

Example 5.6.
$$Q(s) = s^3 + 6s^2 + 12s + 8$$

$$\frac{Q_1(s)}{Q_2(s)} = \frac{s^3 + 12s}{6s^2 + 8} = \frac{1}{6} s + \frac{\dfrac{32}{3} s}{6s^2 + 8}$$

$$= \frac{1}{6} s + \cfrac{1}{\dfrac{9}{16} s + \cfrac{1}{\dfrac{4}{3} s}}$$

Since all the coefficients of s in the continued fraction are positive, i.e. $h_1 = 1/6$, $h_2 = 9/16$ and $h_3 = 4/3$, all the roots of the polynomial $Q(s)$ have negative real parts.

Solved Problems

STABILITY DEFINITIONS

5.1.   The impulse responses of several systems are given below. For each case determine if the impulse response represents a stable or an unstable system.

(a) $h(t) = e^{-t}$,   (b) $h(t) = te^{-t}$,   (c) $h(t) = 1$,   (d) $h(t) = e^{-t} \sin 3t$,   (e) $h(t) = \sin \omega t$.

If the impulse response decays to zero as time approaches infinity, the system is stable. As can be seen in Fig. 5-1, the impulse responses of (a), (b), and (d) decay to zero as time approaches infinity and therefore represent *stable* systems. Since the impulse responses (c) and (e) do not approach zero, they represent *unstable* systems.

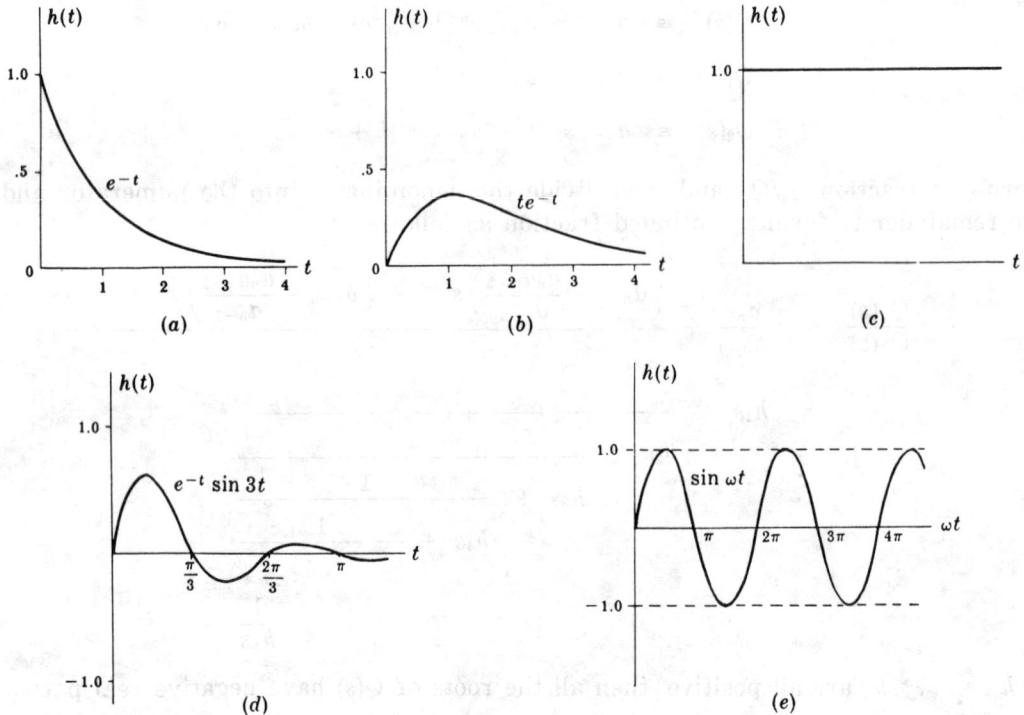

Fig. 5-1

5.2. If a step function is applied to the input of a system and the output remains below a certain level for all time, is the system stable?

The system is not necessarily stable since the output must be bounded for every bounded input. A bounded output to one specific bounded input does not insure stability.

5.3. If a step function is applied to the input of a system and the output is of the form $y = t$, is the system stable or unstable?

The system is unstable since a bounded input produced an unbounded output.

CHARACTERISTIC ROOT LOCATIONS

5.4. The roots of the characteristic equations of several systems are given below. Determine in each case if the set of roots represents stable, marginally stable, or unstable systems.

(a) $-1, -2$	(d) $-1+j, -1-j$	(g) $-6, -4, 7$
(b) $-1, +1$	(e) $-2+j, -2-j, 2j, -2j$	(h) $-2+3j, -2-3j, -2$
(c) $-3, -2, 0$	(f) $2, -1, -3$	(i) $-j, j, -1, 1$

The sets of roots (a), (d) and (h) represent stable systems since all the roots have negative real parts. The sets of roots (c) and (e) represent marginally stable systems since all the roots have non-positive real parts, that is, zero or negative. The sets (b), (f), (g) and (i) represent unstable systems since each has at least one root with a positive real part.

5.5. A system has poles at -1, -5 and zeros at 1 and -2. Is the system stable?

The system is stable since the poles which are the roots of the system characteristic equation (Chapter 4) have negative real parts. The fact that the system has a zero with a positive real part does not affect its stability.

5.6. Determine if the system with the following characteristic equation is stable: $(s+1)(s+2)(s-3) = 0$.

This characteristic equation has the roots -1, -2, and 3 and therefore represents an unstable system since there is a positive real root.

5.7. The differential equation of an integrator may be written as follows: $dy/dt = x$. Determine if an integrator is stable.

The characteristic equation of this system is $s = 0$. Since the root does not have a negative real part, an integrator is not stable. Since it has no roots with positive real parts, an integrator is marginally stable.

5.8. Determine a bounded input which will produce an unbounded output from an integrator.

The input $x = 1$ will produce the output $y = t$ which is unbounded.

ROUTH STABILITY CRITERION

5.9. Determine if the following characteristic equation represents a stable system:

$$s^3 + 4s^2 + 8s + 12 = 0$$

The Routh table for this system is

s^3	1	8
s^2	4	12
s^1	5	0
s^0	12	

Since there are no changes of sign in the first column, all the roots of the characteristic equation have negative real parts and the system is stable.

5.10. Determine if the following characteristic equation has any roots with positive real parts:

$$s^4 + s^3 - s - 1 = 0$$

Note that the coefficient of the s^2 term is zero. The Routh table for this equation is

s^4	1	0	-1
s^3	1	-1	0
s^2	1	-1	
s^1	0	0	
s^0	-1		

The coefficient for the s^0 row was obtained by replacing the 0 of the s^1 row by ϵ and then computing the coefficient of the s^0 row as

$$\frac{\epsilon(-1) - 0}{\epsilon} = -1$$

This procedure is necessary when a zero is obtained in the first column. Since there is one change of sign, the characteristic equation has one root with a positive real part. The presence of the zeros in the s^1 row indicates that the characteristic equation has two roots which satisfy the auxiliary equation formed from the s^2 row as follows: $s^2 - 1 = 0$. The roots of this equation are $+1$ and -1.

5.11. The characteristic equation of a given system is

$$s^4 + 6s^3 + 11s^2 + 6s + K = 0$$

What restrictions must be placed upon the parameter K in order to insure that the system is stable?

The Routh table for this system is

s^4	1	11	K
s^3	6	6	0
s^2	10	K	0
s^1	$\dfrac{60 - 6K}{10}$	0	
s^0	K		

For the system to be stable, the following restrictions must be placed upon the parameter K: $60 - 6K > 0$ or $K < 10$, and $K > 0$. Thus K must be greater than zero and less than 10.

5.12. Construct a Routh table and determine the number of roots with positive real parts for the equation

$$2s^3 + 4s^2 + 4s + 12 = 0$$

The Routh table for this equation is given below. Here the s^2 row was divided by 4 before the s^1 row was computed. The s^1 row was then divided by 2 before the s^0 row was computed.

s^3	2	4
s^2	1	3
s^1	-1	0
s^0	3	

Since there are two changes of sign in the first column of the Routh table, the equation above has two roots with positive real parts.

HURWITZ STABILITY CRITERION

5.13. Determine if the characteristic equation below represents a stable or an unstable system.

$$s^3 + 8s^2 + 14s + 24 = 0$$

The Hurwitz determinants for this system are

$$\Delta_3 = \begin{vmatrix} 8 & 24 & 0 \\ 1 & 14 & 0 \\ 0 & 8 & 24 \end{vmatrix} = 2112, \qquad \Delta_2 = \begin{vmatrix} 8 & 24 \\ 1 & 14 \end{vmatrix} = 88, \qquad \Delta_1 = 8$$

Since each determinant is positive, the system is stable. Note that the general formulation of Example 5.5 could have been used to check the stability in this case by substituting the appropriate values for the coefficients a_0, a_1, a_2, and a_3.

5.14. For what range of values of K is the system with the following characteristic equation stable?

$$s^2 + Ks + 2K - 1 = 0$$

The Hurwitz determinants for this system are

$$\Delta_2 = \begin{vmatrix} K & 0 \\ 1 & 2K-1 \end{vmatrix} = 2K^2 - K = K(2K-1), \qquad \Delta_1 = K$$

In order for these determinants to be positive, it is necessary that $K > 0$ and $2K - 1 > 0$. Thus the system is stable if $K > \frac{1}{2}$.

5.15. A system is designed to give satisfactory performance when a particular amplifier gain K has the value 2. Determine how much this gain can vary before the system becomes unstable if the characteristic equation is

$$s^3 + (4+K)s^2 + 6s + 16 + 8K = 0$$

Substituting the coefficients of the given equation into the general Hurwitz conditions of Example 5.5 results in the following requirements for stability:

$$4 + K > 0, \quad (4+K)6 - (16+8K) > 0, \quad (4+K)(6)(16+8K) - (16+8K)^2 > 0$$

Assuming the amplifier gain K cannot be negative, the first condition is satisfied. The second and third conditions are satisfied if K is less than 4. Hence with an amplifier gain design value of 2, the system could tolerate an increase in gain of a factor of 2 before it would become unstable. The gain could also drop to zero without causing instability.

5.16. Determine the Hurwitz conditions for stability of the following general fourth order characteristic equation, assuming a_4 is positive.

$$a_4 s^4 + a_3 s^3 + a_2 s^2 + a_1 s + a_0 = 0$$

The Hurwitz determinants are

$$\Delta_4 = \begin{vmatrix} a_3 & a_1 & 0 & 0 \\ a_4 & a_2 & a_0 & 0 \\ 0 & a_3 & a_1 & 0 \\ 0 & a_4 & a_2 & a_0 \end{vmatrix} = a_3(a_2 a_1 a_0 - a_3 a_0^2) - a_1^2 a_0 a_4$$

$$\Delta_3 = \begin{vmatrix} a_3 & a_1 & 0 \\ a_4 & a_2 & a_0 \\ 0 & a_3 & a_1 \end{vmatrix} = a_3 a_2 a_1 - a_0 a_3^2 - a_4 a_1^2$$

$$\Delta_2 = \begin{vmatrix} a_3 & a_1 \\ a_4 & a_2 \end{vmatrix} = a_3 a_2 - a_4 a_1$$

$$\Delta_1 = a_3$$

The conditions for stability are then

$$a_3 > 0, \quad a_3 a_2 - a_4 a_1 > 0, \quad a_3 a_2 a_1 - a_0 a_3^2 - a_4 a_1^2 > 0, \quad a_3(a_2 a_1 a_0 - a_3 a_0^2) - a_1^2 a_0 a_4 > 0$$

5.17. Is the system with the following characteristic equation stable?

$$s^4 + 3s^3 + 6s^2 + 9s + 12 = 0$$

Substituting the appropriate values for the coefficients in the general conditions of Problem 5.16, we have

$$3 > 0, \quad 18 - 9 > 0, \quad 162 - 108 - 81 \not> 0, \quad 3(648 - 432) - 972 \not> 0$$

Since the last two conditions are not satisfied, the system is unstable.

CONTINUED FRACTION STABILITY CRITERION

5.18. Repeat Problem 5.9 using the continued fraction stability criterion.

The polynomial $Q(s) = s^3 + 4s^2 + 8s + 12$ is divided into the two parts:

$$Q_1(s) = s^3 + 8s, \quad Q_2(s) = 4s^2 + 12$$

The continued fraction for $Q_1(s)/Q_2(s)$ is

$$\frac{Q_1(s)}{Q_2(s)} = \frac{s^3 + 8s}{4s^2 + 12} = \frac{1}{4}s + \frac{5s}{4s^2 + 12} = \frac{1}{4}s + \cfrac{1}{\frac{4}{5}s + \cfrac{1}{\frac{5}{12}s}}$$

Since all the coefficients of s are positive, the polynomial has all its roots in the left half plane and the system with the characteristic equation $Q(s) = 0$ is stable.

5.19. Determine bounds upon the parameter K for which a system with the following characteristic equation will be stable.

$$s^3 + 14s^2 + 56s + K = 0$$

$$\frac{Q_1(s)}{Q_2(s)} = \frac{s^3 + 56s}{14s^2 + K} = \frac{1}{14}s + \frac{(56 - K/14)s}{14s^2 + K} = \frac{1}{14}s + \cfrac{1}{\left[\dfrac{14}{56 - K/14}\right]s + \cfrac{1}{\left[\dfrac{56 - K/14}{K}\right]s}}$$

For the system to be stable, the following conditions must be satisfied: $56 - K/14 > 0$ and $K > 0$, i.e. $0 < K < 784$.

5.20. Derive the necessary conditions for all the roots of a general third order polynomial to have negative real parts.

For the general third order polynomial $Q(s) = a_3 s^3 + a_2 s^2 + a_1 s + a_0$,

$$\frac{Q_1(s)}{Q_2(s)} = \frac{a_3 s^3 + a_1 s}{a_2 s^2 + a_0} = \frac{a_3}{a_2}s + \frac{[a_1 - a_3 a_0/a_2]s}{a_2 s^2 + a_0} = \frac{a_3}{a_2}s + \cfrac{1}{\left[\dfrac{a_2}{a_1 - a_3 a_0/a_2}\right]s + \cfrac{1}{\left[\dfrac{a_1 - a_3 a_0/a_2}{a_0}\right]s}}$$

The necessary conditions for all the roots of $Q(s)$ to have negative real parts are then

$$\frac{a_3}{a_2} > 0, \qquad \frac{a_2}{a_1 - a_3 a_0/a_2} > 0, \qquad \frac{a_1 - a_3 a_0/a_2}{a_0} > 0$$

Thus if a_3 is positive, the required conditions are $a_2, a_1, a_0 > 0$ and $a_1 a_2 - a_3 a_0 > 0$. Note that if a_3 is not positive, $Q(s)$ should be multiplied by -1 before checking the above conditions.

5.21. Is the system with the following characteristic equation stable?

$$s^4 + 4s^3 + 8s^2 + 16s + 32 = 0$$

$$\frac{Q_1(s)}{Q_2(s)} = \frac{s^4 + 8s^2 + 32}{4s^3 + 16s} = \frac{1}{4}s + \frac{4s^2 + 32}{4s^3 + 16s}$$

$$= \frac{1}{4}s + \cfrac{1}{s + \cfrac{-16s}{4s^2 + 32}} = \frac{1}{4}s + \cfrac{1}{s + \cfrac{1}{-\frac{1}{4}s + \cfrac{1}{-\frac{1}{2}s}}}$$

Since the coefficients of s are not all positive, the system is unstable.

MISCELLANEOUS PROBLEMS

5.22. If a zero appears in the first column of the Routh table, is the system necessarily unstable?

Strictly speaking, a zero in the first column of the Routh table must be interpreted as having no sign, that is, neither positive nor negative. Consequently, all the elements of the first column cannot have the same sign if one of them is zero, and the system is unstable. In some cases, a zero in the first column of the Routh table indicates the presence of two roots of equal magnitude but opposite sign (see Problem 5.10). In other cases, it indicates the presence of one or more roots with zero real parts. Thus a characteristic equation having one or more roots with zero real parts and no roots with positive real parts will produce a Routh table in which all the elements of the first column do not have the same sign and do not have any sign changes.

5.23. Prove that a system is unstable if any coefficients of the characteristic equation are zero.

The characteristic equation may be written in the form

$$(s - s_1)(s - s_2)(s - s_3) \cdots (s - s_n) = 0$$

where $s_1, s_2, \ldots, s_n$ are the roots of the equation. If this equation is multiplied out, n new equations can be obtained relating the roots and the coefficients of the characteristic equation in the usual form. Thus

$$a_n s^n + a_{n-1} s^{n-1} + \cdots + a_0 = 0 \qquad \text{or} \qquad s^n + \frac{a_{n-1}}{a_n} s^{n-1} + \cdots + \frac{a_0}{a_n} = 0$$

and the relations are

$$\frac{a_{n-1}}{a_n} = -\sum_{i=1}^{n} s_i, \quad \frac{a_{n-2}}{a_n} = \sum_{\substack{i=1 \\ i \neq j}}^{n} \sum_{j=1}^{n} s_i s_j, \quad \frac{a_{n-3}}{a_n} = -\sum_{\substack{i=1 \\ i \neq j \neq k}}^{n} \sum_{j=1}^{n} \sum_{k=1}^{n} s_i s_j s_k, \quad \ldots, \quad \frac{a_0}{a_n} = (-1)^n s_1 s_2 \cdots s_n$$

The coefficients $a_{n-1}, a_{n-2}, \ldots, a_0$ all have the same sign as a_n and are nonzero if all the roots $s_1, s_2, \ldots, s_n$ have negative real parts. The only way any one of the coefficients can be zero is for one or more of the roots to have zero or positive real parts. In either case, the system would be unstable.

5.24. Prove that a system is unstable if all the coefficients of the characteristic equation do not have the same sign.

From the relations presented in Problem 5.23, it can be seen that the coefficients $a_{n-1}, a_{n-2}, \ldots, a_0$ have the same sign as a_n if all the roots $s_1, s_2, \ldots, s_n$ have negative real parts. The only way any of these coefficients may differ in sign from a_n is for one or more of the roots to have a positive real part. Thus the system is necessarily unstable if all the coefficients do not have the same sign. Note that a system is *not* necessarily stable if all the coefficients do have the same sign.

5.25. Can the stability criteria presented in this chapter be applied to systems which contain time delays?

No, they cannot be directly applied because systems which contain time delays do not have characteristic equations of the required form, that is, finite polynomials in s. For example, the following characteristic equation represents a system which contains a time delay:

$$s^2 + s + e^{-sT} = 0$$

Strictly speaking, this equation has an infinite number of roots. However, in some cases an approximation may be employed for e^{-sT} to give useful, although not entirely accurate, information concerning a system's stability. To illustrate, let e^{-sT} in the equation above be replaced by the first two terms of its Taylor series. The equation then becomes

$$s^2 + s + 1 - sT = 0 \quad \text{or} \quad s^2 + (1-T)s + 1 = 0$$

One of the stability criteria of this chapter may then be applied to this approximation of the characteristic equation.

5.26. Determine an approximate upper limit on the time delay in order that the system discussed in the solution of Problem 5.25 be stable.

Employing the approximate equation $s^2 + (1-T)s + 1 = 0$, the Hurwitz determinants are $\Delta_1 = \Delta_2 = 1 - T$. Hence for the system to be stable, the time delay T must be less than 1 second.

Supplementary Problems

5.27. For each characteristic polynomial, determine if it represents a stable or an unstable system.

    (a) $2s^4 + 8s^3 + 10s^2 + 10s + 20$ (c) $s^5 + 6s^4 + 10s^2 + 5s + 24$ (e) $s^4 + 8s^3 + 24s^2 + 32s + 16$

    (b) $s^3 + 7s^2 + 7s + 46$ (d) $s^3 - 2s^2 + 4s + 6$ (f) $s^6 + 4s^4 + 8s^2 + 16$

5.28. For what values of K does the polynomial $s^3 + (4+K)s^2 + 6s + 12$ have roots with negative real parts?

5.29. How many roots with positive real parts does each polynomial have?

    (a) $s^3 + s^2 - s + 1$ (b) $s^4 + 2s^3 + 2s^2 + 2s + 1$ (c) $s^3 + s^2 - 2$ (d) $s^4 - s^2 - 2s + 2$ (e) $s^3 + s^2 + s + 6$

5.30. For what positive value of K does the polynomial $s^4 + 8s^3 + 24s^2 + 32s + K$ have roots with zero real parts? What are these roots?

Answers to Supplementary Problems

5.27.   (b) and (e) represent stable systems; (a), (c), (d), and (f) represent unstable systems.

5.28. $K > -2$

5.29.   (a) 2, (b) 0, (c) 1, (d) 2, (e) 2

5.30. $K = 80$; $s = \pm j2$

Chapter 6

Transfer Functions

6.1 DEFINITION OF A TRANSFER FUNCTION

As shown in Chapters 3 and 4, the response of a time-invariant linear system can be separated into two parts: the forced response and the free response. Equation (4.8), Page 63, clearly illustrates this division for the most general constant-coefficient, linear, ordinary differential equation. The forced response includes terms due to initial values x_0^k of the input, and the free response depends only on initial conditions y_0^k on the output. If terms due to *all* initial values, i.e. x_0^k and y_0^k, are lumped together, Equation (4.8) can be written as

$$y(t) \;=\; \mathcal{L}^{-1}\left[\left(\sum_{i=0}^{m} b_i s^i \Big/ \sum_{i=0}^{n} a_i s^i\right) X(s) \;+\; \text{(terms due to \textit{all} initial values } x_0^k, y_0^k)\right]$$

or, in transform notation, as

$$Y(s) \;=\; \left(\sum_{i=0}^{m} b_i s^i \Big/ \sum_{i=0}^{n} a_i s^i\right) X(s) \;+\; \text{(terms due to \textit{all} initial values } x_0^k, y_0^k)$$

The **transfer function** $P(s)$ of a system is defined as that factor in the equation for $Y(s)$ multiplying the transform of the input $X(s)$. For the system described above, the transfer function is

$$P(s) \;=\; \sum_{i=0}^{m} b_i s^i \Big/ \sum_{i=0}^{n} a_i s^i \;=\; \frac{b_m s^m + b_{m-1} s^{m-1} + \cdots + b_0}{a_n s^n + a_{n-1} s^{n-1} + \cdots + a_0}$$

and the transform of the response may be rewritten as

$$Y(s) \;=\; P(s)\,X(s) \;+\; \text{(terms due to \textit{all} initial values } x_0^k, y_0^k)$$

If the quantity (terms due to *all* initial values x_0^k, y_0^k) is zero, the Laplace transform of the output $Y(s)$ in response to an input $X(s)$ is given by

$$Y(s) \;=\; P(s)\,X(s)$$

If the system is at rest prior to application of the input, i.e. $d^k y/dt^k = 0$, $k = 0, 1, \ldots, n-1$ for $t < 0$, then

$$\text{(terms due to \textit{all} initial values } x_0^k, y_0^k) \;=\; 0$$

and the output as a function of time $y(t)$ is simply the inverse transform of $P(s)\,X(s)$.

It is emphasized that not all transfer functions are rational algebraic expressions. The transfer function of a system including time delays contains terms of the form e^{-sT} (e.g. Problem 5.25). In fact, the transfer function of an element representing a pure time delay is $P(s) = e^{-sT}$ where T is the time delay in units of time.

Since the formation of the output transform $Y(s)$ is purely an algebraic multiplication of $P(s)$ and $X(s)$ when (terms due to *all* initial values x_0^k, y_0^k) $= 0$, the multiplication is commutative; that is,

$$Y(s) \;=\; X(s)\,P(s) \;=\; P(s)\,X(s) \qquad (6.1)$$

6.2 PROPERTIES OF A TRANSFER FUNCTION

The transfer function of a system has several useful properties:

1. The transfer function of a system is the Laplace transform of its impulse response. That is, if the input to a system with transfer function $P(s)$ is an impulse and all initial values are zero, the transform of the output is $P(s)$.

2. The system transfer function can be determined from the system differential equation by taking the Laplace transform and ignoring all terms arising from initial values. The transfer function $P(s)$ is then given by

$$P(s) \;=\; \frac{Y(s)}{X(s)}$$

3. The system differential equation can be obtained from the transfer function by replacing the s variable with the differential operator D defined by $D \equiv d/dt$.

4. The stability of a time-invariant linear system can be determined from the characteristic equation (see Chapter 5). The denominator of the system transfer function set equal to zero is the characteristic equation. Consequently if all the roots of the denominator have negative real parts, the system is stable.

5. The roots of the denominator are the system poles and the roots of the numerator are the system zeros (see Chapter 4). The system transfer function can then be specified to within a constant by specifying the system poles and zeros. This constant, usually denoted by K, is the system **gain-factor**. As was described in Chapter 4, the system poles and zeros can be represented schematically by a pole-zero map in the s-plane.

Example 6.1.

Consider the system with the differential equation $\dfrac{dy}{dt} + 2y = \dfrac{dx}{dt} + x$.

The Laplace transform version of this equation with all initial values set equal to zero is $(s+2)\,Y(s) = (s+1)\,X(s)$.

The system transfer function is thus given by $P(s) = \dfrac{Y(s)}{X(s)} = \dfrac{s+1}{s+2}$.

Example 6.2.

Given $P(s) = \dfrac{2s+1}{s^2 + s + 1}$,   the system differential equation is

$$y = \left[\frac{2D+1}{D^2 + D + 1}\right] x \quad \text{or} \quad D^2 y + Dy + y = 2Dx + x \quad \text{or} \quad \frac{d^2 y}{dt^2} + \frac{dy}{dt} + y = 2\frac{dx}{dt} + x$$

Example 6.3.

The transfer function $P(s) = \dfrac{K(s+a)}{(s+b)(s+c)}$ can be specified by giving the zero location $-a$, the pole locations $-b$ and $-c$, and the gain-factor K.

6.3 TRANSFER FUNCTIONS OF CONTROL SYSTEM COMPENSATORS

The transfer functions of three common control system compensators are presented below. Typical mechanizations of these transfer functions, *using R-C networks*, are presented in the solved problems.

Example 6.4.

The transfer function of a **lead compensator** is

$$P_{\text{Lead}}(s) \;=\; \frac{s+a}{s+b} \tag{6.2}$$

This compensator has a zero at $s = -a$ and a pole at $s = -b$. For lead compensation, b is larger than a.

Example 6.5.

The transfer function of a **lag compensator** is

$$P_{\text{Lag}}(s) \;=\; \frac{a(s+b)}{b(s+a)} \tag{6.3}$$

For lag compensation, b is larger than a. However, in this case the zero is at $s = -b$ and the pole is at $s = -a$. The gain factor a/b is included because of the way it is usually mechanized (Problem 6.13).

Example 6.6

The transfer function of a **lag-lead compensator** is

$$P_{\text{LL}}(s) \;=\; \frac{(s + a_1)(s + b_2)}{(s + b_1)(s + a_2)} \tag{6.4}$$

This compensator has two zeros and two poles. For lag-lead compensation, b_1 is greater than a_1 and b_2 is greater than a_2. For mechanization considerations, the restriction $a_1 b_2 = b_1 a_2$ is usually imposed (Problem 6.14).

6.4 SYSTEM TIME RESPONSE

The Laplace transform of the response of a system to a specific input is given by

$$Y(s) \;=\; P(s)\,X(s)$$

when all initial conditions are zero. The inverse transform $y(t) = \mathcal{L}^{-1}[P(s)\,X(s)]$ is the time response and $y(t)$ may be determined by finding the poles of $P(s)\,X(s)$ and evaluating the residues at these poles (when there are no multiple poles). Therefore $y(t)$ depends on both the poles and zeros of the transfer function and the poles and zeros of the input.

The residues can be determined graphically from a *pole-zero map* of $Y(s)$, which is constructed from the pole-zero map of $P(s)$ by simply adding the poles and zeros of $X(s)$. Graphical evaluation of the residues may then be performed as described on Pages 70 and 71 in Chapter 4.

6.5 SYSTEM FREQUENCY RESPONSE

The steady-state response of a system to sinusoidal inputs can be determined from the system transfer function. For the special case of a step function input of amplitude A, often called a **d.c. input**, the Laplace transform of the system output is given by

$$Y(s) \;=\; P(s)\,\frac{A}{s}$$

If the system is stable, the steady-state response is a step function of amplitude $A\,P(0)$ since this is the residue at the input pole. The amplitude of the input signal is thus multiplied by $P(0)$ to determine the amplitude of the output. $P(0)$ is therefore the **d.c. gain** of the system.

Note that for an unstable system such as an integrator with $P(s) = 1/s$, a steady-state response does not always exist. If the input to an integrator is a step function, the output is a ramp which is unbounded (see Problems 5.7 and 5.8). For this reason, integrators are sometimes said to have infinite d.c. gain.

The steady-state response of a stable system to an input $x = A \sin \omega t$ is given by

$$y_{\text{ss}} \;=\; A\,|P(j\omega)|\,\sin\,(\omega t + \phi)$$

where $|P(j\omega)|$ = magnitude of $P(j\omega)$, $\phi = \arg P(j\omega)$, and the complex number $P(j\omega)$ is determined from $P(s)$ by replacing s by $j\omega$ (see Problem 6.20). The system output has the same frequency as the input and can be obtained by multiplying the magnitude of the input by $|P(j\omega)|$ and shifting the phase angle of the input by $\arg P(j\omega)$. The magnitude $|P(j\omega)|$ and angle $\arg P(j\omega)$ for all ω constitute the **system frequency response**. The magnitude $|P(j\omega)|$ represents the *gain* of the system for sinusoidal inputs with frequency ω.

The system frequency response can be determined graphically in the s-plane from a pole-zero map of $P(s)$ in the same manner as the graphical calculation of residues. In this instance, however, the magnitude and phase angle of $P(s)$ are computed at a point on the $j\omega$ axis by measuring the magnitudes and angles of the vectors drawn from the poles and zeros of $P(s)$ to the point on the $j\omega$ axis. The Spirule can be used to rapidly perform the computation of $|P(j\omega)|$ and $\arg P(j\omega)$ (see Section 13.9, Page 241).

Example 6.7.

Consider the system with the transfer function

$$P(s) = \frac{1}{(s+1)(s+2)}$$

Referring to the adjoining figure, the magnitude and angle of $P(j\omega)$ for $\omega = 1$ are computed in the s-plane as follows. The magnitude of $P(j1)$ is

$$|P(j1)| = \frac{1}{\sqrt{5}\cdot\sqrt{2}} = 0.316$$

and the angle is

$$\arg P(j1) = -26.6° - 45° = -71.6°$$

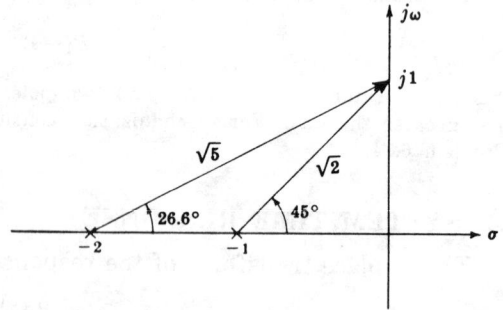

Example 6.8.

The system frequency response is usually represented by two graphs: one of $|P(j\omega)|$ as a function of ω and one of $\arg P(j\omega)$ as a function of ω. For the transfer function of Example 6.7, $P(s) = \dfrac{1}{(s+1)(s+2)}$, these graphs are easily determined by plotting the values of $|P(j\omega)|$ and $\arg P(j\omega)$ for several values of ω as shown below.

ω	0	0.5	1.0	2.0	4.0	8.0
$|P(j\omega)|$	0.5	0.433	0.316	0.158	0.054	0.015
$\arg P(j\omega)$	0	$-40.6°$	$-71.6°$	$-108.5°$	$-139.4°$	$-158.9°$

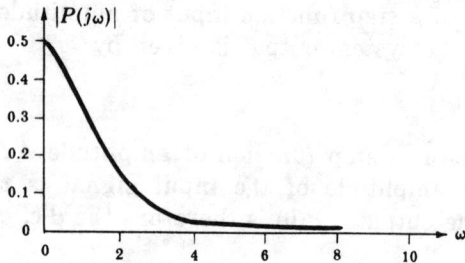

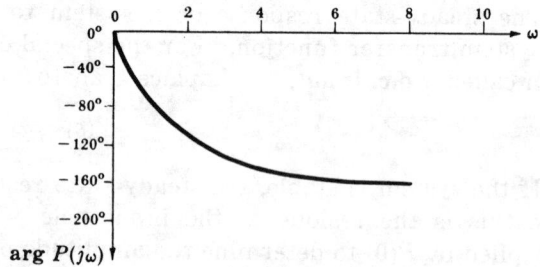

Solved Problems

TRANSFER FUNCTION DEFINITIONS

6.1. What is the transfer function of a system whose input and output are related by the following differential equation?

$$\frac{d^2y}{dt^2} + 3\frac{dy}{dt} + 2y = x + \frac{dx}{dt}$$

Taking the Laplace transform of this equation, ignoring terms due to initial conditions, we obtain

$$s^2 Y(s) + 3s Y(s) + 2 Y(s) = X(s) + s X(s)$$

This equation can be written in the form $Y(s) = \left[\dfrac{s+1}{s^2 + 3s + 2}\right] X(s)$.

The transfer function of this system is therefore given by $P(s) = \dfrac{s+1}{s^2 + 3s + 2}$.

6.2. A particular system containing a time delay has the differential equation $\dfrac{d}{dt}y(t) + y(t) = x(t-T)$. Find the transfer function of this system.

The Laplace transform of the differential equation, ignoring terms due to initial conditions, is $s\,Y(s) + Y(s) = e^{-sT}\,X(s)$. $Y(s)$ and $X(s)$ are related by the following function of s, which is the system transfer function: $P(s) = \dfrac{Y(s)}{X(s)} = \dfrac{e^{-sT}}{s+1}$.

6.3. The position y of a moving object of constant mass M is related to the total force f applied to the object by the differential equation $M\dfrac{d^2y}{dt^2} = f$. Determine the transfer function relating the position to the applied force.

Taking the Laplace transform of the differential equation, we obtain $Ms^2\,Y(s) = F(s)$. The transfer function relating $Y(s)$ to $F(s)$ is therefore $P(s) = \dfrac{Y(s)}{F(s)} = \dfrac{1}{Ms^2}$.

6.4. A motor connected to a load with inertia J and viscous friction B produces a torque which is proportional to the input current i. If the differential equation for the motor and load is $J\dfrac{d^2\theta}{dt^2} + B\dfrac{d\theta}{dt} = Ki$, determine the transfer function between the input current i and the shaft position θ.

The Laplace transform version of the differential equation is $(Js^2 + Bs)\,\Theta(s) = K\,I(s)$, and the required transfer function is $P(s) = \dfrac{\Theta(s)}{I(s)} = \dfrac{K}{s(Js+B)}$.

PROPERTIES OF A TRANSFER FUNCTION

6.5. An impulse is applied at the input of a system and the output is observed to be the time function e^{-2t}. Find the transfer function of this system.

The transfer function is $P(s) = \dfrac{Y(s)}{X(s)}$ and $X(s) = 1$ for $x(t) = \delta(t)$. Therefore,

$$P(s) = Y(s) = \frac{1}{s+2}$$

6.6. The impulse response of a certain system is the sinusoidal signal $\sin t$. Determine the system transfer function and differential equation.

The system transfer function is the Laplace transform of its impulse response, $P(s) = \dfrac{1}{s^2+1}$. Then $P(D) = \dfrac{y}{x} = \dfrac{1}{D^2+1}$, $D^2y + y = x$ or $\dfrac{d^2y}{dt^2} + y = x$.

6.7. The step response of a given system is $y = 1 - \frac{7}{3}e^{-t} + \frac{3}{2}e^{-2t} - \frac{1}{6}e^{-4t}$. What is the transfer function of this system?

Since the derivative of a step is an impulse (see Definition 3.20, Page 38), the impulse response for this system is $p(t) = dy/dt = \frac{7}{3}e^{-t} - 3e^{-2t} + \frac{2}{3}e^{-4t}$.

The Laplace transform of $p(t)$ is the desired transfer function. Thus

$$P(s) = \frac{7/3}{s+1} + \frac{-3}{s+2} + \frac{2/3}{s+4} = \frac{s+8}{(s+1)(s+2)(s+4)}$$

Note that an alternative solution would be to compute the Laplace transform of y and then multiply by s to determine $P(s)$, since a multiplication by s in the s-domain is equivalent to differentiation in the time domain.

6.8. Determine if the transfer function $P(s) = \dfrac{2s+1}{s^2+s+1}$ represents a stable or an unstable system.

The characteristic equation of the system is obtained by setting the denominator polynomial to zero, i.e. $s^2 + s + 1 = 0$. The characteristic equation may then be tested using one of the stability criteria described in Chapter 5. The Routh table for this system is given by

$$
\begin{array}{c|cc}
s^2 & 1 & 1 \\
s^1 & 1 & \\
s^0 & 1 & \\
\end{array}
$$

Since there are no sign changes in the first column, the system is stable.

6.9. Does the transfer function $P(s) = \dfrac{s+4}{(s+1)(s+2)(s-1)}$ represent a stable or an unstable system?

The stability of the system is determined by the roots of the denominator polynomial; that is, the *poles* of the system. Here the denominator is in factored form and the poles are located at $s = -1, -2, +1$. Since there is one pole with a positive real part, the system is unstable.

6.10. What is the transfer function of a system which has a gain-factor of 2 and a pole-zero map in the s-plane as shown in Fig. 6-1?

The transfer function has a zero at -1 and poles at -2 and the origin. Hence the transfer function is $P(s) = \dfrac{2(s+1)}{s(s+2)}$.

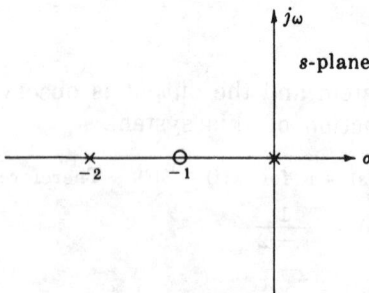

Fig. 6-1

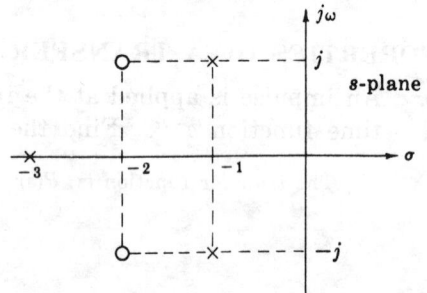

Fig. 6-2

6.11. Determine the transfer function of a system which has a gain factor of 3 and the pole-zero map shown in Fig. 6-2.

The transfer function has zeros at $-2 \pm j$ and poles at -3 and at $-1 \pm j$. The transfer function is therefore $P(s) = \dfrac{3(s+2+j)(s+2-j)}{(s+3)(s+1+j)(s+1-j)}$.

TRANSFER FUNCTIONS OF CONTROL SYSTEM COMPENSATORS

6.12. An R-C network mechanization of a lead compensator is shown. Find its transfer function.

Assuming the circuit is not loaded, i.e. no current flows through the output terminals, Kirchhoff's current law for the output node yields

$$C\frac{d}{dt}(v_i - v_0) + \frac{1}{R_1}(v_i - v_0) = \frac{1}{R_2}v_0$$

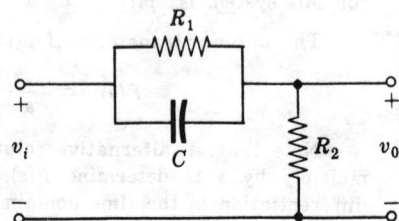

The Laplace transform of this equation (with zero initial conditions) is

$$Cs[V_i(s) - V_0(s)] + \frac{1}{R_1}[V_i(s) - V_0(s)] = \frac{1}{R_2}V_0(s)$$

The transfer function is

$$P_{\text{Lead}} = \frac{V_0(s)}{V_i(s)} = \frac{Cs + 1/R_1}{Cs + 1/R_1 + 1/R_2} = \frac{s+a}{s+b}$$

where $a = 1/R_1 C$ and $b = 1/R_1 C + 1/R_2 C$.

6.13. Determine the transfer function of the R-C network mechanization of the lag compensator shown.

Kirchhoff's voltage law for the loop yields the equation

$$iR_1 + \frac{1}{C}\int_0^t i\,dt + iR_2 = v_i$$

whose Laplace transform is

$$\left(R_1 + R_2 + \frac{1}{Cs}\right)I(s) = V_i(s)$$

The output voltage is given by $V_0(s) = \left(R_2 + \frac{1}{Cs}\right)I(s)$

The transfer function of the lag network is therefore

$$P_{\text{Lag}} = \frac{V_0(s)}{V_i(s)} = \frac{R_2 + 1/Cs}{R_1 + R_2 + 1/Cs} = \frac{a(s+b)}{b(s+a)} \quad \text{where} \quad a = \frac{1}{(R_1+R_2)C}, \quad b = \frac{1}{R_2 C}$$

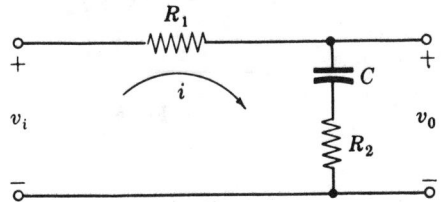

6.14. Derive the transfer function of the R-C network mechanization of the lag-lead compensator shown.

Equating currents at the output node a yields

$$\frac{1}{R_1}(v_i - v_0) + C_1\frac{d}{dt}(v_i - v_0) = i$$

The voltage v_0 and the current i are related by

$$\frac{1}{C_2}\int_0^t i\,dt + iR_2 = v_0$$

Taking the Laplace transform of these two equations (with zero initial conditions) and eliminating $I(s)$ results in the equation

$$\left(\frac{1}{R_1} + C_1 s\right)[V_i(s) - V_0(s)] = \frac{V_0(s)}{1/sC_2 + R_2}$$

The transfer function of the network is therefore

$$P_{\text{LL}} = \frac{V_0(s)}{V_i(s)} = \frac{\left(s + \frac{1}{R_1 C_1}\right)\left(s + \frac{1}{R_2 C_2}\right)}{s^2 + \left(\frac{1}{R_2 C_2} + \frac{1}{R_2 C_1} + \frac{1}{R_1 C_1}\right)s + \frac{1}{R_1 C_1 R_2 C_2}} = \frac{(s+a_1)(s+b_2)}{(s+b_1)(s+a_2)}$$

where $a_1 = \frac{1}{R_1 C_1}$, $b_1 a_2 = a_1 b_2$, $b_1 + a_2 = a_1 + b_2 + \frac{1}{R_2 C_1}$, $b_2 = \frac{1}{R_2 C_2}$.

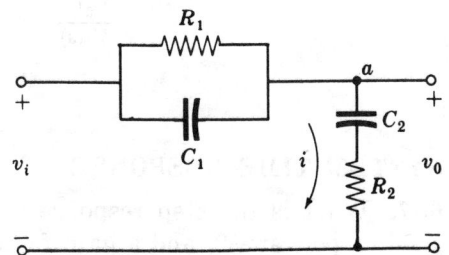

6.15. Find the transfer function of the *simple* lag network shown in Fig. 6-3.

This network is a special case of the lag compensation network of Problem 6.13 with R_2 set equal to zero. Hence the transfer function is given by

$$P(s) \;=\; \frac{V_0(s)}{V_i(s)} \;=\; \frac{1/Cs}{R + 1/Cs} \;=\; \frac{1/RC}{s + 1/RC}$$

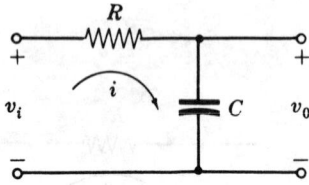

Fig. 6-3

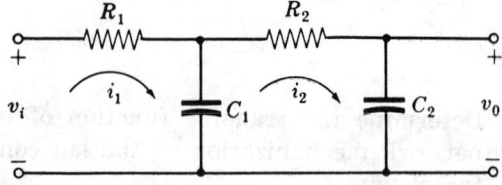

Fig. 6-4

6.16. Determine the transfer function of two simple lag networks connected in series as shown in Fig. 6-4.

The two loop equations are

$$R_1 i_1 \;+\; \frac{1}{C_1} \int_0^t (i_1 - i_2)\, dt \;=\; v_i$$

$$R_2 i_2 \;+\; \frac{1}{C_2} \int_0^t i_2\, dt \;+\; \frac{1}{C_1} \int_0^t (i_2 - i_1)\, dt \;=\; 0$$

Using the Laplace transformation and solving the two loop equations for $I_2(s)$, we obtain

$$I_2(s) \;=\; \frac{C_2 s\, V_i(s)}{R_1 R_2 C_1 C_2 s^2 + (R_1 C_1 + R_1 C_2 + R_2 C_2)s + 1}$$

The output voltage is given by $v_0 = \dfrac{1}{C_2} \displaystyle\int_0^t i_2\, dt$. Thus

$$\frac{V_0(s)}{V_i(s)} \;=\; \frac{1}{R_1 R_2 C_1 C_2 s^2 + (R_1 C_1 + R_1 C_2 + R_2 C_2)s + 1}$$

SYSTEM TIME RESPONSE

6.17. What is the step response of a system whose transfer function has a zero at -1, a pole at -2, and a gain factor of 2?

The Laplace transform of the output is given by $Y(s) = P(s)\, X(s)$. Here

$$P(s) \;=\; \frac{2(s+1)}{s+2}, \qquad X(s) \;=\; \frac{1}{s}, \qquad Y(s) \;=\; \frac{2(s+1)}{s(s+2)} \;=\; \frac{1}{s} + \frac{1}{s+2}$$

Evaluating the inverse transform of the partial fraction expansion of $Y(s)$ gives $y(t) = 1 + e^{-2t}$.

6.18. Graphically evaluate the step response of a system whose transfer function is given by

$$P(s) \;=\; \frac{(s+2)}{(s+0.5)(s+4)}$$

The pole-zero map of the output is obtained by adding the poles and zeros of the input to the pole-zero map of the transfer function. The output pole-zero map therefore has poles at 0, -0.5 and -4 and a zero at -2 as shown below.

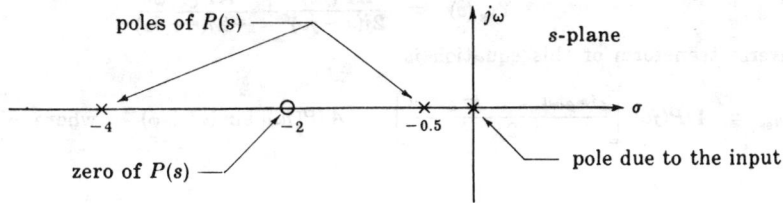

The residue for the pole at the origin is

$$|R_1| = \frac{2}{0.5(4)} = 1 \qquad \arg R_1 = 0°$$

For the pole at -0.5,

$$|R_2| = \frac{1.5}{0.5(3.5)} = 0.857 \qquad \arg R_2 = -180°$$

For the pole at -4,

$$|R_3| = \frac{2}{4(3.5)} = 0.143 \qquad \arg R_3 = -180°$$

The time response is therefore $y(t) = R_1 + R_2 e^{-0.5t} + R_3 e^{-4t} = 1 - 0.857e^{-0.5t} - 0.143e^{-4t}$.

6.19. Evaluate the step response of the system of Problem 6.11.

The Laplace transform of the system output is

$$Y(s) = P(s)\,X(s) = \frac{3(s+2+j)(s+2-j)}{s(s+3)(s+1+j)(s+1-j)}$$

Expanding $Y(s)$ into partial fractions yields

$$Y(s) = \frac{R_1}{s} + \frac{R_2}{s+3} + \frac{R_3}{s+1+j} + \frac{R_4}{s+1-j}$$

where $\quad R_1 = \dfrac{3(2+j)(2-j)}{3(1+j)(1-j)} = \dfrac{5}{2} \qquad R_3 = \dfrac{3(1)(1-2j)}{(-1-j)(2-j)(-2j)} = \dfrac{-3}{20}(7+j)$

$$R_2 = \frac{3(-1+j)(-1-j)}{-3(-2+j)(-2-j)} = \frac{-2}{5} \qquad R_4 = \frac{3(1+2j)(1)}{(2+j)(-1+j)(2j)} = \frac{-3}{20}(7-j)$$

Evaluating the inverse Laplace transform,

$$y = \frac{5}{2} - \frac{2}{5}e^{-3t} - \frac{3\sqrt{2}}{4}e^{-t}\left[e^{-j(t+\theta)} + e^{j(t+\theta)}\right] = \frac{5}{2} - \frac{2}{5}e^{-3t} - \frac{3\sqrt{2}}{2}e^{-t}\cos(t+\theta)$$

where $\quad \theta = -\tan^{-1}[1/7] = -8.13°$.

SYSTEM FREQUENCY RESPONSE

6.20. Prove that the steady-state output of a stable system with transfer function $P(s)$ and input $x = A\sin\omega t$ is given by

$$y_{ss} = A\,|P(j\omega)|\sin(\omega t + \phi) \qquad \text{where } \phi = \arg P(j\omega)$$

The Laplace transform of the output is $\quad Y(s) = P(s)\,X(s) = P(s)\dfrac{A\omega}{s^2 + \omega^2}$.

When this transform is expanded into partial fractions, there will be terms due to the poles of $P(s)$ and two terms due to the poles of the input $(s = \pm j\omega)$. Since the system is stable, all the time functions resulting from the poles of $P(s)$ will decay to zero as time approaches infinity. Thus the steady-state output will contain only the time functions resulting from the terms in the partial fraction expansion due to the poles of the input. The Laplace transform of the steady-state output is therefore

$$Y_{ss}(s) = \frac{AP(j\omega)}{2j(s - j\omega)} + \frac{AP(-j\omega)}{-2j(s + j\omega)}$$

The inverse transform of this equation is

$$y_{ss} = A\,|P(j\omega)|\left[\frac{e^{j\phi}\,e^{j\omega t} - e^{-j\phi}\,e^{-j\omega t}}{2j}\right] = A\,|P(j\omega)|\,\sin(\omega t + \phi) \qquad \text{where } \phi = \arg P(j\omega)$$

6.21. Find the *d.c. gain* of each of the systems represented by the following transfer functions: (a) $P(s) = \dfrac{1}{s+1}$, (b) $P(s) = \dfrac{10}{(s+1)(s+2)}$, (c) $P(s) = \dfrac{(s+8)}{(s+2)(s+4)}$.

The d.c. gain is given by $P(0)$. Then (a) $P(0) = 1$, (b) $P(0) = 5$, (c) $P(0) = 1$.

6.22. Evaluate the gain and phase shift of $P(s) = \dfrac{2}{s+2}$ for $\omega = 1, 2$, and 10.

The gain of $P(s)$ is given by $|P(j\omega)| = \dfrac{2}{\sqrt{\omega^2 + 4}}$. For $\omega = 1$, $|P(j1)| = 2/\sqrt{5} = 0.894$; for $\omega = 2$, $|P(j2)| = 2/\sqrt{8} = 0.707$; for $\omega = 10$, $|P(j10)| = 2/\sqrt{104} = 0.196$.

The phase shift of the transfer function is the phase angle of $P(j\omega)$, $\arg P(j\omega) = -\tan^{-1}\omega/2$. For $\omega = 1$, $\arg P(j1) = -\tan^{-1}\frac{1}{2} = -26.6°$; for $\omega = 2$, $\arg P(j2) = -\tan^{-1}1 = -45°$; for $\omega = 10$, $\arg P(j10) = -\tan^{-1}5 = -78.7°$.

6.23. Sketch the graphs of $|P(j\omega)|$ and $\arg P(j\omega)$ as a function of frequency for the transfer function of Problem 6.22.

In addition to the values calculated in Problem 6.22 for $|P(j\omega)|$ and $\arg P(j\omega)$, the values for $\omega = 0$ will also be useful: $|P(j0)| = 2/2 = 1$, $\arg P(j0) = -\tan^{-1}0 = 0$.

As ω becomes large, $|P(j\omega)|$ asymptotically approaches zero while $\arg P(j\omega)$ asymptotically approaches $-90°$. The graphs representing the frequency response of $P(s)$ are shown below.

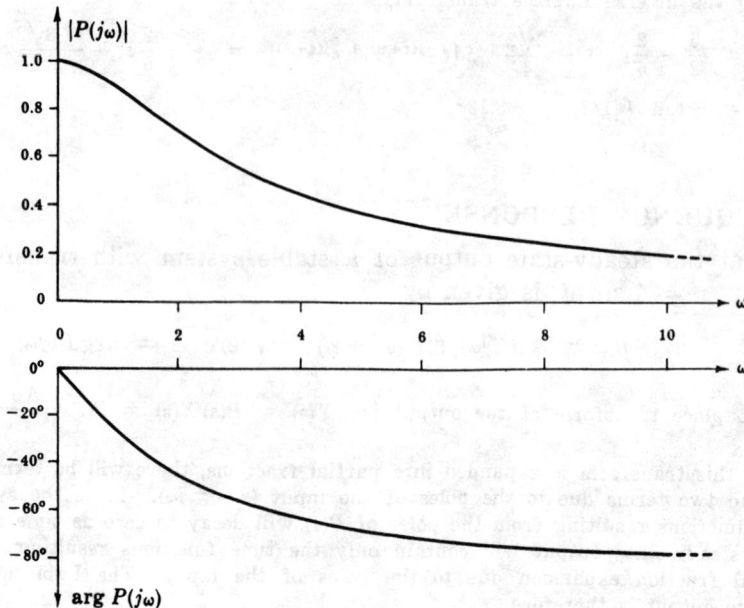

MISCELLANEOUS PROBLEMS

6.24. A *d.c.* (*direct current*) *motor* is shown schematically in the diagram below. L and R represent the inductance and resistance of the motor's armature circuit, and the voltage v_b represents the generated back e.m.f. (electromotive force) which is proportional to the shaft velocity $d\theta/dt$. The torque T generated by the motor is proportional to the armature current i. The inertia J represents the combined inertia of the motor armature and the load, and B is the total viscous friction acting on the output shaft. Determine the transfer function between the input voltage V and the angular position Θ of the output shaft.

Motor Armature Circuit   **Inertial Load**

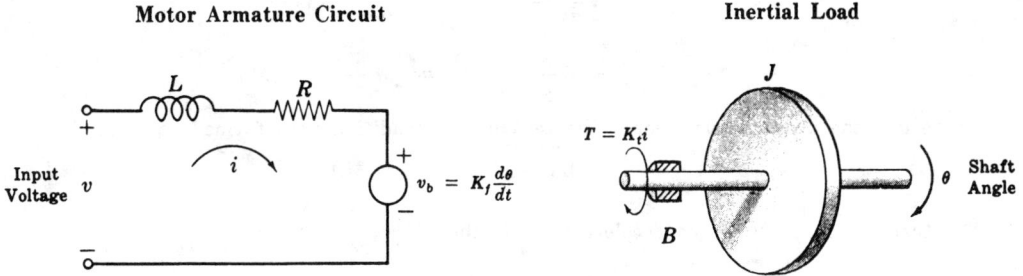

The differential equations of the motor armature circuit and the inertial load are

$$Ri + L\frac{di}{dt} = v - K_f\frac{d\theta}{dt} \quad \text{and} \quad K_t i = J\frac{d^2\theta}{dt^2} + B\frac{d\theta}{dt}$$

Taking the Laplace transform of each equation, ignoring initial conditions,

$$(R + sL)I = V - K_f s\Theta \quad \text{and} \quad K_t I = (Js^2 + Bs)\Theta$$

Solving these equations simultaneously for the transfer function between V and θ, we have

$$\frac{\Theta}{V} = \frac{K_t}{(Js^2 + Bs)(Ls + R) + K_t K_f s} = \frac{K_t/JL}{s[s^2 + (B/J + R/L)s + BR/JL + K_t K_f/JL]}$$

6.25. The back e.m.f. generated by the armature circuit of a d.c. machine is proportional to the angular velocity of its shaft, as noted in the problem above. This principle is utilized in the *d.c. tachometer* shown schematically in the adjoining figure, where v_b is the voltage generated by the armature, L is the armature inductance, R_a is the armature resistance, and v_0 is the output voltage. If K_f is the proportionality constant between v_b and shaft velocity $d\theta/dt$, i.e. $v_b = K_f(d\theta/dt)$, determine the transfer function between the shaft position Θ and the output voltage V_0. The output load is represented by a resistance R_L and $R_L + R_a \equiv R$.

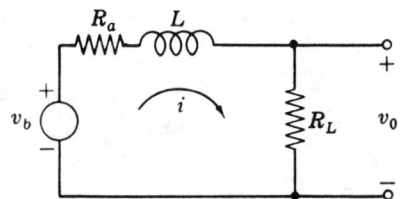

The Laplace transformed equation representing the tachometer is $I(R + sL) = K_f s\Theta$.

The output voltage is given by

$$V_0 = IR_L = \frac{R_L K_f s\Theta}{R + sL}$$

The transfer function of the d.c. tachometer is then

$$\frac{V_0}{\Theta} = \frac{R_L K_f}{L}\left(\frac{s}{s + R/L}\right)$$

6.26. A simple mechanical *accelerometer* is shown
in the adjoining diagram. The position y of
the mass M with respect to the accelerometer
case is proportional to the acceleration of the
case. What is the transfer function between
the input acceleration A $(a = d^2x/dt^2)$ and the
output Y.

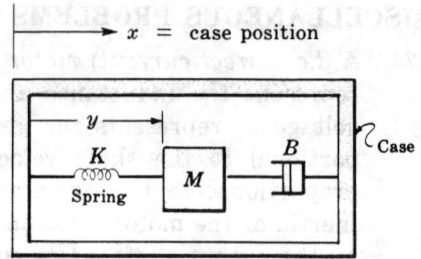

x = case position

Equating the sum of the forces acting on the
mass M to its inertial acceleration, we obtain

$$-B\frac{dy}{dt} - Ky = M\frac{d^2}{dt^2}(y - x)$$

or

$$M\frac{d^2y}{dt^2} + B\frac{dy}{dt} + Ky = M\frac{d^2x}{dt^2} = Ma$$

where a is the input acceleration. The zero initial condition transformed equation is

$$(Ms^2 + Bs + K)Y = MA$$

The transfer function of the accelerometer is therefore $\dfrac{Y}{A} = \dfrac{1}{s^2 + (B/M)s + K/M}$.

6.27. A differential equation describing the dynamic operation of the *one-degree-of-freedom
gyroscope* shown below is

$$J\frac{d^2\theta}{dt^2} + B\frac{d\theta}{dt} + K\theta = H\omega$$

where ω is the angular velocity of the gyroscope about the input axis, θ is the angular
position of the spin axis — the measured output of the gyroscope, H is angular
momentum stored in the spinning wheel, J is the inertia of the wheel about the
output axis, B is the viscous friction coefficient about the output axis, and K is the
spring constant of the restraining spring attached to the spin axis.

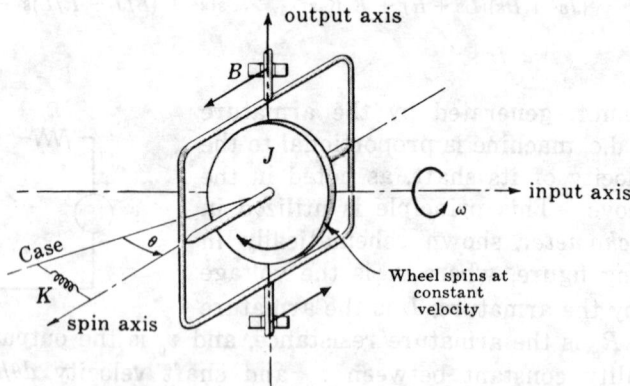

(a) Determine the transfer function relating the Laplace transforms of ω and θ, and
show that the steady-state output is proportional to the magnitude of a constant
rate input. This type of gyroscope is called a *rate gyro*.

(b) Determine the transfer function between ω and θ with the restraining spring
removed ($K = 0$). Since here the output is proportional to the integral of the
input rate, this type of gyroscope is called an *integrating gyro*.

(a) The zero initial condition transform of the gyroscope differential equation is

$$(Js^2 + Bs + K)\Theta = H\Omega$$

where Θ and Ω are the Laplace transforms of θ and ω respectively. The transfer function relating Θ and Ω is therefore

$$\frac{\Theta}{\Omega} = \frac{H}{(Js^2 + Bs + K)}$$

For a constant or d.c. rate input ω_K, the magnitude of the steady-state output θ_{ss} can be obtained by multiplying the input by the d.c. gain of the transfer function, which in this case is H/K. Thus the steady-state output is proportional to the magnitude of the rate input, that is, $\theta_{ss} = (H/K)\omega_K$.

(b) Setting K equal to zero in the transfer function of (a) yields $\dfrac{\Theta}{\Omega} = \dfrac{H}{s(Js + B)}$. This transfer function now has a pole at the origin, so that an integration is obtained between the input Ω and the output Θ. The output is thus proportional to the integral of the input rate or, equivalently, the input angle.

6.28. A differential equation approximating the rotational dynamics of a rigid vehicle moving in the atmosphere is

$$J\frac{d^2\theta}{dt^2} - NL\theta = T$$

where θ is the vehicle attitude angle, J is its inertia, N is the normal-force coefficient, L is the distance from the center of gravity to the center of pressure, and T is any applied torque. Determine the transfer function between an applied torque and the vehicle attitude angle.

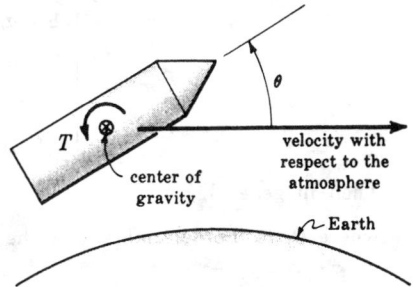

The zero initial condition transformed system differential equation is

$$(Js^2 - NL)\Theta = T$$

The desired transfer function is

$$\frac{\Theta}{T} = \frac{1}{Js^2 - NL} = \frac{1/J}{s^2 - NL/J}$$

Note that if NL is positive (center of pressure forward of the vehicle center of gravity), the system is *unstable* because there is a pole in the right half plane at $s = \sqrt{NL/J}$. If NL is negative, the poles are imaginary and the system is *oscillatory* (marginally stable). However, aerodynamic damping terms not included in the differential equation are actually present and perform the function of damping out any oscillations.

6.29. Pressure receptors called *baroreceptors* measure changes in arterial blood pressure, as outlined in Problem 2.14. They are shown as a block in the feedback path of the block diagram determined in the solution of that problem. The frequency $b(t)$ at which signals (action potentials) move along the vagus and glossopharyngeal nerves from the baroreceptors to the vasomotor center (VMC) in the brain is proportional to arterial blood pressure p plus the time rate of change of blood pressure. Determine the form of the transfer function for the baroreceptors.

From the description given above, the equation for b is

$$b = k_1 p + k_2 \frac{dp}{dt}$$

where k_1 and k_2 are constants, and p is blood pressure. (p should not be confused here with the notation $p(t)$, the inverse Laplace transform of $P(s)$ introduced in this chapter as a general representation for a transfer function.) The Laplace transform of the above equation, with zero initial conditions, is

$$B = k_1 P + k_2 s P = P(k_1 + k_2 s)$$

The transfer function of the baroreceptors is therefore $B/P = k_1 + k_2 s$. We again remind the reader that P represents the transform of arterial blood pressure in this problem.

6.30. Consider the transfer function C_k/R_k for the biological system described in Problem 3.4(a) by the equations

$$c_k(t) \;=\; r_k(t) \;-\; \sum_{i=1}^{n} a_{k-i}\, c_i(t - \Delta t)$$

for $k = 1, 2, \ldots, n$. Explain how C_k/R_k may be computed.

Taking the Laplace transform of the above equations, ignoring initial conditions, yields the following set of equations:

$$C_k \;=\; R_k \;-\; \sum_{i=1}^{n} a_{k-i}\, C_i\, e^{-s\,\Delta t}$$

for $k = 1, 2, \ldots, n$. If all n equations were written down, we would have n equations in n unknowns (C_k for $k = 1, 2, \ldots, n$). The general solution for any C_k in terms of the inputs R_k can then be determined using the standard techniques for solving simultaneous equations. Let D represent the determinant of the coefficient matrix:

$$D \;\equiv\; \begin{vmatrix} 1 + a_0 e^{-s\,\Delta t} & a_{-1} e^{-s\,\Delta t} & \cdot & \cdots & a_{1-n} e^{-s\,\Delta t} \\ a_1 e^{-s\,\Delta t} & 1 + a_0 e^{-s\,\Delta t} & & \cdots & a_{2-n} e^{-s\,\Delta t} \\ \multicolumn{5}{c}{\dotfill} \\ a_{n-1} e^{-s\,\Delta t} & & \cdots & a_1 e^{-s\,\Delta t} & 1 + a_0 e^{-s\,\Delta t} \end{vmatrix}$$

Then in general, $\qquad\qquad\qquad C_k = D_k/D$

where D_k is the determinant of the coefficient matrix with the kth column replaced by

$$\begin{matrix} R_1 \\ R_2 \\ \cdot \\ \cdot \\ \cdot \\ R_n \end{matrix}$$

The transfer function C_k/R_k is then determined by setting all the inputs except R_k equal to zero, computing C_k from the formula above, and dividing C_k by R_k.

Supplementary Problems

6.31. Determine the transfer function of the R-C network shown in Fig. 6-5.

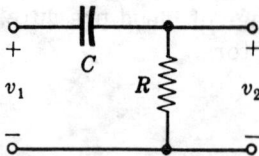

Fig. 6-5

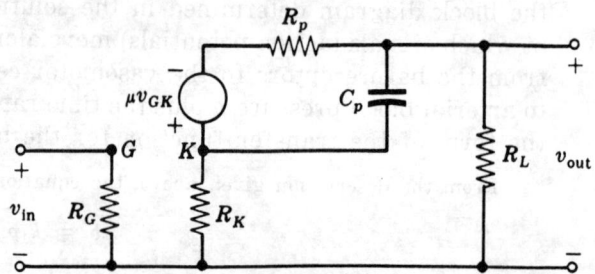

Fig. 6-6

6.32. An equivalent circuit of a vacuum tube amplifier is shown in Fig. 6-6. What is its transfer function?

6.33. Find the transfer function of a system having the impulse response $p(t) = e^{-t}(1 - \sin t)$.

6.34. A sinusoidal input $x = 2 \sin 2t$ is applied to a system with the transfer function $P(s) = \dfrac{2}{s(s+2)}$. Determine the steady-state output y_{ss}.

6.35. Find the step response of a system having the transfer function $P(s) = \dfrac{4}{(s^2-1)(s^2+1)}$.

6.36. Determine which of the following transfer functions represent stable systems and which represent unstable systems:

(a) $P(s) = \dfrac{(s-1)}{(s+2)(s^2+4)}$

(c) $P(s) = \dfrac{(s+2)(s-2)}{(s+1)(s-1)(s+4)}$

(e) $P(s) = \dfrac{5(s+10)}{(s+5)(s^2-s+10)}$

(b) $P(s) = \dfrac{(s-1)}{(s+2)(s+4)}$

(d) $P(s) = \dfrac{6}{(s^2+s+1)(s+1)^2}$

6.37. Use the Final Value Theorem (Chapter 4) to show that the steady-state value of the output of a stable system in response to a unit step input is equal to the d.c. gain of the system.

6.38. Determine the transfer function of two of the networks shown in Problem 6.31 connected in cascade (series).

6.39. Examine the literature for the transfer functions of two- and three-degree-of-freedom gyros and compare them with the one-degree-of-freedom gyro of Problem 6.27.

6.40. Determine the ramp response of a system having the transfer function $P(s) = \dfrac{s+1}{s+2}$.

6.41. Show that if a system described by

$$\sum_{i=0}^{n} a_i \frac{d^i y}{dt^i} = \sum_{i=0}^{m} b_i \frac{d^i x}{dt^i}$$

for $m \le n$ is at rest prior to application of the input, i.e. $d^k y/dt^k = 0$, $k = 0, 1, \ldots, n-1$ for $t < 0$, then (terms due to *all* initial values x_0^k, y_0^k) $= 0$.

(*Hint.* Integrate the differential equation n times from $0^- \equiv \lim_{\substack{\epsilon \to 0 \\ \epsilon < 0}} \epsilon$ to t, and then let $t \to 0^+$.)

Answers to Supplementary Problems

6.31. $\dfrac{V_2}{V_1} = \dfrac{s}{s + 1/RC}$

6.32. $\dfrac{V_{out}}{V_{in}} = \dfrac{-\mu R_L}{(R_k + R_L)R_p C_p s + (\mu + 1)R_k + R_p + R_L}$

6.33. $P(s) = \dfrac{s^2 + s + 1}{(s+1)(s^2 + 2s + 2)}$

6.34. $y_{ss} = 0.707 \sin(2t - 135°)$

6.35. $y = -4 + e^{-t} + e^t + 2 \cos t$

6.36. (b) and (d) represent stable systems; (a), (c), and (e) represent unstable systems.

6.38. $\dfrac{V_2}{V_1} = \dfrac{s^2}{s^2 + (3/RC)s + 1/R^2C^2}$

6.40. $y = \tfrac{1}{4} - \tfrac{1}{4}e^{-2t} + \tfrac{1}{2}t$

Chapter 7

Block Diagram Algebra
and Transfer Functions of Systems

7.1 INTRODUCTION

It is pointed out in Chapters 1 and 2 that the block diagram is a shorthand, graphical representation of a physical system, illustrating the functional relationships among its components. This latter feature permits evaluation of the contributions of the individual elements to the overall performance of the system.

In this chapter we shall first investigate these relationships in more detail, utilizing the frequency domain and transfer function concepts developed in the preceding chapters. Then we will develop methods for reducing complicated block diagrams to manageable forms so that they may be used to predict the overall performance of a system.

7.2 REVIEW OF FUNDAMENTALS

In general, a block diagram consists of a specific configuration of four types of elements — blocks, summing points, takeoff points, and arrows representing unidirectional signal flow:

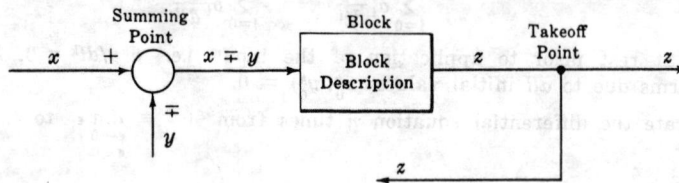

The meaning of each element should be clear from the above diagram.

Time-domain quantities are represented by lower-case letters.

Example 7.1. $r = r(t)$

Capital letters in this chapter are used for Laplace transforms.

Example 7.2. $R = R(s)$

The basic feedback control system configuration presented in Chapter 2 is reproduced here, with all quantities in abbreviated Laplace transform notation:

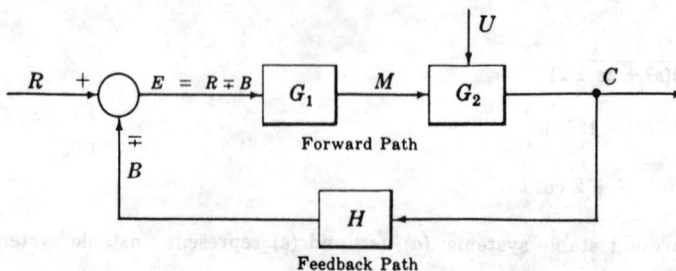

The quantities G_1, G_2, and H are the transfer functions of the components in the blocks.

Example 7.3. $G_1 = M/E$ or $M = G_1 E$

7.3 BLOCKS IN CASCADE

Any finite number of blocks in series may be algebraically combined by multiplication. That is, n components or blocks with transfer functions $G_1, G_2, \ldots, G_n$ connected in cascade are equivalent to a single element G with a transfer function given by

$$G \;=\; G_1 \cdot G_2 \cdot G_3 \cdots G_n \;=\; \prod_{i=1}^{n} G_i \qquad\qquad (7.1)$$

The symbol for multiplication "$\cdot$" is often omitted when no confusion results.

Example 7.4.

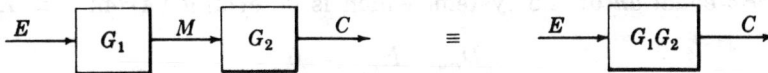

Multiplication of transfer functions is *commutative*; that is,

$$G_i G_j \;=\; G_j G_i \qquad\qquad (7.2)$$

for any i or j.

Example 7.5.

Loading effects (interaction of one transfer function upon its neighbor) must be accounted for in the derivation of the individual transfer functions before blocks can be cascaded.

7.4 CANONICAL FORM OF A FEEDBACK CONTROL SYSTEM

The two blocks in the forward path of the feedback system of Section 7.2 may be combined. Letting $G \equiv G_1 G_2$, the resulting configuration is called the **canonical form** of a feedback control system. G and H are not necessarily unique for a particular system.

The following definitions refer to this block diagram.

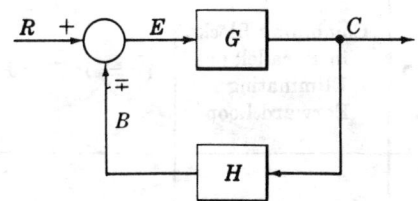

Definition 7.1: $G \equiv$ direct transfer function $\equiv$ forward transfer function

Definition 7.2: $H \equiv$ feedback transfer function

Definition 7.3: $GH \equiv$ loop transfer function $\equiv$ open-loop transfer function

Definition 7.4: $\dfrac{C}{R} \equiv$ closed-loop transfer function $\equiv$ control ratio

Definition 7.5: $\dfrac{E}{R} \equiv$ actuating signal ratio $\equiv$ error ratio

Definition 7.6: $\dfrac{B}{R} \equiv$ primary feedback ratio

In the following equations, the $-$ sign refers to a *positive* feedback system, and the $+$ sign refers to a *negative* feedback system:

$$\frac{C}{R} = \frac{G}{1 \pm GH} \tag{7.3}$$

$$\frac{E}{R} = \frac{1}{1 \pm GH} \tag{7.4}$$

$$\frac{B}{R} = \frac{GH}{1 \pm GH} \tag{7.5}$$

The *characteristic equation* of the system, which is determined from $1 \pm GH = 0$, is

$$D_{GH} \pm N_{GH} = 0 \tag{7.6}$$

where D_{GH} is the denominator and N_{GH} the numerator of GH.

7.5 BLOCK DIAGRAM TRANSFORMATION THEOREMS

Block diagrams of complicated control systems may be simplified using easily derivable transformations. The first important transformation, combining blocks in cascade, has already been presented in Section 7.3. It is repeated for completeness in the following chart illustrating the transformation theorems. The letter P is used to represent any transfer function, and W, X, Y, Z denote any s-domain signals.

	Transformation	Equation	Block Diagram	Equivalent Block Diagram
1	Combining Blocks in Cascade	$Y = (P_1 P_2)X$	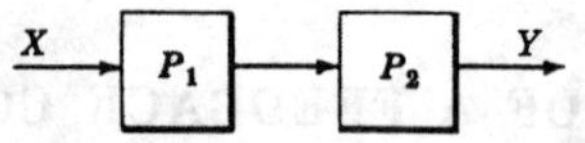	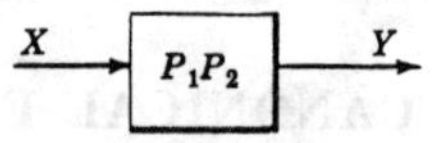
2	Combining Blocks in Parallel; or Eliminating a Forward Loop	$Y = P_1 X \pm P_2 X$	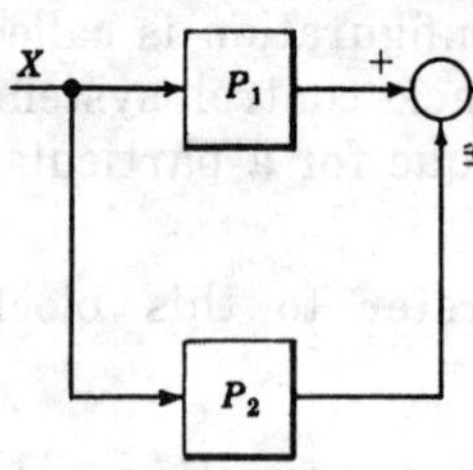	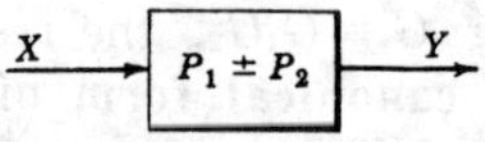
3	Removing a Block from a Forward Path	$Y = P_1 X \pm P_2 X$		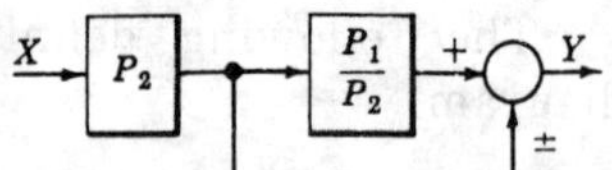
4	Eliminating a Feedback Loop	$Y = P_1(X \mp P_2 Y)$	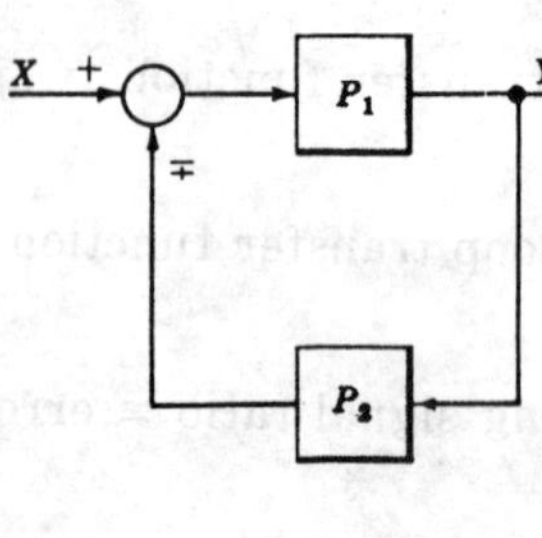	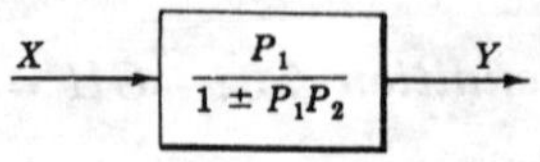
5	Removing a Block from a Feedback Loop	$Y = P_1(X \mp P_2 Y)$		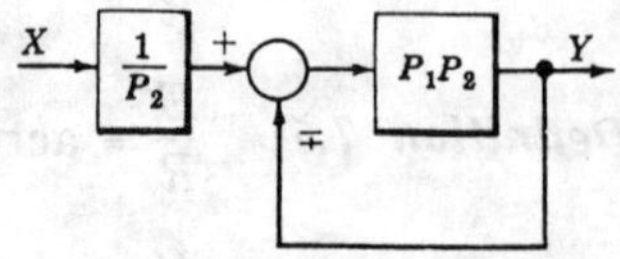

	Transformation	Equation	Block Diagram	Equivalent Block Diagram
6a	Rearranging Summing Points	$Z = W \pm X \pm Y$		
6b	Rearranging Summing Points	$Z = W \pm X \pm Y$		
7	Moving a Summing Point Ahead of a Block	$Z = PX \pm Y$		
8	Moving a Summing Point Beyond a Block	$Z = P[X \pm Y]$		
9	Moving a Takeoff Point Ahead of a Block	$Y = PX$		
10	Moving a Takeoff Point Beyond a Block	$Y = PX$		
11	Moving a Takeoff Point Ahead of a Summing Point	$Z = X \pm Y$		
12	Moving a Takeoff Point Beyond a Summing Point	$Z = X \pm Y$		

7.6 UNITY FEEDBACK SYSTEMS

Definition 7.7:   A **unity feedback system** is a feedback system in which the primary feedback b is identically equal to the controlled output c.

Example 7.6 $H = 1$ for a linear, unity feedback system:

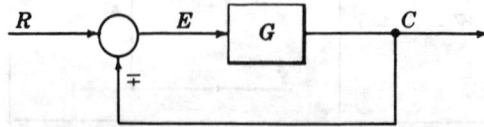

Any feedback system with only linear elements in the feedback loop can be put into the form of a unity feedback system by using Transformation 5.

Example 7.7.

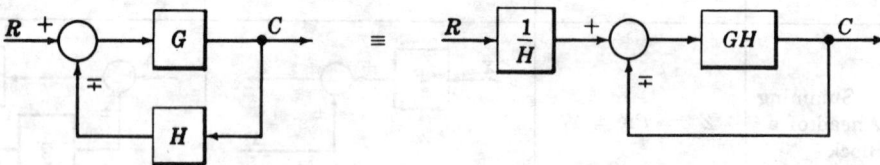

The characteristic equation for the unity feedback system, determined from $1 \pm G = 0$, is

$$D_G \pm N_G = 0 \qquad\qquad (7.7)$$

where D_G is the denominator and N_G the numerator of G.

7.7 MULTIPLE INPUTS

Sometimes it is necessary to evaluate a system's performance when several stimuli are simultaneously applied at different points of the system.

When multiple inputs are present in a *linear* system, each is treated independently of the others. The output due to all stimuli acting together is found in the following manner:

Step 1:   Set all inputs except one equal to zero.

Step 2:   Transform the block diagram to canonical form, using the transformations of Section 7.5.

Step 3:   Calculate the response due to the chosen input acting alone.

Step 4:   Repeat Steps 1 to 3 for each of the remaining inputs.

Step 5:   Algebraically add all of the responses (outputs) determined in Steps 1 to 5. This sum is the total output of the system with all inputs acting simultaneously.

We re-emphasize here that the above superposition process is dependent on the system being linear.

Example 7.8.   We shall determine the output C of the following system:

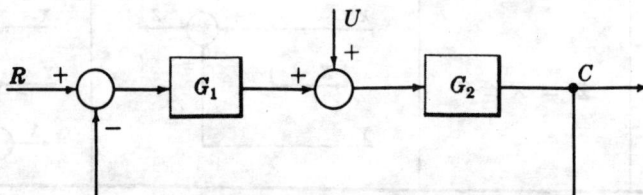

Step 1:   Put $U \equiv 0$.

Step 2:   The system reduces to

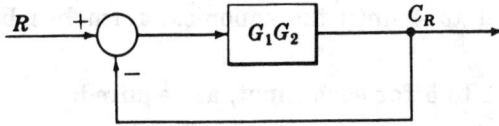

Step 3:   By Equation (7.3), the output C_R due to input R is $C_R = \left[\dfrac{G_1 G_2}{1 + G_1 G_2}\right] R$.

Step 4a: Put $R = 0$.

Step 4b: Put -1 into a block, representing the negative feedback effect:

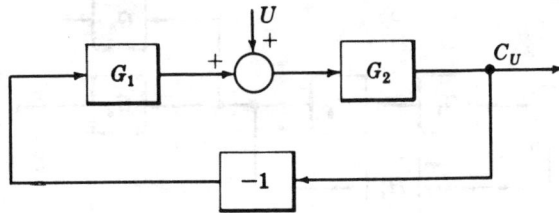

Rearrange the block diagram:

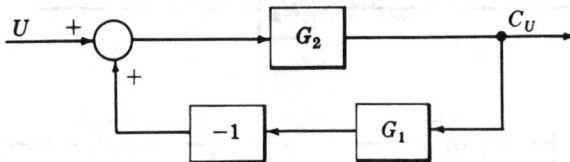

Let the -1 block be absorbed into the summing point:

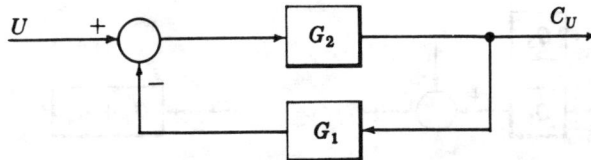

Step 4c: By Equation (7.3), the output C_U due to input U is $C_U = \left[\dfrac{G_2}{1 + G_1 G_2}\right] U$.

Step 5:   The total output is

$$C = C_R + C_U = \left[\frac{G_1 G_2}{1 + G_1 G_2}\right] R + \left[\frac{G_2}{1 + G_1 G_2}\right] U = \left[\frac{G_2}{1 + G_1 G_2}\right] \cdot [G_1 R + U]$$

7.8   REDUCTION OF COMPLICATED BLOCK DIAGRAMS

The block diagram of a practical feedback control system is often quite complicated. It may include several feedback or feedforward loops, and multiple inputs. By means of systematic block diagram reduction, every multiple loop feedback system may be reduced to canonical form. The techniques developed in the preceding paragraphs provide the necessary tools.

The following general steps may be used as a basic approach in the reduction of complicated block diagrams. Each step refers to specific transformations in the chart of Section 7.5.

Step 1:   Combine all cascade blocks using Transformation 1.

Step 2:   Combine all parallel blocks using Transformation 2.

Step 3:   Eliminate all minor feedback loops using Transformation 4.

Step 4: Shift summing points to the left and takeoff points to the right of the major loop, using Transformations 7, 10, and 12.

Step 5: Repeat Steps 1 to 4 until the canonical form has been achieved for a particular input.

Step 6: Repeat Steps 1 to 5 for each input, as required.

Transformations 3, 5, 6, 8, 9 and 11 are sometimes useful, and experience with the reduction technique will determine their application.

Example 7.9.

Let us reduce the following block diagram to canonical form.

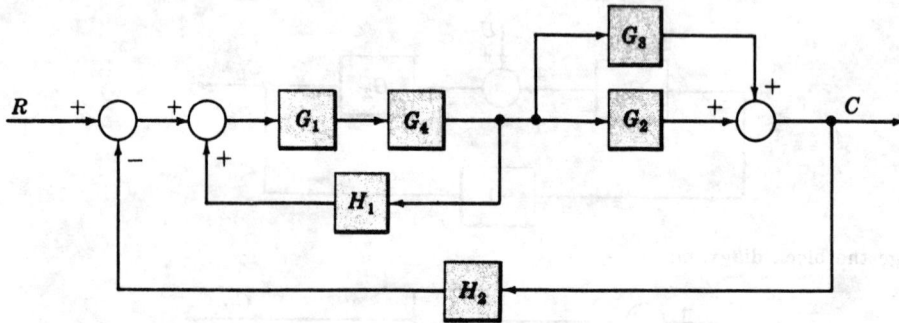

Step 1:

Step 2:

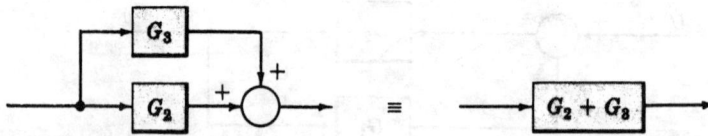

Step 3:

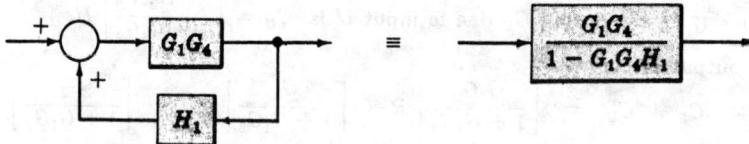

Step 4: Does not apply.

Step 5:

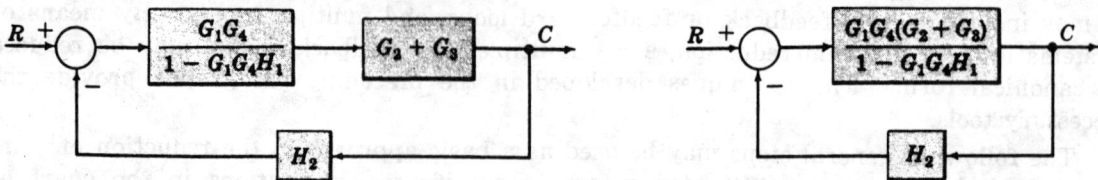

Step 6: Does not apply.

An occasional requirement of block diagram reduction is the isolation of a particular block in a feedback or feedforward loop. This may be desirable in order to more easily examine the effect of a particular block on the overall system.

Isolation of a block may generally be accomplished by applying the same reduction steps to the system, but usually in a different order. Also, the block to be isolated cannot be combined with any others.

Rearranging Summing Points (Transformation 6), and **Transformations 8, 9 and 11** are especially useful for isolating blocks.

Example 7.10.

Let us reduce the block diagram of Example 7.9, isolating block H_1.

Steps 1 and 2:

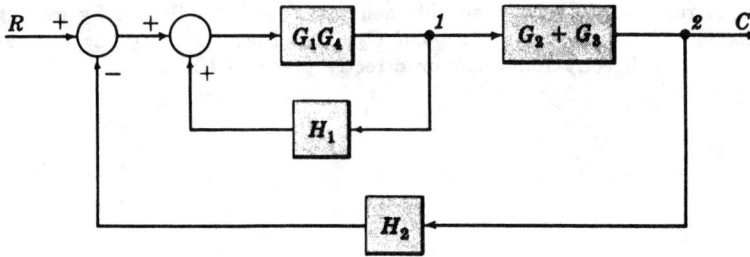

We do not apply Step 3 at this time, but go directly to Step 4, moving takeoff point *1* beyond block $G_2 + G_3$:

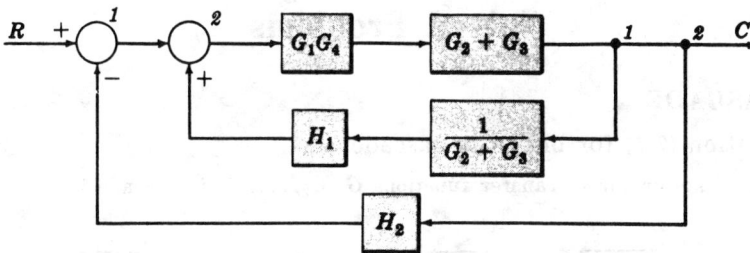

We may now rearrange summing points *1* and *2* and combine the cascade blocks in the forward loop using Transformation 6, then 1:

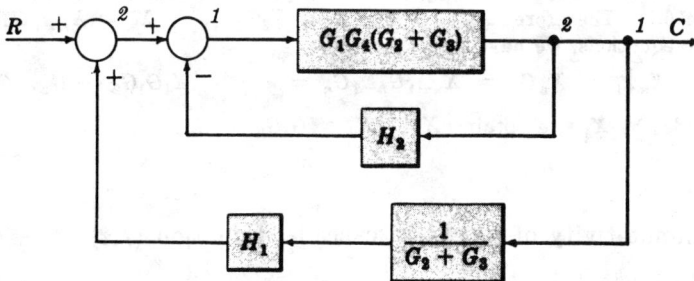

Step 3:

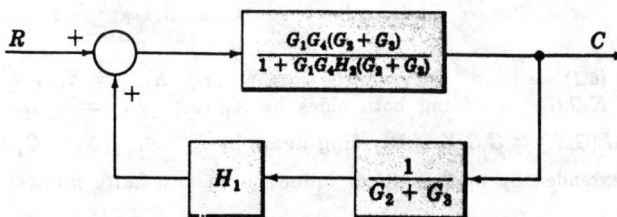

Finally, we apply Transformation 5 to remove $\dfrac{1}{G_2 + G_3}$ from the feedback loop:

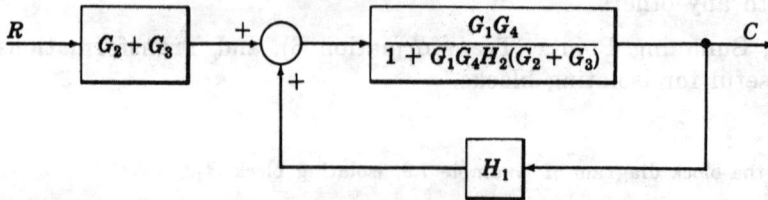

Note that the same result could have been obtained after applying Step 2 by moving takeoff point *2 ahead* of $G_2 + G_3$, instead of takeoff point *1 beyond* $G_2 + G_3$. Block $G_2 + G_3$ has the same effect on the control ratio C/R whether it directly follows R or directly precedes C.

Solved Problems

BLOCKS IN CASCADE

7.1. Prove Equation (*7.1*) for blocks in cascade.

The block diagram for n transfer functions $G_1, G_2, \ldots, G_n$ in cascade is

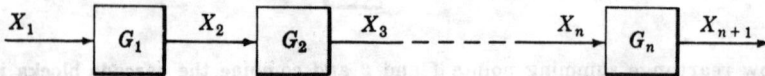

The output transform for any block is equal to the input transform multiplied by the transfer function (see Section 6.1). Therefore $X_2 = X_1 G_1$, $X_3 = X_2 G_2$, $\ldots$, $X_n = X_{n-1} G_{n-1}$, $X_{n+1} = X_n G_n$. Combining these equations, we have

$$X_{n+1} = X_n G_n = X_{n-1} G_{n-1} G_n = \cdots = X_1 G_1 G_2 \cdots G_{n-1} G_n$$

Dividing both sides by X_1, we obtain $X_{n+1}/X_1 = G_1 G_2 \cdots G_{n-1} G_n$.

7.2. Prove the commutativity of blocks in cascade, Equation (*7.2*).

Consider two blocks in cascade:

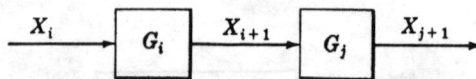

From Equation (*6.1*) we have $X_{i+1} = X_i G_i = G_i X_i$ and $X_{j+1} = X_{i+1} G_j = G_j X_{i+1}$. Therefore $X_{j+1} = (X_i G_i) G_j = X_i G_i G_j$. Dividing both sides by X_i, $X_{j+1}/X_i = G_i G_j$.

Also, $X_{j+1} = G_j(G_i X_i) = G_j G_i X_i$. Dividing again by X_i, $X_{j+1}/X_i = G_j G_i$. Thus $G_i G_j = G_j G_i$.

This result is extended by mathematical induction to any finite number of transfer functions (blocks) in cascade.

7.3. Find X_n/X_1 for each of the following systems:

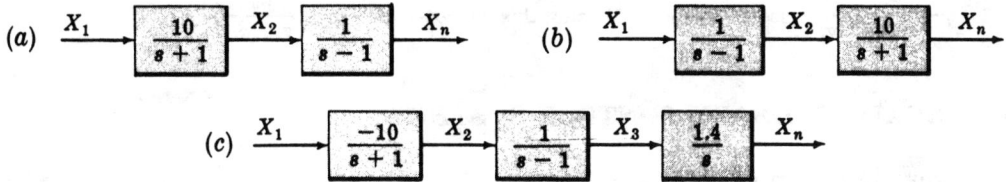

(a) One way to work this problem is to first write X_2 in terms of X_1: $X_2 = \left(\dfrac{10}{s+1}\right)X_1$. Then write X_n in terms of X_2: $X_n = \left(\dfrac{1}{s-1}\right)X_2 = \left(\dfrac{1}{s-1}\right)\left(\dfrac{10}{s+1}\right)X_1$. Multiplying out and dividing both sides by X_1, we have $X_n/X_1 = 10/(s^2-1)$.

    A shorter method is as follows. We know from Equation (7.1) that two blocks can be reduced to one by simply multiplying their transfer functions. Also, the transfer function of a single block is its output-to-input transform. Hence $X_n/X_1 = \left(\dfrac{1}{s-1}\right)\left(\dfrac{10}{s+1}\right) = 10/(s^2-1)$.

(b) This system has the same transfer function determined in Part (a) because multiplication of transfer functions is commutative.

(c) By Equation (7.1), we have $X_n/X_1 = \left(\dfrac{-10}{s+1}\right)\left(\dfrac{1}{s-1}\right)\left(\dfrac{1.4}{s}\right) = \dfrac{-14}{s(s^2-1)}$.

7.4. The transfer function of

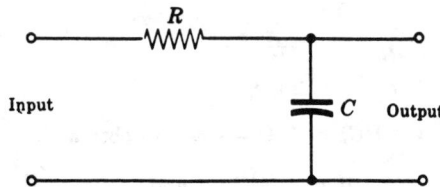

is $\dfrac{\omega_0}{s+\omega_0}$, where $\omega_0 = \dfrac{1}{RC}$. Is the transfer function of

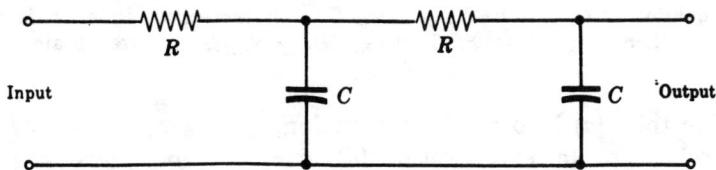

equal to $\left(\dfrac{\omega_0}{s+\omega_0}\right)^2$? Why?

No. If two identical networks are connected in series:

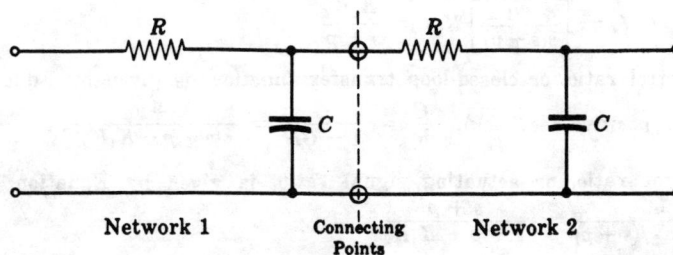

the second loads the first by drawing current from it. Therefore Equation (7.1) cannot be directly applied to the combined system. The correct transfer function for the connected networks is $\dfrac{\omega_0^2}{s^2 + 3\omega_0 s + \omega_0^2}$ (see Problem 6.16), and this is *not* equal to $\left(\dfrac{\omega_0}{s + \omega_0}\right)^2$.

CANONICAL FEEDBACK CONTROL SYSTEMS

7.5. Prove Equation (7.3), $\dfrac{C}{R} = \dfrac{G}{1 \pm GH}$.

The equations describing the canonical feedback system are taken directly from the adjoining block diagram. They are given by $E = R \mp B$, $B = HC$, and $C = GE$. Substituting one into the other, we have

$$C = G(R \mp B) = G(R \mp HC)$$
$$= GR \mp GHC = GR + (\mp GHC)$$

Subtracting ($\mp GHC$) from both sides, we obtain

$$C \pm GHC = GR \quad \text{or} \quad \frac{C}{R} = \frac{G}{1 \pm GH}.$$

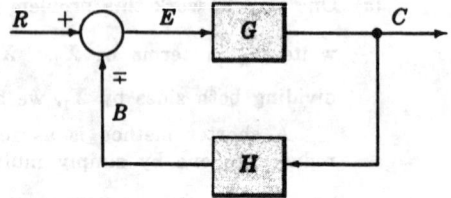

7.6. Prove Equation (7.4), $\dfrac{E}{R} = \dfrac{1}{1 \pm GH}$.

From the preceding problem, we have $E = R \mp B$, $B = HC$, and $C = GE$.

Then $E = R \mp HC = R \mp HGE$, $E \pm GHE = R$, and $\dfrac{E}{R} = \dfrac{1}{1 \pm GH}$.

7.7. Prove Equation (7.5), $\dfrac{B}{R} = \dfrac{GH}{1 \pm GH}$.

From $E = R \mp B$, $B = HC$, and $C = GE$, we obtain $B = HGE = HG(R \mp B) = GHR \mp GHB$.

Then $B \pm GHB = GHR$, $B = \dfrac{GHR}{1 \pm GH}$, and $\dfrac{B}{R} = \dfrac{GH}{1 \pm GH}$.

7.8. Prove Equation (7.6), $D_{GH} \pm N_{GH} = 0$.

By Equation (7.3), $C(1 \pm GH) = GR$. The characteristic equation is obtained by setting the input $R = 0$; then $1 \pm GH = 0$. Putting $GH \equiv N_{GH}/D_{GH}$, we obtain $D_{GH} \pm N_{GH} = 0$.

7.9. Determine the (a) loop transfer function, (b) control ratio, (c) error ratio, (d) primary feedback ratio, and (e) characteristic equation for the adjoining feedback control system in which K_1 and K_2 are constants.

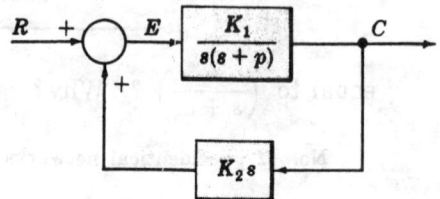

(a) The loop transfer function is equal to GH.

Hence $GH = \left[\dfrac{K_1}{s(s+p)}\right]K_2 s = \dfrac{K_1 K_2}{s + p}$.

(b) The control ratio, or closed-loop transfer function, is given by Equation (7.3) (with a minus sign for positive feedback): $\dfrac{C}{R} = \dfrac{G}{1 - GH} \doteq \dfrac{K_1}{s(s + p - K_1 K_2)}$.

(c) The error ratio, or actuating signal ratio, is given by Equation (7.4): $\dfrac{E}{R} = \dfrac{1}{1 - GH} = \dfrac{1}{1 - K_1 K_2/(s+p)} = \dfrac{s+p}{s + p - K_1 K_2}$.

(d) The primary feedback ratio is given by Equation (7.5): $\dfrac{B}{R} = \dfrac{GH}{1-GH} = \dfrac{K_1 K_2}{s + p - K_1 K_2}$.

(e) The characteristic equation is given by Equation (7.6): $1 - GH = 0, \quad s + p - K_1 K_2 = 0$.

BLOCK DIAGRAM TRANSFORMATIONS

7.10. Prove the equivalence of the block diagrams for Transformation 2 (Section 7.5).

The equation in the second column, $Y = P_1 X \pm P_2 X$, governs the construction of the block diagram in the third column, as shown. Rewrite this equation as $Y = (P_1 \pm P_2)X$. The equivalent block diagram in the last column is clearly the representation of this form of the equation; that is,

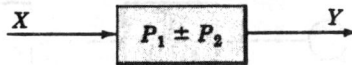

7.11. Repeat Problem 7.10 for Transformation 3.

Rewrite $Y = P_1 X \pm P_2 X$ as $Y = (P_1/P_2)P_2 X \pm P_2 X$. The block diagram for this form of the equation is clearly given by

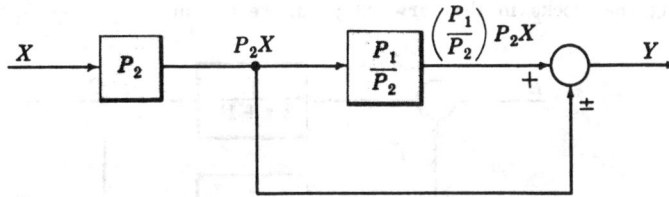

7.12. Repeat Problem 7.10 for Transformation 5.

We have $Y = P_1[X \mp P_2 Y] = P_1 P_2 \left[\left(\dfrac{1}{P_2} \right) X \mp Y \right]$. The block diagram for the latter form is given by

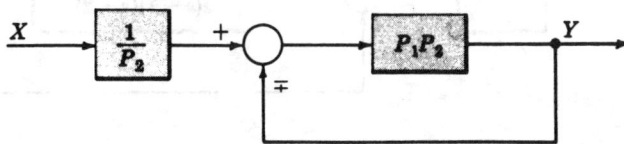

7.13. Repeat Problem 7.10 for Transformation 7.

We have $Z = PX \pm Y = P[X \pm (1/P)Y]$, which yields the block diagram

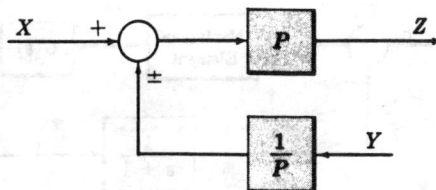

7.14. Repeat Problem 7.10 for Transformation 8.

We have $Z = P(X \pm Y) = PX \pm PY$, whose block diagram is clearly

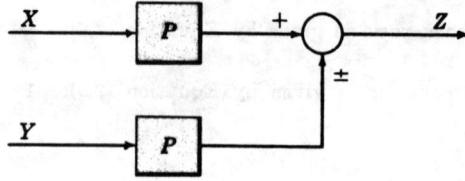

UNITY FEEDBACK SYSTEMS

7.15. Reduce the following block diagram to unity feedback form and find the system characteristic equation.

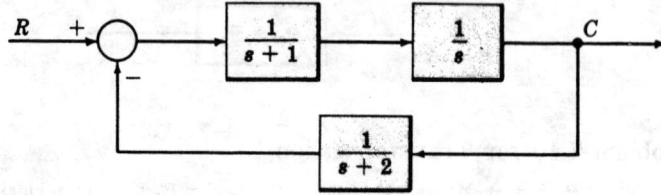

Combining the blocks in the forward path, we obtain

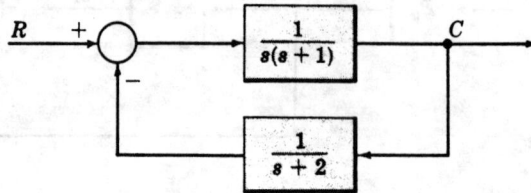

Applying Transformation 5, we have

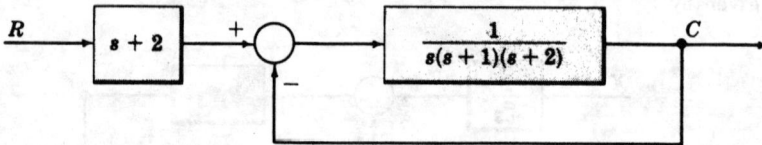

By Equation (7.7), the characteristic equation for this system is $s(s+1)(s+2)+1 = 0$ or $s^3 + 3s^2 + 2s + 1 = 0$.

7.16. Reduce the following system to a unity feedback system.

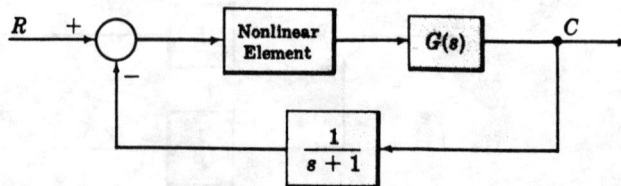

A nonlinear element in the forward path does not restrict the usefulness of Transformation 5. The linear block in the feedback path can be combined with the linear elements of the forward loop, resulting in the unity feedback system given by

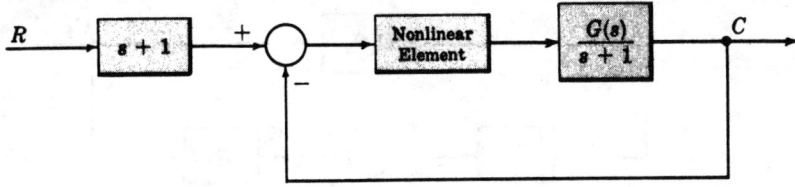

MULTIPLE INPUTS AND OUTPUTS

7.17. Determine the output C for the following system:

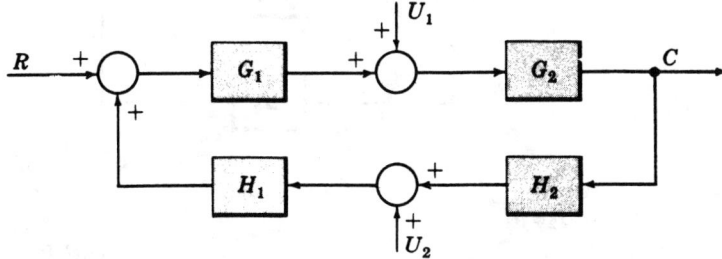

Let $U_1 = U_2 = 0$. After combining the cascaded blocks, the system is given by

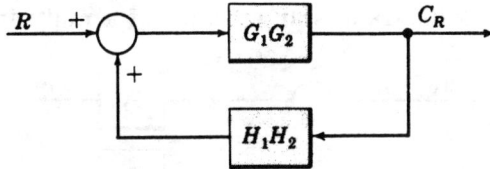

where C_R is the output due to R acting alone. Applying Equation (7.3) to this system,
$$C_R = \left[\frac{G_1 G_2}{1 - G_1 G_2 H_1 H_2} \right] R.$$

Now let $R = U_2 = 0$. The block diagram becomes

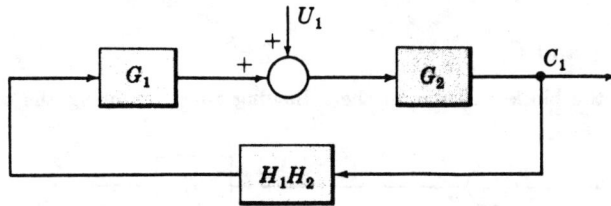

where C_1 is the response due to U_1 acting alone. Rearranging the blocks, we have

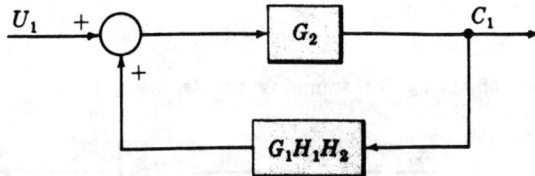

From Equation (7.3), we get $C_1 = \left[\dfrac{G_2}{1 - G_1 G_2 H_1 H_2} \right] U_1.$

Finally, let $R = U_1 = 0$. The block diagram becomes

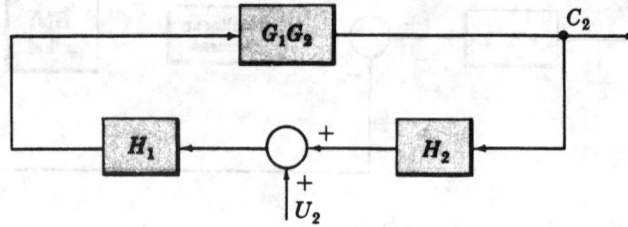

where C_2 is the response due to U_2 acting alone. Rearranging the blocks, we get

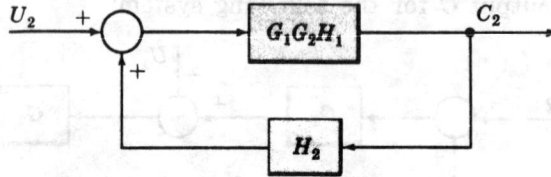

Hence $C_2 = \left[\dfrac{G_1 G_2 H_1}{1 - G_1 G_2 H_1 H_2} \right] U_2$.

By superposition, the total output is $C = C_R + C_1 + C_2 = \dfrac{G_1 G_2 R + G_2 U_1 + G_1 G_2 H_1 U_2}{1 - G_1 G_2 H_1 H_2}$.

7.18. The following block diagram is an example of a multi-input/multi-output system:

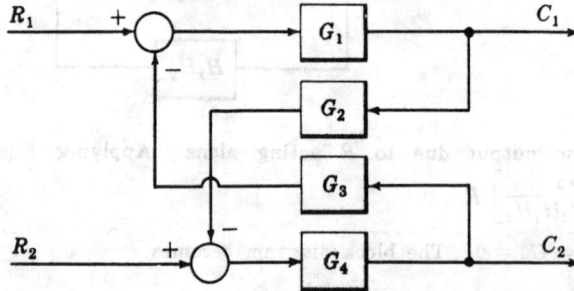

Determine C_1 and C_2.

First put the block diagram in the following form, ignoring the output C_2:

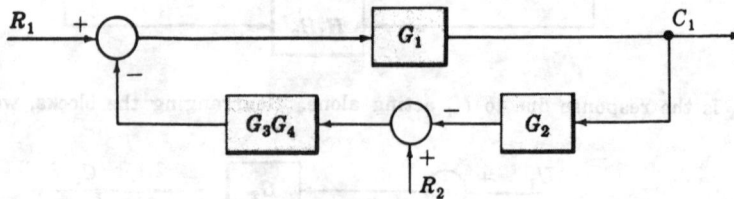

Letting $R_2 = 0$ and combining the summing points, we get

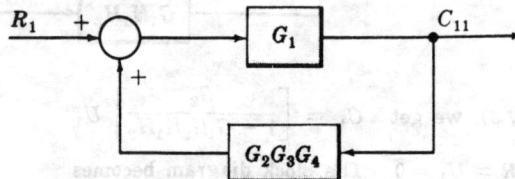

Hence C_{11}, the output at C_1 due to R_1 alone, is $C_{11} = \dfrac{G_1 R_1}{1 - G_1 G_2 G_3 G_4}$.   For $R_1 = 0$, we have

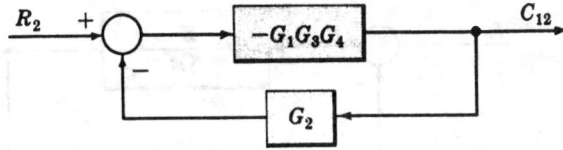

Hence $C_{12} = \dfrac{-G_1 G_3 G_4 R_2}{1 - G_1 G_2 G_3 G_4}$ is the output at C_1 due to R_2 alone.   Thus $C_1 = C_{11} + C_{12} = \dfrac{G_1 R_1 - G_1 G_3 G_4 R_2}{1 - G_1 G_2 G_3 G_4}$.

Now we reduce the original block diagram, ignoring output C_1.   First,

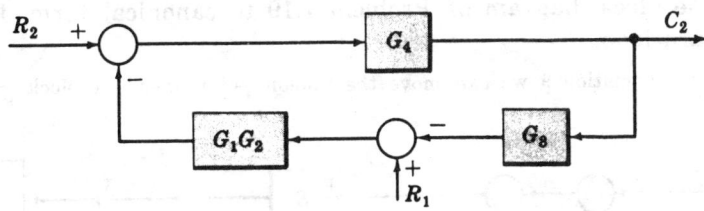

Then,

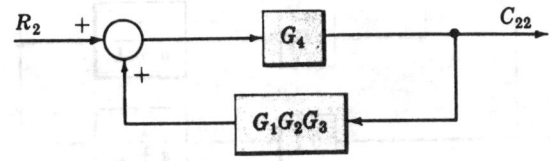

Hence $C_{22} = \dfrac{G_4 R_2}{1 - G_1 G_2 G_3 G_4}$.   Next, let $R_2 = 0$:

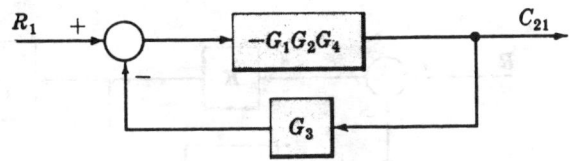

Hence $C_{21} = \dfrac{-G_1 G_2 G_4 R_1}{1 - G_1 G_2 G_3 G_4}$.   Finally, $C_2 = C_{22} + C_{21} = \dfrac{G_4 R_2 - G_1 G_2 G_4 R_1}{1 - G_1 G_2 G_3 G_4}$.

BLOCK DIAGRAM REDUCTION

7.19.   Reduce the following block diagram to canonical form, and find the output transform C. K is a constant.

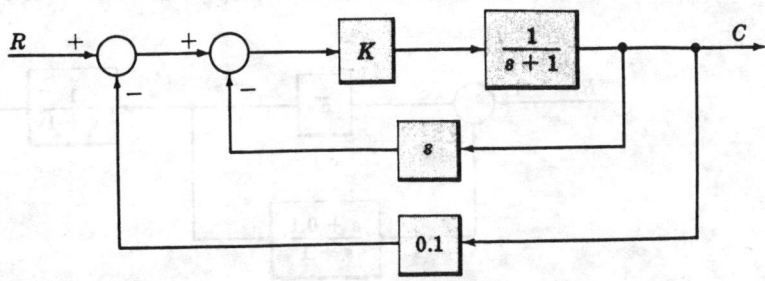

First we combine the cascade blocks of the forward path and apply Transformation 4 to the innermost feedback loop:

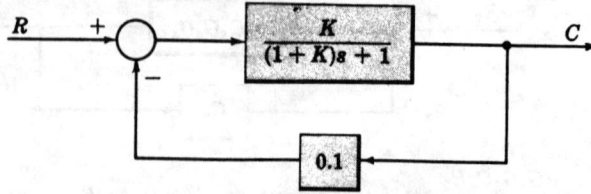

Equation (7.3) or the reapplication of Transformation 4 yields $C = \dfrac{KR}{(1 + K)s + (1 + 0.1K)}$.

7.20. Reduce the block diagram of Problem 7.19 to canonical form, isolating block K in the forward loop.

By Transformation 9 we can move the takeoff point ahead of block $\dfrac{1}{s + 1}$:

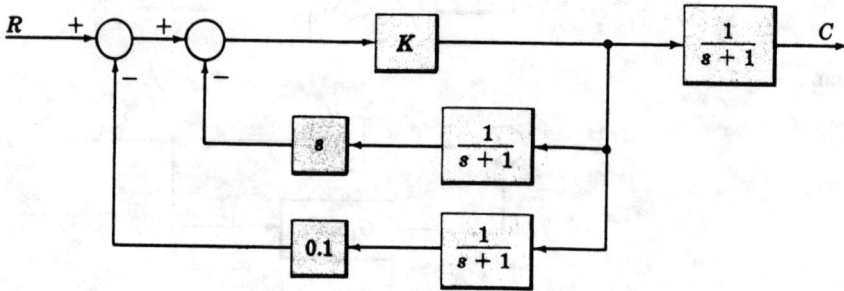

Applying Transformations 1 and 6b, we get

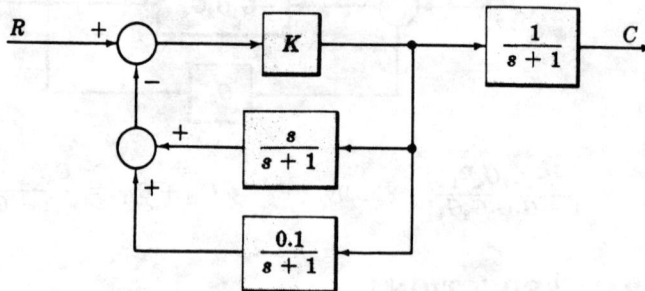

Now we can apply Transformation 2 to the feedback loops, resulting in the final form:

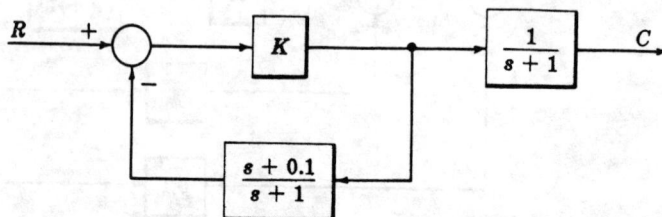

7.21. Reduce the following block diagram to open-loop form.

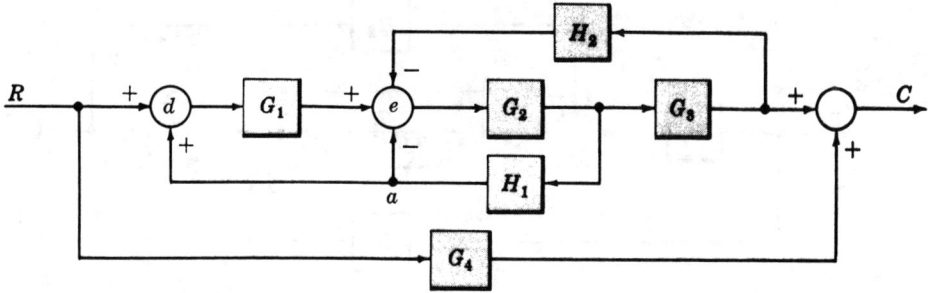

First move the leftmost summing point beyond G_1 (Transformation 8):

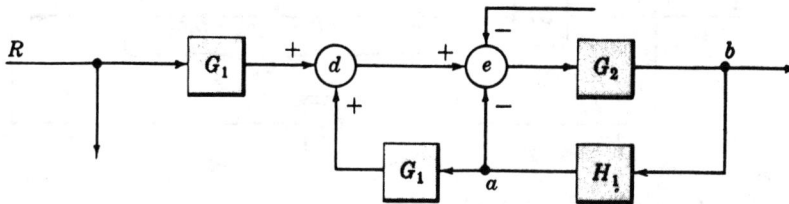

Next move takeoff point a beyond G_1:

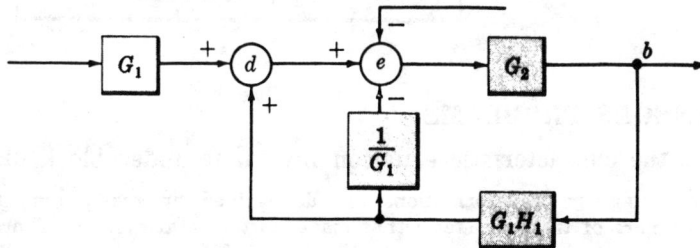

Now use Transformation 6b, and then 2, to combine the two lower feedback loops (from G_1H_1) entering d and e:

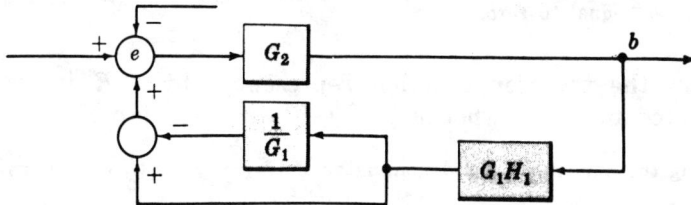

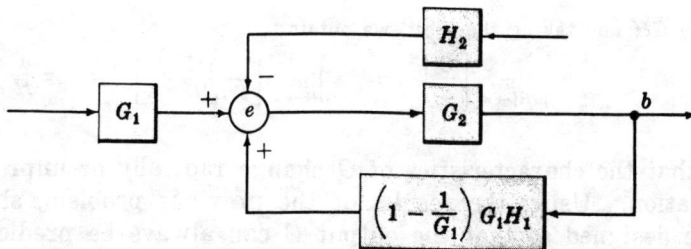

Applying Transformation 4 to this inner loop, the system becomes

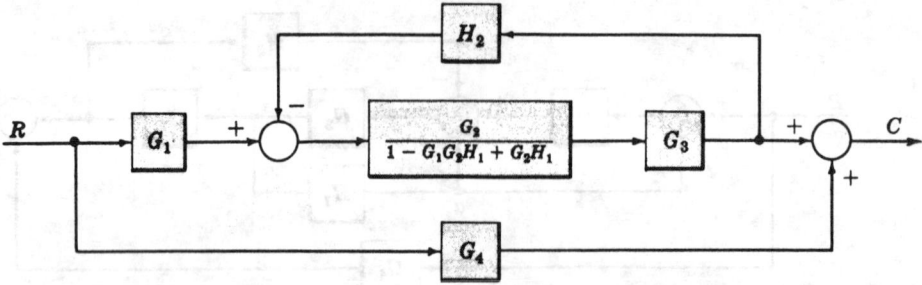

Again, applying Transformation 4 to the remaining feedback loop yields

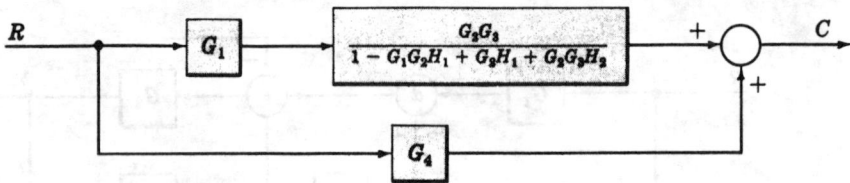

Finally, Transformations 1 and 2 give

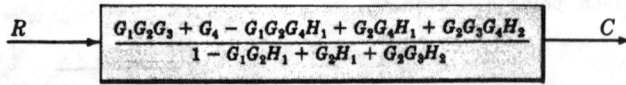

MISCELLANEOUS PROBLEMS

7.22. Why is the characteristic equation invariant under block diagram transformation?

Block diagram transformations are determined by *rearranging* the input-output equations of one or more of the subsystems that make up the total system. Therefore the final transformed system is governed by the same equations, probably arranged in a different manner than those for the original system.

Now, the characteristic equation is determined from the denominator of the overall system transfer function set equal to zero. Factoring or other rearrangement of the numerator and denominator of the system transfer function clearly does not change it, nor does it alter its denominator set equal to zero.

7.23. Prove that the transfer function represented by C/R in Equation (7.3) can be approximated by $\pm 1/H$ when $|G|$ or $|GH|$ are very large.

Dividing the numerator and denominator of $\dfrac{G}{1 \pm GH}$ by G, we get $\dfrac{1}{1/G \pm H}$.   Then

$$\lim_{|G| \to \infty} \left[\frac{C}{R} \right] \;=\; \lim_{|G| \to \infty} \left[\frac{1}{1/G \pm H} \right] \;=\; \pm \frac{1}{H}$$

Dividing by GH and taking the limit, we obtain

$$\lim_{|GH| \to \infty} \left[\frac{C}{R} \right] \;=\; \lim_{|GH| \to \infty} \left[\frac{1/H}{1/GH \pm 1} \right] \;=\; \pm \frac{1}{H}$$

7.24. Assume that the characteristics of G change radically or unpredictably during system operation. Using the results of the previous problem, show how the system should be designed so that the output C can always be predicted reasonably well.

In Problem 7.23 we found that $\lim\limits_{|GH| \to \infty} \left[\dfrac{C}{R} \right] = \pm \dfrac{1}{H}$. Thus $C \to \pm \dfrac{R}{H}$ as $|GH| \to \infty$, or C is independent of G for large $|GH|$. Hence the system should be designed so that $|GH| \gg 1$.

7.25. Determine the transfer function of the system in the following block diagram. *Then* let $H_1 = 1/G_1$ and $H_2 = 1/G_2$.

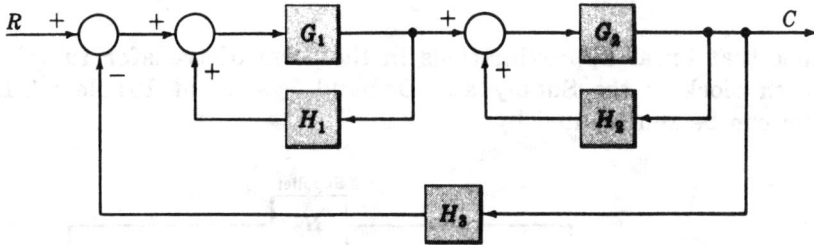

Reducing the inner loops, we have

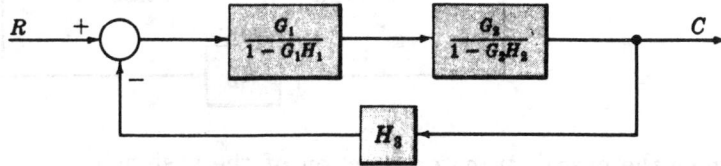

Applying Transformation 4 again, we obtain

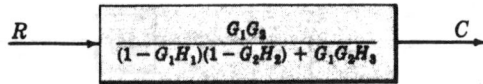

Now put $H_1 = 1/G_1$ and $H_2 = 1/G_2$. This yields $\dfrac{C}{R} = \dfrac{G_1 G_2}{(1-1)(1-1) + G_1 G_2 H_3} = \dfrac{1}{H_3}$.

7.26. Show that

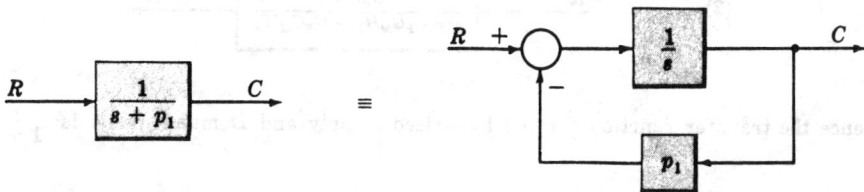

From the open-loop diagram, we have $C = \dfrac{R}{s + p_1}$. Rearranging, $(s + p_1)C = R$ and $C = \dfrac{1}{s}(R - p_1 C)$. The closed-loop diagram follows from this equation.

7.27. Prove that

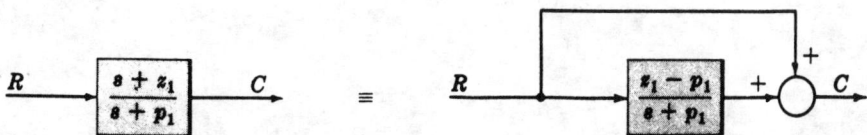

This problem illustrates how a zero may be removed from a block.

From the closed-loop diagram, $C = R + (z_1 - p_1)R/(s + p_1)$. Rearranging,

$$C = \left(1 + \frac{z_1 - p_1}{s + p_1}\right) R = \left(\frac{s + p_1 + z_1 - p_1}{s + p_1}\right) R = \left(\frac{s + z_1}{s + p_1}\right) R$$

This mathematical equivalence clearly proves the equivalence of the block diagrams.

7.28. Assume that linear approximations in the form of transfer functions are available for each block of the Supply and Demand System of Problem 2.13, and that the system can be represented by

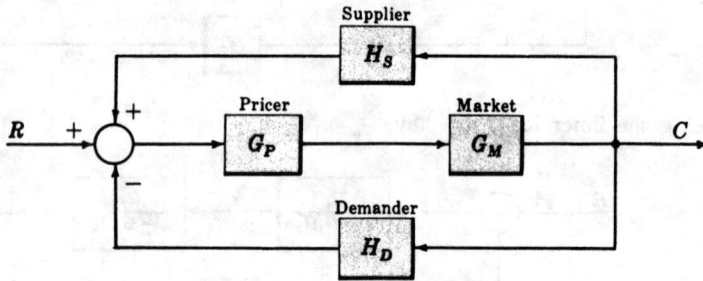

Determine the overall transfer function of the system.

Block diagram Transformation 4, applied twice to this system, gives

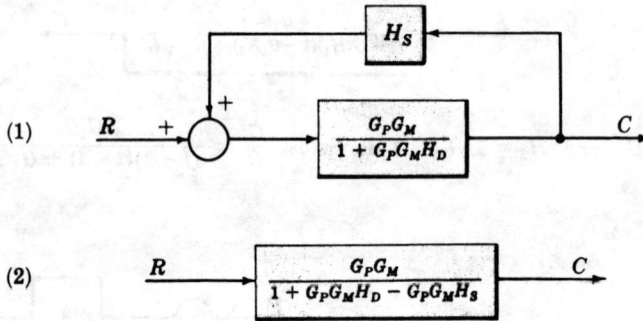

(1)

(2)

Hence the transfer function for the linearized Supply and Demand model is $\dfrac{G_P G_M}{1 + G_P G_M (H_D - H_S)}$.

Supplementary Problems

7.29. Determine C/R for each system.

(a)

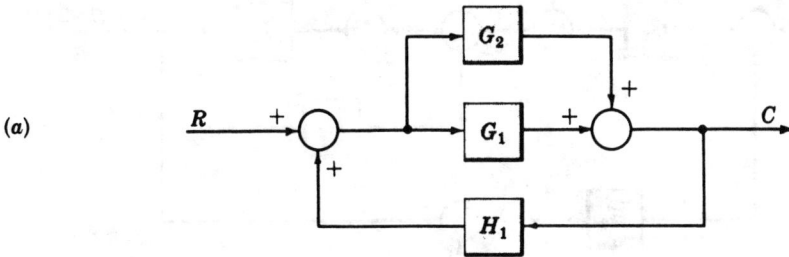

(b)

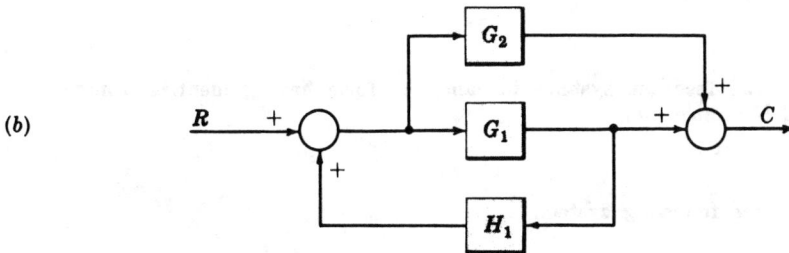

(c)

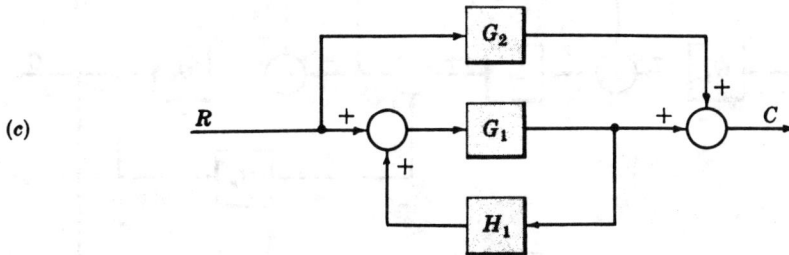

7.30. Consider the blood pressure regulator described in Problem 2.14. Assume the vasomotor center (VMC) can be described by a linear transfer function $G_{11}(s)$, and the baroreceptors by the transfer function $k_1 s + k_2$ (see Problem 6.29). Transform the block diagram into its simplest, unity feedback form.

7.31. Reduce the following block diagram to canonical form.

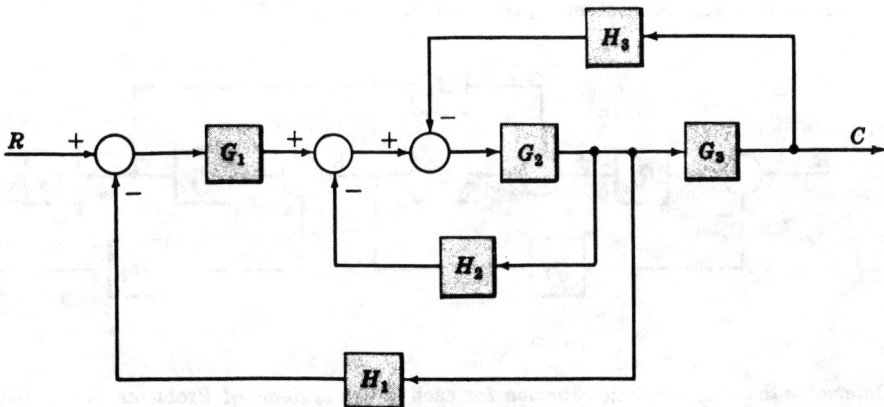

7.32.   Determine C for the following system.

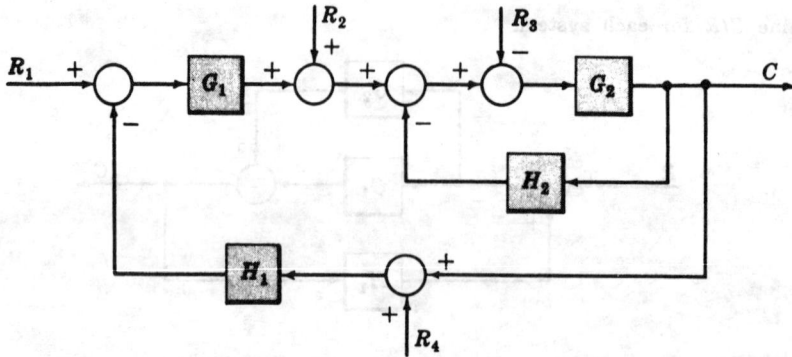

7.33.   Give an example of two feedback systems in canonical form having identical control ratios C/R but different G and H components.

7.34.   Determine C/R_2 for the following system.

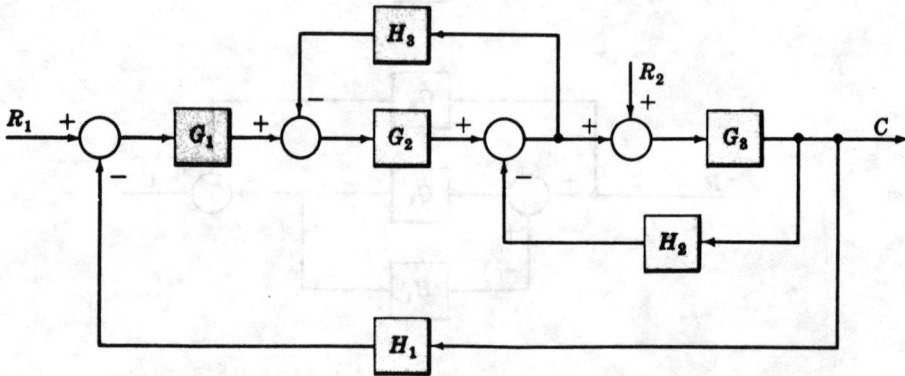

7.35.   Determine the complete output C, with both inputs R_1 and R_2 acting simultaneously, for the system given in the preceding problem.

7.36.   Determine C/R for the system represented by

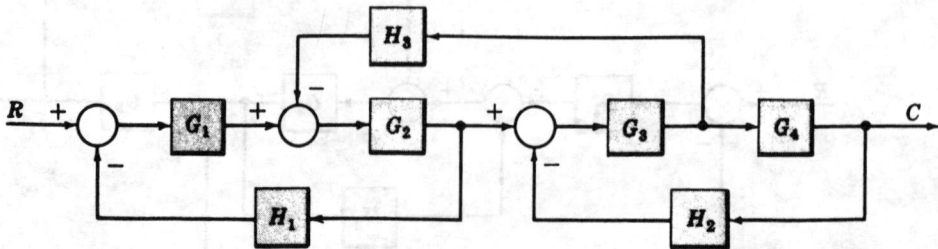

7.37.   Determine the characteristic equation for each of the systems of Problems (a) 7.32, (b) 7.35, (c) 7.36.

Answers to Supplementary Problems

7.29. See Problem 8.15.

7.30.

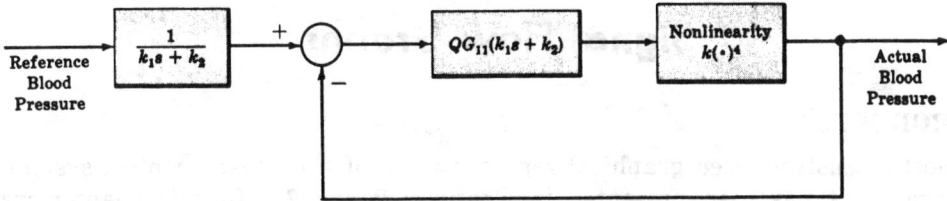

7.31.

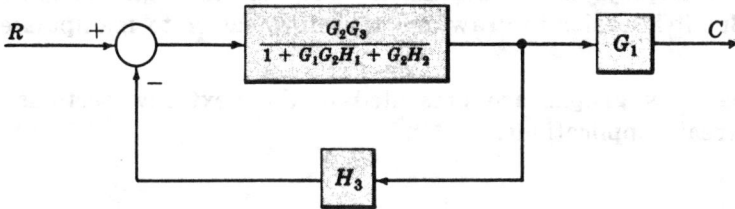

7.32. $C = \dfrac{G_1 G_2 R_1 + G_2 R_2 - G_2 R_3 - G_1 G_2 H_1 R_4}{1 + G_2 H_2 + G_1 G_2 H_1}$

7.34. $\dfrac{C}{R_2} = \dfrac{G_3(1 + G_2 H_3)}{1 + G_3 H_2 + G_2 H_3 + G_1 G_2 G_3 H_1}$

7.35. $C = \dfrac{G_1 G_2 G_3 R_1 + G_3(1 + G_2 H_3) R_2}{1 + G_3 H_2 + G_2 H_3 + G_1 G_2 G_3 H_1}$

7.36. $\dfrac{C}{R} = \dfrac{G_1 G_2 G_3 G_4}{(1 + G_1 G_2 H_1)(1 + G_3 G_4 H_2) + G_2 G_3 H_3}$

7.37. (a) $1 + G_2 H_2 + G_1 G_2 H_1 = 0$

      (b) $1 + G_3 H_2 + G_2 H_3 + G_1 G_2 G_3 H_1 = 0$

      (c) $(1 + G_1 G_2 H_1)(1 + G_3 G_4 H_2) + G_2 G_3 H_3 = 0$

Chapter 8

Signal Flow Graphs

8.1 INTRODUCTION

The most extensively used graphical representation of a feedback control system is the block diagram. It has been presented in Chapters 2 and 7. In this chapter we shall consider another model, the signal flow graph.

A **signal flow graph** is a pictorial representation of the simultaneous equations describing a system. It graphically displays the transmission of signals through the system, as does the block diagram. But it is easier to draw and therefore easier to manipulate than the block diagram.

The properties of signal flow graphs are presented in the next few sections. The remainder of the chapter treats applications.

8.2 FUNDAMENTALS OF SIGNAL FLOW GRAPHS

Let us first consider the simple equation

$$X_i = A_{ij}X_j \tag{8.1}$$

The variables X_i and X_j can be functions of time, complex frequency, or any other quantity. They may even be constants, which are "variables" in the mathematical sense.

For signal flow graphs, A_{ij} is a mathematical operator mapping X_j into X_i, and is called the **transmission function.** For example, A_{ij} may be a constant, in which case X_i is a constant times X_j in Equation (8.1); if X_i and X_j are functions of s, A_{ij} may be a transfer function $A_{ij}(s)$.

The signal flow graph for Equation (8.1) is

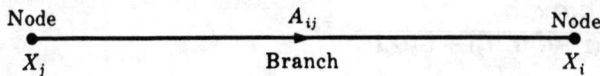

This is the simplest form of a signal flow graph. Note that the variables X_i and X_j are represented by a small dot called a **node,** and the transmission function A_{ij} is represented by a line with an arrow, called a **branch.**

Every variable in a signal flow graph is designated by a node, and every transmission function by a branch. Branches are always unidirectional. The arrow denotes the direction of signal flow.

Example 8.1.

Ohm's Law states that $E = RI$, where E is a voltage, I a current, and R a resistance. The signal flow graph for this equation is

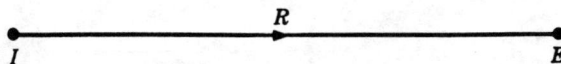

8.3 SIGNAL FLOW GRAPH ALGEBRA

1. The Addition Rule

The value of the variable designated by a node is equal to the **sum of all signals** entering the node. In other words, the equation

$$X_i = \sum_{j=1}^{n} A_{ij} X_j$$

is represented by

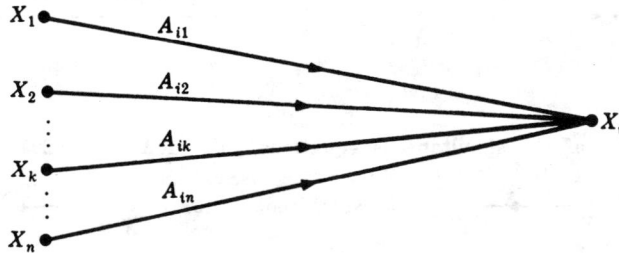

Example 8.2.

The signal flow graph for the equation of a line in rectangular coordinates, $Y = mX + b$, is

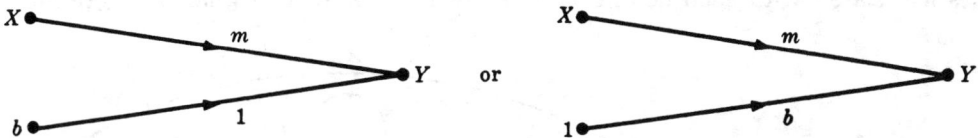

or

Since b, the Y-axis intercept, is a constant it may represent a node (variable) or a **transmission function**.

2. The Transmission Rule

The value of the variable designated by a node is **transmitted** on every branch leaving that node. In other words, the equation

$$X_i = A_{ik} X_k, \qquad i = 1, 2, \ldots, n, \ \ k \text{ fixed}$$

is represented by

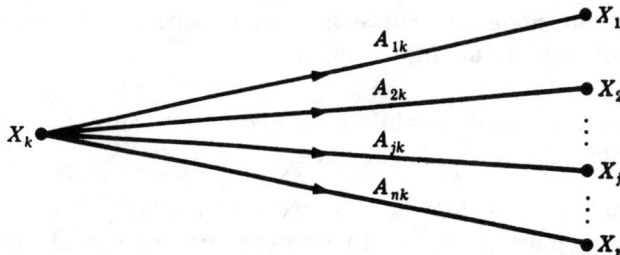

Example 8.3.

The signal flow graph of the simultaneous equations $Y = 3X$, $Z = -4X$ is

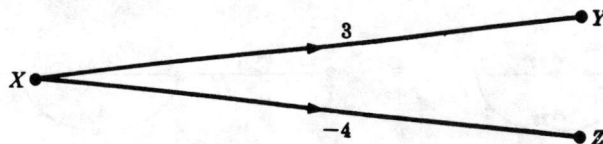

3. The Multiplication Rule

A cascaded (series) connection of $n-1$ branches with transmission functions $A_{21}, A_{32}, A_{43}, \ldots, A_{n(n-1)}$ can be replaced by a single branch with a new transmission function equal to the product of the old ones. That is,

$$X_n = A_{21} \cdot A_{32} \cdot A_{43} \cdots A_{n(n-1)} \cdot X_1$$

The signal flow graph equivalence is represented by

Example 8.4.

The signal flow graph of the simultaneous equations $Y = 10X$, $Z = -20Y$ is

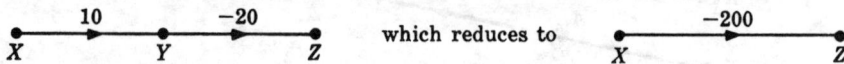

which reduces to

8.4 DEFINITIONS

The following terminology is frequently used in signal flow graph theory. The examples associated with each definition refer to the following signal flow graph:

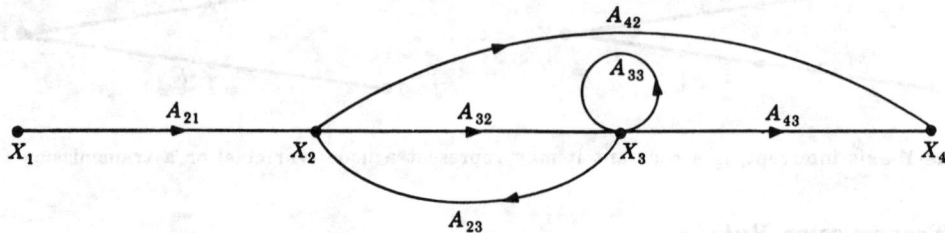

Definition 8.1: A **path** is a continuous, unidirectional succession of branches along which no node is passed more than once. For example, X_1 to X_2 to X_3 to X_4, X_2 to X_3 and back to X_2, and X_1 to X_2 to X_4 are paths.

Definition 8.2: An **input node** or **source** is a node with only outgoing branches. For example, X_1 is an input node.

Definition 8.3: An **output node** or **sink** is a node with only incoming branches. For example, X_4 is an output node.

Very often a variable in a system is a function of the output variable. The canonical feedback system is an obvious example. In this case, if the signal flow graph were to be drawn directly from the equations, the "output node" would require an outgoing branch, contrary to the definition. This problem may be remedied by adding a branch with a transmission function of unity entering a "dummy" node. For example, the following two graphs are equivalent, and Y_4 is an output node. Note that $Y_4 = Y_3$.

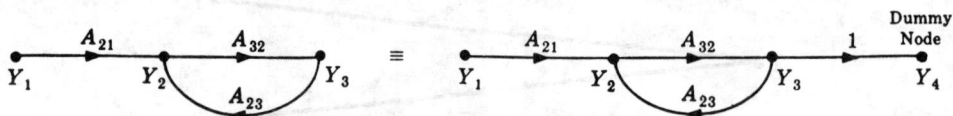

Definition 8.4: A **forward path** is a path from the input node to the output node. For example, X_1 to X_2 to X_3 to X_4, and X_1 to X_2 to X_4 are forward paths.

Definition 8.5: A **feedback path** or **feedback loop** is a path which originates and terminates on the same node. For example, X_2 to X_3 and back to X_2 is a feedback path.

Definition 8.6: A **self-loop** is a feedback loop consisting of a single branch. For example, A_{33} is a self-loop.

Definition 8.7: The **gain** of a branch is the transmission function of that branch when the transmission function is a multiplicative operator. For example, A_{33} is the gain of the self-loop if A_{33} is a constant or transfer function.

Definition 8.8: The **path gain** is the product of the branch gains encountered in traversing a path. For example, the path gain of the forward path from X_1 to X_2 to X_3 to X_4 is $A_{21}A_{32}A_{43}$.

Definition 8.9: The **loop gain** is the product of the branch gains of the loop. For example, the loop gain of the feedback loop from X_2 to X_3 and back to X_2 is $A_{32}A_{23}$.

8.5 CONSTRUCTION OF SIGNAL FLOW GRAPHS

The signal flow graph of a linear feedback control system whose components are specified by non-interacting transfer functions can be constructed by direct reference to the block diagram of the system. Each variable of the block diagram becomes a node and each block becomes a branch.

Example 8.5.

The block diagram of the canonical feedback control system is given by

The signal flow graph is easily constructed from this diagram:

Note that the $-$ or $+$ sign of the summing point is associated with H.

The signal flow graph of a system described by a set of simultaneous equations can be constructed in the following general manner.

1. Write the system equations in the form

$$X_1 = A_{11}X_1 + A_{12}X_2 + \cdots + A_{1n}X_n$$
$$X_2 = A_{21}X_1 + A_{22}X_2 + \cdots + A_{2n}X_n$$
$$\cdots\cdots\cdots\cdots\cdots\cdots\cdots\cdots\cdots\cdots\cdots\cdots\cdots$$
$$X_m = A_{m1}X_1 + A_{m2}X_2 + \cdots + A_{mn}X_n$$

An equation for X_1 is not required if X_1 is an input node.

2. Arrange the m or n (whichever is larger) nodes from left to right. The nodes may be rearranged if the required loops later appear too cumbersome.

3. Connect the nodes by the appropriate branches A_{11}, A_{12}, etc.

4. If the desired output node has outgoing branches, add a dummy node and a unity-gain branch.

5. Rearrange the nodes and/or loops in the graph to achieve maximum pictorial clarity.

Example 8.6.

Let us construct a signal flow graph for the simple resistance network given by

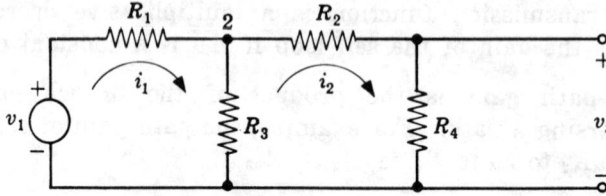

There are five variables, $v_1, v_2, v_3, i_1,$ and i_2. v_1 is known. We can write four independent equations from Kirchhoff's voltage and current laws. Proceeding from left to right in the schematic, we have

$$i_1 = \left(\frac{1}{R_1}\right)v_1 - \left(\frac{1}{R_1}\right)v_2, \quad v_2 = R_3 i_1 - R_3 i_2, \quad i_2 = \left(\frac{1}{R_2}\right)v_2 - \left(\frac{1}{R_2}\right)v_3, \quad v_3 = R_4 i_2$$

Laying out the five nodes in the same order with v_1 as an input node, and connecting the nodes with the appropriate branches, we get

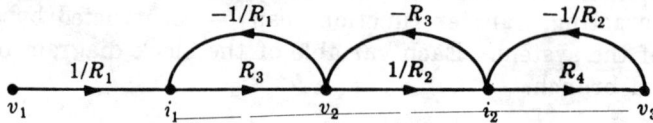

If we wish to consider v_3 as an output node, we must add a unity-gain branch and another node, yielding

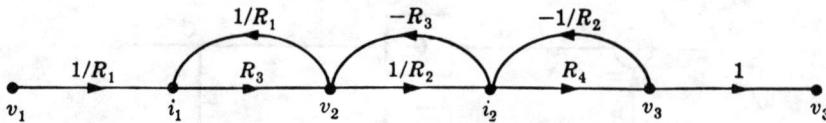

No rearrangement of the nodes is necessary. We have one forward path and three feedback loops clearly in evidence.

Note that signal flow graph representations of equations are not unique. For example, the addition of a unity gain branch followed by a dummy node changes the graph, but not the equations it represents.

8.6 THE GENERAL INPUT-OUTPUT GAIN FORMULA

We found in Chapter 7 that we can reduce complicated block diagrams to canonical form, from which the control ratio is easily written:

$$\frac{C}{R} = \frac{G}{1 \pm GH}$$

It is possible to simplify signal flow graphs in a manner similar to that of block diagram reduction. But it is also possible, and much less time-consuming, to write down the input-output relationship *by inspection* from the original signal flow graph. This can be accomplished using the formula presented below. This formula can also be applied directly to block diagrams, but the signal flow graph representation is easier to read — especially when the block diagram is very complicated.

Let us denote the ratio of the input variable to the output variable by T. For linear feedback control systems, $T = C/R$. For the general signal flow graph presented in preceding paragraphs $T = X_n/X_1$, where X_n is the output and X_1 is the input.

The general formula for any signal flow graph is

$$T = \frac{\sum_i P_i \Delta_i}{\Delta} \qquad (8.2)$$

where P_i = the ith forward path gain

P_{jk} = jth possible product of k non-touching loop gains

$$\Delta = 1 - (-1)^{k+1} \sum_k \sum_j P_{jk}$$

$$= 1 - \sum_j P_{j1} + \sum_j P_{j2} - \sum_j P_{j3} + \cdots$$

$\qquad$ = 1 − (sum of all loop gains) + (sum of all gain-products of 2 non-touching loops) − (sum of all gain-products of 3 non-touching loops) + $\cdots$

Δ_i = Δ evaluated with all loops touching P_i eliminated.

Two loops, paths, or a loop and a path are said to be **non-touching** if they have no nodes in common.

Δ is called the **signal flow graph determinant** or **characteristic function**, since $\Delta = 0$ is the system characteristic equation.

The application of Equation (8.2) is considerably more straightforward than it appears. The following examples illustrate this point.

Example 8.7.

Let us first apply Equation (8.2) to the signal flow graph of the canonical feedback system:

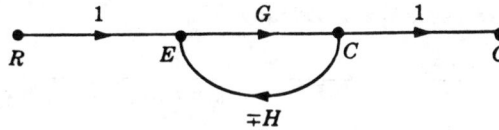

There is only one forward path; hence

$$P_1 = G$$
$$P_2 = P_3 = \cdots = 0$$

There is only one (feedback) loop. Hence

$$P_{11} = \mp GH$$
$$P_{jk} = 0, \quad j \neq 1, \quad k \neq 1$$

Then $\qquad\qquad$ $\Delta = 1 - P_{11} = 1 \pm GH$ $\quad$ and $\quad$ $\Delta_1 = 1 - 0 = 1$

Finally, $\qquad\qquad\qquad$ $T = \dfrac{C}{R} = \dfrac{P_1 \Delta_1}{\Delta} = \dfrac{G}{1 \pm GH}$

Example 8.8.

The signal flow graph of the resistance network of Example 8.6 was shown to be

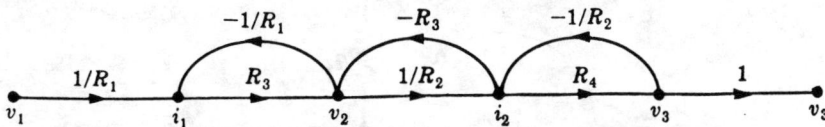

Let us apply Equation (8.2) to this graph and determine the voltage gain $T = v_3/v_1$ of the resistance network.

There is one forward path:

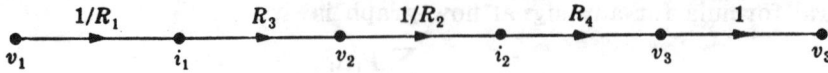

Hence the forward path gain is

$$P_1 = \frac{R_3 R_4}{R_1 R_2}$$

There are three feedback loops:

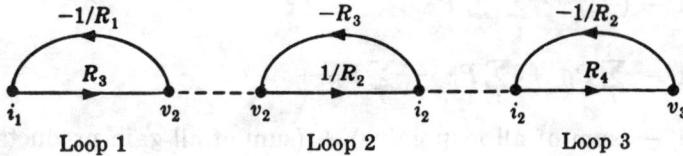

Loop 1   Loop 2   Loop 3

Hence the loop gains are

$$P_{11} = -\frac{R_3}{R_1}, \quad P_{21} = -\frac{R_3}{R_2}, \quad P_{31} = -\frac{R_4}{R_2}$$

There are two non-touching loops, loops one and three. Hence

$$P_{12} = \text{gain-product of the only two non-touching loops} = P_{11} \cdot P_{31} = \frac{R_3 R_4}{R_1 R_2}$$

There are no three loops that do not touch. Therefore

$$\Delta = 1 - (P_{11} + P_{21} + P_{31}) + P_{12} = 1 + \frac{R_3}{R_1} + \frac{R_3}{R_2} + \frac{R_4}{R_2} + \frac{R_3 R_4}{R_1 R_2}$$

$$= \frac{R_1 R_2 + R_1 R_3 + R_1 R_4 + R_2 R_3 + R_3 R_4}{R_1 R_2}$$

Since all loops touch the forward path, $\Delta_1 = 1$. Finally,

$$\frac{v_3}{v_1} = \frac{P_1 \Delta_1}{\Delta} = \frac{R_3 R_4}{R_1 R_2 + R_1 R_3 + R_1 R_4 + R_2 R_3 + R_3 R_4}$$

8.7 TRANSFER FUNCTION COMPUTATION OF CASCADED COMPONENTS

Loading effects of interacting components require little special attention using signal flow graphs. Simply combine the graphs of the components at their normal joining points (output node of one to the input node of another), account for loading by adding new loops at the joined nodes, and compute the overall gain using Equation (8.2). This procedure is best illustrated by example.

Example 8.9.

Assume that two identical resistance networks are to be cascaded and used as the control elements in the forward loop of a control system. The networks are simple voltage dividers of the form

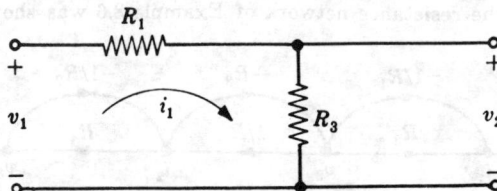

Two independent equations for this network are

$$i_1 = \left(\frac{1}{R_1}\right)v_1 - \left(\frac{1}{R_1}\right)v_2 \quad \text{and} \quad v_2 = R_3 i_1$$

The signal flow graph is easily drawn:

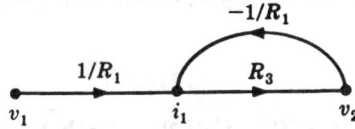

The gain of this network is, by inspection, equal to

$$\frac{v_2}{v_1} = \frac{R_3}{R_1 + R_3}$$

If we were to ignore loading, the overall gain of two cascaded networks would simply be determined by multiplying the individual gains:

$$\left(\frac{v_2}{v_1}\right)^2 = \frac{R_3^2}{R_1^2 + R_3^2 + 2R_1 R_3}$$

This answer is incorrect. We prove this in the following manner. When the two identical networks are cascaded, we note that the result is equivalent to the network of Example 8.6, with $R_2 = R_1$ and $R_4 = R_3$:

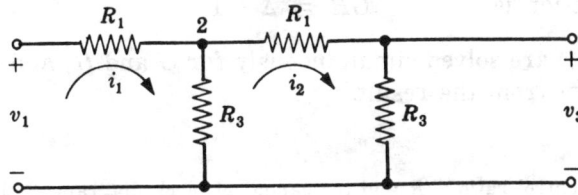

The signal flow graph of this network was also determined in Example 8.6:

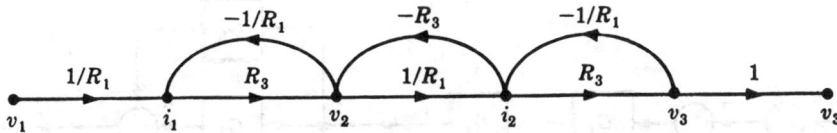

We observe that the feedback branch $-R_3$ in the above graph does not appear in the signal flow graph of the cascaded signal flow graphs of the individual networks connected from node v_2 to v_1':

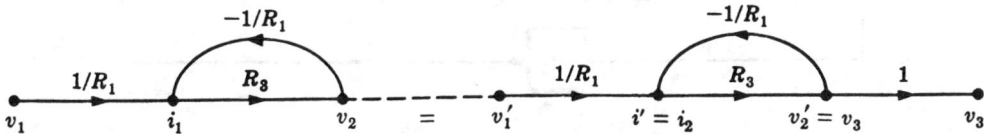

This means that, as a result of connecting the two networks, the second one loads the first, changing the equation for v_2 from

$$v_2 = R_3 i_1 \quad \text{to} \quad v_2 = R_3 i_1 - R_3 i_2$$

This result could also have been obtained by directly writing the equations for the combined networks. In this case, only the equation for v_2 would have changed form.

The gain of the combined networks was determined in Example 8.8 as

$$\frac{v_3}{v_1} = \frac{R_3^2}{R_1^2 + R_3^2 + 3R_1 R_3}$$

when R_2 is set equal to R_1 and R_4 is set equal to R_3. We observe that

$$\left(\frac{v_2}{v_1}\right)^2 = \frac{R_3^2}{R_1^2 + R_3^2 + 2R_1R_3} \neq \frac{v_3}{v_1}$$

It is good general practice to calculate the gain of cascaded networks directly from the *combined* signal flow graph. Most practical control system components load each other when connected in series.

8.8 BLOCK DIAGRAM REDUCTION USING SIGNAL FLOW GRAPHS AND THE GENERAL INPUT-OUTPUT GAIN FORMULA

Often, the easiest way to determine the control ratio of a complicated block diagram is to translate the block diagram into a signal flow graph and apply Equation (8.2). Takeoff points and summing points must be separated by a unity-gain branch in the signal flow graph when using Equation (8.2).

If the elements G and H of a canonical feedback representation are desired, Equation (8.2) also provides this information. The direct transfer function is

$$G = \sum_i P_i \Delta_i \qquad (8.3)$$

The loop transfer function is
$$GH = \Delta - 1 \qquad (8.4)$$

Equations (8.3) and (8.4) are solved simultaneously for G and H, and the canonical feedback control system is drawn from the result.

Example 8.10.

Let us determine the control ratio C/R and the canonical block diagram of the feedback control system of Example 7.9:

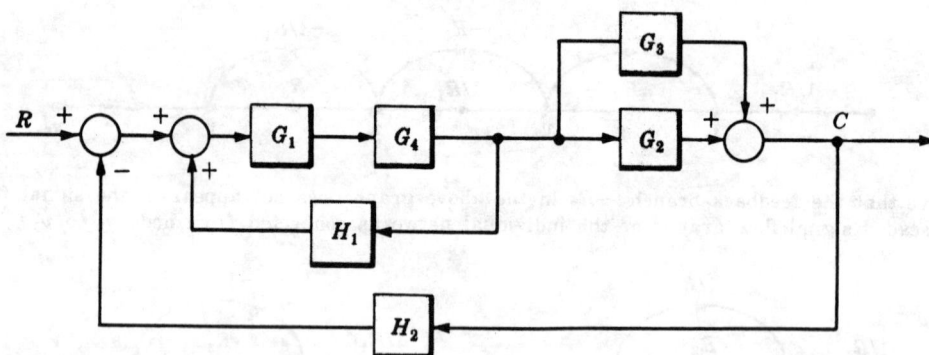

The signal flow graph is

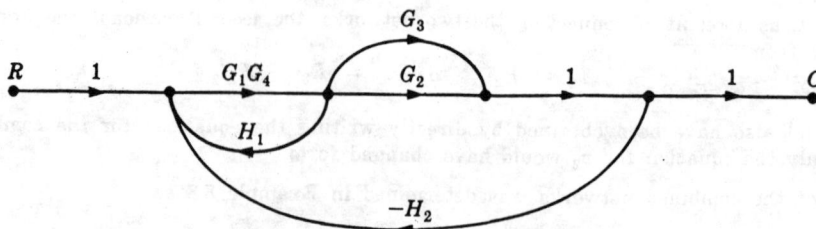

There are two forward paths: $P_1 = G_1G_2G_4, \quad P_2 = G_1G_3G_4$

There are three feedback loops:

$$P_{11} = G_1G_4H_1, \quad P_{21} = -G_1G_2G_4H_2, \quad P_{31} = -G_1G_3G_4H_2$$

There are no non-touching loops, and all loops touch both forward paths; then

$$\Delta_1 = 1, \quad \Delta_2 = 1$$

Therefore the control ratio is

$$T = \frac{C}{R} = \frac{P_1\Delta_1 + P_2\Delta_2}{\Delta} = \frac{G_1G_2G_4 + G_1G_3G_4}{1 - G_1G_4H_1 + G_1G_2G_4H_2 + G_1G_3G_4H_2}$$

$$= \frac{G_1G_4(G_2 + G_3)}{1 - G_1G_4H_1 + G_1G_2G_4H_2 + G_1G_3G_4H_2}$$

From Equations (8.3) and (8.4), we have

$$G = G_1G_4(G_2 + G_3) \quad \text{and} \quad GH = G_1G_4(G_3H_2 + G_2H_2 - H_1)$$

Therefore
$$H = \frac{GH}{G} = \frac{(G_2 + G_3)H_2 - H_1}{G_2 + G_3}$$

The canonical block diagram is therefore given by

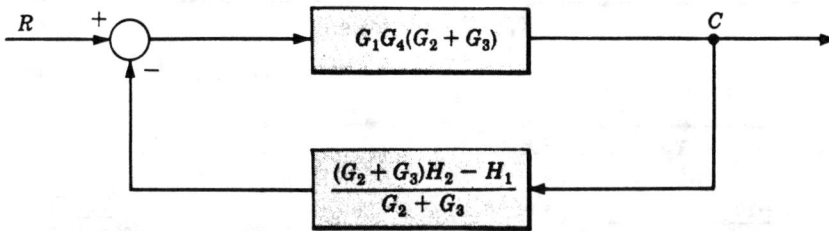

The negative summing point sign for the feedback loop is a result of using a positive sign in the GH formula above. If this is not obvious, refer to Equation (7.3) and its explanation in Section 7.4.

The block diagram above may be put into the final form of Examples 7.9 or 7.10 by using the transformation theorems of Section 7.5.

Solved Problems

SIGNAL FLOW GRAPH ALGEBRA AND DEFINITIONS

8.1. Simplify the following signal flow graphs.

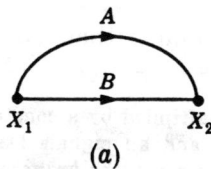

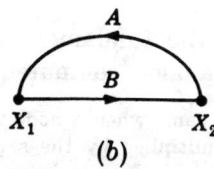

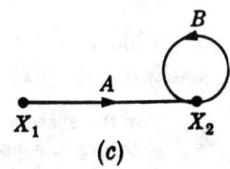

(a) Clearly, $X_2 = AX_1 + BX_1 = (A + B)X_1$. Therefore we have

(b) We have $X_2 = BX_1$ and $X_1 = AX_2$. Hence $X_2 = BAX_2$, or $X_1 = ABX_1$, yielding

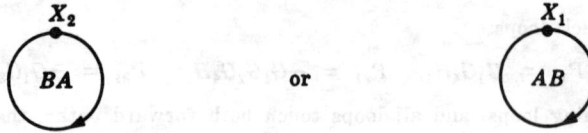

(c) If A and B are multiplicative operators (e.g. constants or transfer functions), we have

$$X_2 = AX_1 + BX_2 = \left(\frac{A}{1-B}\right)X_1. \quad \text{Hence the signal flow graph becomes}$$

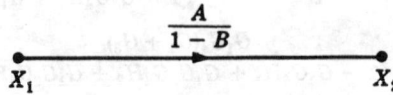

8.2. Draw signal flow graphs for the block diagrams in Problem 7.3 and reduce them by the multiplication rule.

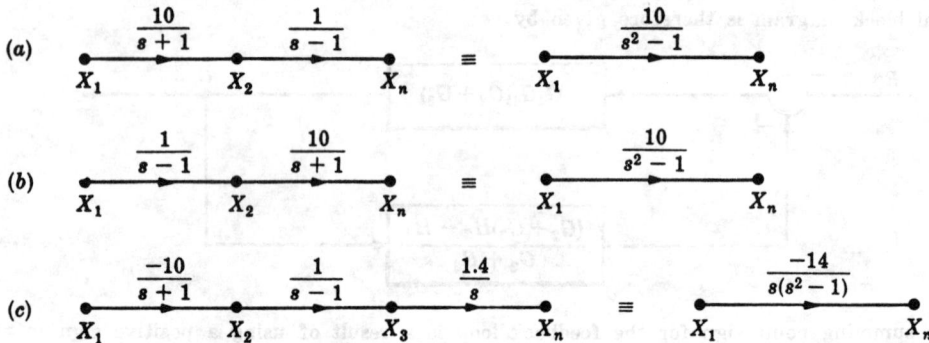

(a)

(b)

(c)

8.3. Consider the following signal flow graph:

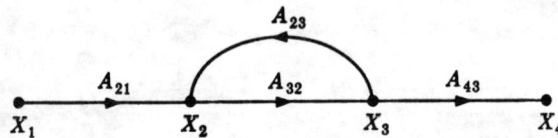

(a) Draw the signal flow graph for the system equivalent to that graphed above, but in which X_3 becomes kX_3 (k constant) and X_1, X_2, X_4 remain the same.

(b) Repeat (a) for the case in which X_2 and X_3 become k_2X_2 and k_3X_3, and X_1, X_4 remain the same (k_2 and k_3 are constants).

This problem illustrates the fundamentals of a technique that can be used for *scaling* the variables in an analog computer program.

(a) For the system to remain the same when a node variable is multiplied by a constant, all signals entering the node must be multiplied by the same constant, and all signals leaving the node divided by that constant. Since X_1, X_2 and X_4 must remain the same, the *branches* are modified:

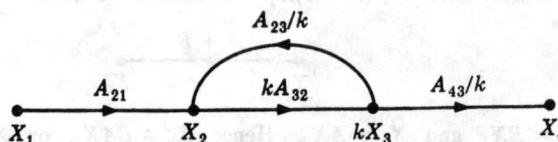

(b) Substitute k_2X_2 for X_2, and k_3X_3 for X_3:

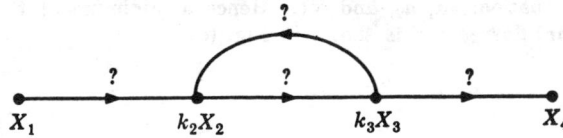

It is clear from the graph that A_{21} becomes k_2A_{21}, A_{32} becomes $(k_3/k_2)A_{32}$, A_{23} becomes $(k_2/k_3)A_{23}$, and A_{43} becomes $(1/k_3)A_{43}$:

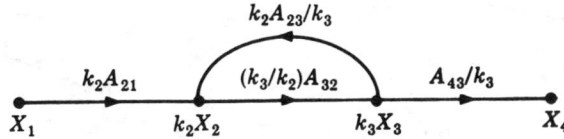

8.4. Consider the signal flow graph given by

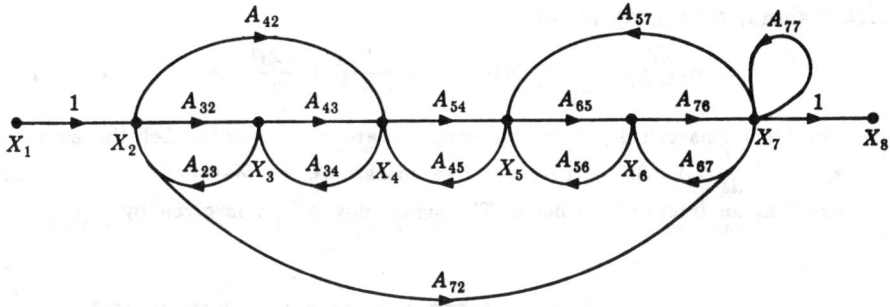

Identify the (a) input node, (b) output node, (c) forward paths, (d) feedback paths, (e) self-loop. Determine the (f) loop gains of the feedback loops, (g) path gains of the forward paths.

(a) X_1

(b) X_8

(c) X_1 to X_2 to X_3 to X_4 to X_5 to X_6 to X_7 to X_8
X_1 to X_2 to X_7 to X_8
X_1 to X_2 to X_4 to X_5 to X_6 to X_7 to X_8

(d) X_2 to X_3 to X_2; X_3 to X_4 to X_3; X_4 to X_5 to X_4; X_2 to X_4 to X_3 to X_2;
X_2 to X_7 to X_5 to X_4 to X_3 to X_2; X_5 to X_6 to X_5; X_6 to X_7 to X_6;
X_5 to X_6 to X_7 to X_5; X_7 to X_7; X_2 to X_7 to X_6 to X_5 to X_4 to X_3 to X_2

(e) X_7 to X_7

(f) $A_{32}A_{23}$; $A_{43}A_{34}$; $A_{54}A_{45}$; $A_{65}A_{56}$; $A_{76}A_{67}$; $A_{65}A_{76}A_{57}$; A_{77}; $A_{42}A_{34}A_{23}$;
$A_{72}A_{57}A_{45}A_{34}A_{23}$; $A_{72}A_{67}A_{56}A_{45}A_{34}A_{23}$

(g) $A_{32}A_{43}A_{54}A_{65}A_{76}$; A_{72}; $A_{42}A_{54}A_{65}A_{76}$

SIGNAL FLOW GRAPH CONSTRUCTION

8.5. Consider the following equations in which $x_1, x_2, \ldots, x_n$ are variables and $a_1, a_2, \ldots, a_n$ are coefficients or mathematical operators:

$$(a) \quad x_3 = a_1x_1 + a_2x_2 - 5 \qquad (b) \quad x_n = \sum_{k=1}^{n-1} a_kx_k + 5$$

What are the minimum number of nodes and the minimum number of branches required to construct the signal flow graphs of these equations? Draw the graphs.

(a) There are four variables in this equation: x_1, x_2, x_3, and ± 5. Therefore a minimum of four nodes are required. There are three coefficients or transmission functions on the right-hand side of the equation: a_1, a_2, and ∓ 1. Hence a minimum of three branches are required. A minimal signal flow graph is shown in Fig. (a).

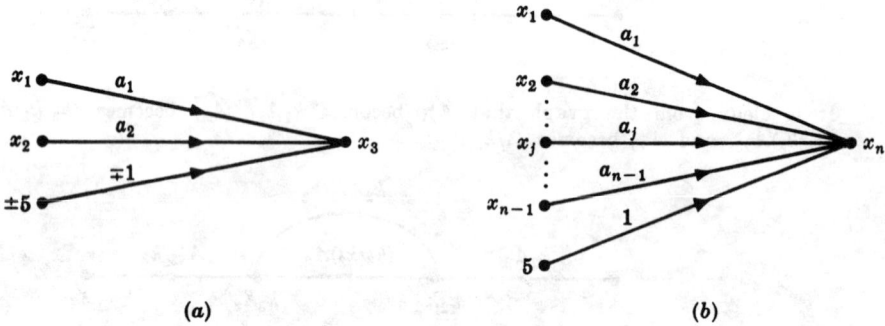

(a)

(b)

(b) There are $n+1$ variables: $x_1, x_2, \ldots, x_n$, and 5; and there are n coefficients: $a_1, a_2, \ldots, a_{n-1}$, and 1. Therefore a minimal signal flow graph is shown in Fig. (b).

8.6. Draw signal flow graphs for

$$(a) \quad x_2 = a_1 \left(\frac{dx_1}{dt} \right) \qquad (b) \quad x_3 = \frac{d^2 x_2}{dt^2} + \frac{dx_1}{dt} - x_1 \qquad (c) \quad x_4 = \int x_3 \, dt$$

(a) The operations called for in this equation are a_1 and d/dt. Let the equation be written as $x_2 = a_1 \cdot \frac{d}{dt} (x_1)$. Since there are two operations, we may define a new variable dx_1/dt and use it as an intermediate node. The signal flow graph is given by

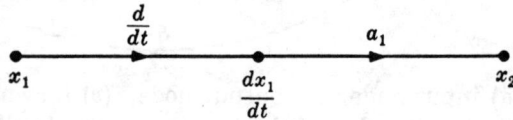

(b) Similarly, $x_3 = \frac{d^2}{dt^2}(x_2) + \frac{d}{dt}(x_1) - x_1$. Therefore

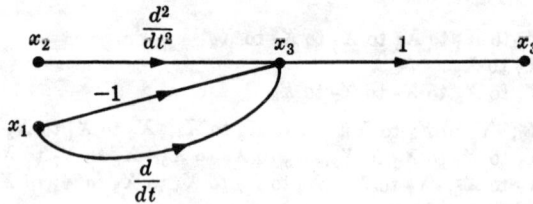

(c) The operation is integration. Let the operator be denoted by $\int dt$. The signal flow graph is

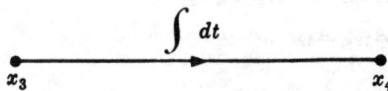

8.7. Construct the signal flow graph for the following set of simultaneous equations:

$$x_2 = A_{21}x_1 + A_{23}x_3, \qquad x_3 = A_{31}x_1 + A_{32}x_2 + A_{33}x_3, \qquad x_4 = A_{42}x_2 + A_{43}x_3$$

There are four variables: $x_1, \ldots, x_4$. Hence four nodes are required. Arranging them from left to right and connecting them with the appropriate branches, we obtain

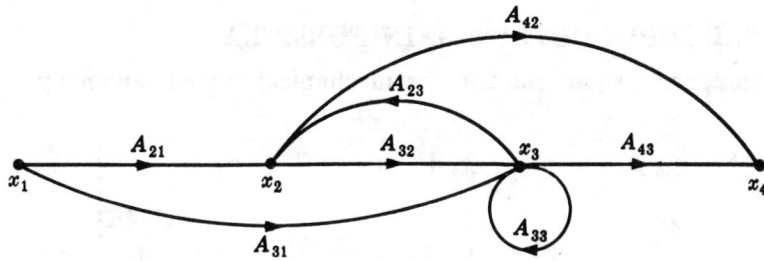

A neater way to arrange this graph is

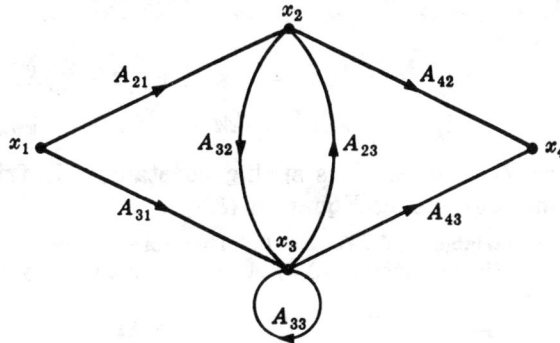

8.8. Draw a signal flow graph for the following resistance network in which $v_2(0) = v_3(0) = 0$. v_2 is the voltage across C_1.

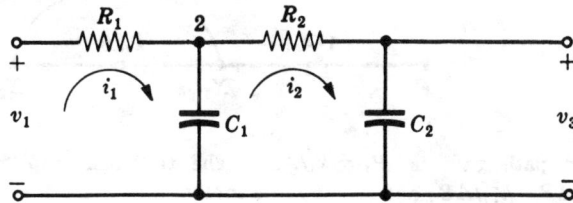

The five variables are v_1, v_2, v_3, i_1, and i_2; and v_1 is the input. The four independent equations derived from Kirchhoff's voltage and current laws are

$$i_1 = \left(\frac{1}{R_1}\right)v_1 - \left(\frac{1}{R_1}\right)v_2, \qquad v_2 = \frac{1}{C_1}\int_0^t i_1\,dt - \frac{1}{C_1}\int_0^t i_2\,dt,$$

$$i_2 = \left(\frac{1}{R_2}\right)v_2 - \left(\frac{1}{R_2}\right)v_3, \qquad v_3 = \frac{1}{C_2}\int_0^t i_2\,dt.$$

The signal flow graph can be drawn directly from these equations:

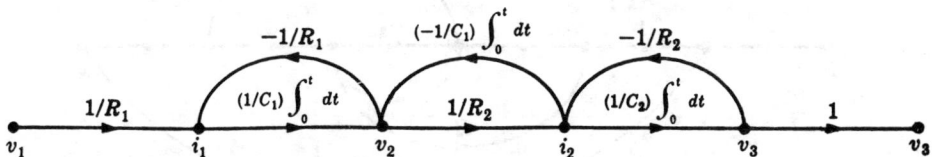

In Laplace transform notation, the signal flow graph is given by

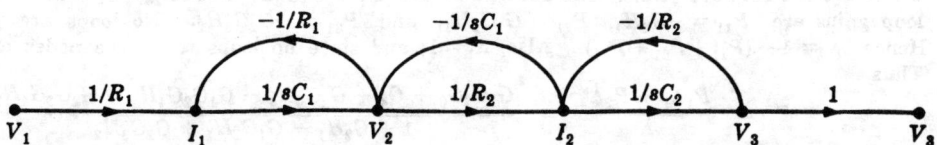

THE GENERAL INPUT-OUTPUT GAIN FORMULA

8.9. The transformed equations for the mechanical system given by

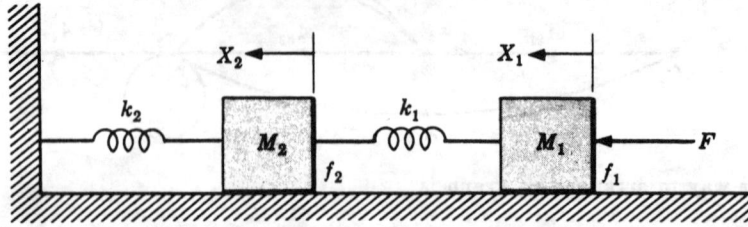

are

$$\text{(i)} \quad F + k_1 X_2 = (M_1 s^2 + f_1 s + k_1) X_1$$

$$\text{(ii)} \quad k_1 X_1 = (M_2 s^2 + f_2 s + k_1 + k_2) X_2$$

where F is force, M is mass, k is spring constant, f is friction, and X is displacement. Determine X_2/F using Equation (8.2).

There are three variables: X_1, X_2 and F. Therefore we need three nodes. In order to draw the signal flow graph, divide Equation (i) by A and Equation (ii) by B, where $A \equiv M_1 s^2 + f_1 s + k_1$, and $B \equiv M_2 s^2 + f_2 s + k_1 + k_2$:

$$\text{(iii)} \quad (1/A)F + (k_1/A)X_2 = X_1$$

$$\text{(iv)} \quad (k_1/B)X_1 = X_2$$

Therefore, the signal flow graph is

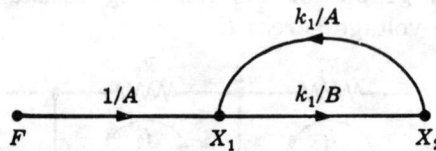

The forward path gain is $P_1 = k_1/AB$. The feedback loop gain is $P_{11} = k_1^2/AB$. Then $\Delta = 1 - P_{11} = (AB - k_1^2)/AB$ and $\Delta_1 = 1$. Finally,

$$X_2/F = P_1 \Delta_1/\Delta = k_1/(AB - k_1^2) = k_1/[(M_1 s^2 + f_1 s + k_1)(M_2 s^2 + f_2 s + k_1 + k_2) - k_1^2]$$

8.10. Determine the transfer function for the block diagram in Problem 7.21 by signal flow graph techniques.

The signal flow graph is drawn directly from the block diagram.

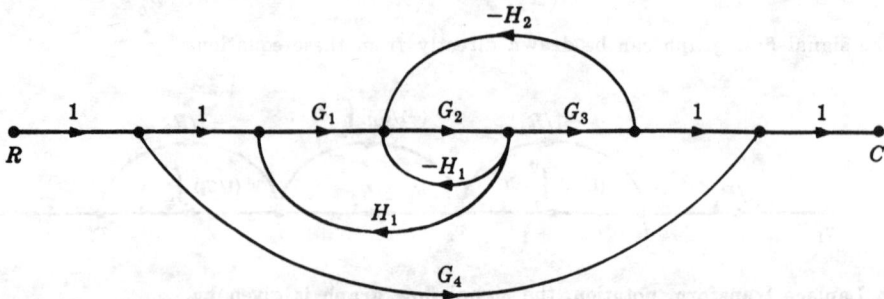

There are two forward paths. The path gains are $P_1 = G_1 G_2 G_3$ and $P_2 = G_4$. The three feedback loop gains are $P_{11} = -G_2 H_1$, $P_{21} = G_1 G_2 H_1$, and $P_{31} = -G_2 G_3 H_2$. No loops are non-touching. Hence $\Delta = 1 - (P_{11} + P_{21} + P_{31})$. Also, $\Delta_1 = 1$; and since no loops touch the nodes of P_2, $\Delta_2 = \Delta$. Thus

$$T = \frac{P_1 \Delta_1 + P_2 \Delta_2}{\Delta} = \frac{G_1 G_2 G_3 + G_4 + G_2 G_4 H_1 - G_1 G_2 G_4 H_1 + G_2 G_3 G_4 H_2}{1 + G_2 H_1 - G_1 G_2 H_1 + G_2 G_3 H_2}$$

8.11. Determine the transfer function V_3/V_1 from the signal flow graph of Problem 8.8.

The single forward path gain is $1/(s^2R_1R_2C_1C_2)$. The loop gains of the three feedback loops are $P_{11} = -1/(sR_1C_1)$, $P_{21} = -1/(sR_2C_1)$ and $P_{31} = -1/(sR_2C_2)$. The gain-product of the only two non-touching loops is $P_{12} = P_{11} \cdot P_{31} = 1/(s^2R_1R_2C_1C_2)$. Hence

$$\Delta = 1 - (P_{11} + P_{21} + P_{31}) + P_{12} = \frac{s^2R_1R_2^2C_1^2C_2 + s(R_2^2C_1C_2 + R_1R_2C_1C_2 + R_1R_2C_1^2) + R_2C_1}{s^2R_1R_2^2C_1^2C_2}$$

Since all loops touch the forward path, $\Delta_1 = 1$. Finally,

$$\frac{V_3}{V_1} = \frac{P_1\Delta_1}{\Delta} = \frac{1}{s^2R_1R_2C_1C_2 + s(R_2C_2 + R_1C_2 + R_1C_1) + 1}$$

8.12. Solve Problem 7.17 with signal flow graph techniques.

The signal flow graph is drawn directly from the block diagram:

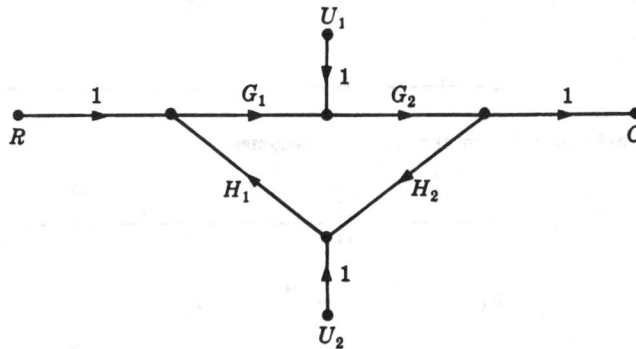

With $U_1 = U_2 = 0$, we have

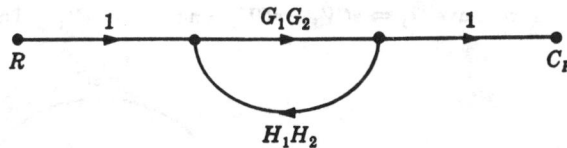

Then $P_1 = G_1G_2$ and $P_{11} = G_1G_2H_1H_2$. Hence $\Delta = 1 - P_{11} = 1 - G_1G_2H_1H_2$, $\Delta_1 = 1$, and

$$C_R = TR = \frac{P_1\Delta_1R}{\Delta} = \frac{G_1G_2R}{1 - G_1G_2H_1H_2}$$

Now put $U_2 = R = 0$:

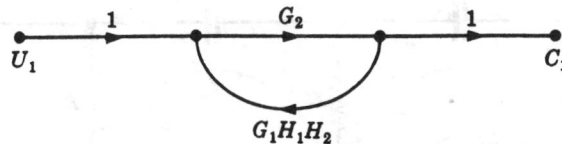

Then $P_1 = G_2$, $P_{11} = G_1G_2H_1H_2$, $\Delta = 1 - G_1G_2H_1H_2$, $\Delta_1 = 1$, and

$$C_1 = TU_1 = \frac{G_2U_1}{1 - G_1G_2H_1H_2}$$

Now put $R = U_1 = 0$:

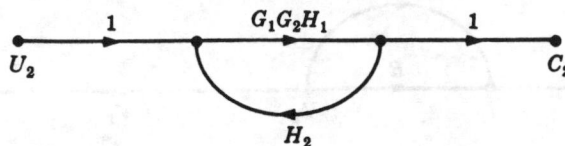

Then $P_1 = G_1G_2H_1$, $P_{11} = G_1G_2H_1H_2$, $\Delta = 1 - G_1G_2H_1H_2$, $\Delta_1 = 1$, and

$$C_2 = TU_2 = \frac{P_1\Delta_1 U_2}{\Delta} = \frac{G_1G_2H_1U_2}{1 - G_1G_2H_1H_2}$$

Finally, we have

$$C = C_R + C_1 + C_2 = \frac{G_1G_2R + G_2U_1 + G_1G_2H_1U_2}{1 - G_1G_2H_1H_2}$$

TRANSFER FUNCTION COMPUTATION OF CASCADED COMPONENTS

8.13. Determine the transfer function for two of the following networks in cascade:

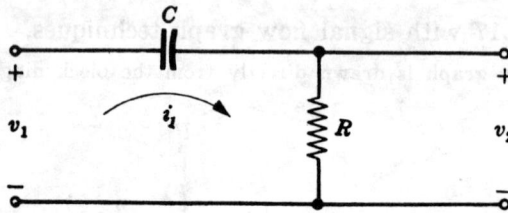

In Laplace transform notation the network becomes

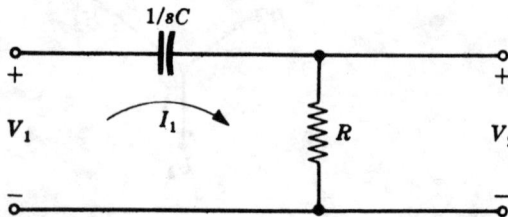

By Kirchhoff's laws, we have $I_1 = sCV_1 - sCV_2$ and $V_2 = RI_1$. The signal flow graph is

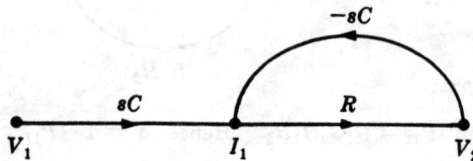

For two networks in cascade,

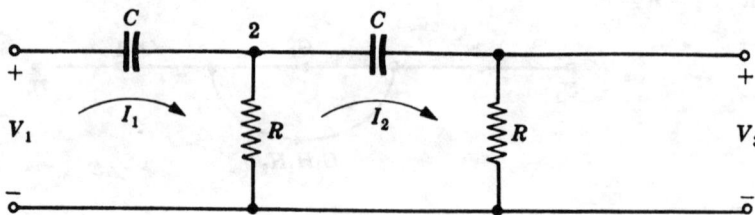

the V_2 equation is also dependent on I_2: $V_2 = RI_1 - RI_2$. Hence two networks are joined at node 2,

and a feedback loop $(-RI_2)$ is added between I_2 and V_2:

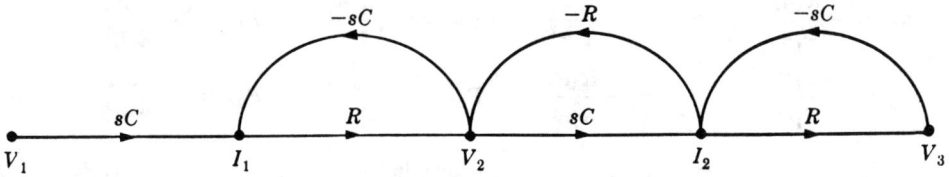

Then $P_1 = s^2R^2C^2$, $P_{11} = P_{21} = P_{31} = -sRC$, $P_{12} = P_{11} \cdot P_{31} = s^2R^2C^2$, $\Delta = 1 - (P_{11} + P_{21} + P_{31}) + P_{12} = 1 + 3sRC + s^2R^2C^2$, $\Delta_1 = 1$, and

$$T = \frac{P_1\Delta_1}{\Delta} = \frac{s^2}{s^2 + (3/RC)s + 1/(RC)^2}$$

8.14. Two resistance networks in the form of that in Example 8.6 are to be used for control elements in the forward path of a control system. They are to be cascaded and shall have identical respective component values:

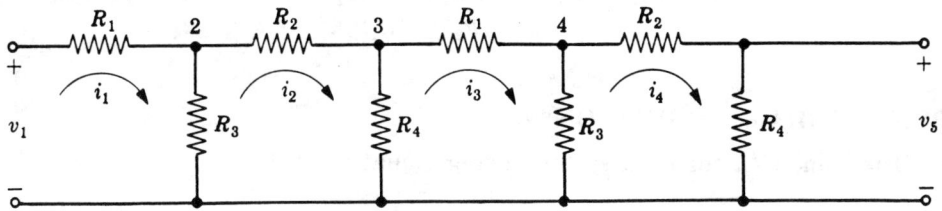

Find v_5/v_1 using Equation (8.2).

There are nine variables: $v_1, v_2, v_3, v_4, v_5, i_1, i_2, i_3,$ and i_4. Eight independent equations are

$$i_1 = (1/R_1)v_1 - (1/R_1)v_2 \qquad\qquad i_3 = (1/R_1)v_3 - (1/R_1)v_4$$

$$v_2 = R_3i_1 - R_3i_2 \qquad\qquad\qquad v_4 = R_3i_3 - R_3i_4$$

$$i_2 = (1/R_2)v_2 - (1/R_2)v_3 \qquad\qquad i_4 = (1/R_2)v_4 - (1/R_2)v_5$$

$$v_3 = R_4i_2 - R_4i_3 \qquad\qquad\qquad v_5 = R_4i_4$$

Only the equation for v_3 is different from those of the single network of Example 8.6; it has an extra term, $(-R_4i_3)$. Therefore the signal flow diagram for each network alone (Example 8.6) may be joined at node v_3, and an extra branch of gain $-R_4$ drawn from i_3 to v_3. The resulting signal flow graph for the double network is

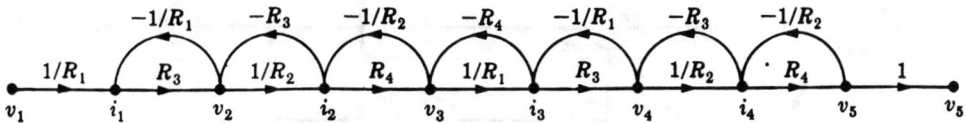

The voltage gain $T = v_5/v_1$ is calculated from Equation (8.2) as follows. One forward path yields $P_1 = \left(\dfrac{R_3R_4}{R_1R_2}\right)^2$. The gains of the seven feedback loops are $P_{11} = -R_3/R_1 = P_{51}$, $P_{21} = -R_3/R_2 = P_{61}$, $P_{31} = -R_4/R_2 = P_{71}$, and $P_{41} = -R_4/R_1$.

There are fifteen gain-products of two non-touching loops. From left to right, we have

$P_{12} = \dfrac{R_3R_4}{R_1R_2}$	$P_{42} = \dfrac{R_3^2}{R_1R_2}$	$P_{72} = \dfrac{R_3^2}{R_1R_2}$	$P_{10,2} = \dfrac{R_3R_4}{R_1R_2}$	$P_{13,2} = \dfrac{R_3R_4}{R_1R_2}$
$P_{22} = \dfrac{R_3R_4}{R_1^2}$	$P_{52} = \dfrac{R_3R_4}{R_1R_2}$	$P_{82} = \left(\dfrac{R_3}{R_2}\right)^2$	$P_{11,2} = \dfrac{R_3R_4}{R_2^2}$	$P_{14,2} = \dfrac{R_4^2}{R_1R_2}$
$P_{32} = \left(\dfrac{R_3}{R_1}\right)^2$	$P_{62} = \dfrac{R_3R_4}{R_1R_2}$	$P_{92} = \dfrac{R_3R_4}{R_2^2}$	$P_{12,2} = \left(\dfrac{R_4}{R_2}\right)^2$	$P_{15,2} = \dfrac{R_3R_4}{R_1R_2}$

There are ten gain-products of three non-touching loops. From left to right we have

$$P_{13} = -\frac{R_3^2 R_4}{R_1^2 R_2} \qquad P_{33} = -\frac{R_3 R_4^2}{R_1 R_2^2} \qquad P_{63} = -\frac{R_3^2 R_4}{R_1^2 R_2} \qquad P_{83} = -\frac{R_3 R_4^2}{R_1 R_2^2} \qquad P_{53} = -\frac{R_3 R_4^2}{R_1^2 R_2}$$

$$P_{23} = -\frac{R_3^2 R_4}{R_1 R_2^2} \qquad P_{43} = -\frac{R_3^2 R_4}{R_1^2 R_2} \qquad P_{73} = -\frac{R_3^2 R_4}{R_1 R_2^2} \qquad P_{93} = -\frac{R_3^2 R_4}{R_1 R_2} \qquad P_{10,3} = -\frac{R_3 R_4^2}{R_1 R_2^2}$$

There is one gain-product of four non-touching loops: $P_{14} = P_{11} P_{31} P_{51} P_{71} = \left(\dfrac{R_3 R_4}{R_1 R_2}\right)^2$.

Therefore the determinant is

$$\Delta = 1 - \sum_{j=1}^{7} P_{j1} + \sum_{j=1}^{15} P_{j2} - \sum_{j=1}^{10} P_{j3} + P_{14}$$

$$= 1 + \frac{R_1 R_3 + R_1 R_4 + R_2 R_3 + R_2 R_4 + 6 R_3 R_4 + 2 R_3^2 + R_4^2}{R_1 R_2} + \frac{R_3 R_4 + R_3^2}{R_1^2} + \frac{R_3^2 + R_4^2 + R_3 R_4}{R_2^2}$$

Since all loops touch the forward path, $\Delta_1 = 1$ and

$$T = \frac{P_1 \Delta_1}{\Delta} = \frac{(R_3 R_4)^2}{(R_1 R_2)^2 + R_1^2 (R_2 R_3 + R_2 R_4 + R_3 R_4 + R_3^2 + R_4^2) + R_2^2 (R_3^2 + R_1 R_3 + R_1 R_4 + R_3 R_4)}{+ 2 R_1 R_2 R_3^2 + R_1 R_2 R_4^2 + 6 R_1 R_2 R_3 R_4}$$

BLOCK DIAGRAM REDUCTION

8.15. Determine C/R for each system, using Equation (8.2).

(a)

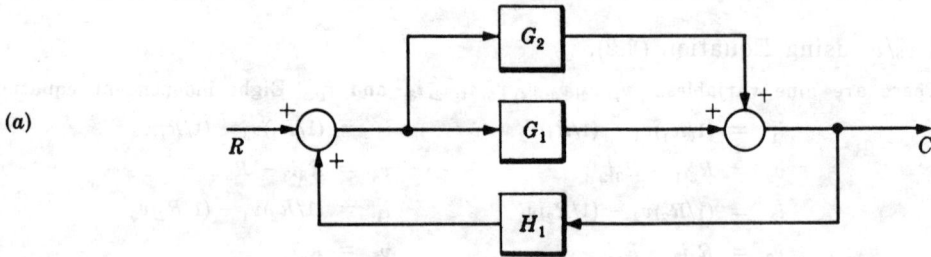

(b)

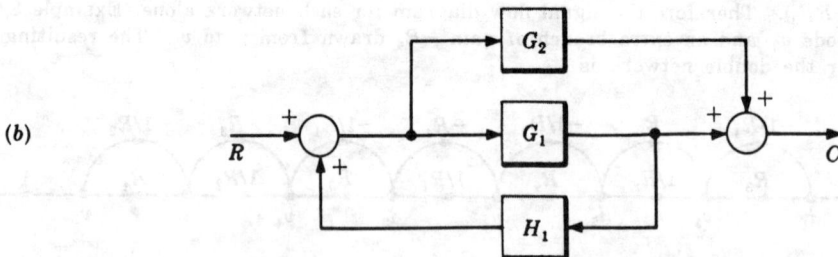

(c)

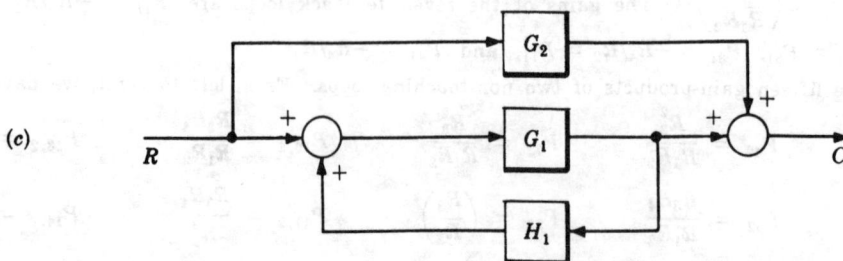

(a) The signal flow graph is

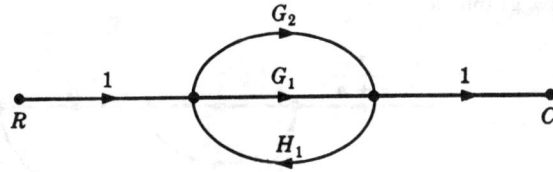

The two forward path gains are: $P_1 = G_1$, $P_2 = G_2$. The two feedback loop gains are: $P_{11} = G_1H_1$, $P_{21} = G_2H_1$. Then

$$\Delta = 1 - (P_{11} + P_{21}) = 1 - G_1H_1 - G_2H_1$$

Now, $\Delta_1 = 1$ and $\Delta_2 = 1$ because both paths touch the feedback loops at both interior nodes. Hence

$$\frac{C}{R} = \frac{P_1\Delta_1 + P_2\Delta_2}{\Delta} = \frac{G_1 + G_2}{1 - G_1H_1 - G_2H_1}$$

(b) The signal flow graph is

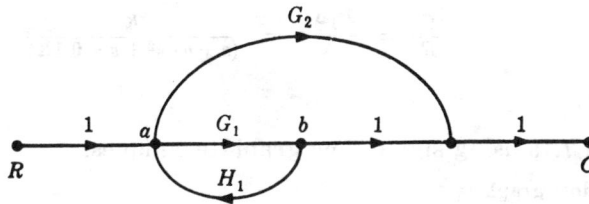

Again, we have $P_1 = G_1$ and $P_2 = G_2$. But now there is only one feedback loop, and $P_{11} = G_1H_1$; then $\Delta = 1 - G_1H_1$. The forward path through G_1 clearly touches the feedback loop at nodes a and b; then $\Delta_1 = 1$. The forward path through G_2 touches the feedback loop at node a; then $\Delta_2 = 1$. Hence

$$\frac{C}{R} = \frac{P_1\Delta_1 + P_2\Delta_2}{\Delta} = \frac{G_1 + G_2}{1 - G_1H_1}$$

(c) The signal flow graph is

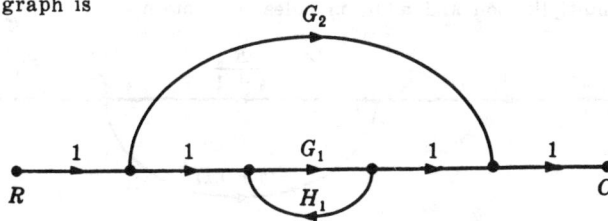

Again, we have $P_1 = G_1$, $P_2 = G_2$, $P_{11} = G_1H_1$, $\Delta = 1 - G_1H_1$, and $\Delta_1 = 1$. But the feedback path does not touch the forward path through G_2 at any node. Therefore $\Delta_2 = \Delta = 1 - G_1H_1$ and

$$\frac{C}{R} = \frac{P_1\Delta_1 + P_2\Delta_2}{\Delta} = \frac{G_1 + G_2(1 - G_1H_1)}{1 - G_1H_1}$$

This problem illustrates the importance of separating summing points and takeoff points with a branch of unity gain when applying Equation (8.2).

8.16. Find the transfer function C/R for the following system in which K is a constant:

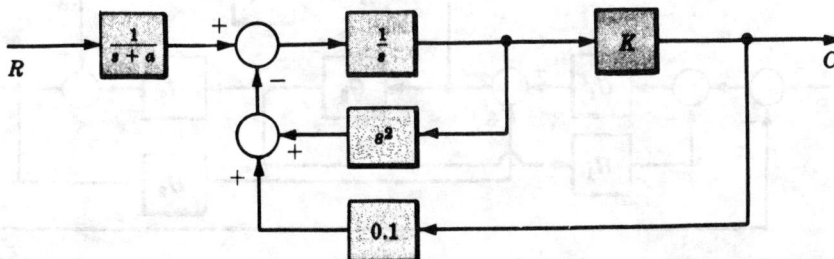

The signal flow graph is

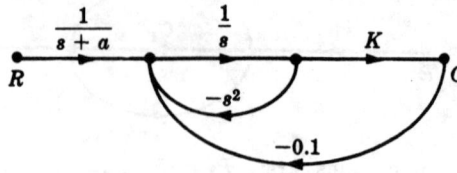

The only forward path gain is $P_1 = \left(\frac{1}{s+a}\right) \cdot \left(\frac{1}{s}\right) K = \frac{K}{s(s+a)}$. The two feedback loop gains are $P_{11} = \left(\frac{1}{s}\right) \cdot (-s^2) = -s$ and $P_{21} = -\frac{0.1K}{s}$. There are no non-touching loops. Hence $\Delta = 1 - (P_{11} + P_{21}) = \frac{s^2 + s + 0.1K}{s}$, $\Delta_1 = 1$ and

$$\frac{C}{R} = \frac{P_1 \Delta_1}{\Delta} = \frac{K}{(s+a)(s^2 + s + 0.1K)}$$

8.17. Solve Problem 7.19 using signal flow graph techniques.

The signal flow graph is

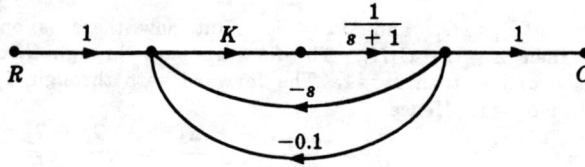

Applying the multiplication and addition rules, we obtain

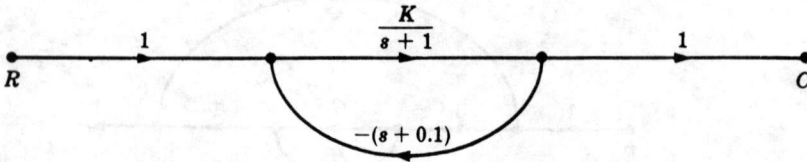

Now $P_1 = \frac{K}{s+1}$, $P_{11} = -\frac{K(s+0.1)}{s+1}$, $\Delta = 1 + \frac{K(s+0.1)}{s+1}$, $\Delta_1 = 1$, and

$$C = TR = \frac{P_1 \Delta_1 R}{\Delta} = \frac{KR}{(1+K)s + 1 + 0.1K}$$

8.18. Find C/R for the control system given by

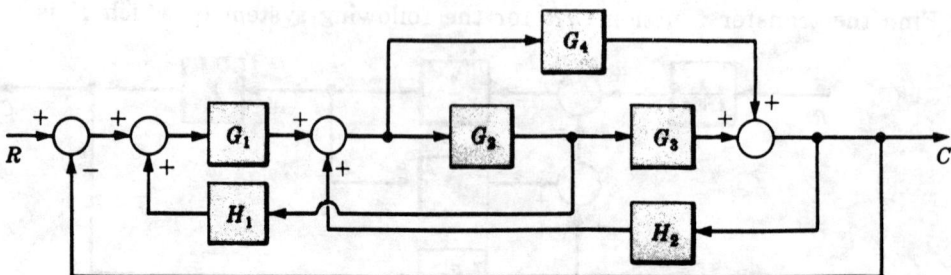

The signal flow graph is

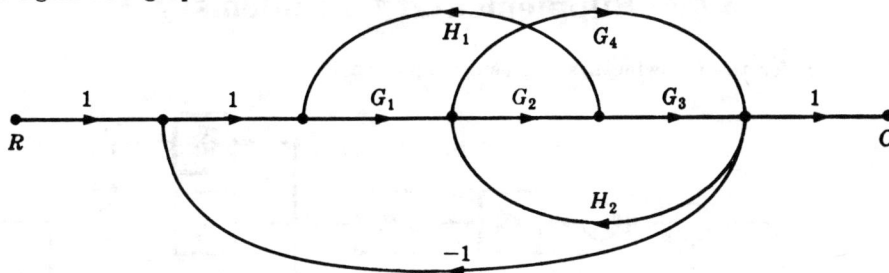

The two forward path gains are $P_1 = G_1G_2G_3$ and $P_2 = G_1G_4$. The five feedback loop gains are $P_{11} = G_1G_2H_1$, $P_{21} = G_2G_3H_2$, $P_{31} = -G_1G_2G_3$, $P_{41} = G_4H_2$, and $P_{51} = -G_1G_4$. Hence

$$\Delta = 1 - (P_{11} + P_{21} + P_{31} + P_{41} + P_{51}) = 1 + G_1G_2G_3 - G_1G_2H_1 - G_2G_3H_2 - G_4H_2 + G_1G_4$$

and $\Delta_1 = \Delta_2 = 1$. Finally,

$$\frac{C}{R} = \frac{P_1\Delta_1 + P_2\Delta_2}{\Delta} = \frac{G_1G_2G_3 + G_1G_4}{1 + G_1G_2G_3 - G_1G_2H_1 - G_2G_3H_2 - G_4H_2 + G_1G_4}$$

8.19. Determine C/R for the system given by

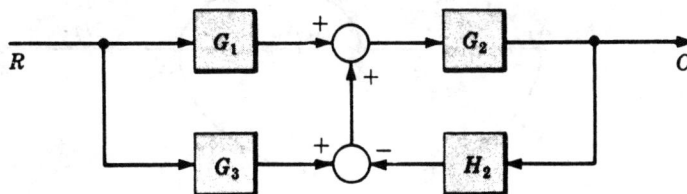

Then put $G_3 = G_1G_2H_2$.

The signal flow graph is

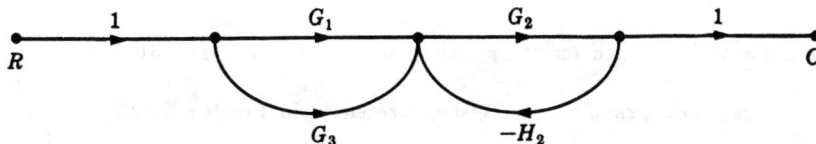

We have $P_1 = G_1G_2$, $P_2 = G_2G_3$, $P_{11} = -G_2H_2$, $\Delta = 1 + G_2H_2$, $\Delta_1 = \Delta_2 = 1$, and

$$\frac{C}{R} = \frac{P_1\Delta_1 + P_2\Delta_2}{\Delta} = \frac{G_2(G_1 + G_3)}{1 + G_2H_2}$$

Putting $G_3 = G_1G_2H_2$, we obtain $C/R = G_1G_2$ and the system transfer function becomes open-loop.

8.20. Determine the elements for a canonical feedback system for the system of Prob. 8.10.

From Problem 8.10, $P_1 = G_1G_2G_3$, $P_2 = G_4$, $\Delta = 1 + G_2H_1 - G_1G_2H_1 + G_2G_3H_2$, $\Delta_1 = 1$, and $\Delta_2 = \Delta$. From Equation (8.3) we have

$$G = \sum_{i=1}^{2} P_i\Delta_i = G_1G_2G_3 + G_4 + G_2G_4H_1 - G_1G_2G_4H_1 + G_2G_3G_4H_2$$

and from Equation (8.4) we obtain

$$H = \frac{\Delta - 1}{G} = \frac{G_2H_1 - G_1G_2H_1 + G_2G_3H_2}{G_1G_2G_3 + G_4 + G_2G_4H_1 - G_1G_2G_4H_1 + G_2G_3G_4H_2}$$

Supplementary Problems

8.21. Find C/R for the following system, using Equation (8.2).

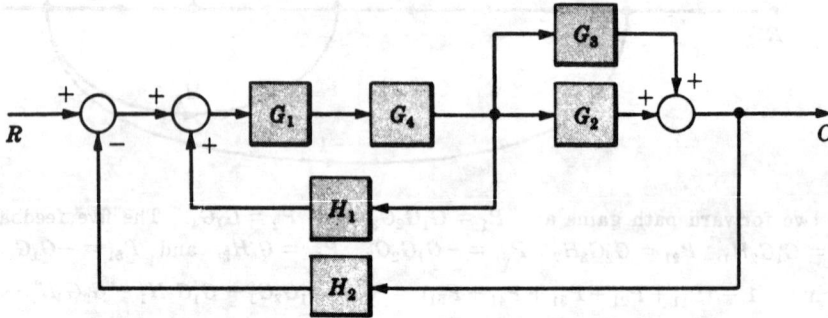

8.22. Determine a set of canonical feedback system transfer functions for the system of the preceding problem, using Equations (8.3) and (8.4).

8.23. Scale the following signal flow graph so that X_3 becomes $X_3/2$ (see Problem 8.3).

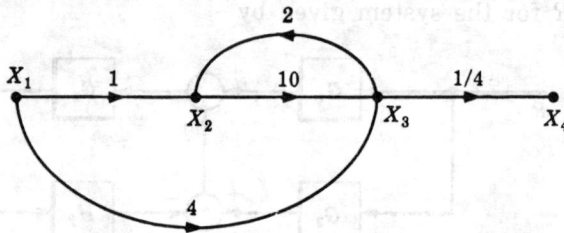

8.24. Draw a signal flow graph for several nodes of the lateral inhibition system described in Problem 3.4 by the equation

$$c_k = r_k - \sum_{i=1}^{n} a_{k-i} c_i$$

8.25. Draw a signal flow graph for the system presented in Problem 7.31.

8.26. Draw a signal flow graph for the system presented in Problem 7.32.

8.27. Determine C/R_4 from Equation (8.2) for the signal flow graph drawn in Problem 8.26.

8.28. Draw a signal flow graph for the following electrical network.

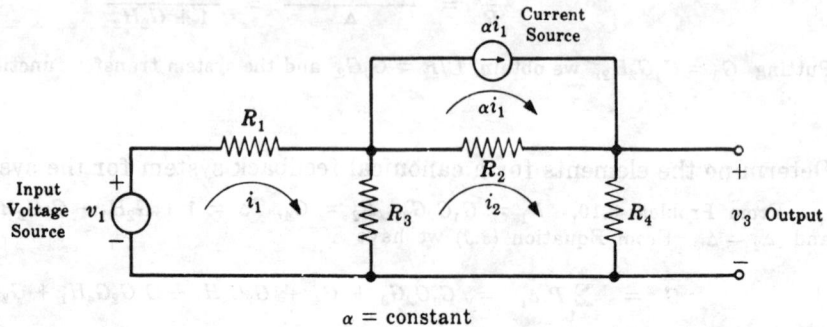

8.29. Determine V_3/V_1 from Equation (8.2) for the network of Problem 8.28.

8.30. Determine the elements for a canonical feedback system for the network of Problem 8.28, using Equations *(8.3)* and *(8.4)*.

8.31. Draw the signal flow graph for the following analog computer circuit.

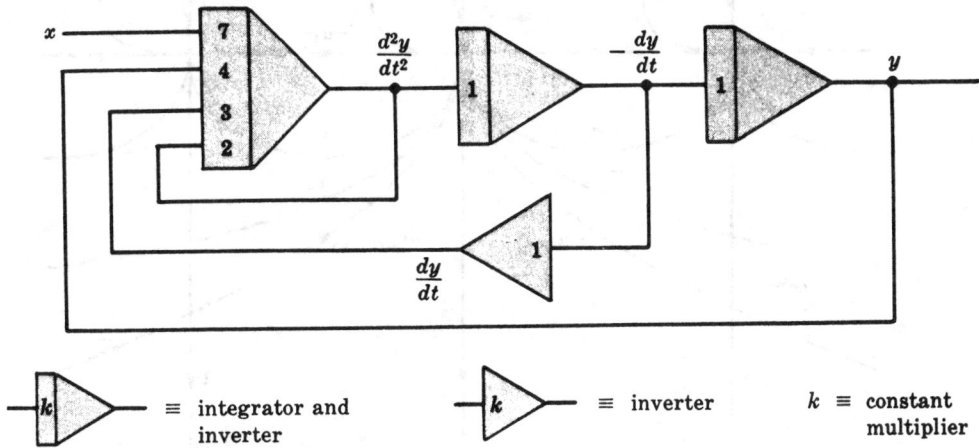

8.32. Scale the analog computer circuit of Problem 8.31 so that y becomes $10y$, $\dfrac{dy}{dt}$ becomes $20\dfrac{dy}{dt}$, and $\dfrac{d^2y}{dt^2}$ becomes $5\dfrac{d^2y}{dt^2}$.

Answers to Supplementary Problems

8.21.

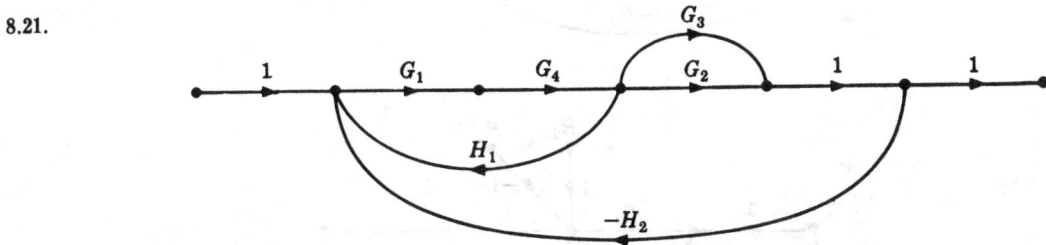

$P_1 = G_1G_2G_4$, $P_2 = G_1G_3G_4$, $P_{11} = G_1G_4H_1$, $P_{21} = -G_1G_2G_4H_2$, $P_{31} = -G_1G_3G_4H_2$, $\Delta = 1 - G_1G_4H_1 + G_1G_2G_4H_2 + G_1G_3G_4H_2$, and $\Delta_1 = \Delta_2 = 1$. Therefore

$$\frac{C}{R} = \frac{P_1\Delta_1 + P_2\Delta_2}{\Delta} = \frac{G_1G_4(G_2 + G_3)}{1 - G_1G_4[H_1 - H_2(G_2 + G_3)]}$$

8.22. $G = P_1\Delta_1 + P_2\Delta_2 = G_1G_4(G_2 + G_3)$, $\quad H = \dfrac{\Delta - 1}{G} = H_2 - \dfrac{H_1}{G_2 + G_3}$

8.23.

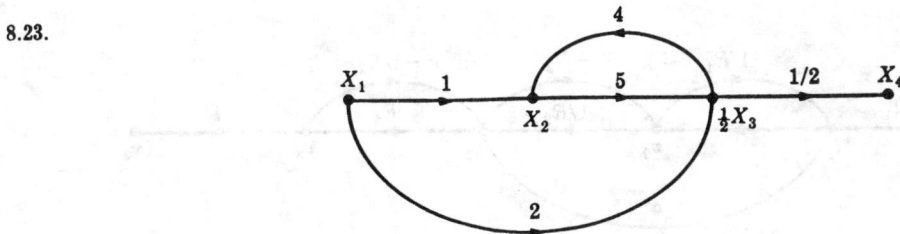

8.24.

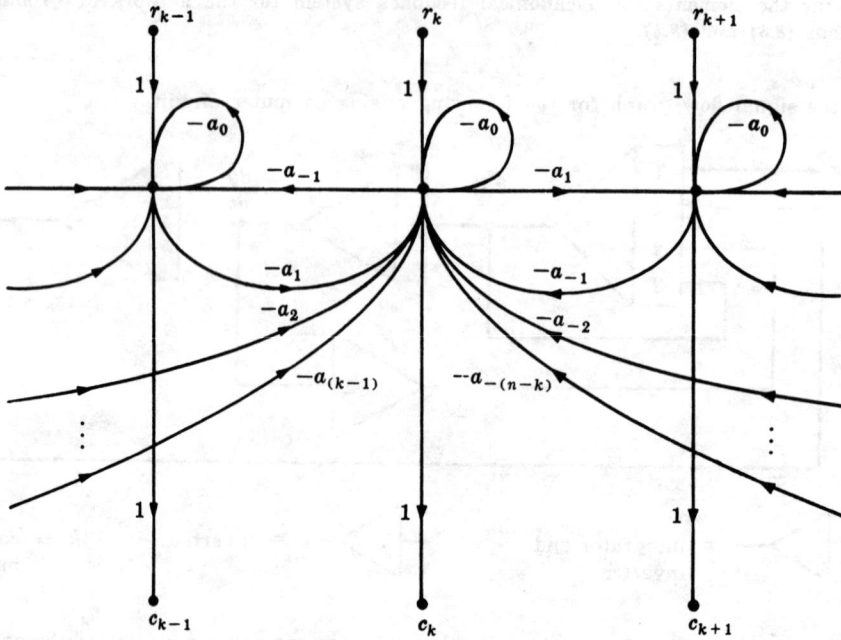

8.25.

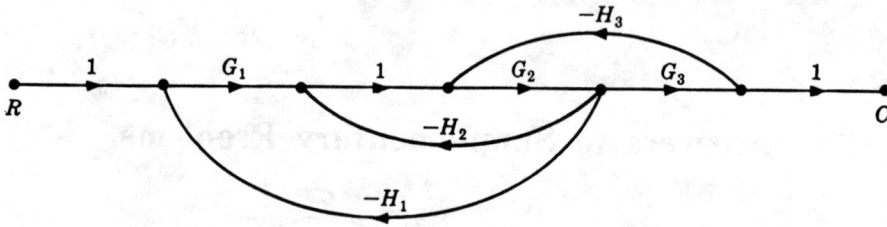

8.26.

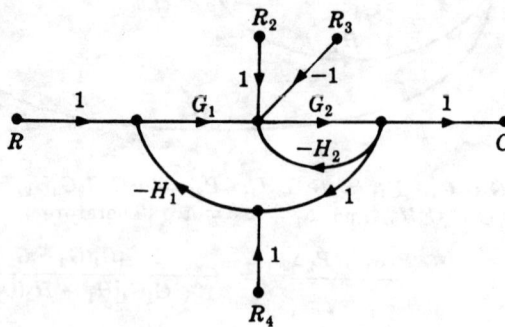

8.27. $\quad \dfrac{C}{R_4} = \dfrac{-G_1 G_2 H_1}{1 + G_2 H_2 + G_1 G_2 H_1}$

8.28.

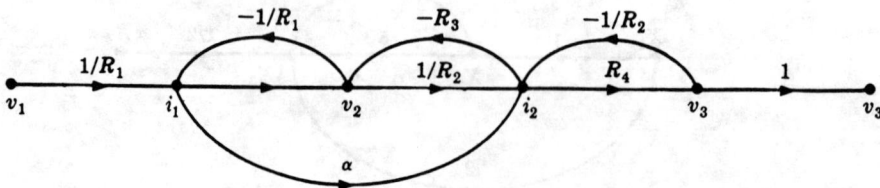

8.29. $\dfrac{V_3}{V_1} = \dfrac{R_3R_4 + \alpha R_2R_4}{R_1R_2 + R_1R_3 + R_1R_4 + R_2R_3 + R_3R_4 - \alpha R_2R_3}$

8.30. $G = R_4(R_3 + \alpha R_2)$

$H = \dfrac{R_1(R_2 + R_3 + R_4) + R_3R_4 + R_2R_3(1 - \alpha)}{R_4(R_3 + \alpha R_2)}$

8.31.

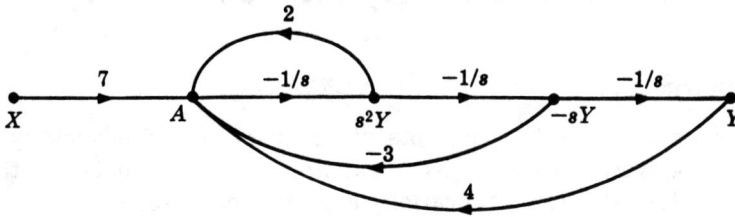

8.32.

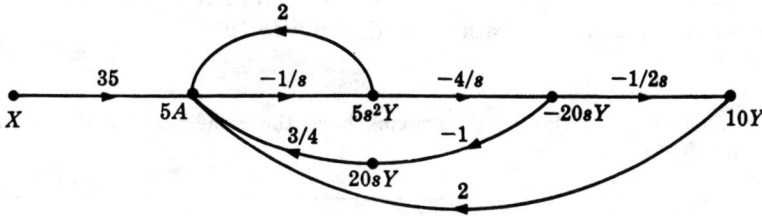

Chapter 9

System Classification, Error Constants, and Sensitivity

9.1 INTRODUCTION

In several earlier chapters the concepts of feedback and feedback systems have been emphasized. Since a system with a given transfer function can be synthesized in either an open-loop or a closed-loop configuration, a closed-loop (feedback) configuration must have some desirable properties which an open-loop configuration does not have.

In this chapter some of the properties of feedback and feedback systems are further discussed, and quantitative measures of the effectiveness of feedback are developed in terms of the concepts of *Error Constants* and *Sensitivity*.

Example 9.1.

The following open-loop and closed-loop systems have the same plant and the same overall system transfer function for $K = 2$.

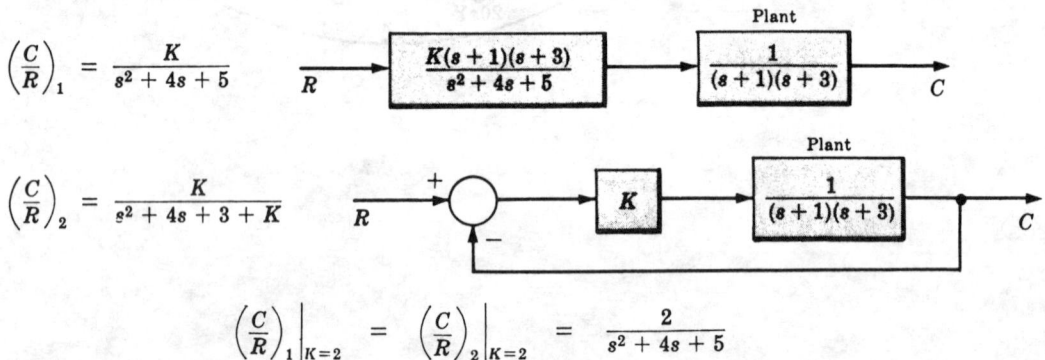

$$\left(\frac{C}{R}\right)_1 = \frac{K}{s^2 + 4s + 5}$$

$$\left(\frac{C}{R}\right)_2 = \frac{K}{s^2 + 4s + 3 + K}$$

$$\left(\frac{C}{R}\right)_1\bigg|_{K=2} = \left(\frac{C}{R}\right)_2\bigg|_{K=2} = \frac{2}{s^2 + 4s + 5}$$

Although these systems are precisely equivalent for $K = 2$, their properties differ significantly for small (and large) deviations of K from $K = 2$. This situation is clarified in Section 9.8.

9.2 CLASSIFICATION OF FEEDBACK SYSTEMS BY TYPE

Consider the class of canonical feedback systems defined by the adjacent block diagram. The open-loop transfer function of the system may be written as:

$$GH = \frac{K \prod_{i=1}^{m} (s + z_i)}{\prod_{i=1}^{n} (s + p_i)}$$

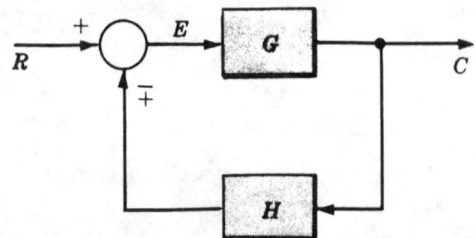

where K is a constant, $m \leq n$, and $-z_i$ and $-p_i$ are the finite zeros and poles, respectively, of GH. If there are a zeros and b poles at the origin, then

162

$$GH = \frac{Ks^a \prod_{i=1}^{m-a} (s + z_i)}{s^b \prod_{i=1}^{n-b} (s + p_i)}$$

In the remainder of this chapter, only systems for which $b \geq a$ are considered, and $l \equiv b - a$.

Definition 9.1:   A canonical feedback system whose open-loop transfer function can be written in the form

$$GH = \frac{K \prod_{i=1}^{m-a} (s + z_i)}{s^l \prod_{i=1}^{n-a-l} (s + p_i)} \equiv \frac{K B_1(s)}{s^l B_2(s)} \qquad (9.1)$$

where $l \geq 0$ and $-z_i$ and $-p_i$ are the nonzero finite zeros and poles of GH, respectively, is called a **type l system**.

Example 9.2.

The system defined by the block diagram

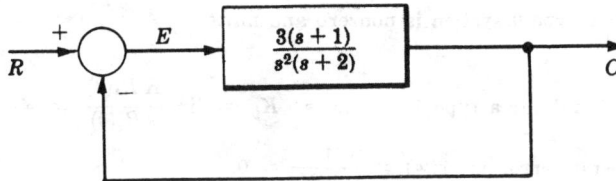

is a *type 2 system*.

Example 9.3.

The system defined by the block diagram

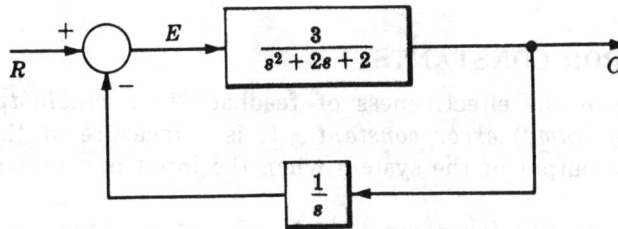

is a *type 1 system*.

Example 9.4.

The system defined by the block diagram

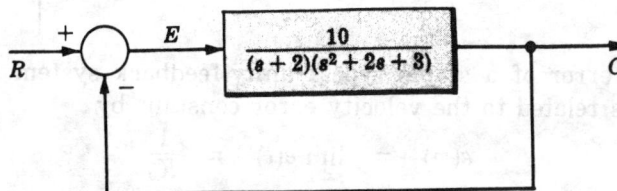

is a *type 0 system*.

9.3 POSITION ERROR CONSTANTS

One criterion of the effectiveness of feedback in a *stable type l unity-feedback system* is the *position (step) error constant*. It is a measure of the steady-state error between the input and output when the input is a unit-step function.

Definition 9.2:   The **position error constant** K_p of a type l unity-feedback system is defined as

$$K_p \equiv \lim_{s \to 0} G(s) = \lim_{s \to 0} \frac{K B_1(s)}{s^l B_2(s)} = \begin{cases} \dfrac{K B_1(0)}{B_2(0)} & l = 0 \\[2mm] \infty & l > 0 \end{cases} \qquad (9.2)$$

The steady-state error of a stable type l unity-feedback system when the input is a unit-step function $(e(\infty) = 1 - c(\infty))$ is related to the position error constant by

$$e(\infty) = \lim_{t \to \infty} e(t) = \frac{1}{1 + K_p} \qquad (9.3)$$

Example 9.5.

The position error constant for a type 0 system is finite. That is,

$$|K_p| = \left| \frac{K B_1(0)}{B_2(0)} \right| < \infty$$

The steady-state error for a type 0 system is nonzero and finite.

Example 9.6.

The position error constant for a type 1 system is $K_p = \lim_{s \to 0} \dfrac{K B_1(0)}{s B_2(0)} = \infty$.

Therefore the steady-state error is $e(\infty) = \dfrac{1}{1 + K_p} = 0$.

Example 9.7.

The position error constant for a type 2 system is $K_p = \lim_{s \to 0} \dfrac{K B_1(s)}{s^2 B_2(s)} = \infty$.

Therefore the steady-state error is $e(\infty) = \dfrac{1}{1 + K_p} = 0$.

9.4 VELOCITY ERROR CONSTANTS

Another criterion of the effectiveness of feedback in a *stable type l unity-feedback system* is the *velocity (ramp) error constant*. It is a measure of the steady-state error between the input and output of the system when the input is a unit-ramp function.

Definition 9.3:   The **velocity error constant** K_v of a stable type l unity-feedback system is defined as

$$K_v \equiv \lim_{s \to 0} s\, G(s) = \lim_{s \to 0} \frac{K B_1(s)}{s^{l-1} B_2(s)} = \begin{cases} 0 & l = 0 \\[2mm] \dfrac{K B_1(0)}{B_2(0)} & l = 1 \\[2mm] \infty & l > 1 \end{cases} \qquad (9.4)$$

The steady-state error of a stable type l unity-feedback system when the input is a unit-ramp function is related to the velocity error constant by

$$e(\infty) = \lim_{t \to \infty} e(t) = \frac{1}{K_v} \qquad (9.5)$$

Example 9.8.

The velocity error constant for a type 0 system is $K_v = 0$. Hence the steady-state error is infinite.

Example 9.9.

The velocity error constant for a type 1 system, $K_v = \dfrac{K B_1(0)}{B_2(0)}$, is finite. Therefore the **steady-state error** is nonzero and finite.

Example 9.10.

The velocity error constant for a type 2 system is infinite. Therefore the **steady-state error is zero**.

9.5 ACCELERATION ERROR CONSTANTS

A third criterion of the effectiveness of feedback in a *stable type l unity-feedback* system is the *acceleration (parabolic) error constant*. It is a measure of the steady-state error of the system when the input is a unit-parabolic function; that is, $r = t^2/2$ and $R = 1/s^3$.

Definition 9.4: The **acceleration error constant** K_a of a stable type l unity-feedback system is defined as

$$K_a \equiv \lim_{s \to 0} s^2 G(s) = \lim_{s \to 0} \frac{K B_1(s)}{s^{l-2} B_2(s)} = \begin{cases} 0 & l = 0, 1 \\ \dfrac{K B_1(0)}{B_2(0)} & l = 2 \\ \infty & l > 2 \end{cases} \tag{9.6}$$

The steady-state error of a stable type l unity-feedback system when the input is a unit-parabolic function is related to the acceleration error constant by

$$e(\infty) = \lim_{t \to \infty} e(t) = \frac{1}{K_a} \tag{9.7}$$

Example 9.11.

The acceleration error constant for a type 0 system is $K_a = 0$. Hence the steady-state error is infinite.

Example 9.12.

The acceleration error constant for a type 1 system is $K_a = 0$. Hence the steady-state error is infinite.

Example 9.13.

The acceleration error constant for a type 2 system, $K_a = \dfrac{K B_1(0)}{B_2(0)}$, is finite. Hence the steady-state error is nonzero and finite.

9.6 TABULATED SUMMARY OF RESULTS

The error constants and steady-state errors for stable type 0, type 1, and type 2 unity-feedback systems, which were developed in Sections 9.3, 9.4, and 9.5, are tabulated below for easy reference.

Input	Unit Step		Unit Ramp		Unit Parabola	
System Type	K_p	Steady State Error	K_v	Steady State Error	K_a	Steady State Error
Type 0	$\dfrac{K B_1(0)}{B_2(0)}$	$\dfrac{1}{1 + K_p}$	0	∞	0	∞
Type 1	∞	0	$\dfrac{K B_1(0)}{B_2(0)}$	$\dfrac{1}{K_v}$	0	∞
Type 2	∞	0	∞	0	$\dfrac{K B_1(0)}{B_2(0)}$	$\dfrac{1}{K_a}$

9.7 ERROR CONSTANTS FOR GENERAL SYSTEMS

The results of Sections 9.3-9.6 are only applicable to stable unity-feedback systems. They can be readily extended, however, to general stable systems. In the figure below, T_d represents the transfer function of a desired (ideal) system, and C/R represents the transfer function of the actual system (an approximation of T_d). R is the input to both systems, and E is the difference (the error) between the desired output and the actual output. For this general system three error constants are defined below and are related to the steady-state error.

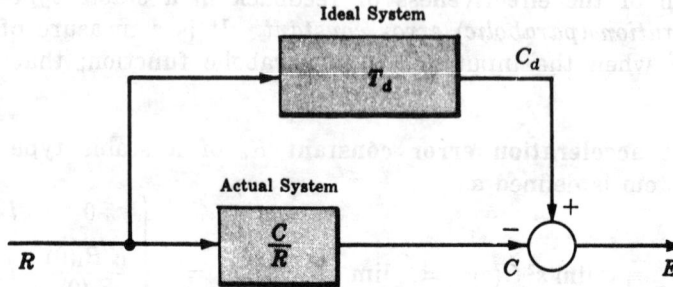

Definition 9.5: The **step error constant** K_s is defined as

$$K_s \equiv \frac{1}{\lim_{s \to 0} [T_d - C/R]} \qquad (9.8)$$

The steady-state error for the general system when the input is a unit-step function is related to K_s by

$$e(\infty) = \lim_{t \to \infty} e(t) = \frac{1}{K_s} \qquad (9.9)$$

Definition 9.6: The **ramp error constant** K_r is defined as

$$K_r \equiv \frac{1}{\lim_{s \to 0} \frac{1}{s} [T_d - C/R]} \qquad (9.10)$$

The steady-state error for the general system when the input is a unit-ramp function is related to K_r by

$$e(\infty) = \lim_{t \to \infty} e(t) = \frac{1}{K_r} \qquad (9.11)$$

Definition 9.7: The **parabolic error constant** K_{pa} is defined as

$$K_{pa} \equiv \frac{1}{\lim_{s \to 0} \frac{1}{s^2} [T_d - C/R]} \qquad (9.12)$$

The steady-state error for the general system when the input is a unit-parabolic function is related to K_{pa} by

$$e(\infty) = \lim_{t \to \infty} e(t) = \frac{1}{K_{pa}} \qquad (9.13)$$

Example 9.14.

The non-unity feedback system

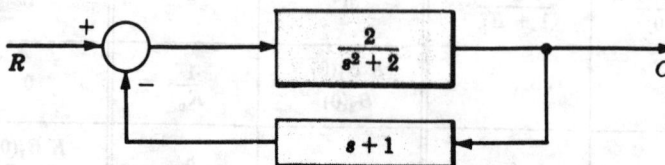

has the transfer function $\dfrac{C}{R} = \dfrac{2}{s^2 + 2s + 4}$. If the desired transfer function which C/R approximates is $T_d = \frac{1}{2}$, then $T_d - \dfrac{C}{R} = \dfrac{s(s+2)}{2(s^2+2s+4)}$.   Therefore

$$K_s = \frac{1}{\displaystyle\lim_{s\to 0}\left[\frac{s(s+2)}{2(s^2+2s+4)}\right]} = \infty \qquad K_r = \frac{1}{\displaystyle\lim_{s\to 0}\frac{1}{s}\left[\frac{s(s+2)}{2(s^2+2s+4)}\right]} = 4$$

$$K_{pa} = \frac{1}{\displaystyle\lim_{s\to 0}\frac{1}{s^2}\left[\frac{s(s+2)}{2(s^2+2s+4)}\right]} = 0$$

Example 9.15.

   For the system of Example 9.14 the steady-state errors due to a unit-step input, a unit-ramp input, and a unit-parabolic input can be found using the results of that example. For a unit-step input, $e(\infty) = 1/K_s = 0$. For a unit-ramp input, $e(\infty) = 1/K_r = \frac{1}{4}$. For a unit-parabolic input, $e(\infty) = 1/K_{pa} = \infty$.

   To establish relationships between the general error constants K_s, K_r, and K_{pa} and the error constants K_p, K_v and K_a for unity-feedback systems, we let the actual system be a unity-feedback system and let the desired system have a unity transfer function. That is, we let

$$T_d = 1 \qquad \text{and} \qquad \frac{C}{R} = \frac{G}{1+G}$$

Therefore

$$K_s = \frac{1}{\displaystyle\lim_{s\to 0}\left[\frac{1}{1+G}\right]} = 1 + \lim_{s\to 0} G(s) = 1 + K_p \qquad (9.14)$$

$$K_r = \frac{1}{\displaystyle\lim_{s\to 0}\left[\frac{1}{s}\left(\frac{1}{1+G}\right)\right]} = \lim_{s\to 0} s\,G(s) = K_v \qquad (9.15)$$

$$K_{pa} = \frac{1}{\displaystyle\lim_{s\to 0}\left[\frac{1}{s^2}\left(\frac{1}{1+G}\right)\right]} = \lim_{s\to 0} s^2\,G(s) = K_a \qquad (9.16)$$

9.8 SENSITIVITY

   A first step in the analysis or design of a control system is the generation of models for the various elements in the system. If the system is assumed to be linear and time-invariant, two important mathematical models which can be generated are the *transfer function* and the *frequency response function* (see Chapter 6).

   The transfer function is fixed when a finite number of constant parameters have been chosen. The values given to these parameters are called the **nominal values** and the corresponding transfer function is called the **nominal transfer function**. The accuracy of the model apparently depends on how closely these nominal parameter values approximate the actual parameter values, and also how much these parameters deviate from the nominal values during the course of system operation. The *sensitivity* of a system can then be defined as a measure of the amount by which the system transfer function differs from its nominal value when one of its parameters differs from the number chosen as its nominal value.

   The frequency response function of a system can be determined directly from the transfer function of the system, if it is known, by replacing the complex variable s in the transfer function by $j\omega$. In this case the frequency response function is defined by the same parameters as those of the transfer function, and its accuracy is determined by the accuracy of these parameters. The frequency response function can alternatively be defined by graphs of its magnitude and phase-angle, both plotted as a function of the real frequency ω. These graphs are often determined experimentally, and in many cases cannot be defined by a finite number of parameters. Hence an infinite number of values of amplitude and phase angle (values for all frequencies) define the frequency response function.   The accuracy of the model then depends on how closely the amplitude and phase angle graphs approximate the actual frequency response function.   The *sensitivity* of the system is in this case a measure of the amount by which its frequency response function differs from its nominal value when the frequency response function of an element of the system differs from its nominal value.

Consider the mathematical model $T(k)$ (transfer function or frequency response function) of a linear time-invariant system, written in polar form as

$$T(k) = |T(k)| e^{j\phi_T} \qquad (9.17)$$

where k is a parameter upon which $T(k)$ depends.   Usually both $T(k)$ and ϕ_T depend on k, and k is a real or complex quantity representing some identifiable parameter of the system.

Definition 9.8:   For the mathematical model $T(k)$, where k is regarded as a variable, the **sensitivity of $T(k)$ with respect to the parameter k** is defined by

$$S_k^{T(k)} \equiv \frac{d \ln T(k)}{d \ln k} = \frac{dT(k)/T(k)}{dk/k} = \frac{dT(k)}{dk} \frac{k}{T(k)} \qquad (9.18)$$

Definition 9.9:   The **sensitivity of the magnitude of $T(k)$ with respect to parameter k** is defined by

$$S_k^{|T(k)|} \equiv \frac{d \ln |T(k)|}{d \ln k} = \frac{d|T(k)|/|T(k)|}{dk/k} = \frac{d|T(k)|}{dk} \frac{k}{|T(k)|} \qquad (9.19)$$

Definition 9.10:   The **sensitivity of the phase angle ϕ_T of $T(k)$ with respect to the parameter k** is defined by

$$S_k^{\phi_T} \equiv \frac{d \ln \phi_T}{d \ln k} = \frac{d\phi_T/\phi_T}{dk/k} = \frac{d\phi_T}{dk} \frac{k}{\phi_T} \qquad (9.20)$$

The sensitivities of $T(k) = |T(k)| e^{j\phi_T}$, the magnitude $|T(k)|$, and the phase angle ϕ_T with respect to parameter k are related by the expression

$$S_k^{T(k)} = S_k^{|T(k)|} + j\phi_T S_k^{\phi_T} \qquad (9.21)$$

Note that, in general, $S_k^{|T(k)|}$ and $S_k^{\phi_T}$ are complex numbers.   In the special but very important case where k is real, then both $S_k^{|T(k)|}$ and $S_k^{\phi_T}$ are real.

Example 9.16.

Consider the frequency response function

$$T(\mu) = e^{-j\omega\mu}$$

where $\mu \equiv k$.   The magnitude of $T(\mu)$ is $|T(\mu)| = 1$, and the phase angle of $T(\mu)$ is $\phi_T = -\omega\mu$.

The sensitivity of $T(\mu)$ with respect to the parameter μ is

$$S_\mu^{T(\mu)} = \frac{d(e^{-j\omega\mu})}{d\mu} \frac{\mu}{e^{-j\omega\mu}} = -j\omega\mu$$

The sensitivity of the magnitude of $T(\mu)$ with respect to the parameter μ is

$$S_\mu^{|T(\mu)|} = \frac{d|T(\mu)|}{\mu} \frac{\mu}{|T(\mu)|} = 0$$

The sensitivity of the phase angle of $T(\mu)$ with respect to the parameter μ is

$$S_\mu^{\phi_T} = \frac{d\phi_T}{d\mu} \frac{\mu}{\phi_T} = -\omega \cdot \frac{\mu}{-\omega\mu} = 1$$

Note that $\qquad\qquad S_\mu^{|T(\mu)|} + j\phi_T S_\mu^{\phi_T} = -j\omega\mu = S_\mu^{T(\mu)}$

Example 9.17.

The frequency response function for the unity-feedback system

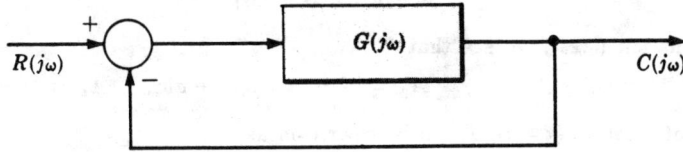

is related to the forward-loop frequency response function $G(j\omega)$ by

$$\frac{C}{R}(j\omega) \;=\; \left|\frac{C}{R}(j\omega)\right| e^{j\phi_{C/R}} \;=\; \frac{G(j\omega)}{1 + G(j\omega)} \;=\; \frac{|G(j\omega)|e^{j\phi_G}}{1 + |G(j\omega)|e^{j\phi_G}}$$

where $\phi_{C/R}$ is the phase-angle of $(C/R)(j\omega)$ and ϕ_G is the phase angle of $G(j\omega)$. *The sensitivity of $(C/R)(j\omega)$ with respect to $|G(j\omega)|$ is given by*

$$S_{|G(j\omega)|}^{(C/R)(j\omega)} \;=\; \frac{d\left(\dfrac{C}{R}(j\omega)\right)}{d|G(j\omega)|} \cdot \frac{|G(j\omega)|}{\dfrac{C}{R}(j\omega)} \;=\; \frac{e^{j\phi_G}}{(1 + |G(j\omega)|e^{j\phi_G})^2} \cdot \frac{|G(j\omega)|}{\dfrac{|G(j\omega)|e^{j\phi_G}}{1 + |G(j\omega)|e^{j\phi_G}}}$$

$$\;=\; \frac{1}{1 + |G(j\omega)|e^{j\phi_G}} \;=\; \frac{1}{1 + G(j\omega)} \qquad\qquad (9.22)$$

Example 9.18.

For the system of Example 9.17 assume $\omega = 1$ and $G(j\omega) = G(j1) = 1 + j$. Then $|G(j\omega)| = \sqrt{2}$, $\phi_G = \pi/4$ radians, $(C/R)(j\omega) = \frac{3}{5} + j\frac{1}{5}$, $|(C/R)(j\omega)| = \sqrt{10}/5$, and $\phi_{C/R} = 0.3215$ radians.

Using the result of the previous example the *sensitivity of $(C/R)(j\omega)$ with respect to $|G(j\omega)|$ is* $S_{|G(j\omega)|}^{(C/R)(j\omega)} = \dfrac{1}{2+j} = \dfrac{2}{5} - j\dfrac{1}{5}$. Then from Equation (9.21) we have

$$S_{|G(j\omega)|}^{|(C/R)(j\omega)|} \;=\; \frac{2}{5} \;=\; 0.4, \qquad \phi_{C/R}\,S_{|G(j\omega)|}^{\phi_{C/R}} \;=\; -\frac{1}{5}, \qquad S_{|G(j\omega)|}^{\phi_{C/R}} \;=\; -\frac{1}{5(0.3215)} \;=\; -0.622$$

These real values of sensitivity mean that a 10% change in $|G(j\omega)|$ will **approximately produce a 4%** change in $|(C/R)(j\omega)|$ and a -6.22% change in $\phi_{C/R}$.

A special but very important class of system transfer functions has the form

$$T \;=\; \frac{A_1 + kA_2}{A_3 + kA_4} \qquad\qquad (9.23)$$

where k is a parameter and A_1, A_2, A_3, A_4 are polynomials in s. This type of **dependence** between a parameter k and a transfer function T is general enough to include most of the systems considered in this book.

Note that while the following development is in terms of transfer functions, everything is applicable to frequency response functions by simply replacing s in all equations by $j\omega$.

Example 9.19.

The transfer function of the following system is

$$T \;\equiv\; \frac{C}{R} \;=\; \frac{K}{s^3 + (a+b)s^2 + abs + K}$$

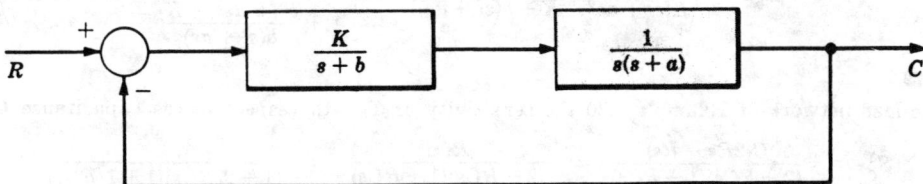

If K is the parameter of interest ($k \equiv K$), we group terms in T as follows:

$$T = \frac{K}{[s^3 + (a+b)s^2 + abs] + K}$$

Comparing T with Equation (9.23), we see that

$$A_1 = 0, \quad A_2 = 1, \quad A_3 = s^3 + (a+b)s^2 + abs, \quad A_4 = 1$$

If a is the parameter of interest ($k \equiv a$), T can be rewritten as

$$T = \frac{K}{[s^3 + bs^2 + K] + a[s^2 + bs]}$$

Comparing this expression with Equation (9.23), we see that

$$A_1 = K, \quad A_2 = 0, \quad A_3 = s^3 + bs^2 + K, \quad A_4 = s^2 + bs$$

If b is the parameter of interest ($k \equiv b$), T can be rewritten as

$$T = \frac{K}{[s^3 + as^2 + K] + b[s^2 + as]}$$

Again comparing this expression with Equation (9.23), we see that

$$A_1 = K, \quad A_2 = 0, \quad A_3 = s^3 + as^2 + K, \quad A_4 = s^2 + as$$

Example 9.20.

For the lead network shown in the adjoining figure the transfer function is $T \equiv \dfrac{E_0}{E_i} = \dfrac{1 + RCs}{2 + RCs}$. If C (capacitance) is the parameter of interest, we write $T = \dfrac{1 + C(Rs)}{2 + C(Rs)}$. Comparing this expression with Equation (9.23), we see that $A_1 = 1$, $A_2 = Rs$, $A_3 = 2$, $A_4 = Rs$.

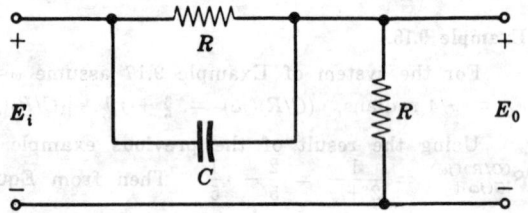

For a transfer function with the form of Equation (9.23), the *sensitivity of T with respect to parameter k* is given by

$$S_k^T \equiv \frac{dT}{dk} \cdot \frac{k}{T} = \frac{k(A_2 A_3 - A_1 A_4)}{(A_3 + kA_4)(A_1 + kA_2)} \tag{9.24}$$

In general, S_k^T is a function of the complex variable s.

Example 9.21.

For the system of Example 9.19 the *sensitivity of T with respect to K* is

$$S_K^T = \frac{K[s^3 + (a+b)s^2 + abs]}{K[s^3 + (a+b)s^2 + abs + K]} = \frac{1}{1 + \dfrac{K}{s^3 + (a+b)s^2 + abs}}$$

The *sensitivity of T with respect to the parameter a* is

$$S_a^T = \frac{-aK(s^2 + bs)}{K[s^3 + bs^2 + K + a(s^2 + bs)]} = \frac{-1}{1 + \dfrac{s^3 + bs^2 + K}{a(s^2 + bs)}}$$

The *sensitivity of T with respect to the parameter b* is

$$S_b^T = \frac{-bK(s^2 + as)}{K[s^3 + as^2 + K + b(s^2 + as)]} = \frac{-1}{1 + \dfrac{s^3 + as^2 + K}{b(s^2 + as)}}$$

Example 9.22.

For the lead network of Example 9.20 the sensitivity of T with respect to the capacitance C is

$$S_C^T = \frac{C(2Rs - Rs)}{(2 + RCs)(1 + RCs)} = \frac{RCs}{(2 + RCs)(1 + RCs)} = \frac{1}{(1 + 2/RCs)(1 + 1/RCs)}$$

Solved Problems

SYSTEM CONFIGURATIONS

9.1. A given plant has the transfer function G_2. A system is desired which includes G_2 as the output element and has a transfer function C/R. Show that, if no constraints (such as stability) are placed on the compensating elements, then such a system can be synthesized as either an open-loop or a unity-feedback system.

If the system can be synthesized as an open-loop system, then it will have the following configuration:

where G_1 is an unknown compensating element. The system transfer function is $C/R = G_1' G_2$, from which $G_1' = (C/R)/G_2$. This value for G_1' permits synthesis of C/R as an open-loop system.

If the system can be synthesized as a unity-feedback system, then it will have the following configuration:

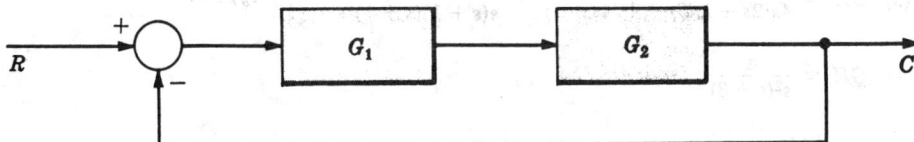

The system transfer function is $\dfrac{C}{R} = \dfrac{G_1 G_2}{1 + G_1 G_2}$ from which $G_1 = \dfrac{1}{G_2}\left(\dfrac{C/R}{1 - C/R}\right)$. This value for G_1 permits synthesis of C/R as a unity-feedback system.

9.2. Using the results of Problem 9.1, show how the system transfer function $\dfrac{C}{R} = \dfrac{2}{s^2 + s + 2}$ which includes as its output element the plant $G_2 = \dfrac{1}{s(s+1)}$ can be synthesized as (a) an open-loop system, (b) a unity-feedback system.

(a) For the open-loop system, $G_1' = \dfrac{C/R}{G_2} = \dfrac{2s(s+1)}{s^2+s+2}$ and the system block diagram is

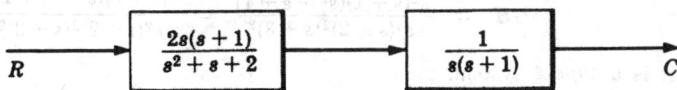

(b) For the unity-feedback system, $G_1 = \dfrac{1}{G_2}\left(\dfrac{C/R}{1-C/R}\right) = s(s+1)\left[\dfrac{2/(s^2+s+2)}{(s^2+s+2-2)/(s^2+s+2)}\right] = 2$ and the system block diagram is

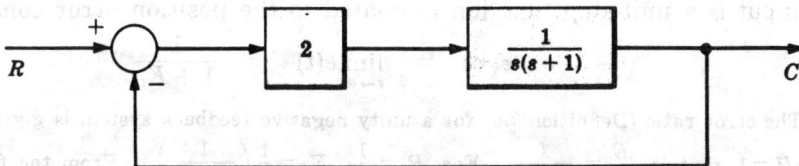

SYSTEMS CLASSIFICATION BY TYPE

9.3. The canonical feedback system is represented by

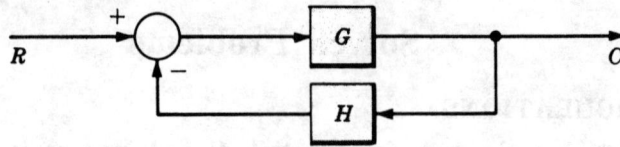

Classify this system according to type if

(a) $G = \dfrac{1}{s}$ $H = 1$

(b) $G = \dfrac{5}{s(s+3)}$ $H = \dfrac{s+1}{s+2}$

(c) $G = \dfrac{2}{s^2 + 2s + 5}$ $H = s + 5$

(d) $G = \dfrac{24}{(2s+1)(4s+1)}$ $H = \dfrac{4}{4s(3s+1)}$

(e) $G = \dfrac{4}{s(s+3)}$ $H = \dfrac{1}{s}$

(a) $GH = \dfrac{1}{s}$; *type 1*.   (b) $GH = \dfrac{5(s+1)}{s(s+2)(s+3)}$; *type 1*.   (c) $GH = \dfrac{2(s+5)}{s^2 + 2s + 5}$; *type 0*.

(d) $GH = \dfrac{96}{4s(2s+1)(3s+1)(4s+1)} = \dfrac{1}{s(s+\frac{1}{2})(s+\frac{1}{3})(s+\frac{1}{4})}$; *type 1*.

(e) $GH = \dfrac{4}{s^2(s+3)}$; *type 2*.

9.4. Classify the following system by type:

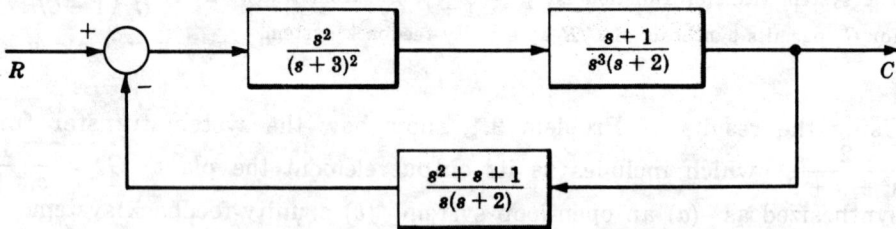

The open-loop transfer function of this system is

$$GH = \frac{s^2(s+1)(s^2+s+1)}{s^4(s+2)^2(s+3)^2} = \frac{(s+1)(s^2+s+1)}{s^2(s+2)^2(s+3)^2}$$

Therefore it is a *type 2* system.

ERROR CONSTANTS AND STEADY STATE ERRORS

9.5. Show that the steady-state error $e(\infty)$ of a stable type l unity-feedback system when the input is a unit-step function is related to the position error constant by

$$e(\infty) = \lim_{t \to \infty} e(t) = \frac{1}{1 + K_p}$$

The error ratio (Definition 7.5) for a unity negative feedback system is given by Equation (7.4) with $H = 1$, that is, $\dfrac{E}{R} = \dfrac{1}{1+G}$. For $R = \dfrac{1}{s}$, $E = \dfrac{1}{s}\left(\dfrac{1}{1+G}\right)$. From the final value theorem, Page 58, we obtain

$$e(\infty) = \lim_{s \to 0} s E(s) = \lim_{s \to 0}\left(\frac{s}{s[1 + G(s)]}\right) = \frac{1}{1 + \lim_{s \to 0} G(s)} = \frac{1}{1 + K_p}$$

where we have used the definition $K_p \equiv \lim_{s \to 0} G(s)$.

9.6.   Show that the steady-state error $e(\infty)$ of a stable type l unity-feedback system with a unit-ramp function input is related to the velocity error constant by $e(\infty) = \lim\limits_{t \to \infty} e(t) = 1/K_v$.

We have $\dfrac{E}{R} = \dfrac{1}{1+G}$, and $E = \dfrac{1}{s^2}\left(\dfrac{1}{1+G}\right)$ for $R = \dfrac{1}{s^2}$. Since $G = \dfrac{K B_1(s)}{s^l B_2(s)}$ by Definition 9.1, then

$$E = \frac{1}{s^2}\left[\frac{s^l B_2(s)}{s^l B_2(s) + K B_1(s)}\right]$$

For $l > 0$, we have

$$s E(s) = \frac{B_2(s)}{s B_2(s) + K B_1(s)/s^{l-1}}$$

where $l - 1 \geq 0$. Now we can use the final value theorem, as was done in the previous problem, because the condition for the application of this theorem is satisfied. That is, for $l > 0$ we have

$$e(\infty) = \lim_{s \to 0} s E(s) = \begin{cases} 0 & \text{for } l > 1 \\[2mm] \dfrac{B_2(0)}{K B_1(0)} & \text{for } l = 1 \end{cases}$$

$B_1(0)$ and $B_2(0)$ are nonzero and finite by Definition 9.1; hence the limit exists (i.e. it is finite).

We cannot evoke the final value theorem for the case $l = 0$ because

$$s E(s)\Big|_{l=0} = \frac{1}{s}\left[\frac{B_2(s)}{B_2(s) + K B_1(s)}\right]$$

and the limit as $s \to 0$ of the quantity on the right does not exist. However, we may use the following argument for $l = 0$. Since the system is stable, $B_2(s) + K B_1(s) = 0$ has roots only in the left-half plane. Therefore E can be written with its denominator in the general factored form

$$E = \frac{B_2(s)}{s^2 \prod\limits_{i=1}^{r} (s + p_i)^{n_i}}$$

where $\text{Re}(p_i) > 0$ and $\sum\limits_{i=1}^{r} n_i = n - a$ (see Definition 9.1), i.e. some roots may be repeated. Expanding E into partial fractions (Equation (4.10a), Page 65), we obtain

$$E = \frac{c_{20}}{s^2} + \frac{c_{10}}{s} + \sum_{i=1}^{r}\sum_{k=1}^{n_i}\frac{c_{ik}}{(s + p_i)^k}$$

where b_n in Equation (4.10a) is zero because the degree of the denominator is greater than that of the numerator ($m < n$). Inverting $E(s)$ (Section 4.8, Page 66), we get

$$e(t) = c_{20}t + c_{10} + \sum_{i=1}^{r}\sum_{k=1}^{n_i}\frac{c_{ik}}{(k-1)!}\,t^{k-1}\,e^{-p_i t}$$

Since $\text{Re}(p_i) > 0$ and c_{20} and c_{10} are finite nonzero constants (E is a rational algebraic expression), then

$$e(\infty) = \lim_{t \to \infty} e(t) = \lim_{t \to \infty}(c_{20}t) + c_{10} = \infty$$

Collecting results, we have

$$e(\infty) = \begin{cases} \infty & \text{for } l = 0 \\[2mm] \dfrac{B_2(0)}{K B_1(0)} & \text{for } l = 1 \\[2mm] 0 & \text{for } l > 1 \end{cases} \qquad \text{Equivalently,} \quad \frac{1}{e(\infty)} = \begin{cases} 0 & \text{for } l = 0 \\[2mm] \dfrac{K B_1(0)}{B_2(0)} & \text{for } l = 1 \\[2mm] \infty & \text{for } l > 1 \end{cases}$$

These three values for $1/e(\infty)$ define K_v; thus

$$e(\infty) = 1/K_v$$

9.7. For the stable system

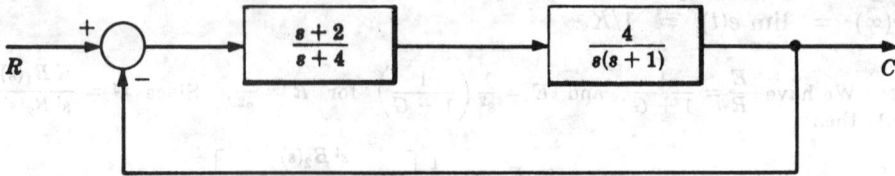

find the position, velocity and acceleration error constants.

Position error constant $K_p = \lim_{s \to 0} G(s) = \lim_{s \to 0} \dfrac{4(s+2)}{s(s+1)(s+4)} = \infty.$

Velocity error constant $K_v = \lim_{s \to 0} s\,G(s) = \lim_{s \to 0} \dfrac{4(s+2)}{(s+1)(s+4)} = 2.$

Acceleration error constant $K_a = \lim_{s \to 0} s^2\,G(s) = \lim_{s \to 0} \dfrac{4s(s+2)}{(s+1)(s+4)} = 0.$

9.8. For the system in Problem 9.7, find the steady-state error for (a) a unit-step input,
(b) a unit-ramp input, (c) a unit-parabolic input.

(a) The steady-state error for a unit-step input is given by $e(\infty) = 1/(1 + K_p)$. Using the result
of Problem 9.7 yields $e(\infty) = 1/(1 + \infty) = 0.$

(b) The steady-state error for a unit-ramp input is given by $e(\infty) = 1/K_v$. Again using the
result of Problem 9.7 we get $e(\infty) = 1/2.$

(c) The steady-state error for a unit-parabolic input is given by $e(\infty) = 1/K_a$.   Then
$e(\infty) = 1/0 = \infty.$

9.9. The following system approximately represents a differentiator:

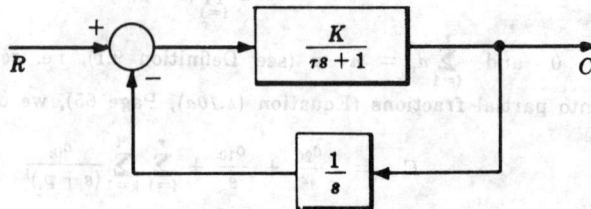

Its transfer function is $\dfrac{C}{R} = \dfrac{Ks}{s(\tau s + 1) + K}.$ Note that $\lim\limits_{\substack{\tau \to 0 \\ K \to \infty}} C/R = s,$ that is,

C/R is a pure differentiator in the limit. Find the step, ramp and parabolic error
constants for this system, where the ideal system T_d is assumed to be a differentiator.

Using the notation of Section 9.7, Page 166, $T_d = s$ and $T_d - \dfrac{C}{R} = \dfrac{s^2(\tau s + 1)}{s(\tau s + 1) + K}.$

Applying Definitions 9.5, 9.6 and 9.7 yields

$$K_s = \frac{1}{\lim\limits_{s \to 0} [T_d - C/R]} = \frac{1}{\lim\limits_{s \to 0}\left[\dfrac{s^2(\tau s + 1)}{s(\tau s + 1) + K}\right]} = \infty$$

$$K_r = \frac{1}{\lim\limits_{s \to 0}\dfrac{1}{s}[T_d - C/R]} = \frac{1}{\lim\limits_{s \to 0}\left[\dfrac{s(\tau s + 1)}{s(\tau s + 1) + K}\right]} = \infty$$

$$K_{pa} = \frac{1}{\lim\limits_{s \to 0}\dfrac{1}{s^2}[T_d - C/R]} = \frac{1}{\lim\limits_{s \to 0}\left[\dfrac{\tau s + 1}{s(\tau s + 1) + K}\right]} = K$$

9.10. Find the steady-state value of the difference (error) between the outputs of a pure differentiator and the approximate differentiator of the previous problem for (a) a unit-step input, (b) a unit-ramp input, (c) a unit-parabolic input.

From Problem 9.10, $K_s = \infty$, $K_r = \infty$ and $K_{pa} = K$.

(a) The steady-state error for a unit-step input is $e(\infty) = 1/K_s = 0$.

(b) The steady-state error for a unit-ramp input is $e(\infty) = 1/K_r = 0$.

(c) The steady state error for a unit-parabolic input is $e(\infty) = 1/K_{pa} = 1/K$.

9.11. Given the stable type 2 unity-feedback system find (a) the position, velocity and acceleration error constants, (b) the steady-state error when the input is $R = \dfrac{3}{s} - \dfrac{1}{s^2} + \dfrac{1}{2s^3}$.

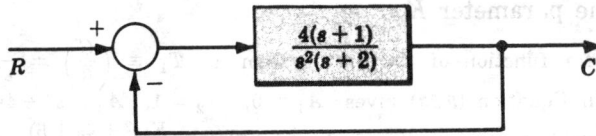

(a) Using the last row of the table on Page 165 (*type 2* systems), the error constants are $K_p = \infty$, $K_v = \infty$, $K_a = (4)(1)/2 = 2$.

(b) The steady-state errors for unit-step, unit-ramp and unit-parabolic inputs are obtained from the same row of the table and are given by: $e_1(\infty) = 0$ for a unit-step; $e_2(\infty) = 0$ for a unit-ramp; $e_3(\infty) = \frac{1}{2}$ for a unit-parabola.

Since the system is linear, the errors can be superimposed. Thus the steady-state error when the input is $R = 3/s - 1/s^2 + 1/2s^3$ is given by $e(\infty) = 3e_1(\infty) - e_2(\infty) + \frac{1}{2}e_3(\infty) = \frac{1}{4}$.

SENSITIVITY

9.12. The following two systems have the same transfer function when $K_1 = K_2 = 100$:

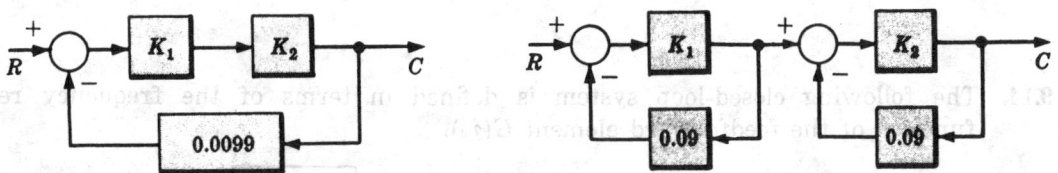

$$T_1 = \left(\frac{C}{R}\right)_1\Bigg|_{\substack{K_1=100 \\ K_2=100}} = \frac{K_1 K_2}{1 + 0.0099 K_1 K_2} = 100$$

$$T_2 = \left(\frac{C}{R}\right)_2\Bigg|_{\substack{K_1=100 \\ K_2=100}} = \left(\frac{K_1}{1 + 0.09 K_1}\right)\left(\frac{K_2}{1 + 0.09 K_2}\right) = 100$$

Compare the sensitivities of these two systems with respect to parameter K_1 for nominal values $K_1 = K_2 = 100$.

For the first system $T_1 = \dfrac{K_1 K_2}{1 + K_1(0.0099 K_2)}$. Comparing this expression with Equation (9.23) yields $A_1 = 0$, $A_2 = K_2$, $A_3 = 1$, $A_4 = 0.0099 K_2$. Substituting these values into Equation (9.24), we obtain

$$S_{K_1}^{T_1} = \frac{K_1 K_2}{(1 + 0.0099 K_1 K_2)(K_1 K_2)} = \frac{1}{1 + 0.0099 K_1 K_2} = 0.01 \quad \text{for } K_1 = K_2 = 100$$

For the second system

$$T_2 = \left(\frac{K_1}{1 + 0.09K_1}\right)\left(\frac{K_2}{1 + 0.09K_2}\right) = \frac{K_1K_2}{1 + 0.09K_1 + 0.09K_2 + 0.0081K_1K_2}$$

Comparing this expression with Equation (9.23) yields $A_1 = 0$, $A_2 = K_2$, $A_3 = 1 + 0.09K_2$, $A_4 = 0.09 + 0.0081K_2$. Substituting these values into Equation (9.24), we have

$$S_{K_1}^{T_2} = \frac{K_1K_2(1 + 0.09K_2)}{(1 + 0.09K_1)(1 + 0.09K_2)(K_1K_2)} = \frac{1}{1 + 0.09K_1} = 0.1 \quad \text{for } K_1 = K_2 = 100$$

A 10% variation in K_1 will approximately produce a 0.1% variation in T_1 and a 1% variation in T_2. Thus the second system T_2 is 10 times more sensitive to variations in K_1 than is the first system T_1.

9.13. Determine the sensitivities of the two systems of Example 9.1, Page 162, with respect to the parameter K.

The transfer function of the first system is $T_1 \equiv \left(\frac{C}{R}\right) = \frac{K}{s^2 + 4s + 5}$. Comparing this expression with Equation (9.23) gives $A_1 = 0$, $A_2 = 1$, $A_3 = s^2 + 4s + 5$, $A_4 = 0$. Substituting these values into Equation (9.24), we obtain $S_K^{T_1} = \frac{K(s^2 + 4s + 5)}{(s^2 + 4s + 5)K} = 1$ for all K.

The transfer function of the second system is $T_2 \equiv \left(\frac{C}{R}\right)_2 = \frac{K}{s^2 + 4s + 3 + K}$. Comparing this expression with Equation (9.23) yields $A_1 = 0$, $A_2 = 1$, $A_3 = s^2 + 4s + 3$, $A_4 = 1$. Substituting these values into Equation (9.24), we obtain

$$S_K^{T_2} = \frac{K(s^2 + 4s + 3)}{(s^2 + 4s + 3 + K)(K)} = \frac{1}{1 + K/(s^2 + 4s + 3)}$$

For $K = 2$, $S_K^{T_2} = \frac{1}{1 + 2/(s^2 + s + 3)}$.

Note that the sensitivity of the open-loop system T_1 is fixed at 1 for all values of gain K. On the other hand, the closed-loop sensitivity is a function of K and the complex variable s. Thus $S_K^{T_2}$ may be adjusted in a design problem by varying K or maintaining the frequencies of the input function within an appropriate range.

9.14. The following closed-loop system is defined in terms of the frequency response function of the feedforward element $G(j\omega)$.

$$\frac{C}{R}(j\omega) = \frac{G(j\omega)}{1 + G(j\omega)}$$

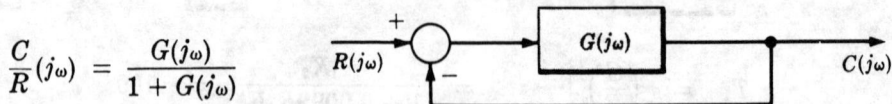

Suppose that $G(j\omega) = 1/(j\omega + 1)$. In Chapter 15 it is shown that the frequency response functions $1/(j\omega + 1)$ can be approximated by the following straight line graphs of magnitude and phase of $G(j\omega)$.

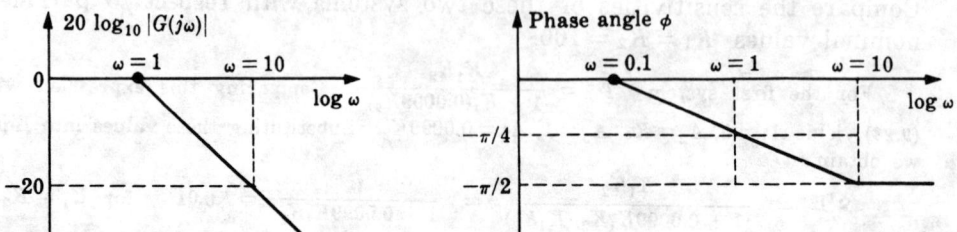

At $\omega = 1$ the true values of $20 \log_{10} |G(j\omega)|$ and ϕ are -3 and $-\pi/4$, respectively. For $\omega = 1$, find:

(a) The sensitivity of $|(C/R)(j\omega)|$ with respect to $|G(j\omega)|$.

(b) Using the result of (a), determine an approximate value for the error in $|(C/R)(j\omega)|$ caused by using the straight line approximations for $1/(j\omega + 1)$.

(a) Using Equation (9.22), Page 169, the sensitivity of $(C/R)(j\omega)$ with respect to $|G(j\omega)|$ is given by

$$S_{|G(j\omega)|}^{(C/R)(j\omega)} = \frac{1}{1 + G(j\omega)} = \frac{1}{2 + j\omega} = \frac{2 - j\omega}{4 + \omega^2}$$

Since $|G(j\omega)|$ is real, $S_{|G(j\omega)|}^{|(C/R)(j\omega)|} = \mathrm{Re}\, S_{|G(j\omega)|}^{(C/R)(j\omega)} = \frac{2}{4 + \omega^2}$. For $\omega = 1$, $S_{|G(j\omega)|}^{|(C/R)(j\omega)|} = 0.4$.

(b) For $\omega = 1$, the exact value of $|G(j\omega)|$ is $|G(j\omega)| = 1/\sqrt{2} = 0.707$. The approximate value taken from the graph is $|G(j\omega)| = 1$. Then the percentage error in the approximation is $100(1 - 0.707)/0.707 = 41.4\%$. The approximate percentage error in $|(C/R)(j\omega)|$ is $41.4\, S_{|G(j\omega)|}^{|(C/R)(j\omega)|} = 16.6\%$.

9.15. Show that the sensitivities of $T(k) = |T(k)|\, e^{j\phi_T}$, the magnitude $|T(k)|$, and the phase angle ϕ_T with respect to parameter k are related by

$$S_k^{T(k)} = S_k^{|T(k)|} + j\phi_T \cdot S_k^{\phi_T} \qquad \text{(Equation (9.21), Page 168)}$$

Using Equation (9.18), Page 168,

$$S_k^{T(k)} = \frac{d \ln T(k)}{d \ln k} = \frac{d \ln [|T(k)|\, e^{j\phi_T}]}{d \ln k} = \frac{d [\ln |T(k)| + j\phi_T]}{d \ln k}$$

$$= \frac{d \ln |T(k)|}{d \ln k} + j\frac{d\phi_T}{d \ln k} = \frac{d \ln |T(k)|}{d \ln k} + j\phi_T \frac{d \ln \phi_T}{d \ln k} = S_k^{|T(k)|} + j\phi_T S_k^{\phi_T}$$

Note that if k is real, then $S_k^{|T(k)|}$ and $S_k^{\phi_T}$ are both real, and

$$S_k^{|T(k)|} = \mathrm{Re}\, S_k^{T(k)}, \qquad \phi_T S_k^{\phi_T} = \mathrm{Im}\, S_k^{T(k)}$$

9.16. Show that the sensitivity of the transfer function $T = \dfrac{A_1 + kA_2}{A_3 + kA_4}$ with respect to the parameter k is given by $S_k^T = \dfrac{k(A_2 A_3 - A_1 A_4)}{(A_3 + kA_4)(A_1 + kA_2)}$.

By definition, the sensitivity of T with respect to the parameter k is

$$S_k^T = \frac{d \ln T}{d \ln k} = \frac{dT}{dk} \cdot \frac{k}{T}$$

Now $\qquad \dfrac{dT}{dk} = \dfrac{A_2(A_3 + kA_4) - A_4(A_1 + kA_2)}{(A_3 + kA_4)^2} = \dfrac{A_2 A_3 - A_1 A_4}{(A_3 + kA_4)^2}$

Thus $\qquad S_k^T = \dfrac{A_2 A_3 - A_1 A_4}{(A_3 + kA_4)^2} \cdot \dfrac{k(A_3 + kA_4)}{A_1 + kA_2} = \dfrac{k(A_2 A_3 - A_1 A_4)}{(A_3 + kA_4)(A_1 + kA_2)}$

Supplementary Problems

9.17. Prove the validity of Equation (*9.9*), Page 166. (*Hint.* See Problems 9.5 and 9.6.)

9.18. Prove the validity of Equation (*9.11*), Page 166. (*Hint.* See Problems 9.5 and 9.6.)

9.19. Prove the validity of Equation (*9.13*), Page 166. (*Hint.* See Problems 9.5 and 9.6.)

9.20. Determine the sensitivity of the system in Problem 7.9, Page 122, to variations in each of the parameters K_1, K_2 and p individually.

9.21. Generate an expression, in terms of the sensitivities determined in Problem 9.20, which relates the total variation in the transfer function of the system in Problem 7.9, Page 122, to variations in K_1, K_2 and p.

9.22. Show that the steady-state error $e(\infty)$ of a stable type l unity feedback system with a unit-parabolic input is related to the acceleration error constant by $e(\infty) = \lim_{t \to \infty} e(t) = 1/K_a$. (*Hint.* See Problem 9.6, Page 173.)

Answers to Supplementary Problems

9.20. $S_{K_1}^{C/R} = \dfrac{s + p}{s + p - K_1 K_2}$ $S_{K_2}^{C/R} = \dfrac{K_1 K_2}{s + p - K_1 K_2}$ $S_p^{C/R} = \dfrac{-p}{s + p - K_1 K_2}$

9.21. $\Delta \dfrac{C}{R} = \dfrac{(s + p)\, \Delta K_1 + (K_1 K_2)\, \Delta K_2 - p\, \Delta p}{s + p - K_1 K_2}$

Chapter 10

The Analysis and Design of Feedback Control Systems: Objectives and Methods

10.1 INTRODUCTION

The basic concepts, the mathematical tools, and the properties of feedback control systems have been presented in the first nine chapters. Attention is now focused on our major goal: the *analysis and design* of feedback control systems.

We have considered only linear models and linear mathematics. In actuality, linear systems do not exist. All physical systems are nonlinear to some extent. Fortunately, a large percentage of systems can be represented by linear models over a limited operating range. Many always operate within this *linear range*. Others exceed the limits of linear operation, but may be *approximated* by linear systems.

The analysis and design methods presented in the next eight chapters are linear techniques, applicable to linear models. Techniques for solving control system problems represented by nonlinear models are quite involved and are not treated here.

This chapter is devoted to making explicit the objectives and to briefly describing the methodology of analysis and design.

10.2 OBJECTIVES OF ANALYSIS

The three predominant objectives of feedback control systems analysis are the determination of the following system characteristics:

1. The degree or extent of system stability
2. The steady-state performance
3. The transient response

Knowing whether a system is absolutely stable or not is insufficient information for most purposes. If a system is stable, we usually want to know how close the system is to being unstable. We need to determine its *relative stability*.

In Chapter 3 we learned that the complete solution of the differential equations describing a system may be split into two parts. The first, the steady-state solution, is that part of the complete solution which does not approach zero as time approaches infinity. The second, the transient response, is that part of the complete solution which does approach zero (or decays) as time approaches infinity.

There is a strong correlation between relative stability and transient response of feedback control systems.

10.3 METHODS OF ANALYSIS

The general procedure for analyzing a control system is the following:

1. Determine the equations or transfer function for each system component.
2. Choose a model for representing the system (block diagram or signal flow graph).

3. Formulate the system model by appropriately connecting the components (blocks or nodes and branches).

4. Determine the system characteristics.

There are several methods available for determining the characteristics of linear systems. Direct solution of the system differential equation may be employed to find the steady-state and transient solutions (Chapters 3 and 4). This technique is cumbersome for higher than second-order systems, and relative stability is difficult to study in the time-domain.

Four primarily graphical methods are available to the control systems analyst which are simpler and more direct than the time-domain method for practical linear models of feedback control systems. They are:

1. The Root-Locus Method
2. Bode-Plot Representations
3. Nyquist Diagrams
4. Nichols Charts

The latter three are frequency-domain techniques. All four are considered in detail in Chapters 13, 15, 11, and 17, respectively.

10.4 DESIGN OBJECTIVES

The basic goal of control system design is meeting *performance specifications*. Performance specifications are the constraints put on the mathematical functions describing system characteristics. They may be stated in any number of ways. Generally they take two forms:

1. Frequency-domain specifications (pertinent quantities expressed as functions of frequency)
2. Time-domain specifications (in terms of time-response)

The desired system characteristics may be prescribed in either or both of the above forms. In general, they specify three important properties of dynamic systems:

1. Speed of response
2. Relative stability
3. System accuracy or allowable error

Frequency domain specifications are usually stated in the following terms:

1. Gain Margin

Gain margin, a measure of relative stability, is defined as the magnitude of the reciprocal of the open-loop transfer function, evaluated at the frequency ω_π at which the phase angle (see Chapter 6) is -180 degrees. That is,

$$\text{gain margin} \equiv \frac{1}{|GH(j\omega_\pi)|} \qquad (10.1)$$

where $\arg GH(j\omega_\pi) = -180$ degrees $= -\pi$ radians   and ω_π is called the **phase crossover** frequency.

2. Phase Margin ϕ_{PM}

Phase margin ϕ_{PM}, a measure of relative stability, is defined as 180 degrees plus the phase angle ϕ_1 of the open-loop transfer function at unity gain. That is,

$$\phi_{PM} \equiv [180 + \arg GH(j\omega_1)] \text{ degrees} \qquad (10.2)$$

where $|GH(j\omega_1)| = 1$ and ω_1 is called the **gain crossover** frequency.

Example 10.1.

The gain and phase margins of a typical feedback control system are illustrated in the following graphs:

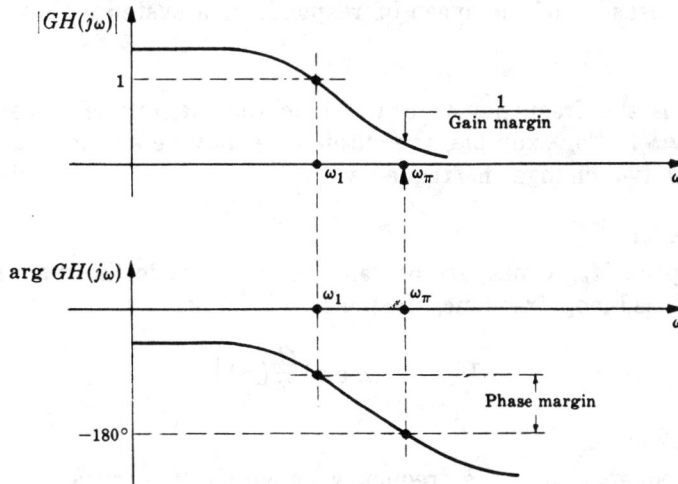

3. Delay Time T_d

Delay time T_d, a measure of the speed of response, is given by

$$T_d(\omega) = -\frac{d\gamma}{d\omega} \qquad (10.3)$$

where $\gamma = \arg \dfrac{C}{R}(j\omega)$. The average value of $T_d(\omega)$ over the frequencies of interest is usually specified.

4. Bandwidth (BW)

The bandwidth of a system was defined in Chapter 1 as that range of frequencies (of the input) over which the system will respond satisfactorily.

Satisfactory performance is determined by the application and the characteristics of the particular system. For example, audio amplifiers are often compared on the basis of their bandwidth. An ideal high-fidelity audio amplifier has a *flat frequency response* from 20 Hz to 20 kHz. That is, it has a pass-band or bandwidth of 19,980 Hz (usually rounded-off to 20 kHz). Flat frequency response means that the *magnitude ratio* of output to input is essentially constant over the bandwidth. Hence signals in the audio spectrum are faithfully reproduced by a 20 kHz bandwidth amplifier. The magnitude ratio is the absolute value of the system transfer function with $s = j\omega$.

The frequency response of a high-fidelity audio amplifier is shown in the adjoining figure. The magnitude ratio is 0.707 of, or approximately 3 db below, its maximum at the *cutoff frequencies*

$$f_{c1} = 20 \text{ Hz}, \quad f_{c2} = 20 \text{ kHz}$$

"db" is the abbreviation for *decibel*, defined by the following equation:

$$\text{db} \equiv 20 \log_{10}(\text{magnitude ratio}) \qquad (10.4)$$

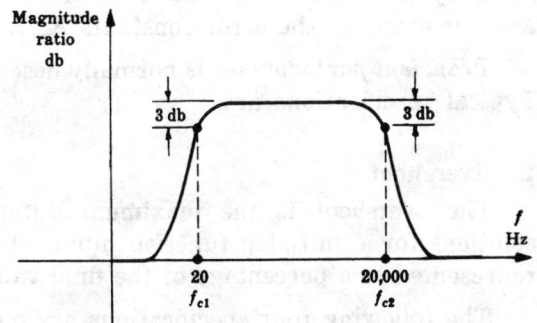

Often the bandwidth of a system is defined as that range of frequencies over which the magnitude ratio does not differ by more than -3 db from its value at a specified frequency. For many feedback control systems this frequency is zero. The bandwidth is in this case equal to the cutoff frequency ω_c (in radians per second).

In general, the precise meaning of bandwidth is made clear by the problem description.

Bandwidth is a measure of the speed of response of a system.

5. Cutoff Rate

The cutoff rate is the frequency rate at which the magnitude ratio decreases beyond the cutoff frequency ω_c. For example, the cutoff rate may be specified as 6 db/octave. An octave is a factor of two change in frequency.

6. Resonance Peak M_p

The resonance peak M_p, a measure of relative stability, is the maximum value of the magnitude of the closed-loop frequency response. That is,

$$M_p \equiv \max_{\omega} \left| \frac{C}{R}(j\omega) \right| \tag{10.5}$$

7. Resonant Frequency ω_p

The resonant frequency ω_p is the frequency at which M_p occurs.

Example 10.2.

The bandwidth BW, cutoff frequency ω_c, resonance peak M_p, and resonant frequency ω_p for an underdamped second-order system are illustrated below.

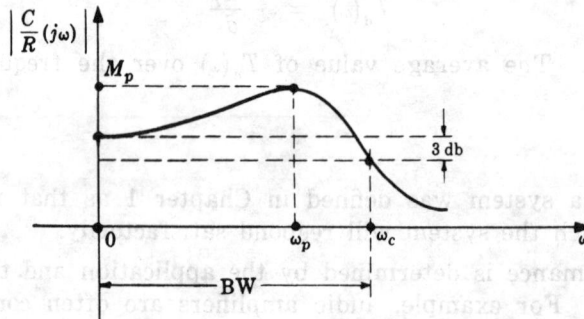

Time-domain specifications are customarily defined in terms of unit-step function, ramp function, and parabolic function responses. Each response has a steady-state and a transient component.

Steady-state performance, in terms of steady-state error, is a measure of system accuracy when a specific input is applied. Figures of merit for steady-state performance are, for example, the error constants K_p, K_v, K_a defined in Chapter 9.

Transient performance is normally described in terms of the unit-step function response. Typical specifications are:

1. Overshoot

The overshoot is the maximum difference between the transient and steady-state solutions for a unit-step function input. It is a measure of relative stability and is often represented as a percentage of the final value of the output (steady-state solution).

The following four specifications are measures of the speed of response.

2. Delay Time T_d

The delay time T_d, interpreted as a time-domain specification, is often defined as the time required for the response to a unit-step function input to reach 50 percent of its final value.

3. Rise Time T_r

The rise time T_r is customarily defined as the time required for the response to a unit-step function input to rise from 10 to 90 percent of its final value.

4. Settling Time T_s

The settling time T_s is most often defined as the time required for the response to a unit-step function input to reach and remain within a specified percentage (frequently 2 or 5 percent) of its final value.

5. Predominant Time Constant τ

The predominant time constant τ is an alternative measure for settling time. The envelope of the transient response decays to 37 percent of its initial value in τ seconds.

The time constant of a stable system is defined in terms of the exponentially decaying character of the transient response. For first-order and second-order underdamped systems the transient terms have the form $Ae^{-\alpha t}$ and $Ae^{-\alpha t} \cos{(\omega_d t + \theta)}$ respectively $(\alpha > 0)$. In each case the decay is governed by $e^{-\alpha t}$. The time constant τ is defined as the time required to make the exponent $-\alpha t$ equal to -1. Hence $\tau = 1/\alpha$.

For feedback control systems of order higher than two, the predominant time constant can often be estimated from the time constant of an underdamped second-order system which approximates the higher order system, since

$$\tau \leq \frac{1}{\zeta \omega_n} \qquad\qquad (10.6)$$

ζ and ω_n (Chapter 3) are the two most significant figures of merit, defined for second-order but often useful for higher-order systems. Specifications are often given in terms of ζ and ω_n.

Example 10.3.

The following plot of the unit-step response of an underdamped second-order system illustrates time-domain specifications:

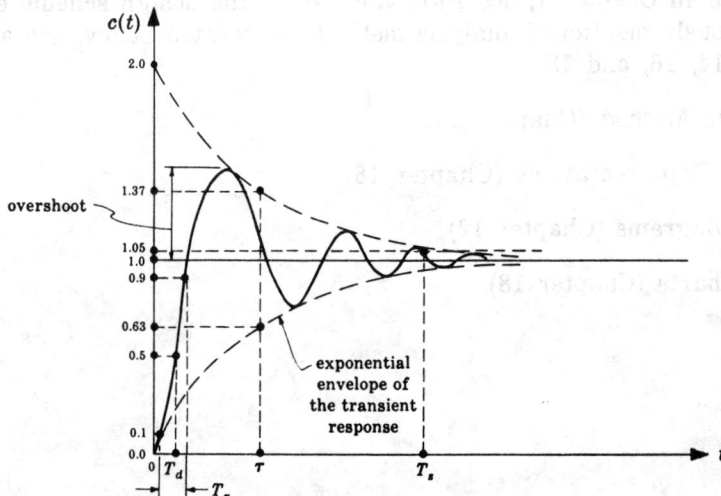

10.5 DESIGN METHODS

In order to meet the performance specifications for feedback control systems, appropriate *compensation* networks must usually be introduced into the system. (We assume here that G and H are fixed configurations of components over which the designer has no control.) Compensation networks may consist of either passive or active elements, several of which have already been discussed, especially in Chapter 6. They may be introduced into the forward path (cascade compensation), or the feedback path (feedback compensation):

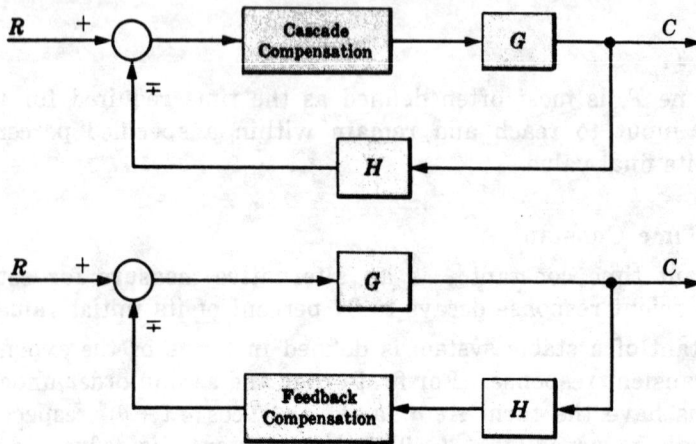

Feedback compensation may also occur in minor feedback loops:

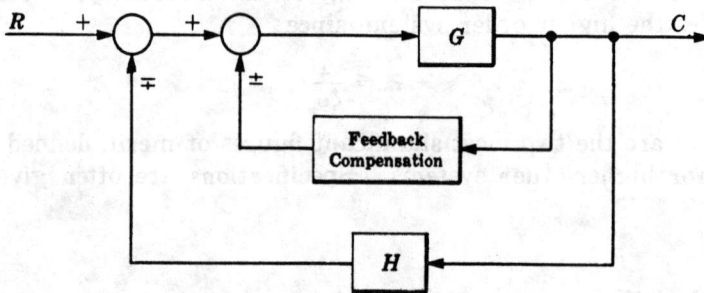

As pointed out in Chapter 1, design-by-analysis is the design scheme employed in this book. The previously mentioned analysis methods, reiterated below, are applied to design in Chapters 12, 14, 16, and 18.

1. Root-Locus Method (Chapter 14)

2. Bode-Plot Representations (Chapter 16)

3. Nyquist Diagrams (Chapter 12)

4. Nichols Charts (Chapter 18)

Solved Problems

10.1. The following graph represents the input-output characteristic of a controller-amplifier for a feedback control system whose other components are linear. What is the linear range of $e(t)$ for this system?

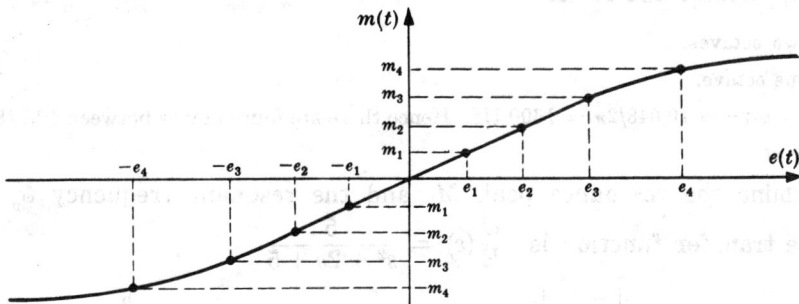

The amplifier-controller operates linearly over the approximate range $-e_3 \leq e \leq e_3$.

10.2. Determine the gain margin for the system in which $GH(j\omega) = \dfrac{1}{(j\omega + 1)^3}$.

Writing $GH(j\omega)$ in polar form, we have

$$GH(j\omega) = \frac{1}{(\omega^2 + 1)^{3/2}} \underline{/-3 \tan^{-1} \omega}, \quad \arg GH(j\omega) = -3 \tan^{-1} \omega$$

Then $-3 \tan^{-1} \omega_\pi = -\pi$, $\omega_\pi = \tan(\pi/3) = 1.732$. Hence, by Equation (10.1), gain margin $= \dfrac{1}{|GH(j\omega_\pi)|} = 8$.

10.3. Determine the phase margin for the system of Problem 10.2.

We have $|GH(j\omega)| = \dfrac{1}{(\omega^2 + 1)^{3/2}} = 1$ only when $\omega = \omega_1 = 0$. Therefore

$$\phi_{PM} = 180° + (-3 \tan^{-1} 0) = 180° = \pi \text{ rad}$$

10.4. Determine the average value of $T_d(\omega)$ over the frequency range $0 \leq \omega \leq 10$ for $\dfrac{C}{R} = \dfrac{j\omega}{j\omega + 1}$. $T_d(\omega)$ is given by Equation (10.3).

$$\gamma = \arg \frac{C}{R}(j\omega) = \frac{\pi}{2} - \tan^{-1} \omega \quad \text{and} \quad T_d(\omega) = \frac{-d\gamma}{d\omega} = \frac{d}{d\omega}[\tan^{-1} \omega] = \frac{1}{1 + \omega^2}$$

Therefore

$$\text{Avg } T_d(\omega) = \frac{1}{10} \int_0^{10} \frac{d\omega}{1 + \omega^2} = 0.147 \text{ s}$$

10.5. Determine the bandwidth for the system whose transfer function is $\dfrac{C}{R}(s) = \dfrac{1}{s + 1}$.

We have $\left| \dfrac{C}{R}(j\omega) \right| = \dfrac{1}{\sqrt{\omega^2 + 1}}$. A sketch of $\left| \dfrac{C}{R}(j\omega) \right|$ versus ω is given by

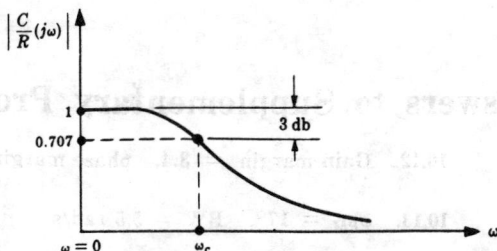

ω_c is determined from $1/\sqrt{\omega_c^2 + 1} = 0.707$. Since $\left|\dfrac{C}{R}(j\omega)\right|$ is a strictly decreasing function of positive frequency, we have BW $= \omega_c = 1$ rad.

10.6. How many octaves are between (a) 200 Hz and 800 Hz, (b) 200 Hz and 100 Hz, (c) 10,048 rad/s and 100 Hz?

(a) Two octaves.

(b) One octave.

(c) $f = \omega/2\pi = 10,048/2\pi = 1600$ Hz. Hence there are four octaves between 10,048 rad/s and 100 Hz.

10.7. Determine the resonance peak M_p and the resonant frequency ω_p for the system whose transfer function is $\dfrac{C}{R}(s) = \dfrac{5}{s^2 + 2s + 5}$.

$$\left|\frac{C}{R}(j\omega)\right| = \frac{5}{|-\omega^2 + 2j\omega + 5|} = \frac{5}{\sqrt{\omega^4 - 6\omega^2 + 25}}$$

Setting the derivative of $\left|\dfrac{C}{R}(j\omega)\right|$ equal to zero, we get $\omega_p = \pm\sqrt{3}$. Therefore

$$M_p = \max_\omega \left|\frac{C}{R}(j\omega)\right| = \left|\frac{C}{R}(j\sqrt{3})\right| = \frac{5}{4}$$

10.8. The output in response to a unit step function input for a particular control system is $c(t) = 1 - e^{-t}$. What is the delay time T_d?

The output is given as a function of time. Therefore the time-domain definition of T_d presented in Section 10.4 is applicable. The final value of the output is $\lim_{t \to \infty} c(t) = 1$. Hence T_d (at 50 percent of the final value) is the solution of $0.5 = 1 - e^{-T_d}$, and is equal to $\log_e(2)$, or 0.693.

10.9. Find the rise time T_r for $c(t) = 1 - e^{-t}$.

At 10 percent of the final value, $0.1 = 1 - e^{-t_1}$; hence $t_1 = 0.104$ s. At 90 percent of the final value, $0.9 = 1 - e^{-t_2}$; thus $t_2 = 2.302$ s. Then $T_r = 2.302 - 0.104 = 2.198$ s.

Supplementary Problems

10.10. Determine the phase margin for $GH = 2(s+1)/s^2$.

10.11. Find the bandwidth for $GH = \dfrac{60}{s(s+2)(s+6)}$ for the closed-loop system.

10.12. Calculate the gain and phase margin for $GH = \dfrac{432}{s(s^2 + 13s + 115)}$.

10.13. Calculate the phase margin and bandwidth for $GH = \dfrac{640}{s(s+4)(s+16)}$ for the closed-loop system.

Answers to Supplementary Problems

10.10. $\phi_{PM} = 65.5°$ **10.12.** Gain margin $= 3.4$, phase margin $= 65°$

10.11. BW $= 3$ rad/s   **10.13.** $\phi_{PM} = 17°$, BW $= 5.5$ rad/s

Chapter 11

Nyquist Analysis

11.1 INTRODUCTION

Nyquist analysis, a frequency response method, is essentially a graphical procedure for determining absolute and relative stability of closed-loop control systems. Information about stability is available directly from a graph of the sinusoidal open-loop transfer function $GH(j\omega)$, once the feedback system has been put into canonical form.

There are several reasons why the Nyquist method may be chosen to determine information about system stability. The methods of Chapter 5 (Routh, Hurwitz, etc.) are often inadequate because, with few exceptions, they can only be used for determining *absolute* stability, and are only applicable to systems whose characteristic equation is a *finite polynomial* in s. When a signal is delayed by T seconds somewhere in the loop of a system, exponential terms of the form e^{-Ts} appear in the characteristic equation. The methods of Chapter 5 can be applied to such systems only if e^{-Ts} is approximated by a few terms of the power series

$$e^{-Ts} = 1 - Ts + \frac{T^2 s^2}{2!} - \frac{T^3 s^3}{3!} + \cdots$$

and this technique yields only *approximate* stability information. The Nyquist method handles systems with time delays without the necessity of approximations, and hence yields *exact* results about both absolute and relative stability of the system.

Nyquist techniques are also useful for obtaining information about transfer functions of components or systems from experimental frequency response data. The Polar Plot (Section 11.5) may be directly graphed from sinusoidal steady-state measurements on the components making up the open-loop transfer function. This feature is very useful in the determination of system stability characteristics when transfer functions of loop components are not available in analytic form, or when physical systems are to be tested and evaluated experimentally.

In the next several sections the mathematical preliminaries and techniques necessary for generating Polar Plots and Nyquist Stability Plots of feedback control systems, and the mathematical basis and properties of the Nyquist Stability Criterion, are presented. The remaining sections of this chapter deal with the interpretation and uses of Nyquist analysis for the determination of *relative* stability and evaluation of the closed-loop frequency response.

11.2 PLOTTING COMPLEX FUNCTIONS OF A COMPLEX VARIABLE

A real function of a real variable is easily graphed on a single set of coordinate axes. For example, the real function $f(x)$, x real, is easily plotted in rectangular coordinates with x as the abscissa and $f(x)$ as the ordinate. A complex function of a complex variable, such as the transfer function $P(s)$ with $s = \sigma + j\omega$, cannot be plotted on a single set of coordinates.

The complex variable $s = \sigma + j\omega$ is dependent upon two independent quantities, the real and imaginary parts of s. Hence s cannot be represented by a line. The complex function $P(s)$ also has real and imaginary parts. It too cannot be graphed in a single dimension.

In general, in order to plot $P(s)$ with $s = \sigma + j\omega$, two 2-dimensional graphs are required. The first is a graph of $j\omega$ versus σ called the **s-plane**, the same set of coordinates as those used for plotting pole-zero maps in Chapter 4. The second is the imaginary part of $P(s)$ (Im P) versus the real part of $P(s)$ (Re P) called the **$P(s)$-plane**.

There is a correspondence between points in the two planes that is called a **mapping** or **transformation**. Points in the s-plane are *mapped* into points of the $P(s)$-plane by the function P:

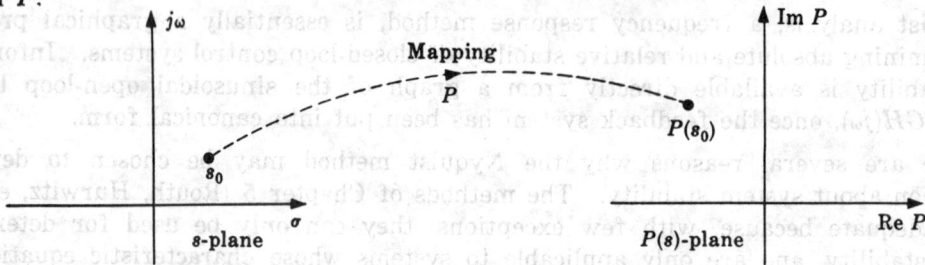

In general, only a very specific locus of points in the s-plane is mapped into the $P(s)$-plane. For Nyquist Stability Plots this locus is called the *Nyquist Path*, and is the subject of Section 11.7.

For the special case $\sigma = 0$, $s = j\omega$, the s-plane degenerates into a line and $P(j\omega)$ may be represented in a $P(j\omega)$-plane with ω as a parameter. *Polar Plots* are constructed in the $P(j\omega)$-plane from this line $(s = j\omega)$ in the s-plane.

Example 11.1.

Consider the complex function $P(s) = s^2 + 1$. The point $s_0 = 2 + j4$ is mapped into the point $P(s_0) = P(2 + j4) = (2 + j4)^2 + 1 = -11 + j16$.

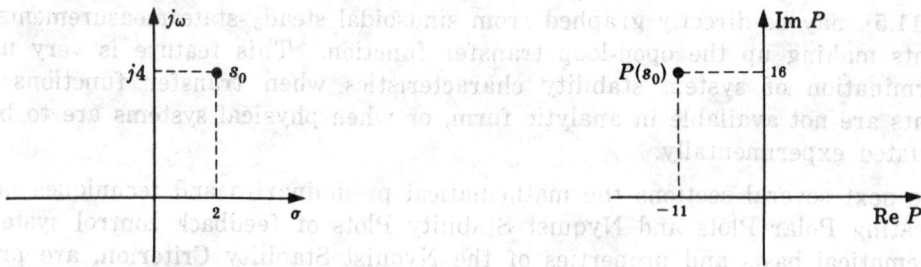

11.3 DEFINITIONS

The following definitions are required in subsequent sections.

Definition 11.1: If the *derivative* of P at s_0 defined by

$$\left.\frac{dP}{ds}\right|_{s=s_0} \equiv \lim_{s \to s_0}\left[\frac{P(s) - P(s_0)}{s - s_0}\right]$$

exists at all points in a region of the s-plane (that is, if the limit is finite and unique), then P is **analytic** in that region.

Transfer functions of all practical physical systems (those considered in this book) are analytic in the finite s-plane except at the poles of $P(s)$.

Definition 11.2: A point at which $P(s)$ is not analytic is a **singular point** or **singularity** of $P(s)$.

A *pole* of $P(s)$ is a singular point.

Definition 11.3: A **closed contour** in a complex plane is a continuous curve beginning and ending at the same point.

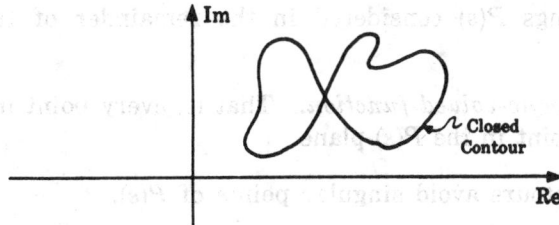

Definition 11.4: All points to the right of a contour as it is traversed in a prescribed direction are said to be **enclosed** by it.

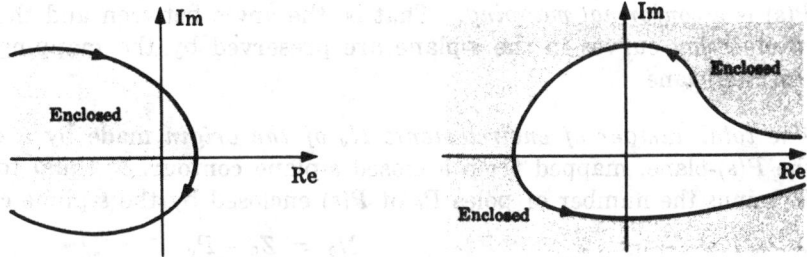

Definition 11.5: A *clockwise* (CW) traverse around a contour is defined as the **positive direction.**

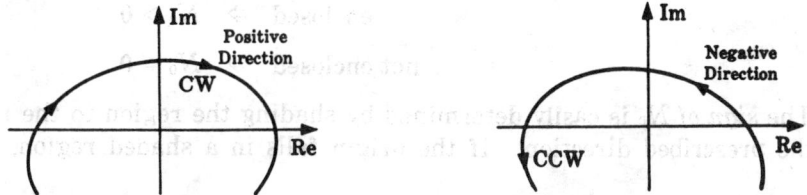

Definition 11.6: A closed contour in the $P(s)$-plane is said to make n **positive encirclements** of the origin if a radial line drawn from the origin to a point on the $P(s)$ curve rotates in a clockwise (CW) direction through $360n$ degrees in going completely around the closed path. If the path is traversed in a counterclockwise (CCW) direction, a **negative encirclement** is obtained. The **total number** of encirclements N_0 is equal to the CW minus the CCW encirclements.

Example 11.2.

The following $P(s)$-plane contour encircles the origin once. That is, $N_0 = 1$.

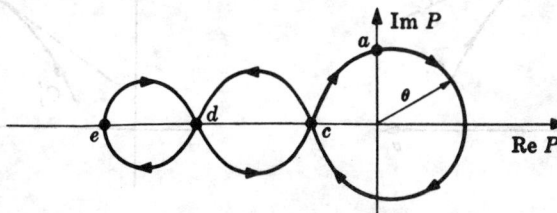

Beginning at point a, we rotate a radial line from the origin to the contour in a CW direction to point c. The angle subtended is $+270°$. From c to d the angle increases, then decreases, and the sum total is $0°$. From d to e and back to d again, the angle swept out by the radial line is again $0°$. d to c is $0°$ and c to a is clearly $+90°$. Hence the total angle is $270° + 90° = 360°$. Therefore $N_0 = 1$.

11.4 PROPERTIES OF THE MAPPING $P(s)$

All mappings $P(s)$ considered in the remainder of this chapter have the following properties.

1. $P(s)$ is a *single-valued function*. That is, every point in the s-plane maps into one and only one point in the $P(s)$-plane.

2. s-plane contours avoid singular points of $P(s)$.

3. $P(s)$ is *analytic* except possibly at a finite number of points (singularities) in the s-plane.

4. Every closed contour in the s-plane maps into a closed contour in the $P(s)$-plane.

5. $P(s)$ is a *conformal mapping*. That is, the angle between and the direction of any two intersecting curves in the s-plane are preserved by the mapping of these curves into the $P(s)$-plane.

6. The *total number of encirclements N_0 of the origin* made by a closed $P(s)$ contour in the $P(s)$-plane, mapped from a closed s-plane contour, is equal to the number of zeros Z_0 minus the number of poles P_0 of $P(s)$ enclosed by the s-plane contour. That is,

$$N_0 = Z_0 - P_0 \tag{11.1}$$

7. If the origin is *enclosed* by the $P(s)$ contour, then $N_0 > 0$. If the origin is *not enclosed* by the $P(s)$ contour, then $N_0 \lesseqgtr 0$. That is

$$\text{enclosed} \;\Rightarrow\; N_0 > 0$$

$$\text{not enclosed} \;\Rightarrow\; N_0 \lesseqgtr 0$$

The *sign of N_0* is easily determined by shading the region to the right of the contour in the prescribed direction. If the origin falls in a shaded region, $N_0 > 0$; if not, $N_0 \lesseqgtr 0$.

Example 11.3.

The principle of conformal mapping is illustrated in the following graphs. Curves C_1 and C_2 are mapped into C_1' and C_2'. The angle between the tangents to these curves at s_0 and $P(s_0)$ is equal to α, and the direction indicated by the arrows is CW in both graphs.

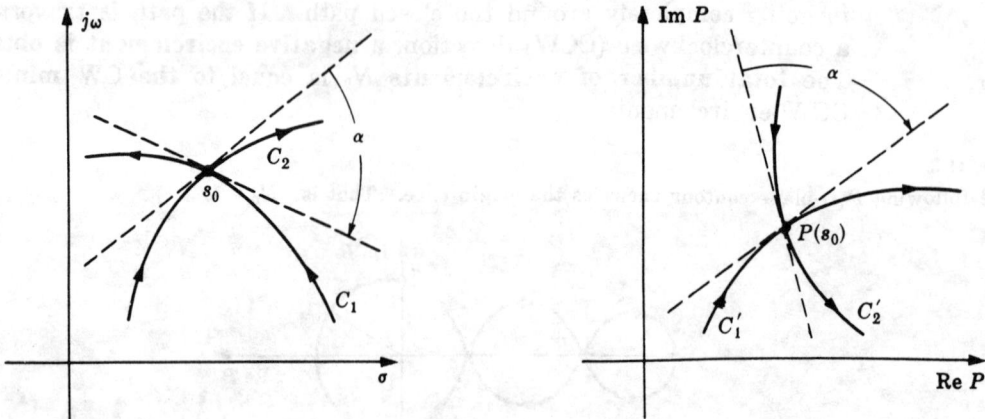

Example 11.4.

A certain transfer function $P(s)$ is known to have one zero in the right half of the s-plane, and this zero is enclosed by the s-plane contour mapped into the $P(s)$-plane as shown below. Points s_1, s_2, s_3 and $P(s_1), P(s_2), P(s_3)$ determine the directions of their respective contours.

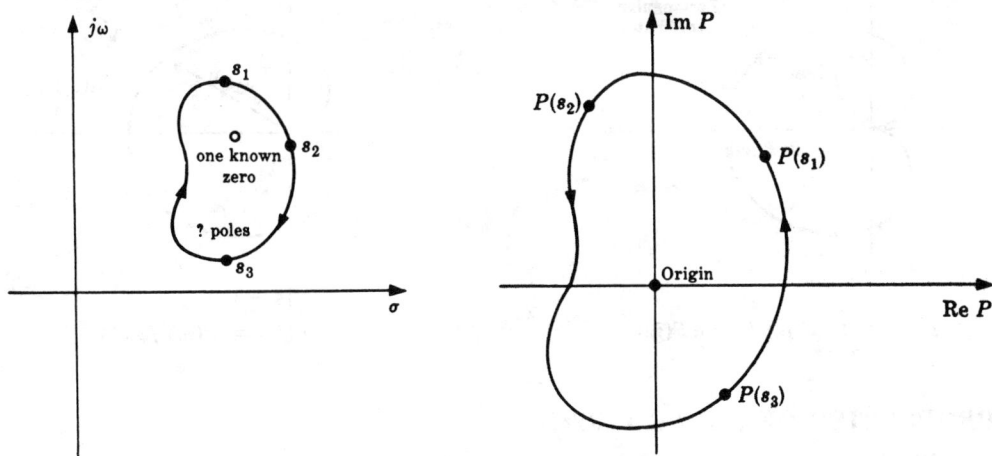

The shaded region to the right of the $P(s)$-plane contour indicates that $N_0 \leqq 0$, since the origin does not lie in the shaded region. But, clearly, the $P(s)$ contour encircles the origin once in a CCW direction. Hence $N_0 = -1$. Thus the number of poles of $P(s)$ enclosed by the s-plane contour is $P_0 = Z_0 - N_0 = 1 - (-1) = 2$.

11.5 POLAR PLOTS

A transfer function $P(s)$ may be represented in the frequency domain as a sinusoidal transfer function by substituting $j\omega$ for s in the expression for $P(s)$. The resulting form $P(j\omega)$ is a complex function of the single variable ω. Therefore it may be plotted in two dimensions with ω as a parameter and written in the following equivalent forms:

$$\text{Polar Form:} \qquad P(j\omega) \;=\; |P(j\omega)|\, \underline{/\phi(\omega)} \qquad\qquad (11.2)$$

$$\text{Euler Form:} \qquad P(j\omega) \;=\; |P(j\omega)|\,(\cos \phi(\omega) + j \sin \phi(\omega)) \qquad (11.3)$$

$|P(j\omega)|$ is the magnitude of the complex function $P(j\omega)$, and $\phi(\omega)$ is its phase angle, $\arg P(\omega)$.

$|P(j\omega)| \cos \phi(\omega)$ is the real part, and $|P(j\omega)| \sin \phi(\omega)$ is the imaginary part of $P(j\omega)$. Therefore $P(j\omega)$ may also be written as

$$\text{Rectangular or Complex Form:} \qquad P(j\omega) \;=\; \operatorname{Re} P(j\omega) + j \operatorname{Im} P(j\omega) \qquad (11.4)$$

A **Polar Plot** of $P(j\omega)$ is a graph of $\operatorname{Im} P(j\omega)$ versus $\operatorname{Re} P(j\omega)$ in the finite portion of the $P(j\omega)$-plane for $-\infty < \omega < \infty$. At singular points of $P(j\omega)$ (poles on the $j\omega$-axis), $|P(j\omega)| \to \infty$. A Polar Plot may also be generated on polar coordinate paper. The magnitude and phase angle of $P(j\omega)$ are plotted with ω varying from $-\infty$ to $+\infty$.

The locus of $P(j\omega)$ is identical on either rectangular or polar coordinates. The choice of coordinate system may depend on whether $P(j\omega)$ is available in analytic form or as experimental data. If $P(j\omega)$ is expressed analytically, the choice of coordinates depends on whether it is easier to write $P(j\omega)$ in the form of Equation (11.2), in which case polar coordinates are used, or in the form of Equation (11.4) for rectangular coordinates. Experimental data on $P(j\omega)$ are usually expressed in terms of magnitude and phase angle. In this case, polar coordinates are the natural choice.

Example 11.5.

The following Polar Plots are identical; only the coordinate systems are different.

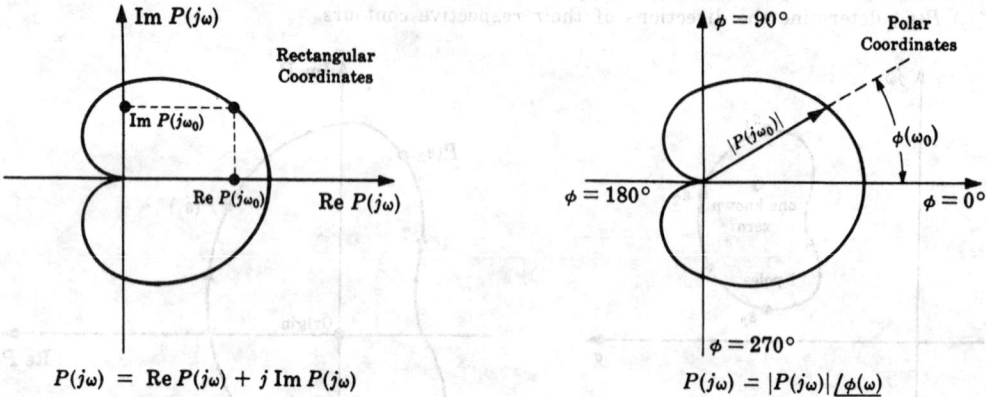

$$P(j\omega) = \operatorname{Re} P(j\omega) + j \operatorname{Im} P(j\omega)$$

$$P(j\omega) = |P(j\omega)|\,\underline{/\phi(\omega)}$$

11.6 PROPERTIES OF POLAR PLOTS

The following are several useful properties of Polar Plots.

1. The Polar Plot for $$P(j\omega) + a$$

   where a is any complex constant, is identical to the plot for $P(j\omega)$ with the origin of coordinates shifted to the point $-a = -(\operatorname{Re} a + j \operatorname{Im} a)$.

2. The Polar Plot of the transfer function of a time-invariant, constant-coefficient, linear system exhibits conjugate symmetry. That is, the graph for $-\infty < \omega < 0$ is the mirror image about the horizontal axis of the graph for $0 \leq \omega < \infty$.

3. The Polar Plot may be constructed directly from a Bode Plot (Chapter 15), if one is available. Values of magnitude and phase angle at various frequencies ω on the Bode Plot represent points along the locus of the Polar Plot.

4. Constant increments of frequency are not generally separated by equal intervals along the Polar Plot.

Example 11.6.

For $a = 1$ and $P = GH$, the Polar Plot of the function $1 + GH$ is given by the plot for GH, with the origin of coordinates shifted to the point $-1 + j0$ in rectangular coordinates:

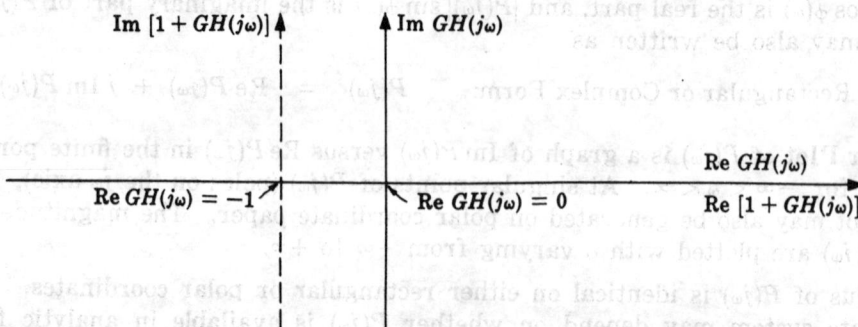

Example 11.7.

In order to illustrate the plotting of a transfer function, consider the open-loop transfer function GH given by

$$GH(s) = \frac{1}{s+1}$$

Letting $s = j\omega$ and rewriting $GH(j\omega)$ in the form of Equation (11.2) (polar form), we have

$$GH(j\omega) \;=\; \frac{1}{j\omega + 1} \;=\; \frac{1}{\sqrt{\omega^2 + 1}} \; \underline{/ - \tan^{-1} \omega}$$

For $\omega = 0$, $\omega = 1$, and $\omega \to \infty$:

$$GH(j0) \;=\; 1 \; \underline{/0^\circ}$$

$$GH(j1) \;=\; (1/\sqrt{2}) \; \underline{/-45^\circ}$$

$$\lim_{\omega \to \infty} GH(j\omega) \;=\; 0 \; \underline{/-90^\circ}$$

Substitution of several other positive values of ω yields a semicircular locus for $0 \leq \omega < \infty$. The graph for $-\infty < \omega < 0$ is the mirror image about the diameter of this semicircle. It is shown in the adjoining complete Polar Plot by a broken line. Note the strikingly unequal increments of frequency between the arcs $\overline{ab}$ and $\overline{bc}$.

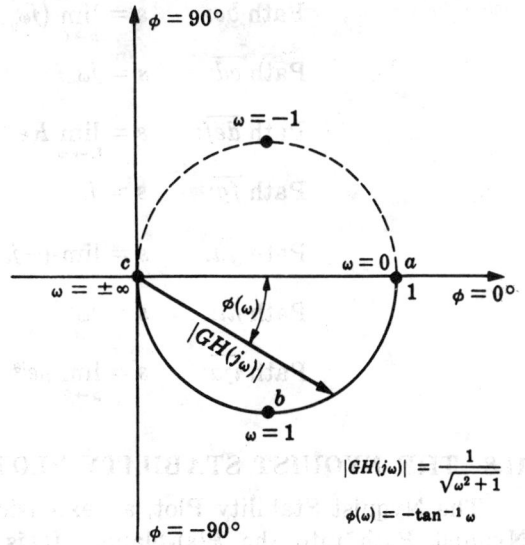

11.7 THE NYQUIST PATH

The **Nyquist Path** is a closed contour in the s-plane which completely encloses the entire right half of the s-plane (RHP).

In order that the Nyquist Path should not pass through any poles of $P(s)$, small semicircles along the imaginary axis or at the origin of $P(s)$ are required in the path if $P(s)$ has poles on the $j\omega$ axis or at the origin. The radii ρ of these small circles are interpreted as approaching zero in the limit.

In order to enclose the RHP at infinity, a large semicircular path is drawn in the RHP and the radius R of this semicircle is interpreted as being infinite in the limit.

The generalized Nyquist Path is illustrated by the following s-plane contour:

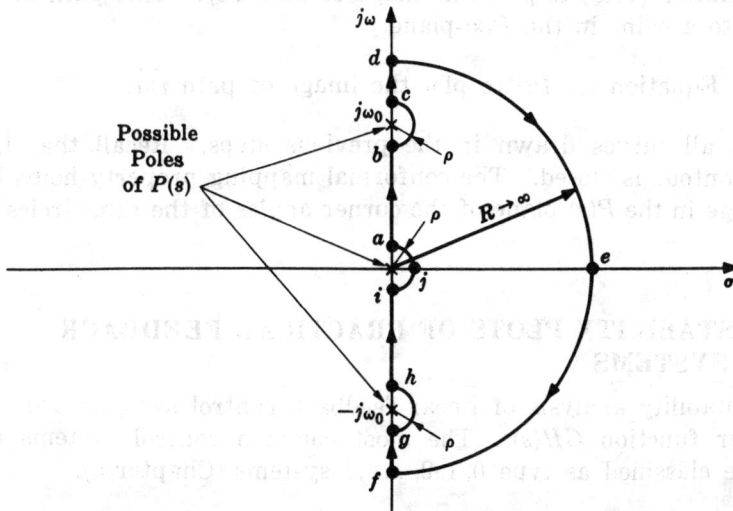

It is apparent that *every pole and zero of $P(s)$ in the RHP is enclosed by the Nyquist Path* when it is mapped into the $P(s)$-plane.

The various portions of the Nyquist Path can be described analytically in the following manner.

$$\text{Path } \overline{ab}: \qquad s = j\omega \qquad\qquad 0 < \omega < \omega_0 \qquad\qquad (11.5)$$

Path $\overline{bc}$:	$s = \lim_{\rho \to 0} (j\omega_0 + \rho e^{j\theta})$	$-90° \leqq \theta \leqq 90°$	(11.6)
Path $\overline{cd}$:	$s = j\omega$	$\omega_0 \leqq \omega < \infty$	(11.7)
Path $\overline{def}$:	$s = \lim_{R \to \infty} R e^{j\theta}$	$+90° \leqq \theta \leqq -90°$	(11.8)
Path $\overline{fg}$:	$s = j\omega$	$-\infty < \omega < -\omega_0$	(11.9)
Path $\overline{gh}$:	$s = \lim_{\rho \to 0} (-j\omega_0 + \rho e^{j\theta})$	$-90° \leqq \theta \leqq 90°$	(11.10)
Path $\overline{hi}$:	$s = j\omega$	$-\omega_0 < \omega < 0$	(11.11)
Path $\overline{ija}$:	$s = \lim_{\rho \to 0} \rho e^{j\theta}$	$-90° \leqq \theta \leqq 90°$	(11.12)

11.8 THE NYQUIST STABILITY PLOT

The Nyquist Stability Plot, an extension of the Polar Plot, is a mapping of the *entire* Nyquist Path into the $P(s)$-plane. It is constructed using the mapping properties of Sections 11.4 and 11.6, Equations (11.5) through (11.8) and Equation (11.12) of Section 11.7. A carefully drawn sketch is sufficient for most purposes.

A general construction procedure is outlined in the following steps.

Step 1: Check $P(s)$ for poles on the $j\omega$ axis and at the origin.

Step 2: Using Equations (11.5) through (11.7), sketch the image of path $\overline{ad}$ in the $P(s)$-plane. If there are no poles on the $j\omega$-axis, Equation (11.6) need not be employed. In this case, Step 2 should read: sketch the Polar Plot of $P(j\omega)$.

Step 3: Draw the mirror image about the real axis $\text{Re}\,P$ of the sketch resulting from Step 2. This is the mapping of path $\overline{fi}$.

Step 4: Use Equation (11.8) to plot the image of path $\overline{def}$. This path at infinity usually plots into a point in the $P(s)$-plane.

Step 5: Employ Equation (11.12) to plot the image of path $\overline{ija}$.

Step 6: Connect all curves drawn in the previous steps. Recall that the image of a closed contour is closed. The conformal mapping property helps by determining the image in the $P(s)$-plane of the corner angles of the semicircles in the Nyquist Path.

11.9 NYQUIST STABILITY PLOTS OF PRACTICAL FEEDBACK CONTROL SYSTEMS

For Nyquist stability analysis of linear feedback control systems, $P(s)$ is equal to the open-loop transfer function $GH(s)$. The most common control systems encountered in practice are those classified as type $0, 1, 2, \ldots, l$ systems (Chapter 9).

Example 11.8.

Type 0 system

$$GH(s) = \frac{1}{s+1}$$

By definition, a *type 0 system* has no poles at the origin. This particular system has no poles on the $j\omega$ axis. The Nyquist Path is given in Fig. 11-1.

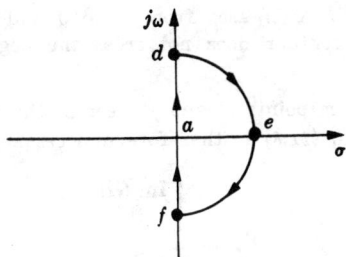

Fig. 11-1

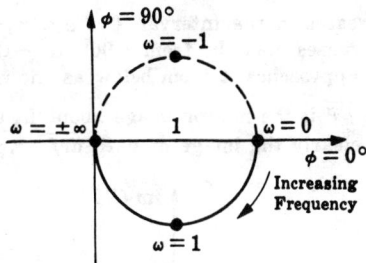

Fig. 11-2

The Polar Plot for this loop transfer function was constructed in Example 11.7, and it is shown in Fig. 11-2 above. This plot is the image of the $j\omega$ axis, or path $\overline{fad}$ of the Nyquist Path, in the $GH(s)$-plane. The semicircular path $\overline{def}$ at infinity is mapped into the $GH(s)$-plane in the following manner. Equation (11.8) implies substitution of $s = \lim_{R \to \infty} Re^{j\theta}$ into the expression for $GH(s)$, where $90° \leqq \theta \leqq -90°$. Hence

$$GH(s)\bigg|_{\substack{\text{path} \\ \overline{def}}} \equiv GH(\infty) = \frac{1}{\lim_{R \to \infty} Re^{j\theta} + 1}$$

By the elementary properties of limits,

$$GH(\infty) = \lim_{R \to \infty} \left[\frac{1}{Re^{j\theta} + 1} \right]$$

But since $|a + b| \geqq |\,|a| - |b|\,|$, then

$$|GH(\infty)| = \lim_{R \to \infty} \left| \frac{1}{Re^{j\theta} + 1} \right| \leqq \lim_{R \to \infty} \left(\frac{1}{R - 1} \right) = 0$$

and the infinite semicircle plots into a point at the origin. Of course, this computation was unnecessary for this simple example because the Polar Plot produces a completely closed contour in the $GH(s)$-plane. In fact, Polar Plots of all *type 0 systems* exhibit this property. The Nyquist Stability Plot is a replica of the Polar Plot with the axes relabeled, and is given by

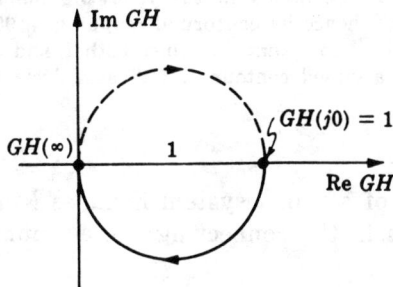

Example 11.9.

Type 1 system

$$GH(s) = \frac{1}{s(s + 1)}$$

There is one pole at the origin. The Nyquist Path is given in the adjacent diagram.

Path $\overline{ad}$: $s = j\omega$ for $0 < \omega < \infty$, and

$$GH(j\omega) = \frac{1}{j\omega(j\omega + 1)} = \frac{1}{\omega\sqrt{\omega^2 + 1}} \underline{/-90° - \tan^{-1}\omega}$$

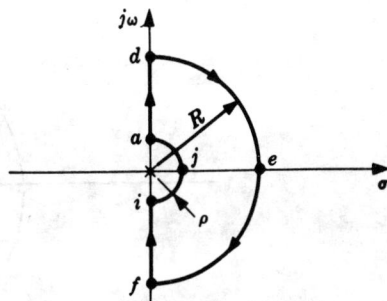

At extreme values of ω we have

$$\lim_{\omega \to 0} GH(j\omega) = \infty \underline{/-90°}$$

$$\lim_{\omega \to \infty} GH(j\omega) = 0 \underline{/-180°}$$

As ω increases in the interval $0 < \omega < \infty$, the magnitude of GH decreases from ∞ to 0 and the phase angle decreases steadily from $-90°$ to $-180°$. Therefore the contour does not cross the negative real axis, but approaches it from below as shown in Fig. 11-3.

Path $\overline{f'i'}$ is the mirror image about Re GH of path $\overline{a'd'}$. Since points d' and f' meet at the origin, the origin is clearly the image of path $\overline{def}$. Application of Equation (11.8) is therefore unnecessary.

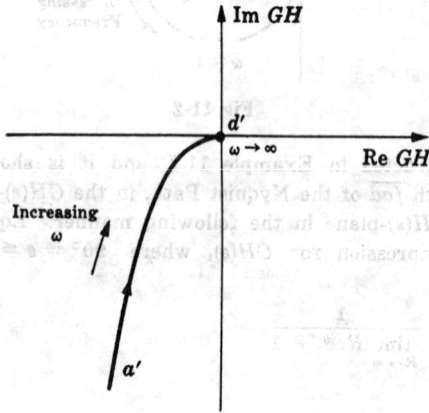

Fig. 11-3

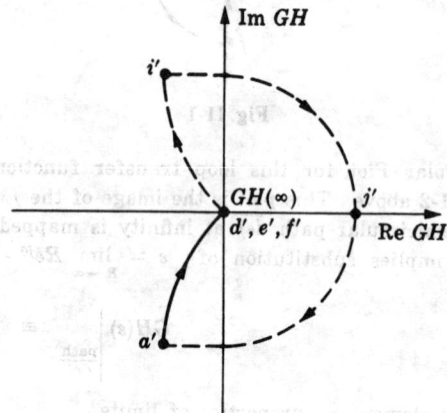

Fig. 11-4

Path $\overline{ija}$: $s = \lim_{\rho \to 0} \rho e^{j\theta}$ for $-90° \leq \theta \leq 90°$,   and

$$\lim_{\rho \to 0} GH(\rho e^{j\theta}) = \lim_{\rho \to 0}\left[\frac{1}{\rho e^{j\theta}(\rho e^{j\theta}+1)}\right] = \lim_{\rho \to 0}\left[\frac{1}{\rho e^{j\theta}}\right] = \infty \cdot e^{-j\theta} = \infty \underline{/-\theta}$$

where we have used the fact that $(\rho e^{j\theta}+1) \to 1$ as $\rho \to 0$. Hence path $\overline{ija}$ maps into a semicircle of infinite radius. For point i, $GH = \infty \underline{/90°}$; for point j, $GH = \infty \underline{/0°}$; and for point a, $GH = \infty \underline{/-90°}$. The resulting Nyquist Stability Plot is given in Fig. 11-4.

Path $\overline{i'j'a'}$ could also have been determined in the following manner. The Nyquist Path makes a 90 degree turn to the right at point i; hence by conformal mapping, a 90 degree right turn must be made at i' in the $GH(s)$-plane. The same goes for point a'. Since both i' and a' are points at infinity, and since the Nyquist Stability Plot must be a closed contour, a CW semicircle of infinite radius must join point i' to point a'.

Type l systems

The Nyquist Stability Plot of a type l system includes l infinite semicircles in its path. That is, there are $180l$ degrees in the connecting arc at infinity of the $GH(s)$-plane.

Example 11.10.

The type 3 system with $GH(s) = \dfrac{1}{s^3(s+1)}$

has three infinite semicircles in its Nyquist Stability Plot:

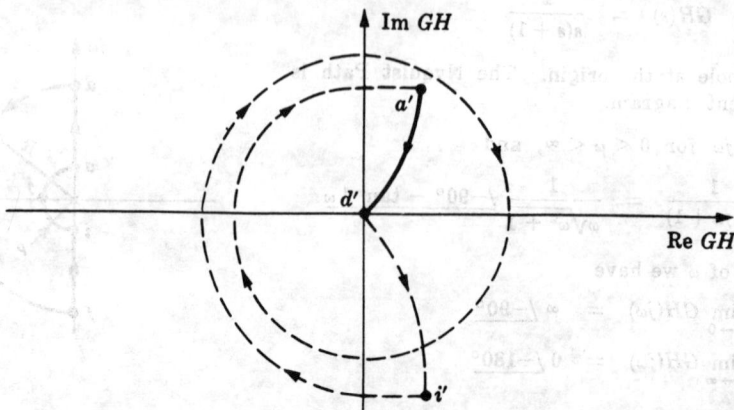

11.10 THE NYQUIST STABILITY CRITERION

A closed-loop control system is absolutely stable if the roots of the characteristic equation have negative real parts (Section 5.2). Equivalently, the poles of the closed-loop transfer function, or the *zeros* of the denominator

$$1 + GH(s)$$

of the closed-loop transfer function, must lie in the left-half plane (LHP). The Nyquist Stability Criterion establishes the number of zeros of $1 + GH(s)$ in the RHP directly from the Nyquist Stability Plot of $GH(s)$. It may be stated in the following manner:

Nyquist Stability Criterion

The closed-loop control system whose open-loop transfer function is $GH(s)$ is stable if and only if

$$N = -P_0 \leqq 0 \tag{11.13}$$

where $P_0 \equiv$ number of *poles* of GH in the RHP $\geqq$ 0

$N \equiv$ total number of CW encirclements of the $(-1, 0)$ point (i.e. $GH(s) = -1$) in the $GH(s)$-plane.

If $N > 0$, *the number of zeros Z_0 of $1 + GH$ in the RHP* is determined by

$$Z_0 = N + P_0 \tag{11.14}$$

If $N \leqq 0$, the $(-1, 0)$ point is not enclosed by the Nyquist Stability Plot. Therefore $N \leqq 0$ if the region to the right of the contour in the prescribed direction does not include the $(-1, 0)$ point. Shading of this region helps significantly in determining whether $N \leqq 0$.

If $N \leqq 0$ *and* $P_0 = 0$, then the system is absolutely stable if and only if $N = 0$; that is, if and only if the $(-1, 0)$ point *does not* lie in the shaded region.

Example 11.11.

The Nyquist Stability Plot for $GH(s) = \dfrac{1}{s(s + 1)}$ was determined in Example 11.9 as shown in Fig. 11-5. The region to the right of the contour has been shaded. Clearly, the $(-1, 0)$ point is not in the shaded region; therefore it is not enclosed by the contour and so $N \leqq 0$. The poles of $GH(s)$ are at $s = 0$ and $s = -1$, neither of which are in the RHP; hence $P_0 = 0$. Thus

$$N = -P_0 = 0$$

and the system is absolutely stable.

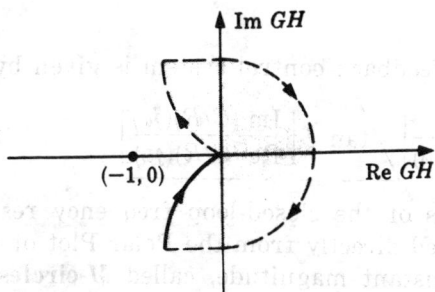

Fig. 11-5.

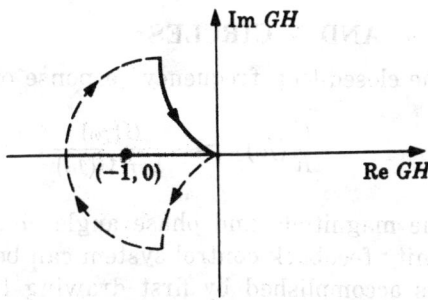

Fig. 11-6.

Example 11.12.

The Nyquist Stability Plot for $GH(s) = \dfrac{1}{s(s - 1)}$ is given in Fig. 11-6. The region to the right of the contour has been shaded and the $(-1, 0)$ point is enclosed; then $N > 0$. (It is clear that $N = 1$.) The poles of GH are at $s = 0$ and $s = +1$, the latter pole being in the RHP. Hence $P_0 = 1$.

$N \neq -P_0$ indicates that the system is *unstable*. From Equation (*11.14*) we have

$$Z_0 = N + P_0 = 2$$

zeros of $1 + GH$ in the RHP.

11.11 RELATIVE STABILITY

The relative stability of a feedback control system is easily determined from the Polar or Nyquist Stability Plot.

The **phase crossover frequency** ω_π is that frequency at which the phase angle of $GH(j\omega)$ is -180 degrees, that is, the frequency at which the Polar Plot crosses the negative real axis. The **gain margin** is given by

$$\text{gain margin} = \frac{1}{|GH(j\omega_\pi)|}$$

These quantities are illustrated in Fig. 11-7.

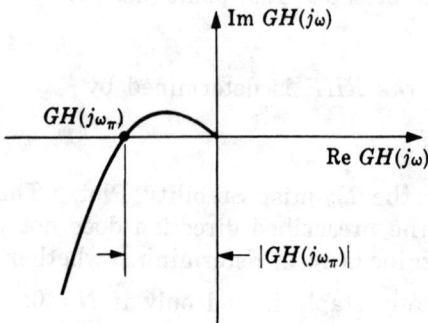

Fig. 11-7

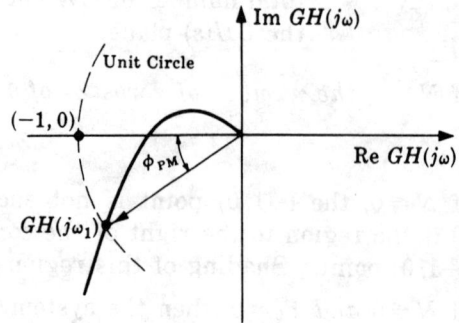

Fig. 11-8

The **gain crossover frequency** ω_1 is that frequency at which $|GH(j\omega)| = 1$. The **phase margin** ϕ_{PM} is the angle by which the Polar Plot must be rotated to cause it to pass through the $(-1, 0)$ point. It is given by

$$\phi_{\text{PM}} = [180 + \arg GH(j\omega_1)] \text{ degrees}$$

These quantities are illustrated in Fig. 11-8.

11.12 *M* AND *N* CIRCLES*

The closed-loop frequency response of a unity feedback control system is given by

$$\frac{C}{R}(j\omega) = \frac{G(j\omega)}{1 + G(j\omega)} = \left| \frac{G(j\omega)}{1 + G(j\omega)} \right| \bigg/ \tan^{-1}\left[\frac{\text{Im}\,(C/R)(j\omega)}{\text{Re}\,(C/R)(j\omega)} \right] \qquad (11.15)$$

The magnitude and phase angle characteristics of the closed-loop frequency response of a unity feedback control system can be determined directly from the Polar Plot of $G(j\omega)$. This is accomplished by first drawing lines of constant magnitude, called **M-circles**, and lines of constant phase angle, called **N-circles**, directly onto the $G(j\omega)$-plane, where

---

*The letter symbols M, N used in this section for M and N circles are not equal to and should not be confused with the manipulated variable $M = M(s)$ defined in Chapter 2 and with the number of encirclements N of the $(-1, 0)$ point of Section 11.10. It is unfortunate that the same symbols have been used to signify more than one quantity. But in the interest of being consistent with most other control system texts, we have maintained the terminology of the classical literature and have now pointed this out to the reader.

$$M \equiv \left| \frac{G(j\omega)}{1 + G(j\omega)} \right| \qquad (11.16)$$

$$N \equiv \frac{\text{Im}\,(C/R)(j\omega)}{\text{Re}\,(C/R)(j\omega)} \qquad (11.17)$$

The intersection of the Polar Plot with a particular M-circle yields the value of M at the frequency ω of $G(j\omega)$ at the point of intersection. The intersection of the Polar Plot with a particular N-circle yields the value of N at the frequency ω of $G(j\omega)$ at the intersection point. M versus ω and N versus ω are easily plotted from these points.

Several M-circles are superimposed on a typical Polar Plot in the $G(j\omega)$-plane shown below. The **radius of an M-circle** is given by

$$\text{radius of } M\text{-circle} \;=\; \left| \frac{M}{M^2 - 1} \right| \qquad (11.18)$$

The **center of an M-circle** always lies on the $\text{Re}\,G(j\omega)$-axis. The center point is given by

$$\text{center of } M\text{-circle} \;=\; \left(\frac{-M^2}{M^2 - 1}, 0 \right) \qquad (11.19)$$

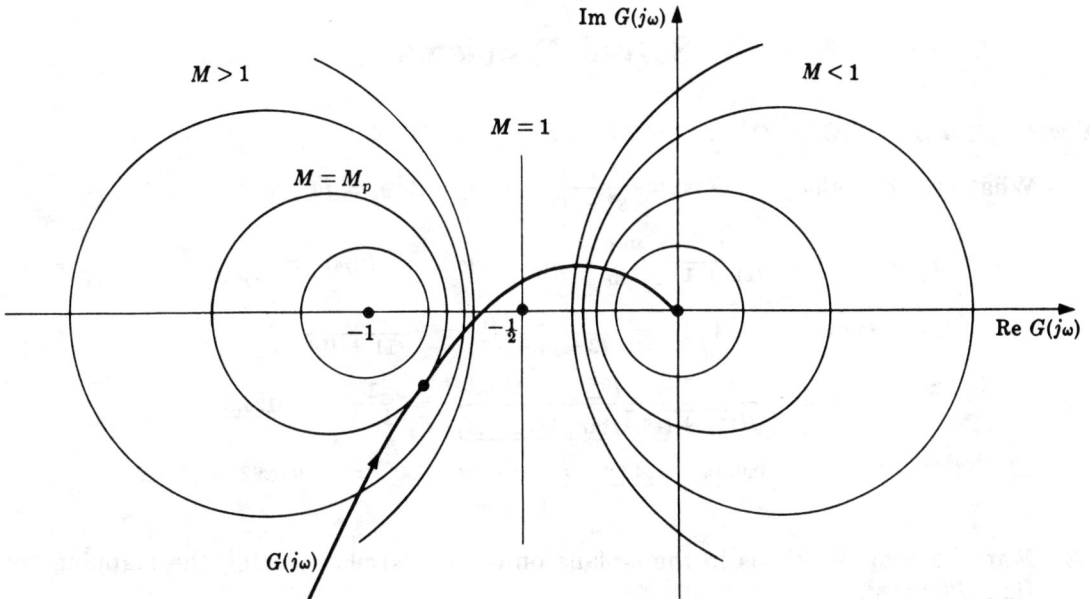

The **resonance peak M_p** is given by the largest value of M of the M-circle(s) tangent to the Polar Plot. (There may be more than one tangency.)

The **damping ratio ζ** for a second-order system with $0 \leq \zeta \leq 0.707$ is related to M_p by

$$M_p \;=\; \frac{1}{2\zeta \cdot \sqrt{1 - \zeta^2}} \qquad (11.20)$$

Several N-circles are superimposed on the Polar Plot shown below. The **radius of an N-circle** is given by

$$\text{radius of } N\text{-circle} \;=\; \sqrt{\tfrac{1}{4} + (1/2N)^2} \qquad (11.21)$$

The **center of an N-circle** always falls on the line $\text{Re}\,G(j\omega) = -\tfrac{1}{2}$. The center point is given by

$$\text{center of } N\text{-circle} \;=\; (-\tfrac{1}{2},\, 1/2N) \qquad (11.22)$$

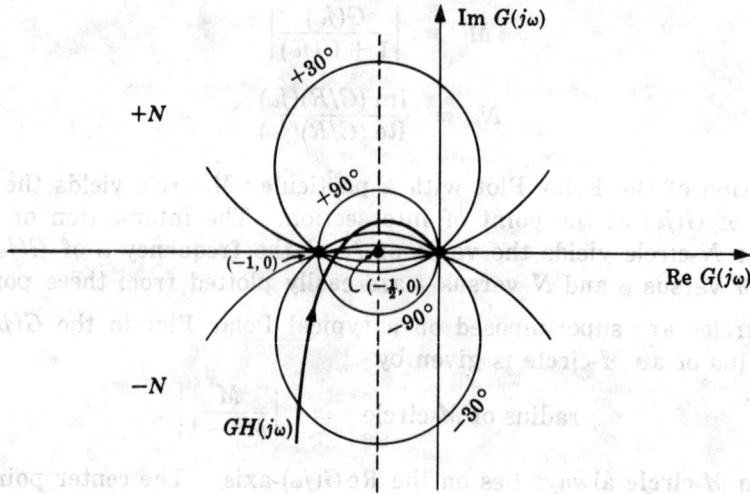

Solved Problems

COMPLEX FUNCTIONS OF A COMPLEX VARIABLE

11.1. What are the values of $P(s) = \dfrac{1}{s^2+1}$ for $s_1 = 2$, $s_2 = j4$, and $s_3 = 2 + j4$?

$$P(s_1) \;=\; P(2) \;=\; \frac{1}{(2)^2+1} \;=\; \frac{1}{5} + j0, \qquad P(s_2) \;=\; P(j4) \;=\; \frac{1}{(j4)^2+1} \;=\; \frac{1}{17} + j0$$

$$P(s_3) \;=\; P(2+j4) \;=\; \frac{1}{(2+j4)^2+1} \;=\; \frac{1}{-11+j16}$$

$$=\; \frac{1\,\underline{/0^\circ}}{\sqrt{(11)^2+(16)^2}\,\underline{/\tan^{-1}(16/-11)}} \;=\; \frac{1}{19.4}\,\underline{/0^\circ - 124.6^\circ}$$

$$=\; 0.0514\,\underline{/-124.6^\circ} \;=\; -0.0514\,\underline{/55.4^\circ} \;=\; -0.0292 - j0.0423$$

11.2. Map the imaginary axis in the s-plane onto the $P(s)$-plane, using the mapping function $P(s) = s^2$.

We have $s = j\omega$, $-\infty < \omega < \infty$. Therefore, $P(j\omega) = (j\omega)^2 = -\omega^2$. Now when $\omega \to -\infty$, $P(j\omega) \to -\infty$ (or $-\infty^2$, if you prefer). When $\omega \to +\infty$, $P(j\omega) \to -\infty$; and when $\omega = 0$, $P(j0) = 0$. Thus as $j\omega$ increases along the negative *imaginary* axis from $-j\infty$ toward $j0$, $P(j\omega)$ increases along the negative *real* axis from $-\infty$ to 0. When $j\omega$ increases from $j0$ to $+j\infty$, $P(j\omega)$ decreases back to $-\infty$, again along the negative *real* axis. The mapping is plotted in the following manner:

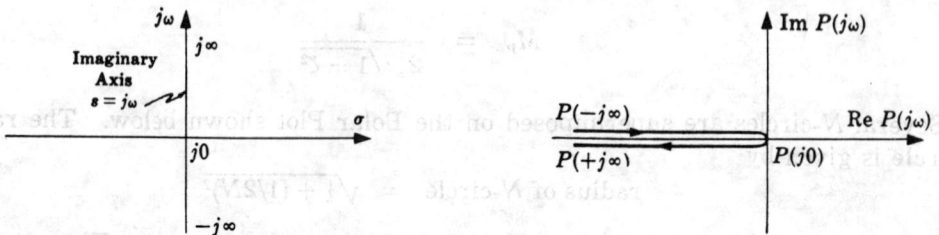

The two lines in the $P(j\omega)$-plane are actually superimposed, but they are shown here separated for clarity.

11.3. Map the rectangular region in the s-plane bounded by the lines $\omega = 0$, $\sigma = 0$, $\omega = 1$, and $\sigma = 2$ onto the $P(s)$-plane using the transformation $P(s) = s + 1 - j2$.

We have

$$\omega = 0: \quad P(\sigma) = (\sigma + 1) - j2 \qquad\qquad \omega = 1: \quad P(\sigma + j1) = (\sigma + 1) - j1$$

$$\sigma = 0: \quad P(j\omega) = 1 + j(\omega - 2) \qquad\qquad \sigma = 2: \quad P(2 + j\omega) = 3 + j(\omega - 2)$$

Since σ varies over all real numbers $(-\infty < \sigma < \infty)$ on the line $\omega = 0$, so does $\sigma + 1$ on $P(\sigma) = (\sigma + 1) - j2$. Therefore $\omega = 0$ maps onto the line $-j2$ in the $P(s)$-plane. Similarly $\sigma = 0$ maps onto the line $P(s) = 1$, $\omega = 1$ maps onto the line $P(s) = -j1$, and $\sigma = 2$ onto the line $P(s) = 3$. The resulting transformation is illustrated as follows:

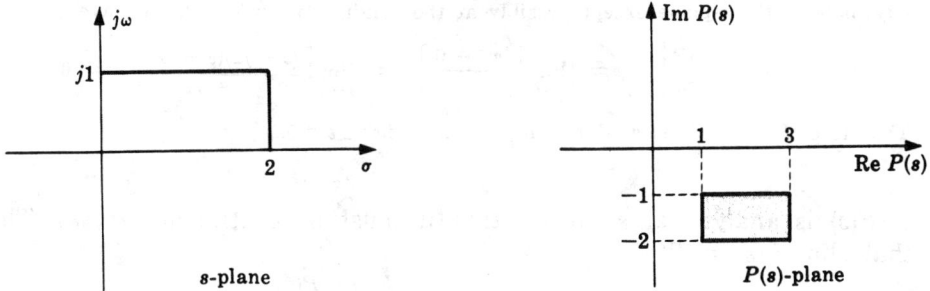

This type of mapping is called a **translation mapping**.

11.4. Find the derivative of $P(s) = s^2$ at the points $s = s_0$ and $s_0 = 1$.

$$\left.\frac{dP}{ds}\right|_{s=s_0} = \lim_{s \to s_0}\left[\frac{P(s) - P(s_0)}{s - s_0}\right] = \lim_{s \to s_0}\left[\frac{s^2 - s_0^2}{s - s_0}\right] = \lim_{s \to s_0}(s + s_0) = 2s_0$$

At $s_0 = 1$, we have $(dP/ds)|_{s=1} = 2$.

ANALYTIC FUNCTIONS AND SINGULARITIES

11.5. Is $P(s) = s^2$ an analytic function in any region of the s-plane? If so, which region?

From the preceding problem $(dP/ds)|_{s=s_0} = 2s_0$. Hence s^2 is analytic wherever $2s_0$ is finite (Definition 11.1). Thus s^2 is analytic in the entire finite region of the s-plane. Such functions are often called **entire functions**.

11.6. Is $P(s) = 1/s$ analytic in any region of the s-plane?

$$\left.\frac{dP}{ds}\right|_{s=s_0} = \lim_{s \to s_0}\left[\frac{1/s - 1/s_0}{s - s_0}\right] = \lim_{s \to s_0}\left[\frac{-(s - s_0)}{ss_0(s - s_0)}\right] = \frac{-1}{s_0^2}$$

This derivative is unique and finite for all $s_0 \neq 0$. Hence $1/s$ is analytic at all points in the s-plane except the origin, $s = s_0 = 0$. The point $s = 0$ is a *singularity* (pole) of $1/s$. Singularities other than poles exist, but not in the **transfer functions** of ordinary control system components.

11.7. Is $P(s) = |s|^2$ analytic in any region of the s-plane?

First put $s = \sigma + j\omega$, $s_0 = \sigma_0 + j\omega_0$. Then

$$\left.\frac{dP}{ds}\right|_{s=s_0} = \lim_{(s - s_0) \to 0}\left[\frac{|\sigma + j\omega|^2 - |\sigma_0 + j\omega_0|^2}{(\sigma + j\omega) - (\sigma_0 + j\omega_0)}\right]$$

$$= \lim_{[(\sigma - \sigma_0) + j(\omega - \omega_0)] \to 0}\left[\frac{(\sigma - \sigma_0)(\sigma + \sigma_0) + (\omega - \omega_0)(\omega + \omega_0)}{(\sigma - \sigma_0) + j(\omega - \omega_0)}\right]$$

If the limit exists it must be unique and should not depend on how s approaches s_0, or equivalently how $[(\sigma - \sigma_0) + j(\omega - \omega_0)]$ approaches zero. So first let $s \to s_0$ along the $j\omega$ axis and obtain

$$\left.\frac{dP}{ds}\right|_{s=s_0} = \lim_{\substack{\omega \to \omega_0 \\ \sigma = \sigma_0}} \left[\frac{(\omega - \omega_0)(\omega + \omega_0)}{j(\omega - \omega_0)} \right] = -j2\omega_0$$

Now let $s \to s_0$ along the σ-axis; that is,

$$\left.\frac{dP}{ds}\right|_{s=s_0} = \lim_{\substack{\sigma \to \sigma_0 \\ \omega = \omega_0}} \left[\frac{(\sigma - \sigma_0)(\sigma + \sigma_0)}{\sigma - \sigma_0} \right] = 2\sigma_0$$

Hence the limit *does not* exist for any nonzero values of σ_0 and ω_0, and $|s|^2$ is not analytic anywhere in the s-plane except possibly at the origin. When $s_0 = 0$, we have

$$\left.\frac{dP}{ds}\right|_{s=0} = \lim_{s \to 0} \left[\frac{|s|^2 - 0}{s} \right] = \lim_{s \to 0} \left[\frac{(\sigma + j\omega)(\sigma - j\omega)}{\sigma + j\omega} \right] = 0$$

Therefore $P(s) = |s|^2$ is analytic only at the origin, $s = 0$.

11.8. If $P(s)$ is analytic at s_0, prove that it must be continuous at s_0. That is, show that $\lim_{s \to s_0} P(s) = P(s_0)$.

Since
$$P(s) - P(s_0) = \frac{P(s) - P(s_0)}{(s - s_0)} \cdot (s - s_0)$$

for $s \neq s_0$, then

$$\lim_{s \to s_0} [P(s) - P(s_0)] = \lim_{s \to s_0} \left[\frac{P(s) - P(s_0)}{(s - s_0)} \right] \cdot \lim_{s \to s_0} (s - s_0) = \left[\left.\frac{dP}{ds}\right|_{s=s_0} \right] \cdot 0 = 0$$

because $\left.\dfrac{dP}{ds}\right|_{s=s_0}$ exists by hypothesis (i.e. $P(s)$ is analytic). Therefore

$$\lim_{s \to s_0} [P(s) - P(s_0)] = 0 \quad \text{or} \quad \lim_{s \to s_0} P(s) = P(s_0)$$

11.9. **Polynomial functions** are defined by $Q(s) \equiv a_n s^n + a_{n-1} s^{n-1} + \cdots + a_1 s + a_0$. where $a_n \neq 0$, n is a positive integer called the **degree of the polynomial**, and $a_0, a_1, \ldots, a_n$ are complex constants. Prove that $Q(s)$ is analytic in every bounded (finite) region of the s-plane.

First consider s^n:

$$\left.\frac{d}{ds}[s^n]\right|_{s=s_0} = \lim_{s \to s_0} \left[\frac{s^n - s_0^n}{s - s_0} \right] = \lim_{s \to s_0} (s^{n-1} + s^{n-2} s_0 + \cdots + s s_0^{n-2} + s_0^{n-1}) = n s_0^{n-1}$$

Thus s^n is analytic in every finite region of the s-plane. By mathematical induction s^{n-1}, s^{n-2}, $\ldots$, s are also analytic. Hence, by the elementary theorems on limits of sums and products, we see that $Q(s)$ is analytic in every finite region of the s-plane.

11.10. **Rational algebraic functions** are defined by $P(s) \equiv N(s)/D(s)$, where $N(s)$ and $D(s)$ are polynomials. Show that $P(s)$ is analytic at every point s where $D(s) \neq 0$; that is, prove that the transfer functions of control system elements that take the form of rational algebraic functions are analytic except at their poles.

The overwhelming majority of control system elements are in this category. The fundamental theorem of algebra, "a polynomial of degree n has n zeros and can be expressed as a product of n linear factors", helps to put $P(s)$ in a form more recognizable as a control system transfer function; that is, $P(s)$ can be written in the familiar form

$$P(s) \equiv \frac{N(s)}{D(s)} = \frac{b_m s^m + b_{m-1} s^{m-1} + \cdots + b_0}{a_n s^n + a_{n-1} s^{n-1} + \cdots + a_0} = \frac{b_m(s + z_1)(s + z_2) \cdots (s + z_m)}{a_n(s + p_1)(s + p_2) \cdots (s + p_n)}$$

where $-z_1, -z_2, \ldots, -z_n$ are zeros, $-p_1, -p_2, \ldots, -p_n$ are poles, and $m \leq n$.

From the identity given by

$$\frac{N(s)}{D(s)} - \frac{N(s_0)}{D(s_0)} \equiv \frac{1}{D(s)\,D(s_0)} [D(s_0)\,(N(s) - N(s_0)) - N(s_0)\,(D(s) - D(s_0))]$$

where $D(s) \neq 0$, we get

$$\frac{dP}{ds}\bigg|_{s=s_0} = \lim_{s \to s_0} \left[\frac{\dfrac{N(s)}{D(s)} - \dfrac{N(s_0)}{D(s_0)}}{s - s_0} \right]$$

$$= \lim_{s \to s_0} \left[\frac{1}{D(s)\,D(s_0)} \left(D(s_0)\left[\frac{N(s) - N(s_0)}{s - s_0}\right] - N(s_0)\left[\frac{D(s) - D(s_0)}{s - s_0}\right] \right) \right]$$

$$= \lim_{s \to s_0} \left[\frac{1}{D(s)}\left(\frac{N(s) - N(s_0)}{s - s_0}\right) \right] - \lim_{s \to s_0} \left[\frac{N(s_0)}{D(s)\,D(s_0)}\left(\frac{D(s) - D(s_0)}{s - s_0}\right) \right]$$

$$= \lim_{s \to s_0} \left[\frac{1}{D(s)}\right] \cdot \lim_{s \to s_0} \left[\frac{N(s) - N(s_0)}{s - s_0}\right] - \lim_{s \to s_0} \left[\frac{N(s_0)}{D(s)\,D(s_0)}\right] \cdot \lim_{s \to s_0} \left[\frac{D(s) - D(s_0)}{s - s_0}\right]$$

$$= \frac{1}{D(s_0)} \cdot \frac{dN}{ds}\bigg|_{s=s_0} - \frac{N(s_0)}{D(s_0)^2} \cdot \frac{dD}{ds}\bigg|_{s=s_0}$$

where we have used the results of Problems 11.8, 11.9, and Definition 11.1. Therefore the derivative of $P(s)$ exists ($P(s)$ is analytic) for all points s where $D(s) \neq 0$.

Note that we have computed the derivative of a rational algebraic function (the last part of the above equation) in terms of the derivatives of its numerator and denominator in addition to solving the required problem.

11.11. Prove that e^{-sT} is analytic in every bounded region of the s-plane.

In complex variable theory e^{-sT} is defined by the power series

$$e^{-sT} = \sum_{k=0}^{\infty} \frac{(-sT)^k}{k!}$$

By the ratio test, as $k \to \infty$ we have

$$\left| \frac{(-sT)^k / k!}{(-sT)^{k+1}/(k+1)!} \right| = \left| \frac{k+1}{-sT} \right| \to \infty$$

Hence the radius of convergence of this power series is infinite. The sum of a power series is analytic within its radius of convergence. Thus e^{-sT} is analytic in every bounded region of the s-plane.

11.12. Prove that $e^{-sT} \cdot P(s)$ is analytic wherever $P(s)$ is analytic. Hence systems containing a combination of rational algebraic transfer functions and time delay operators (i.e. e^{-sT}) are analytic except at the poles of the system.

By Problem 11.11, e^{-sT} is analytic in every bounded region of the s-plane; and by Problem 11.10, $P(s)$ is analytic except at its poles. Now

$$\frac{d}{ds}\left[e^{-sT} \cdot P(s)\right]\bigg|_{s=s_0} = \lim_{s \to s_0} \left[\frac{e^{-sT} \cdot P(s) - e^{-s_0 T} \cdot P(s_0)}{s - s_0} \right]$$

$$= \lim_{s \to s_0} \left[e^{-sT}\left(\frac{P(s) - P(s_0)}{s - s_0}\right) + P(s_0)\left(\frac{e^{-sT} - e^{-s_0 T}}{s - s_0}\right) \right]$$

$$= e^{-s_0 T} \cdot \frac{dP}{ds}\bigg|_{s=s_0} + P(s_0) \cdot \frac{d}{ds}(e^{-sT})\bigg|_{s=s_0}$$

Therefore $e^{-sT} \cdot P(s)$ is analytic wherever $P(s)$ is analytic.

11.13. Consider the function given by $P(s) = e^{-sT}(s^2 + 2s + 3)/(s^2 - 2s + 2)$. Where are the singularities of this function? Where is $P(s)$ analytic?

The singular points are at the poles of $P(s)$. Since $s^2 - 2s + 2 = (s - 1 + j1)(s - 1 - j1)$, the two poles are given by $-p_1 = 1 - j1$ and $-p_2 = 1 + j1$. $P(s)$ is analytic in every bounded region of the s-plane except at the points $s = -p_1$ and $s = -p_2$.

CONTOURS AND ENCIRCLEMENTS

11.14. What points are *enclosed* by the following contours?

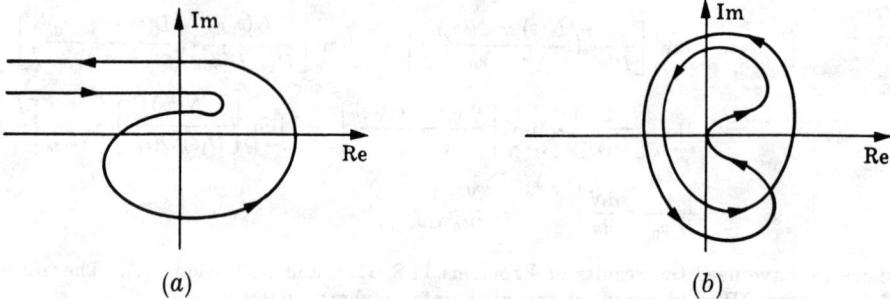

$$(a) \qquad\qquad\qquad (b)$$

By shading the region to the right of each contour as it is traversed in the prescribed direction, we get

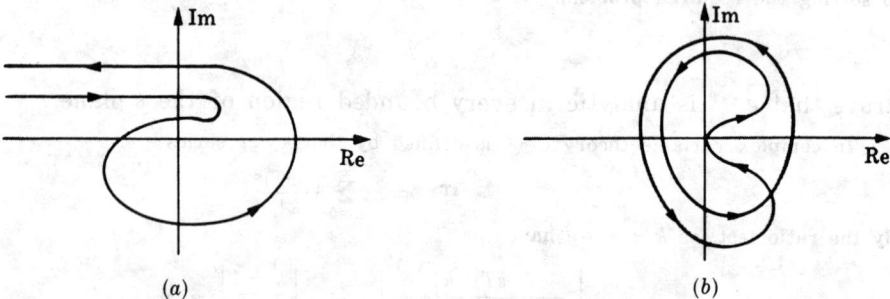

$$(a) \qquad\qquad\qquad (b)$$

All points in the shaded regions are enclosed.

11.15. Which contours of Problem 11.14 are *closed*?

Clearly, the contour of Part (b) is closed. The contour of Part (a) may or may not close upon itself at infinity in the complex plane. This cannot be determined from the given graph.

11.16. What is the *direction* (positive or negative) of each contour in Problem 11.14(a) and (b)?

Using the origin as a base, each contour is directed in the counterclockwise, negative direction about the origin.

11.17. Determine the number of encirclements N_0 of the origin for the adjoining contour.

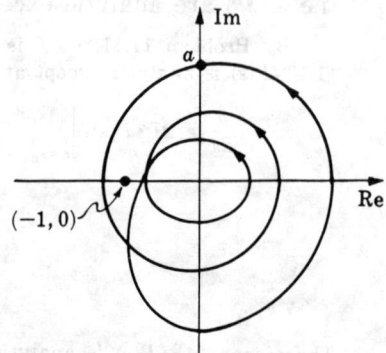

Beginning at the point a, we rotate a radial line from the origin to the contour in the direction of the arrows. Three counterclockwise rotations of $360°$ result in the radial line returning to the point a. Hence $N_0 = -3$.

11.18. Determine the number of encirclements N_0 of the origin for the contour in Fig. 11-9.

Beginning at point a, $+180°$ is swept out by the contour when b is reached for the first time. In going from b to c and back to b, the net angular gain is zero. Returning to a from b yields $+180°$. Thus $N_0 = +1$.

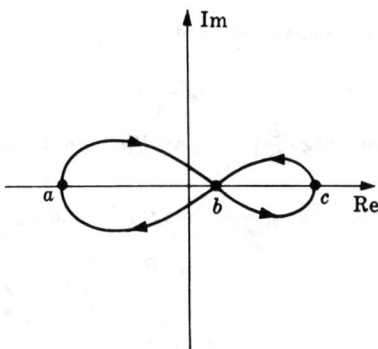

Fig. 11-9

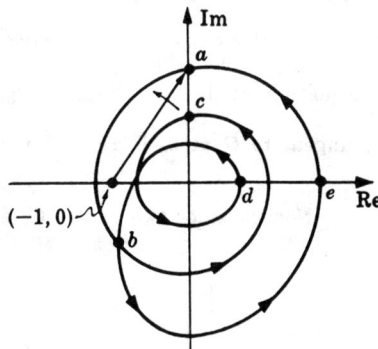

Fig. 11-10

11.19. Determine the number of encirclements N of the $(-1, 0)$ point (i.e. the -1 point on the real axis) for the contour of Problem 11.17.

Again beginning at point a, we rotate a radial line *from the* $(-1, 0)$ *point* to the contour in the direction of the arrows as shown in Fig. 11-10. In going from a to b to c, the radial line sweeps out somewhat less than $-360°$. But from c to d and back to b, the angle increases again toward the value reached in going only from a to b. Then from b to e to a the resultant angle is $-360°$. Thus $N = -1$.

PROPERTIES OF THE MAPPING $P(s)$

11.20. Are the following functions single-valued: (a) $P(s) = s^2$,   (b) $P(s) = s^{1/2}$?

(a) Substitution of any complex number s into $P(s) = s^2$ yields a unique value for $P(s)$. Hence $P(s) = s^2$ is a single-valued function.

(b) In polar form we have $s = |s|e^{j\theta}$, where $\theta = \arg(s)$. Therefore $s^{1/2} = |s|^{1/2}e^{j\theta/2}$. Now if we increase θ by 2π we return to the same point s. But

$$P(s) \;=\; |s|^{1/2}e^{j(\theta + 2\pi)/2} \;=\; |s|^{1/2}e^{j\theta/2} \cdot e^{j\pi} \;=\; P(s) \cdot e^{j\pi}$$

which is *another* point in the $P(s)$-plane. Hence $P(s) = s^{1/2}$ has two points in the $P(s)$-plane for every point in the s-plane. It is not a single-valued function; it is a **multiple-valued function** (with two values).

11.21. Prove that every closed contour containing no singular points of $P(s)$ in the s-plane maps into a closed contour in the $P(s)$-plane.

Suppose not. Then at some point s_0 where the s-plane contour closes upon itself the $P(s)$-plane contour is not closed. This means that one (non-singular) point s_0 in the s-plane is mapped into more than one point in the $P(s)$-plane (the images of the point s_0). This contradicts the fact that $P(s)$ is a single-valued function (Property 1, Section 11.4).

11.22. Prove that P is a conformal mapping wherever P is analytic and $dP/ds \neq 0$.

Consider two curves: C in the s-plane and C', the image of C, in the $P(s)$-plane. Let the curve in the s-plane be described by a parameter t; that is, each t corresponds to a point $s = s(t)$ along the curve C. Hence C' is described by $P[s(t)]$ in the $P(s)$-plane. The derivatives ds/dt and dP/dt represent tangent vectors to corresponding points on C and C'. Now

$$\frac{dP[s(t)]}{dt}\bigg|_{P(s)=P(s_0)} = \frac{ds}{dt} \cdot \frac{dP(s)}{ds}\bigg|_{s=s_0}$$

where we have used the fact that P is analytic at some point $s_0 \equiv s(t_0)$. Put $dP/dt \equiv r_1 e^{j\phi}$, $dP/ds \equiv r_2 e^{j\alpha}$, and $ds/dt \equiv r_3 e^{j\theta}$. Then

$$r_1(s_0)e^{j\phi(s_0)} = r_2(s_0) \cdot r_3(s_0)e^{j[\theta(s_0)+\alpha(s_0)]}$$

Equating angles, we have $\phi(s_0) = \theta(s_0) + \alpha(s_0) = \theta(s_0) + \arg \dfrac{dP}{ds}\bigg|_{s=s_0}$, and we see that the tangent to C at s_0 is rotated through an angle $\arg \dfrac{dP}{ds}\bigg|_{s=s_0}$ at $P(s_0)$ on C' in the $P(s)$-plane.

Now consider two curves C_1 and C_2 intersecting at s_0, with images C_1' and C_2' in the $P(s)$-plane:

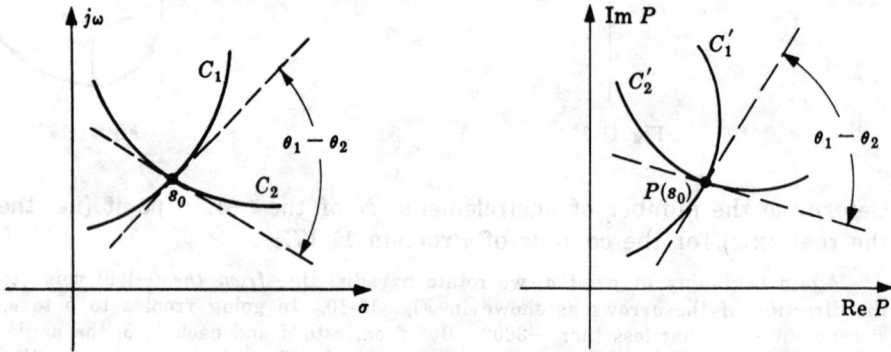

Let θ_1 be the angle of inclination of the tangent to C_1, and θ_2 for C_2. Then the angles of inclination for C_1' and C_2' are $\theta_1 + \arg \dfrac{dP}{ds}\bigg|_{s=s_0}$, and $\theta_2 + \arg \dfrac{dP}{ds}\bigg|_{s=s_0}$. Therefore the angle $(\theta_1 - \theta_2)$ between C_1 and C_2 is equal in magnitude and sense to the angle between C_1' and C_2',

$$\theta_1 + \arg \frac{dP}{ds}\bigg|_{s=s_0} - \theta_2 - \arg \frac{dP}{ds}\bigg|_{s=s_0} = \theta_1 - \theta_2.$$

Note that $\arg \dfrac{dP}{ds}\bigg|_{s=s_0}$ is indeterminate if $\dfrac{dP}{ds}\bigg|_{s=s_0} = 0$.

11.23. Show that $P(s) = e^{-sT}$ is conformal in every bounded region of the s-plane.

e^{-sT} is analytic (Problem 11.11). Moreover, $\dfrac{d}{ds}(e^{-sT}) = -Te^{-sT} \neq 0$ in any bounded (finite) region of the s-plane. Then by Problem 11.22, $P(s) = e^{-sT}$ is conformal.

11.24. Show that $P(s) \cdot e^{-sT}$ is conformal for rational $P(s)$ and $dP/ds \neq 0$.

By Problem 11.12, Pe^{-sT} is analytic except at the poles of P. By Problem 11.12,

$$\frac{d}{ds}[Pe^{-sT}] = e^{-sT}\frac{dP}{ds} - P \cdot Te^{-sT} = e^{-sT}\left(\frac{dP}{ds} - TP\right)$$

Suppose $\dfrac{d}{ds}[Pe^{-sT}] = 0$. Then since $e^{-sT} \neq 0$ for any finite s, we have $dP/ds - TP = 0$ whose general solution is $P(s) = ke^{sT}$, k constant. But P is rational and e^{sT} is not. Hence $\dfrac{d}{ds}[Pe^{-sT}] \neq 0$.

11.25. Two s-plane contours C_1 and C_2 intersect in a 90 degree angle as shown. The analytic function $P(s)$ maps these contours into the $P(s)$-plane and $dP/ds \neq 0$ at s_0. Sketch the image of contour C_2 in a neighborhood of $P(s_0)$. The image of C_1 is also given.

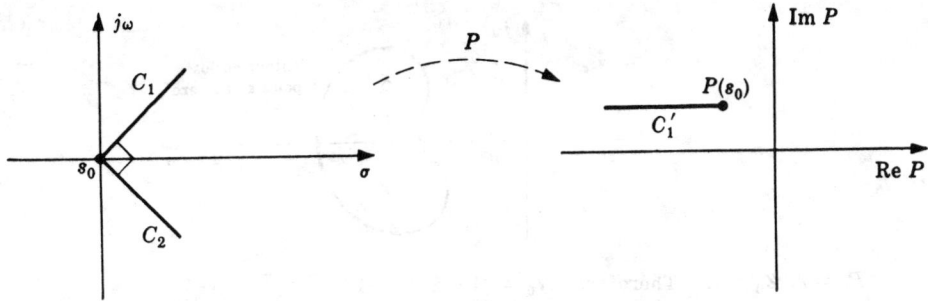

By Problem 11.22, P is conformal; hence the angle between C_1' and C_2' is 90 degrees. Since C_1 makes a left turn onto C_2 at s_0, then C_1' must also turn left at $P(s_0)$:

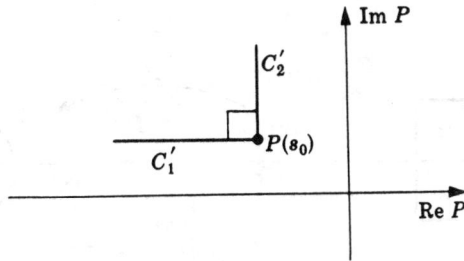

11.26. Prove Equation *(11.1)*, Page 190: $N_0 = Z_0 - P_0$.

The bulk of the proof is somewhat more involved than can be handled with the complex variable theory presented in this book. So we shall assume knowledge of a well-known theorem of functions of a complex variable and continue from there. The theorem states that if C is a closed contour in the s-plane, $P(s)$ an analytic function on C and within C except for possible poles, and $P(s) \neq 0$ on C, then

$$\frac{1}{2\pi j} \int_C \frac{P'(s)}{P(s)} ds \; = \; Z_0 - P_0$$

where Z_0 is the total number of zeros inside C, P_0 the total number of poles inside C, and $P' \equiv dP/ds$. Multiple poles and zeros are counted one for one; that is, a double pole at a point is two poles of the total, a triple zero is three zeros of the total.

Now since $d[\ln P(s)] = [P'(s)/P(s)]ds$ and $\ln P(s) \equiv \ln |P(s)| + j \arg P(s)$, we have

$$\frac{1}{2\pi j} \int_C [P'(s)/P(s)] \, ds \; = \; \frac{1}{2\pi j} \int_C d[\ln P(s)] \; = \; \frac{1}{2\pi j} [\ln P(s)] \Big|_C \; = \; \frac{1}{2\pi j} [\ln |P(s)| + j \arg P(s)] \Big|_C$$

$$= \; \frac{1}{2\pi j} [\ln |P(s)|] \Big|_C + \frac{1}{2\pi j} [j \arg P(s)] \Big|_C$$

Now since $\ln |P(s)|$ returns to its original value when we go once around C, the first term in the last equation is zero. Hence

$$Z_0 - P_0 \; = \; \frac{1}{2\pi} [\arg P(s)] \Big|_C$$

Since C is closed the image of C in the $P(s)$-plane is closed, and the net change in the angle $\arg P(s)$ around the $P(s)$ contour is 2π times the number of encirclements N_0 of the origin in the $P(s)$-plane. Then $Z_0 - P_0 = 2N_0\pi/2\pi = N_0$. This result is often called *the principle of the argument*.

11.27. Determine the number N_0 of $P(s)$-plane contour encirclements for the following s-plane contour mapped into the $P(s)$-plane.

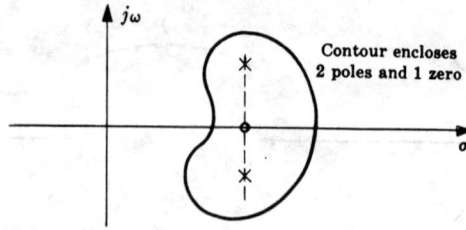

$P_0 = 2$, $Z_0 = 1$.   Therefore $N_0 = 1 - 2 = -1$.

11.28. Determine the number of zeros Z_0 enclosed by the s-plane contour, where $P_0 = 5$.

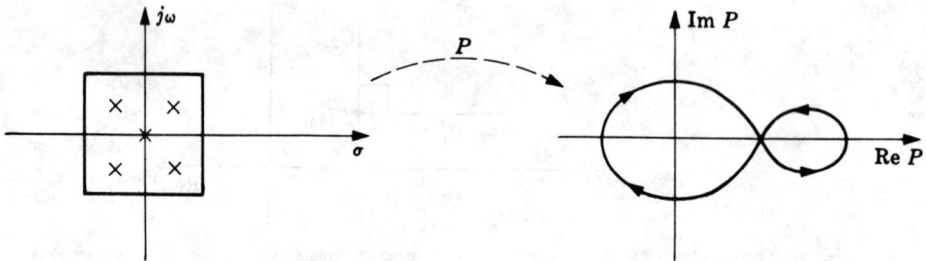

$N_0 = 1$ was computed in Problem 11.18 for the given $P(s)$-plane contour. Since $P_0 = 5$, then $Z_0 = N_0 + P_0 = 1 + 5 = 6$.

11.29. Determine the number of poles P_0 enclosed by the s-plane contour, where $Z_0 = 0$.

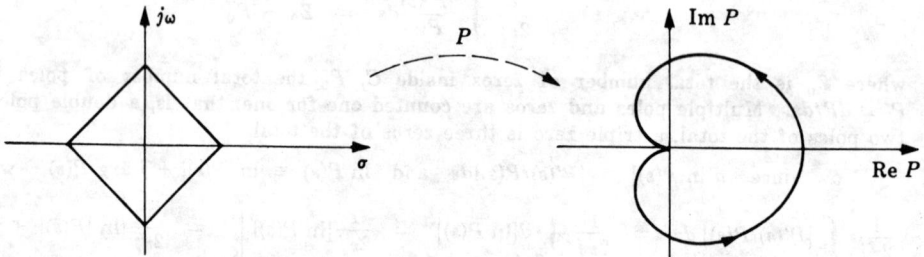

Clearly, $N_0 = -1$. Hence $P_0 = Z_0 - N_0 = 0 + 1 = 1$.

11.30. Determine N_0 (Equation (11.1), Page 190) for the following transfer function (transformation) and s-plane contour:

$$P(s) \;=\; \frac{K(s+5)(s-2.1)}{s^2(s+1)}$$

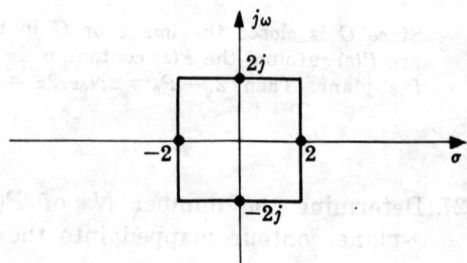

The pole-zero map of $P(s)$ is

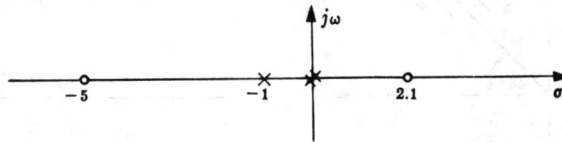

Hence three poles (two at $s = 0$ and one at $s = -1$) and no zeros are enclosed by the contour. Thus $P_0 = 3$, $Z_0 = 0$, and $N_0 = -3$.

11.31. Is the origin *enclosed* by the following contour?

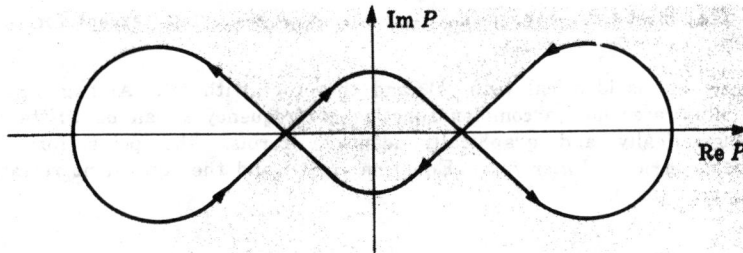

The region to the right of the contour has been shaded. The origin falls in a shaded region and is therefore *enclosed* by the contour.

11.32. What is the sign of N_0 in Problem 11.31?

Since the origin is enclosed, $N_0 > 0$.

POLAR PLOTS

11.33. Prove Property 1 of Section 11.6, Page 192.

Let $P(j\omega) \equiv P_1(\omega) + jP_2(\omega)$ and $a \equiv a_1 + ja_2$, where $P_1(\omega)$, $P_2(\omega)$, a_1 and a_2 are real. Then

$$P(j\omega) + a = (P_1(\omega) + a_1) + j(P_2(\omega) + a_2)$$

and the image of any point $(P_1(\omega), P_2(\omega))$ in the $P(j\omega)$-plane is $(P_1(\omega) + a_1, P_2(\omega) + a_2)$ in the $P(j\omega) + a$ plane. Hence the image of a $P(j\omega)$ contour is simply a *translation* (see **Problem 11.3**, Page 201). Clearly, translation of the contour by a units is equivalent to translation of the axes (origin) by $-a$ units.

11.34. Prove Property 2 of Section 11.6.

The transfer function $P(s)$ of a constant-coefficient linear system is, in general, a ratio of polynomials with constant coefficients. The roots of such polynomials occur in conjugate pairs; that is, if $a + jb$ is a root, then $a - jb$ is also a root. If we let an asterisk (*) represent complex conjugation, then $a + jb = (a - jb)^*$, and if $a = 0$, then $jb = (-jb)^*$. Therefore $P(j\omega) = P(-j\omega)^*$ or $P(-j\omega) = P(j\omega)^*$. Graphically this means that the plot for $P(-j\omega)$ is the mirror image about the real axis of the plot for $P(j\omega)$ since only the imaginary part of $P(j\omega)$ changes sign.

11.35. Sketch the Polar Plot of each of the following complex functions:

(a) $P(j\omega) = \omega^2 \underline{/45°}$, (b) $P(j\omega) = \omega^2(\cos 45° + j \sin 45°)$, (c) $P(j\omega) = 0.707\omega^2 + 0.707j\omega^2$.

(a) $\omega^2\underline{/45°}$ is in the form of Equation (11.2), Page 191. Hence polar coordinates are used in Fig. 11-11 below.

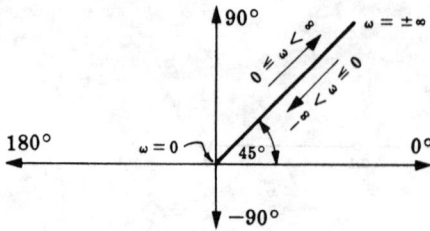

Fig. 11-11

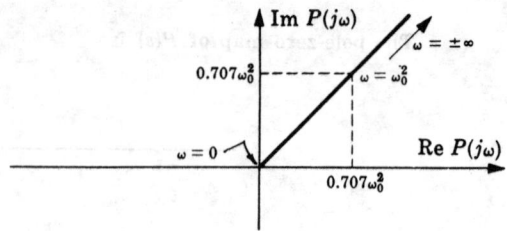

Fig. 11-12

(b)
$$P(j\omega) = \omega^2(\cos 45° + j \sin 45°) = \omega^2(0.707 + 0.707j)$$

That is, $P(j\omega)$ is in the form of Equation (11.3) or (11.4), Page 191. Hence rectangular coordinates is the natural choice as shown in Fig. 11-12.

Note that this graph is identical with that of part (a) except for the coordinates. In fact, $\omega^2(0.707 + 0.707j) = \omega^2 \underline{/45°}$.

(c) Clearly, (c) is identical with (b), and therefore with (a). Among other things, this problem has illustrated how a complex function of frequency ω can be written in three different but mathematically and graphically identical forms: the polar form, Equation (11.2); the trigonometric or *Euler form*, Equation (11.3); and the equivalent rectangular (complex) form, Equation (11.4).

11.36. Sketch the Polar Plot of

$$P(j\omega) = 0.707\omega^2(1 + j) + 1$$

The Polar Plot of $0.707\omega^2(1+j)$ was drawn in Problem 11.35(b). By Property 1 of Section 11.6, Page 192, the required Polar Plot is given by that of Problem 11.35(b) with its origin shifted to $-a = -1$ as shown in the adjoining figure.

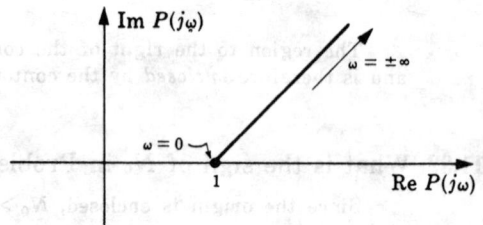

11.37. Construct a Polar Plot from the following set of graphs of the magnitude and phase angle of $P(j\omega)$, representing the frequency response of a linear constant-coefficient system:

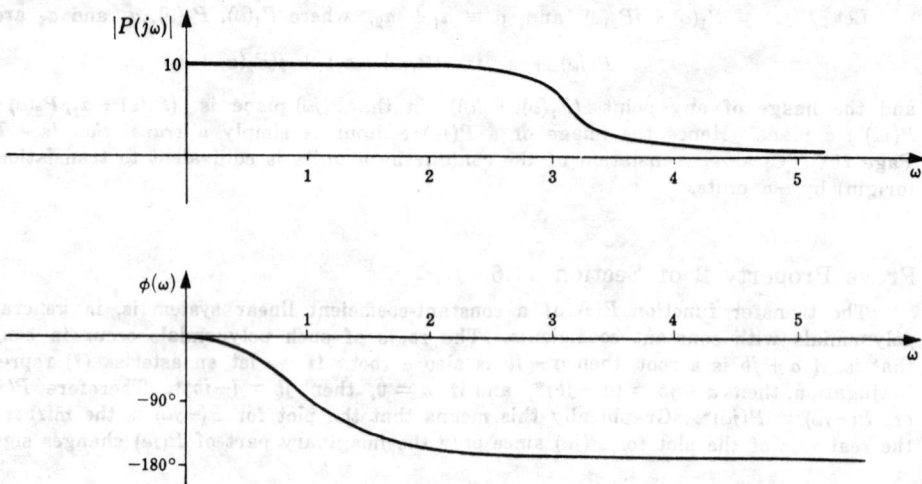

The graphs shown above differ little from *Bode representations*, discussed in detail in Chapter 15. The Polar Plot is constructed by mapping this set of graphs into the $P(j\omega)$-plane. It is only necessary to choose values of ω and corresponding values of $|P(j\omega)|$ and $\phi(\omega)$ from the graphs and plot these points in the $P(j\omega)$-plane. For example at $\omega = 0$, $|P(j\omega)| = 10$ and $\phi(\omega) = 0$. The resulting Polar Plot is given by

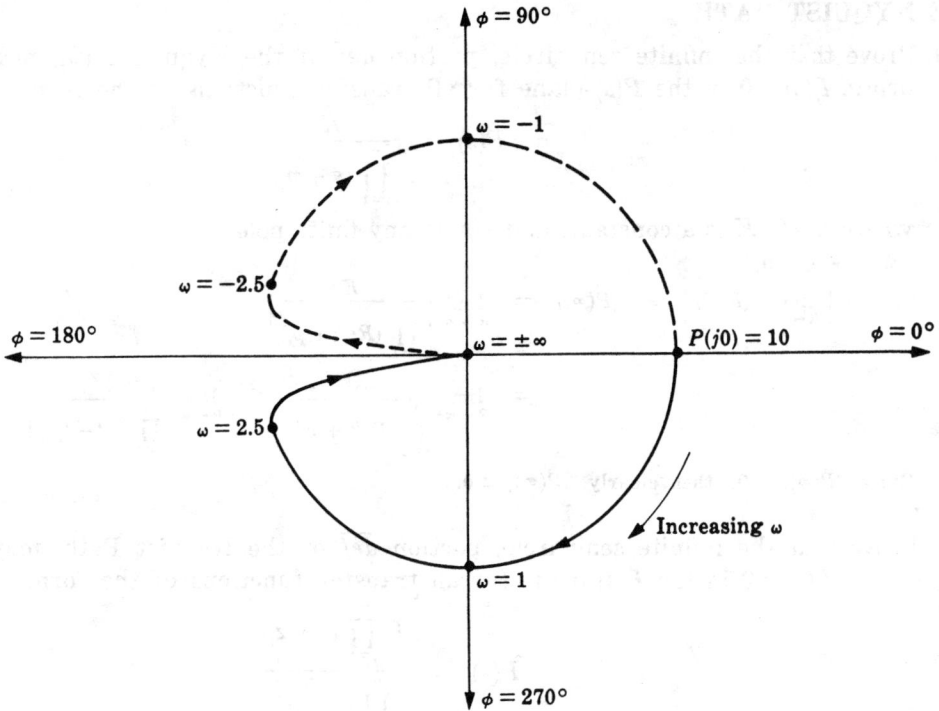

The portion of the graph for $-\infty < \omega < 0$ has been drawn using the property of conjugate symmetry (Section 11.6, Page 192).

11.38. Sketch the Polar Plot for

$$GH(s) = \frac{1}{s^4(s+p)}, \quad p > 0$$

Substituting $j\omega$ for s, and applying Equation (*11.2*), Page 191, we obtain

$$GH(j\omega) = \frac{1}{j^4\omega^4(j\omega + p)}$$

$$= \frac{1}{\omega^4\sqrt{\omega^2 + p^2}} \underline{/-\tan^{-1}(\omega/p)}$$

For $\omega = 0$ and $\omega \to \infty$, we have

$$GH(j0) = \infty \underline{/0°}$$

$$\lim_{\omega \to \infty} GH(j\omega) = 0 \underline{/-90°}$$

Clearly, as ω increases from zero to infinity, the phase angle remains negative and decreases to $-90°$, and the magnitude decreases monotonically to zero. Thus the Polar Plot may be sketched as shown in the adjoining figure. The dashed line represents the mirror image of the plot for $0 < \omega < \infty$ (Section 11.6, Property 2). Hence it is the Polar Plot for $-\infty < \omega < 0$.

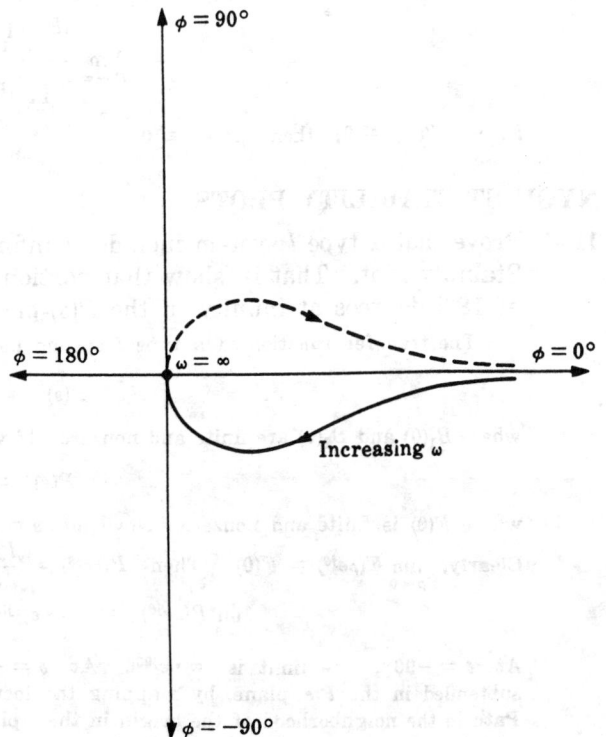

THE NYQUIST PATH

11.39. Prove that the infinite semicircle, portion $\overline{def}$ of the Nyquist Path, maps into the origin $P(s) = 0$ in the $P(s)$-plane for all transfer functions of the form

$$P(s) = \frac{K}{\displaystyle\prod_{i=1}^{n}(s+p_i)}$$

where $n > 0$, K is a constant, and $-p_i$ is any finite pole.

For $n > 0$,

$$\left| \lim_{R \to \infty} P(Re^{j\theta}) \right| \equiv |P(\infty)| = \lim_{R \to \infty} \left| \frac{K}{\displaystyle\prod_{i=1}^{n}(Re^{j\theta}+p_i)} \right|$$

$$= \lim_{R \to \infty} \frac{|K|}{\displaystyle\prod_{i=1}^{n}|Re^{j\theta}+p_i|} \leq \lim_{R \to \infty} \frac{|K|}{\displaystyle\prod_{i=1}^{n}\left| R-|p_i| \right|} = 0$$

Since $|P(\infty)| \leq 0$, then clearly $|P(\infty)| \equiv 0$.

11.40. Prove that the infinite semicircle, portion $\overline{def}$ of the Nyquist Path, maps into the origin $P(s) = 0$ in the $P(s)$-plane for all transfer functions of the form

$$P(s) = \frac{K\displaystyle\prod_{i=1}^{m}(s+z_i)}{\displaystyle\prod_{i=1}^{n}(s+p_i)}$$

where $m < n$, K is a constant, and $-p_i$ and $-z_i$ are finite poles and zeros, respectively.

For $m < n$,

$$\left| \lim_{R \to \infty} P(Re^{j\theta}) \right| \equiv |P(\infty)| = \lim_{R \to \infty} \left| \frac{K\displaystyle\prod_{i=1}^{m}(Re^{j\theta}+z_i)}{\displaystyle\prod_{i=1}^{n}(Re^{j\theta}+p_i)} \right|$$

$$= \lim_{R \to \infty} \frac{|K|\displaystyle\prod_{i=1}^{m}|Re^{j\theta}+z_i|}{\displaystyle\prod_{i=1}^{n}|Re^{j\theta}+p_i|} \leq \lim_{R \to \infty} \frac{|K|\displaystyle\prod_{i=1}^{m}\left| R+|z_i| \right|}{\displaystyle\prod_{i=1}^{n}\left| R-|p_i| \right|} = 0$$

Since $|P(\infty)| \leq 0$, then $|P(\infty)| \equiv 0$.

NYQUIST STABILITY PLOTS

11.41. Prove that a type l system includes l infinite semicircles in the locus of its Nyquist Stability Plot. That is, show that portion $\overline{ija}$ of the Nyquist Path maps into an arc of $180l$ degrees at infinity in the $P(s)$-plane.

The transfer function of a type l system has the form

$$P(s) = \frac{B_1(s)}{s^l B_2(s)}$$

where $B_1(0)$ and $B_2(0)$ are finite and nonzero. If we let $B_1(s)/B_2(s) \equiv F(s)$, then

$$P(s) = \frac{F(s)}{s^l}$$

where $F(0)$ is finite and nonzero. Now put $s = \rho e^{j\theta}$, as required by Equation (11.12), Page 194. Clearly, $\lim\limits_{\rho \to 0} F(\rho e^{j\theta}) = F(0)$. Then $P(\rho e^{j\theta}) = \dfrac{F(\rho e^{j\theta})}{\rho^l e^{jl\theta}}$ and

$$\lim_{\rho \to 0} P(\rho e^{j\theta}) = \infty \cdot e^{-jl\theta} \qquad -90^\circ \leq \theta \leq +90^\circ$$

At $\theta = -90^\circ$, the limit is $\infty \cdot e^{j90l}$. At $\theta = +90^\circ$, the limit is $\infty \cdot e^{-j90l}$. Hence the angle subtended in the $P(s)$-plane, by mapping the locus of the infinitesimal semicircle of the Nyquist Path in the neighborhood of the origin in the s-plane, is $90l - (-90l) = 180l$ degrees, which represents l infinite semicircles in the $P(s)$-plane.

11.42. Sketch the Nyquist Stability Plot for the open-loop transfer function given by

$$GH(s) = \frac{1}{(s+p_1)(s+p_2)}, \quad p_1, p_2 > 0$$

The Nyquist Path for this type 0 system is shown in Fig. 11-13.

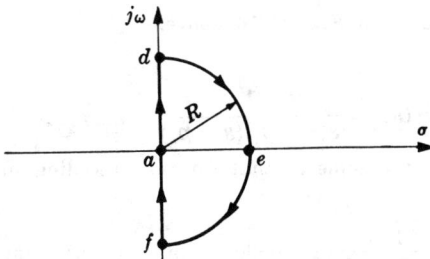

Fig. 11-13

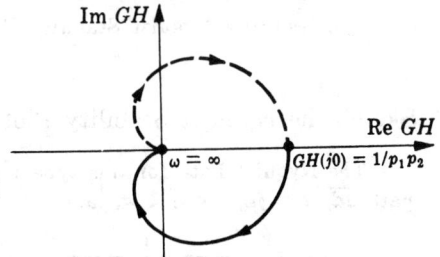

Fig. 11-14

Since there are no poles on the $j\omega$-axis, Step 2 of Section 11.8, Page 194, indicates that the Polar Plot of $GH(j\omega)$ yields the image of path $\overline{ad}$ (and hence $\overline{fad}$) in the $GH(s)$-plane. Letting $s = j\omega$ for $0 < \omega < \infty$, we get

$$GH(j\omega) = \frac{1}{(j\omega + p_1)(j\omega + p_2)} = \frac{1}{\sqrt{(\omega^2 + p_1^2)(\omega^2 + p_2^2)}}\underline{/-\tan^{-1}(\omega/p_1) - \tan^{-1}(\omega/p_2)}$$

$$GH(j0) = \frac{1}{p_1 p_2}\underline{/0^\circ}, \quad \lim_{\omega \to \infty} GH(j\omega) = 0\underline{/180^\circ}$$

For $0 < \omega < \infty$, the Polar Plot passes through the third and fourth quadrants because $\phi = -[\tan^{-1}(\omega/p_1) + \tan^{-1}(\omega/p_2)]$ varies from 0° to 180° when ω increases.

From Problem 11.39, path $\overline{def}$ plots into the origin $P(s) = 0$. Therefore the Nyquist Stability Plot is a replica of the Polar Plot. This is easily sketched from the above derivations, and is shown in Fig. 11-14 above.

11.43. Sketch the Nyquist Stability Plot for $GH(s) = 1/s$.

The Nyquist Path for this simple type 1 system is shown in Fig. 11-15.

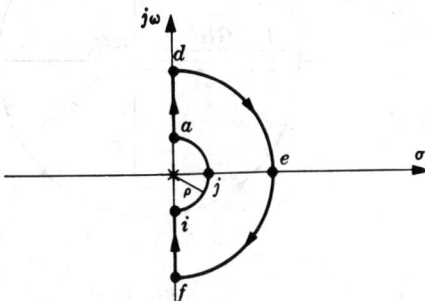

Fig. 11-15

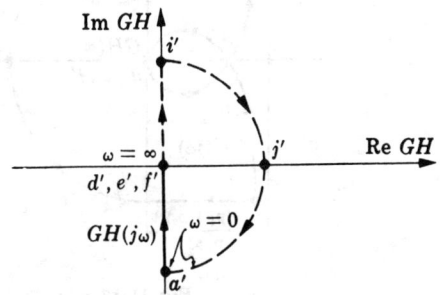

Fig. 11-16

For path $\overline{ad}$, $s = j\omega$, $0 < \omega < \infty$, and

$$GH(j\omega) = \frac{1}{j\omega} = \frac{1}{\omega}\underline{/-90^\circ}, \quad \lim_{\omega \to 0} GH(j\omega) = \infty\underline{/-90^\circ}, \quad \lim_{\omega \to \infty} GH(j\omega) = 0\underline{/-90^\circ}$$

Path $\overline{def}$ maps into the origin (see Problem 11.39).

Path $\overline{f'i'}$ is the mirror image of $\overline{a'd'}$ about the real axis.

The image of path $\overline{ija}$ is determined from Equation (11.12), Page 194, by letting $s = \lim_{\rho \to 0} \rho e^{j\theta}$, where $-90^\circ \leq \theta \leq 90^\circ$:

$$\lim_{\rho \to 0} GH(\rho e^{j\theta}) = \lim_{\rho \to 0} \left[\frac{1}{\rho} e^{-i\theta} \right] = \infty \cdot e^{-i\theta} = \infty \underline{/-\theta}$$

For point i, $\theta = -90°$; then i maps into i' at $\infty \underline{/90°}$. At point j, $\theta = 0°$; then j maps into j' at $\infty \underline{/0°}$. Similarly, a maps into a' at $\infty \underline{/-90°}$. Path $\overline{i'j'a'}$ could also have been obtained from the conformal mapping property of the transformation as explained in Example 11.9, Page 195, plus the statement proved in Problem 11.41.

The resulting Nyquist Stability Plot is shown in Fig. 11-16 above.

11.44. Sketch the Nyquist Stability Plot for $GH(s) = \dfrac{1}{s(s + p_1)(s + p_2)}$, $p_1, p_2 > 0$.

The Nyquist Path for this type 1 system is the same as that for the preceding problem. For path $\overline{ad}$, $s = j\omega$, $0 < \omega < \infty$, and

$$GH(j\omega) = \frac{1}{j\omega(j\omega + p_1)(j\omega + p_2)} = \frac{1}{\omega \sqrt{(\omega^2 + p_1^2)(\omega^2 + p_2^2)}} \underline{/-90° - \tan^{-1}(\omega/p_1) - \tan^{-1}(\omega/p_2)}$$

$$\lim_{\omega \to 0} GH(j\omega) = \infty \underline{/-90°}, \quad \lim_{\omega \to \infty} GH(j\omega) = 0 \underline{/-270°} = 0 \underline{/+90°}$$

Since the phase angle changes sign as ω increases, the plot crosses the real axis. At intermediate values of frequency, the phase angle ϕ is within the range $-90° < \phi < -270°$. Hence the plot is in the second and third quadrants. An asymptote of $GH(j\omega)$ for $\omega \to 0$ is found by writing $GH(j\omega)$ as a real plus an imaginary part, and *then* taking the limit as $\omega \to 0$:

$$GH(j\omega) = \frac{-(p_1 + p_2)}{(\omega^2 + p_1^2)(\omega^2 + p_2^2)} - \frac{j(p_1 p_2 - \omega^2)}{\omega(\omega^2 + p_1^2)(\omega^2 + p_2^2)} \quad \lim_{\omega \to 0} GH(j\omega) = \frac{-(p_1 + p_2)}{p_1^2 p_2^2} - j\infty$$

Hence the line $GH = -(p_1 + p_2)/p_1^2 p_2^2$ is an asymptote of the Polar Plot.

Path $\overline{def}$ maps into the origin (see Problem 11.39). Path $\overline{f'i'}$ is the mirror image of $\overline{a'd'}$ about the real axis. Path $\overline{i'j'a'}$ is most easily determined by the conformal mapping property and the fact that a type 1 system has *one* infinite semicircle in its path (Problem 11.41). The resulting Nyquist Stability Plot is shown in Fig. 11-17.

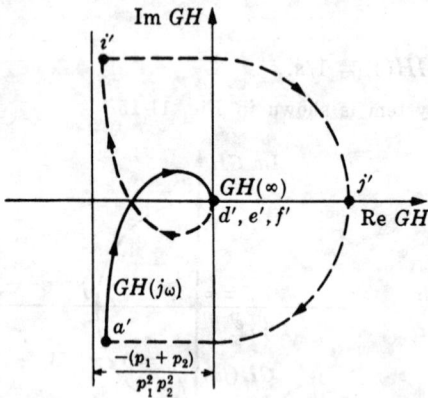

Fig. 11-17

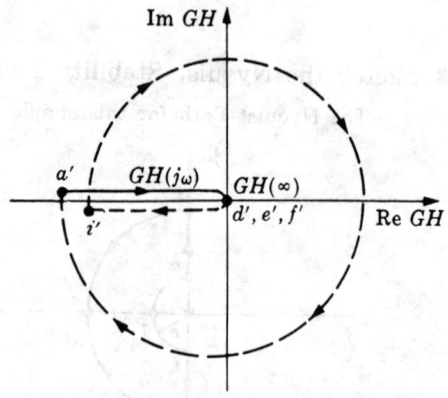

Fig. 11-18

11.45. Sketch the Nyquist Stability Plot for $GH(s) = 1/s^2$.

The Nyquist Path for this type 2 system is the same as that for the preceding problem, except there are two poles at the origin instead of one. For $\overline{ad}$,

$$GH(j\omega) = \frac{1}{j^2 \omega^2} = \frac{1}{\omega^2} \underline{/180°}, \quad \lim_{\omega \to 0} GH(j\omega) = \infty \underline{/180°}, \quad \lim_{\omega \to \infty} GH(j\omega) = 0 \underline{/180°}$$

The Polar Plot clearly lies along the negative real axis, increasing from $-\infty$ to 0 as ω increases. Path $\overline{def}$ maps into the origin and path $\overline{ija}$ maps into *two* infinite semicircles at infinity (see Problem 11.41). Since the Nyquist Path makes right turns at i and a, so does the Nyquist Stability Plot at i' and a'. The resulting locus is shown in Fig. 11-18.

11.46. Sketch the Nyquist Stability Plot for $GH(s) = \dfrac{1}{s^2(s+p)}$, $p > 0$.

The Nyquist Path for this type 2 system is the same as that for the previous problem. For $\overline{ad}$,

$$GH(j\omega) = \frac{1}{j^2\omega^2(j\omega + p)}$$

$$= \frac{1}{\omega^2\sqrt{\omega^2 + p^2}} \big/\underline{-180° - \tan^{-1}(\omega/p)}$$

$$\lim_{\omega \to 0} GH(j\omega) = \infty \big/\underline{-180°}$$

$$\lim_{\omega \to \infty} GH(j\omega) = 0 \big/\underline{-270°}$$

For $0 < \omega < \infty$ the phase angle varies continuously from $-180°$ to $-270°$; thus the plot lies in the second quadrant. The remainder of the Nyquist Path is mapped into the GH-plane as in the preceding problem. The resulting Nyquist Stability Plot is shown in the adjoining figure.

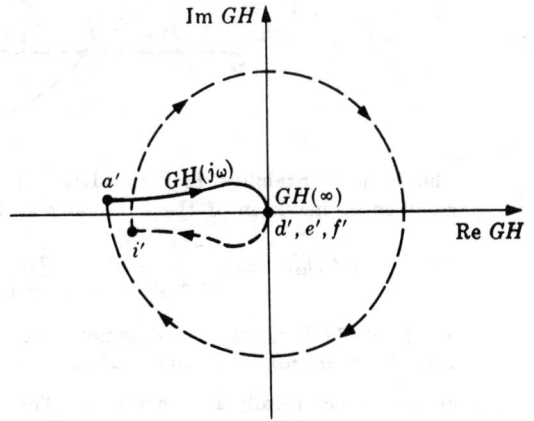

11.47. Sketch the Nyquist Stability Plot for $GH(s) = \dfrac{1}{s^4(s+p)}$, $p > 0$.

There are four poles at the origin in the s-plane, and the Nyquist Path is the same as that of the previous problem. The Polar Plot for this system was determined in Problem 11.38, Page 211. The remainder of the Nyquist Path is mapped using the results of Problems 11.39 and 11.41, and the conformal mapping property. The resulting Nyquist Stability Plot is given by

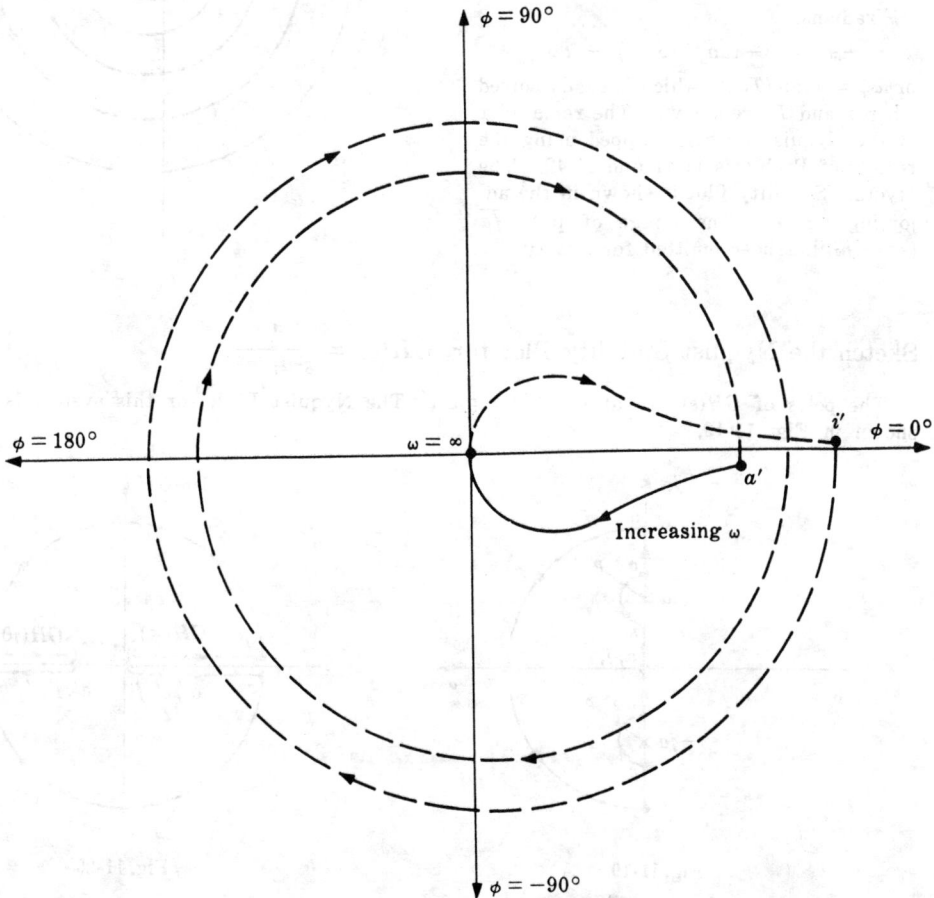

11.48. Sketch the Nyquist Stability Plot for $GH(s) = \dfrac{e^{-Ts}}{s+p}$, $p > 0$.

The e^{-Ts} term represents a time-delay of T seconds in the forward or feedback path. For example, a signal flow graph of such a system can be represented by

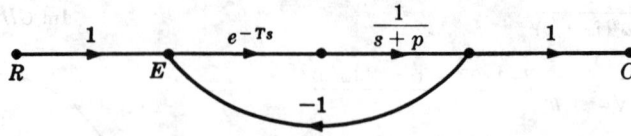

The Nyquist Stability Plot for $1/(s+1)$ was drawn in Example 11.8, Page 195. The plot is modified by inclusion of the e^{-Ts} term in the following manner. For path $\overline{ad}$,

$$GH(j\omega) = \frac{e^{-Tj\omega}}{j\omega + p} = \frac{1}{\sqrt{\omega^2 + p^2}}\,\underline{/-\tan^{-1}(\omega/p) - T\omega}, \qquad GH(j0) = (1/p)\,\underline{/0^\circ}$$

The limit of $GH(j\omega)$ as $\omega \to \infty$ does not exist. But $\lim\limits_{\omega \to \infty} |GH(j\omega)| = 0$ and $|GH(j\omega)|$ decreases monotonically as ω increases. The phase angle term

$$\phi(\omega) = -\tan^{-1}(\omega/p) - T\omega$$

revolves repeatedly about the origin between 0° and -360° as ω increases. Therefore the Polar Plot is a decreasing spiral, beginning at $(1/p)\,\underline{/0^\circ}$ and approaching the origin in a CW direction. The points where the locus crosses the negative real axis are determined by letting $\phi = -180^\circ = -\pi$ radians:

$$-\pi = -\tan^{-1}(\omega_\pi/p) - T\omega_\pi$$

or $\omega_\pi = p \tan(T\omega_\pi)$, which is easily solved when p and T are known. The remainder of the Nyquist Path is mapped using the results of Problems 11.41 and 11.42. The Nyquist Stability Plot is shown in the adjoining figure. The image of path $\overline{fa}$ ($s = -j\omega$) has been omitted for clarity.

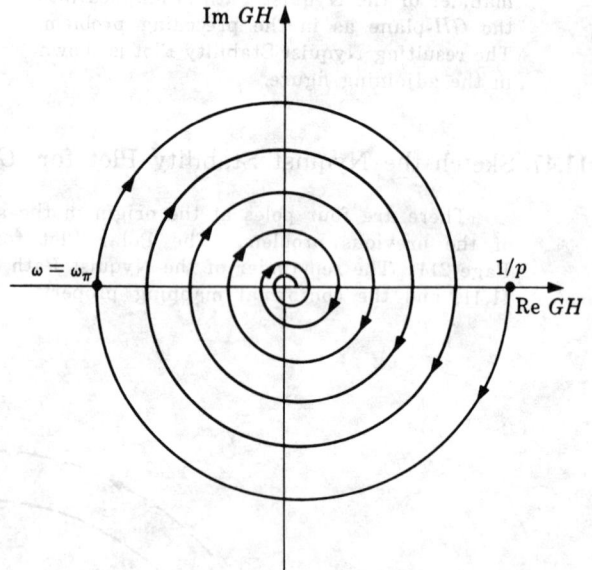

11.49. Sketch the Nyquist Stability Plot for $GH(s) = \dfrac{1}{s^2 + a^2}$.

The poles of $GH(s)$ are at $s = \pm ja \equiv \pm j\omega_0$. The Nyquist Path for this system is therefore as shown in Fig. 11-19.

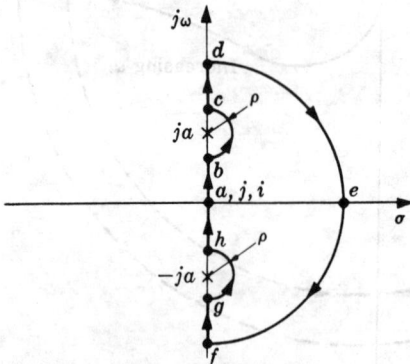

Fig. 11-19

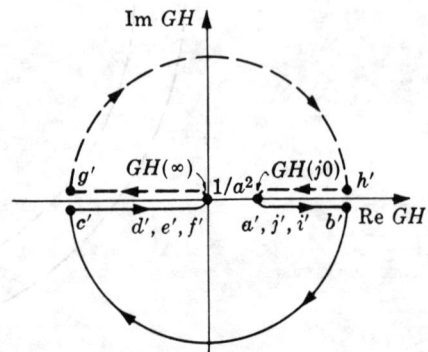

Fig. 11-20

For path $\overline{ab}$, $\omega < a$ and

$$GH(j\omega) = [1/(a^2 - \omega^2)] \underline{/0°}, \quad GH(j0) = (1/a^2) \underline{/0°}, \quad \lim_{\omega \to a} GH(j\omega) = \infty \underline{/0°}$$

For path $\overline{bc}$, let $s \equiv ja + \rho e^{j\theta}$, $-90° \leq \theta \leq 90°$; then

$$\lim_{\rho \to 0} GH(ja + \rho e^{j\theta}) = \lim_{\rho \to 0} \left[\frac{1}{\rho e^{j\theta}(2ja + \rho e^{j\theta})} \right] = -j\infty \cdot e^{-j\theta} = \infty \underline{/-\theta° - 90°}$$

At $\theta = -90°$ the limit is $\infty \underline{/0°}$; at $\theta = 0°$ it is $\infty \underline{/-90°}$; at $\theta = 90°$ it is $\infty \underline{/-180°}$.

For path $\overline{cd}$, $\omega > a$ and

$$\lim_{\omega \to a} GH(j\omega) = \infty \underline{/180°}, \quad \lim_{\omega \to \infty} GH(j\omega) = 0 \underline{/180°}$$

Path $\overline{def}$ maps into the origin by Problem 11.39, and $\overline{f'g'h'a'}$ is the mirror image of $\overline{a'b'c'd'}$ about the real axis. The resulting Nyquist Stability Plot is shown in Fig. 11-20 above.

11.50. Sketch the Nyquist Stability Plot for $GH(s) = \dfrac{s - z}{s(s + p)}$, $z, p > 0$.

The Nyquist Path for this type 1 system is the same as that for Problem 11.43. For path $\overline{ad}$,

$$GH(j\omega) = \frac{j\omega - z}{j\omega(j\omega + p)} = \frac{\sqrt{\omega^2 + z^2}}{\omega^2 \sqrt{\omega^2 + p^2}} \underline{\left/ 90° - \tan^{-1} \left[\frac{\omega(p + z)}{pz - \omega^2} \right] \right.}$$

where we have used $\tan^{-1} x \pm \tan^{-1} y \equiv \tan^{-1}\left[\dfrac{x \pm y}{1 \mp xy} \right]$. Now

$$\lim_{\omega \to 0} GH(j\omega) = \infty \underline{/+90°}$$

$$GH(j\sqrt{pz}) = \frac{1}{p\sqrt{pz}} \underline{/0°}$$

$$\lim_{\omega \to \infty} GH(j\omega) = 0 \underline{/-90°}$$

Thus the locus comes down in the first quadrant, crosses the positive real axis into the fourth quadrant, and approaches the origin from an angle of $-90°$.

Path $\overline{def}$ maps into the origin, and $\overline{ija}$ maps into one semicircle at infinity. The resulting plot is shown in the adjoining figure.

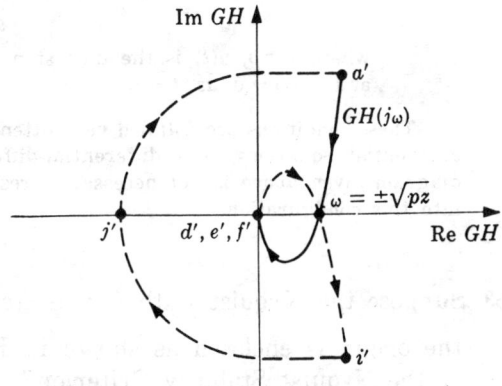

NYQUIST STABILITY CRITERION

11.51. Prove the Nyquist Stability Criterion.

Equation (11.1), Page 190, states that the number of CW encirclements N_0 of the origin made by a closed $P(s)$ contour in the P-plane, mapped from a closed s-plane contour, is equal to the number of zeros Z_0 minus the number of poles P_0 of $P(s)$ enclosed by the s-plane contour: $N_0 = Z_0 - P_0$. This has been proven in Problem 11.26, Page 207.

Now let $P(s) \equiv 1 + GH(s)$. Then the origin for $1 + GH(s)$ in the GH-plane is at $GH = -1$. (See Example 11.6 and Problem 11.33.) Hence let N denote the number of CW encirclements of this $-1 + j0 \equiv (-1, 0)$ point, and let the s-plane contour be the Nyquist Path defined in Section 11.7, Page 193. Then $N = Z_0 - P_0$, where Z_0 and P_0 are the number of zeros and poles of $1 + GH$ enclosed by the Nyquist Path. P_0 is also the number of poles of GH enclosed, since if $GH \equiv N/D$, then $1 + GH = 1 + N/D = (D + N)/D$. That is, GH and $1 + GH$ have the same denominator.

We know from Chapter 5 that a feedback system is absolutely stable if and only if the zeros of the characteristic polynomial $1 + GH$ (the roots of the characteristic equation $1 + GH = 0$) are in the LHP, that is, $Z_0 = 0$. Therefore $N = -P_0$, and clearly $P_0 \geq 0$.

11.52. Extend the Nyquist Stability Criterion so that it can be applied to a larger class of linear systems than those already considered in this chapter.

The Criterion has been extended by C. A. Desoer [7] to include a large class of causal (physically realizable) linear time-invariant systems. The following statement is a modification of this generalization, which can be found with its proof in the reference.

A Generalized Nyquist Stability Criterion: Consider the linear time-invariant system described by the following block diagram:

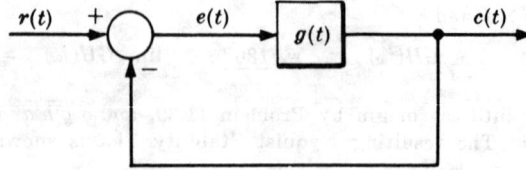

If $g(t)$ satisfies the conditions given below and the Nyquist Stability Plot of $G(s)$ *does not* enclose the $(-1, 0)$ point, then the system is *stable*. If the $(-1, 0)$ point is enclosed, the system is unstable.

1. $G(s)$ represents a causal, linear time-invariant system element.

2.   The input-output relationship for $g(t)$ is

$$c(t) = c_a(t) + \int_0^t g(t-\tau)\, e(\tau)\, d\tau \qquad t \geq 0$$

where $c_a(t)$, the free response of the system $g(t)$, is bounded for all $t \geq 0$ and all initial conditions, and approaches a finite value dependent upon the initial conditions as $t \to \infty$.

3.   The unit impulse response $g(t)$ is

$$g(t) = [k + g_1(t)]\, u(t)$$

where $k \geq 0$, $u(t)$ is the unit step function, $g_1(t)$ is bounded and integrable for all $t \geq 0$, and $g_1(t) \to 0$ as $t \to \infty$.

These conditions are fulfilled very often by physical systems described by ordinary and partial differential equations, and differential-difference equations. The form of the closed-loop block diagram given above is not necessarily restrictive. Most systems of interest can be transformed into this configuration.

11.53. Suppose the Nyquist Path for $GH(s) = \dfrac{1}{s(s+p)}$ were modified so that the pole at the origin is enclosed as shown in Fig. 11-21. How does this modify application of the Nyquist Stability Criterion?

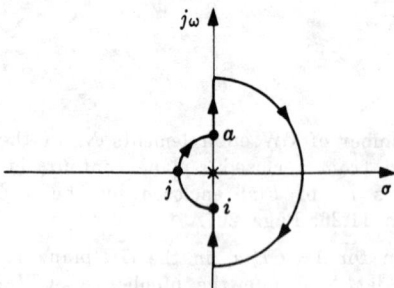

Fig. 11-21

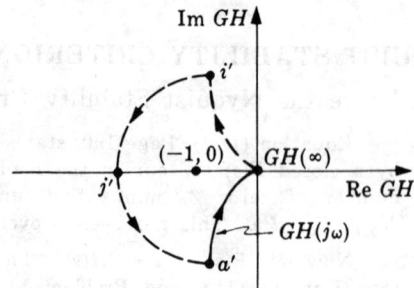

Fig. 11-22

The Polar Plot remains the same, but the image of path $\overline{ija}$ makes *left* instead of right turns at i' and a', just as in the Nyquist Path. The Nyquist Stability Plot is therefore given by Fig. 11-22. Clearly, $N = -1$. But since the pole of GH at the origin is enclosed by the Nyquist Path, then $P_0 = 1$, and $Z_0 = N + P_0 = -1 + 1 = 0$. Therefore the system is stable. Application of the Nyquist Stability Criterion does not depend on the path chosen in the s-plane.

11.54. Is the system of Problem 11.42 stable or unstable?

Shading the region to the right of the contour in the prescribed direction yields Fig. 11-23. It is clear that $N = 0$. The $(-1, 0)$ point is not in the shaded region. Now, since $p_1 > 0$ and $p_2 > 0$, then $P_0 = 0$. Therefore $N = -P_0 = 0$, or $Z_0 = N + P_0 = 0$, and the system is stable.

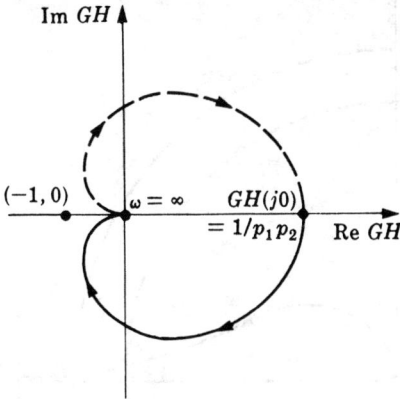

Fig. 11-23

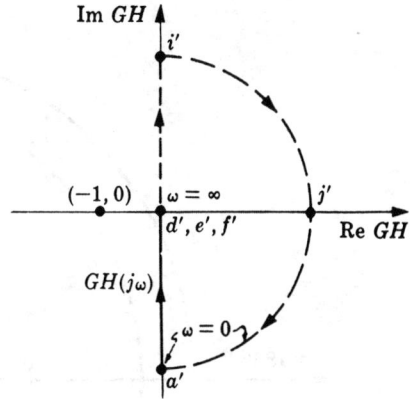

Fig. 11-24

11.55. Is the system of Problem 11.43 stable or unstable?

The region to the right of the contour has been shaded in Fig. 11-24. The $(-1, 0)$ point is not enclosed, and $N = 0$. Since $P_0 = 0$, then $Z_0 = P_0 + N = 0$, and the system is stable.

11.56. Determine the stability of the system of Problem 11.44.

The region to the right of the contour has been shaded in Fig. 11-25. If the $(-1, 0)$ point lies to the left of point k, then $N = 0$; if it lies to the right, then $N = 1$. Since $P_0 = 0$, then $Z_0 = 0$ or 1. Hence the system is stable if and only if the $(-1, 0)$ point lies to the left of point k. Point k can be determined by solving for $GH(j\omega_\pi)$, where

$$-\pi = -\pi/2 - \tan^{-1}(\omega_\pi/p_1) - \tan^{-1}(\omega_\pi/p_2)$$

ω_π is easily determined from this equation when p_1 and p_2 are given.

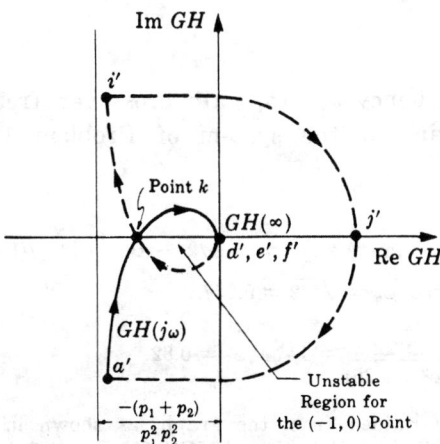

Fig. 11-25

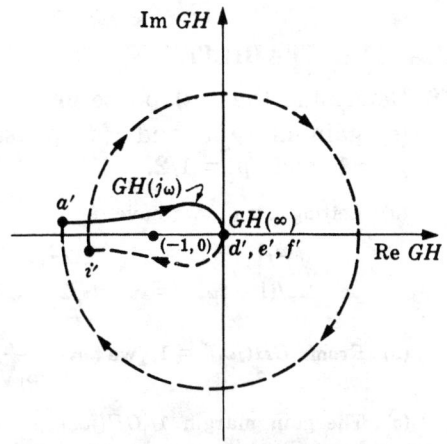

Fig. 11-26

11.57. Determine the stability of the system of Problem 11.46.

The region to the right of the contour has been shaded in Fig. 11-26. Clearly, $N = 1$, $P_0 = 0$, and $Z_0 = 1 + 0 = 1$. Hence the system is unstable for all $p > 0$.

11.58. Determine the stability of the system of Problem 11.47.

The region to the right of the contour has been shaded:

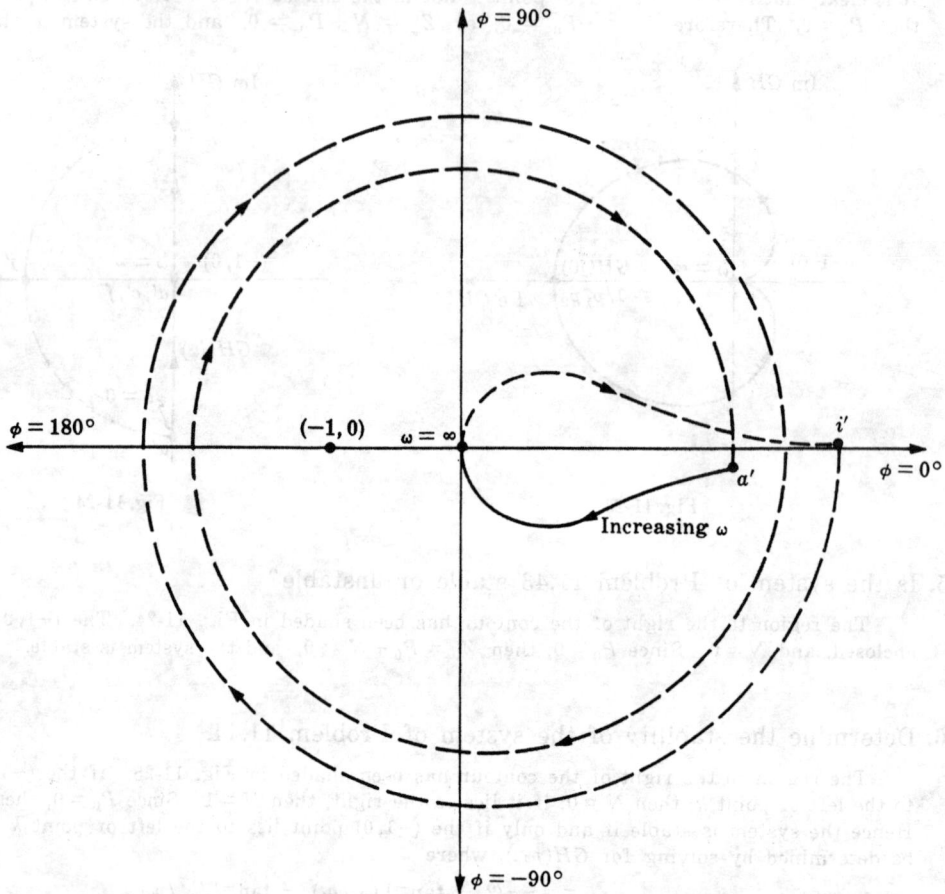

It is clear that $N > 0$. Since $P_0 = 0$ for $p > 0$, then $N \neq -P_0$. Hence the system is unstable.

RELATIVE STABILITY

11.59. Determine the (a) phase crossover frequency ω_π, (b) gain crossover frequency ω_1, (c) gain margin, and (d) phase margin for the system of Problem 11.44 with $p_1 = 1$ and $p_2 = 1/2$.

(a) Letting $\omega = \omega_\pi$, we have

$$\phi(\omega_\pi) = -\pi = -\pi/2 - \tan^{-1}\omega_\pi - \tan^{-1} 2\omega_\pi = -\pi/2 - \tan^{-1}[3\omega_\pi/(1 - 2\omega_\pi^2)]$$

or $3\omega_\pi/(1 - 2\omega_\pi^2) = \tan(\pi/2) = \infty$. Hence $\omega_\pi = \sqrt{1/2} = 0.707$.

(b) From $|GH(j\omega_1)| = 1$, we have $\dfrac{1}{\omega_1 \sqrt{(\omega_1^2 + 1)(\omega_1^2 + 0.25)}} = 1$ or $\omega_1 = 0.82$.

(c) The gain margin $1/|GH(j\omega_\pi)|$ is easily determined from the graph, as shown in Fig. 11-27 below. It can also be calculated analytically: $|GH(j\omega_\pi)| = |GH(j0.707)| = 4/3$; hence gain margin $= 3/4$.

(d) The phase margin is easily determined from the graph, or calculated analytically:

$$\arg GH(j\omega_1) = \arg GH(j0.82) = -90° - \tan^{-1}(0.82) - \tan^{-1}(1.64) = -187.8°$$

Hence $\phi_{PM} = 180° + \arg GH(j\omega_1) = -7.8°$. Negative phase margin means that the system is unstable.

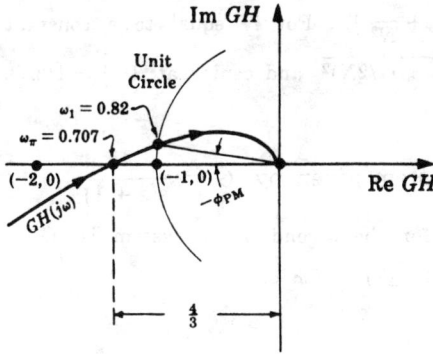

Fig. 11-27

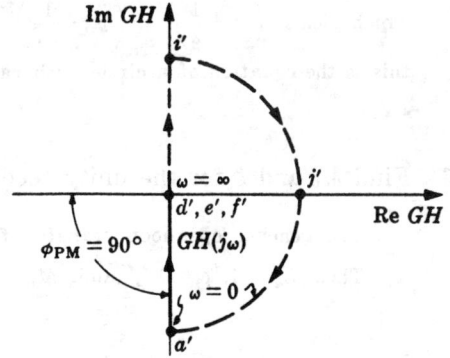

Fig. 11-28

11.60. Determine the gain and phase margins for the system of Problem 11.43 ($GH = 1/s$).

The Nyquist Stability Plot of $1/s$ never crosses the negative real axis as shown in Fig. 11-28; hence the gain margin is undefined for this system. The phase margin is $\phi_{PM} = 90°$.

M AND N CIRCLES

11.61. Prove Equations (*11.18*) and (*11.19*), Page 199, which give the radius and center of an M-circle, respectively.

Let $G(j\omega) \equiv x + jy$. Then $M \equiv \left| \dfrac{G(j\omega)}{1 + G(j\omega)} \right| = \left| \dfrac{x + jy}{1 + x + jy} \right|$. Squaring both sides and rearranging yields

$$\left[x - \left(\frac{M^2}{1 - M^2} \right) \right]^2 + y^2 = \left(\frac{M}{1 - M^2} \right)^2, \quad M < 1; \qquad \left[x + \left(\frac{M^2}{M^2 - 1} \right) \right]^2 + y^2 = \left(\frac{M}{M^2 - 1} \right)^2, \quad M > 1$$

For $M = $ constant, these are equations of circles with radii $|M/(M^2 - 1)|$ and centers at $(-M^2/(M^2 - 1), 0)$.

11.62. Prove Equation (*11.20*), Page 199.

The transfer function G for the second order system whose signal flow graph is shown in the adjoining figure is given by $G = \dfrac{\omega_n^2}{s(s + 2\zeta\omega_n)}$. Now

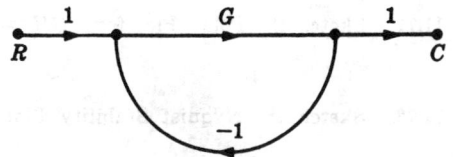

$$M^2 = \left| \frac{G}{1 + G} \right|^2 = \frac{\omega_n^4}{(\omega_n^2 - \omega^2)^2 + 4\zeta^2\omega_n^2\omega^2}$$

To find ω_p, we maximize the above expression:

$$\frac{d}{d\omega}(M^2) = \frac{\omega_n^4[2(\omega_n^2 - \omega^2)(-2\omega) + 8\zeta^2\omega_n^2\omega]}{[(\omega_n^2 - \omega^2)^2 + 4\zeta^2\omega_n^2\omega^2]^2} = 0$$

from which $\omega = \omega_p = \pm\omega_n \sqrt{1 - 2\zeta^2}$. Hence for $0 \le \zeta \le 0.707$,

$$M_p = \left[\frac{\omega_n^4}{[\omega_n^2 - \omega_n^2(1 - 2\zeta^2)]^2 + 4\zeta^2\omega_n^4(1 - 2\zeta^2)} \right]^{1/2} = \frac{1}{2\zeta\sqrt{1 - \zeta^2}}$$

11.63. Prove Equations (*11.21*) and (*11.22*), Page 199, which give the radius and center of an N-circle.

Let $G(j\omega) \equiv x + jy$. Then

$$\frac{C(j\omega)}{R(j\omega)} = \frac{x^2 + x + y^2 + jy}{(1 + x)^2 + y^2} \quad \text{and} \quad N \equiv \frac{\text{Im } (C/R)(j\omega)}{\text{Re } (C/R)(j\omega)} = \frac{y}{x^2 + x + y^2}$$

which yields $\left(x + \dfrac{1}{2}\right)^2 + \left(y - \dfrac{1}{2N}\right)^2 = \dfrac{1}{4}\left(1 + \dfrac{1}{N^2}\right)$. For N equal to a constant parameter, this is the equation of a circle with radius $\sqrt{\tfrac{1}{4} + (1/2N)^2}$ and center at $(-\tfrac{1}{2}, -1/2N)$.

11.64. Find M_p and ζ for the unity feedback system given by $G = \dfrac{1}{s(s+1)}$.

The general open-loop transfer function for the second order system is $G = \dfrac{\omega_n^2}{s(s + 2\zeta\omega_n)}$. Then $\omega_n = 1$, $\zeta = 0.5$, and $M_p = 1/(2\zeta\sqrt{1-\zeta^2}) = 0.866$.

Supplementary Problems

11.65. Let $T = 2$ and $p = 5$ in the system of Problem 11.48, Page 216. Is this system stable?

11.66. Is the system of Problem 11.49, Page 216, stable or unstable?

11.67. Is the system of Problem 11.50, Page 217, stable or unstable?

11.68. Sketch the Polar Plot for $GH = \dfrac{K(s+z_1)(s+z_2)}{s^3(s+p_1)(s+p_2)}$, $z_i, p_i > 0$.

11.69. Sketch the Polar Plot for $GH = \dfrac{K}{(s+p_1)(s+p_2)(s+p_3)}$, $p_i > 0$.

11.70. Find the closed-loop frequency response of the unity feedback system described by $G = \dfrac{10(s+0.5)}{s^2(s+1)(s+10)}$, using M and N circles.

11.71. Sketch the Polar Plot for $GH = \dfrac{K(s+z)}{s^2(s+p_1)(s+p_2)(s+p_3)}$, $z, p_i > 0$.

11.72. Sketch the Nyquist Stability Plot for $GH = \dfrac{Ke^{-Ts}}{s(s+1)}$.

11.73. Sketch the Polar Plot for $GH = \dfrac{s+z_1}{s(s+p_1)}$, $z_1, p_1 > 0$.

11.74. Sketch the Polar Plot for $GH = \dfrac{s+z_1}{s(s+p_1)(s+p_2)}$, $z_1, p_i > 0$.

11.75. Sketch the Polar Plot for $GH = \dfrac{K}{s^2(s+p_1)(s+p_2)}$, $p_i > 0$.

11.76. Sketch the Polar Plot for $GH = \dfrac{s+z_1}{s^2(s+p_1)}$, $z_1, p_1 > 0$.

11.77. Sketch the Polar Plot for $GH = \dfrac{s+z_1}{s^2(s+p_1)(s+p_2)}$, $z_1, p_i > 0$.

11.78. Sketch the Polar Plot for $GH = \dfrac{(s+z_1)(s+z_2)}{s^2(s+p_1)(s+p_2)(s+p_3)}$, $z_i, p_i > 0$.

11.79. Sketch the Polar Plot for $GH = \dfrac{K}{s^3(s + p_1)(s + p_2)}$, $p_i > 0$.

11.80. Sketch the Polar Plot for $GH = \dfrac{(s + z_1)}{s^3(s + p_1)(s + p_2)}$, $z_1, p_i > 0$.

11.81. Sketch the Polar Plot for $GH = \dfrac{s + z_1}{s^4(s + p_1)}$, $z_1, p_1 > 0$.

11.82. Sketch the Polar Plot for $GH = \dfrac{e^{-Ts}(s + z_1)}{s^2(s + p_1)}$, $z_1, p_1 > 0$.

11.83. Sketch the Polar Plot for $GH = \dfrac{e^{-Ts}(s + z_1)}{s^2(s^2 + a)(s^2 + b)}$, $z_1, a, b > 0$.

11.84. Sketch the Polar Plot for $GH = \dfrac{(s - z_1)}{s^2(s + p_1)}$, $z_1, p_1 > 0$.

11.85. Sketch the Polar Plot for $GH = \dfrac{s}{(s + p_1)(s - p_2)}$, $p_i > 0$.

Answers to Supplementary Problems

11.65. Yes

11.66. Unstable

11.67. Unstable

11.68.

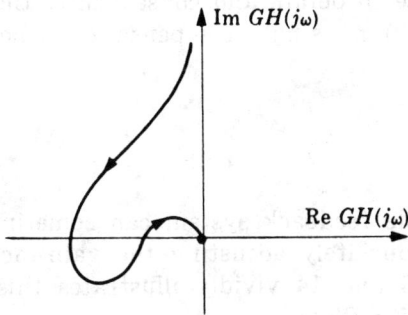

11.69.

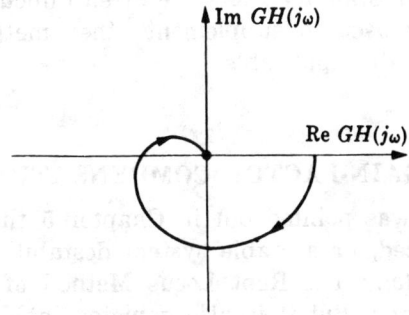

11.71.

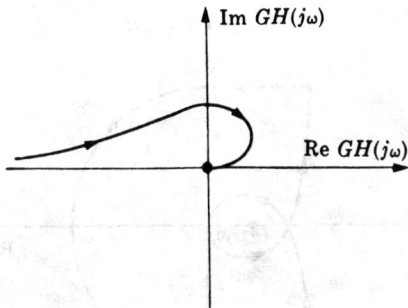

11.72.

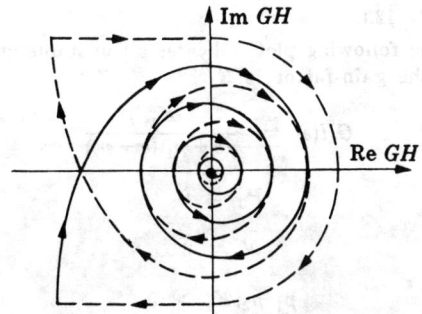

Chapter 12

Nyquist Design

12.1 DESIGN PHILOSOPHY

Design by analysis in the frequency domain using Nyquist techniques is performed in the same general manner as all other design methods in this book: appropriate compensation networks are introduced in the forward or feedback paths and the behavior of the resulting system is critically analyzed. In this manner, the Polar Plot is shaped and reshaped until performance specifications are met.

Since the Polar Plot is a graph of the open-loop transfer function $GH(j\omega)$, many types of compensation networks can be used in either the forward or feedback path, becoming part of either G or H. Often only one, or a combination of both cascade and feedback compensation, can be used to satisfy specifications. But, for illustrative purposes only, cascade compensation is employed in this chapter.

We emphasize that: (1) Nyquist design techniques are rarely used exclusively because Nyquist Stability Plots are often difficult to generate in detail, and consequently they are usually used to supplement other methods; and (2) no single compensation scheme is universally applicable.

12.2 GAIN-FACTOR COMPENSATION

It was pointed out in Chapter 5 that an unstable feedback system can sometimes be stabilized, or a stable system destabilized, by appropriately adjusting the gain-factor K of $GH(s)$. The Root Locus Method of Chapters 13 and 14 vividly illustrates this phenomenon. But it is also depicted in Nyquist Stability Plots.

Example 12.1.

The following plot indicates an unstable system when the gain-factor is K_1:

$$GH(s) = \frac{K_1}{s(s + p_1)(s + p_2)}$$

$$P_0 = 0$$

$$N = 2$$

$$p_1, p_2, K_1 > 0$$

A sufficient decrease in the gain-factor to K_2 $(K_2 < K_1)$ stabilizes the system:

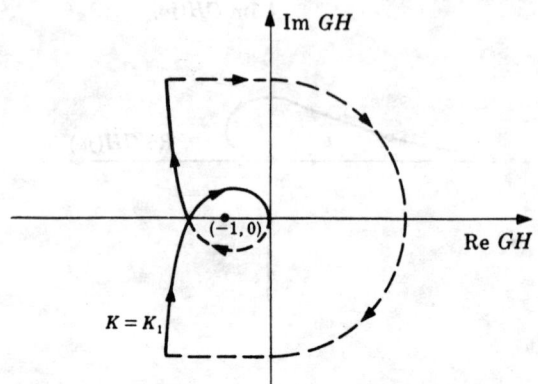

224

$$GH(s) = \frac{K_2}{s(s + p_1)(s + p_2)}$$

$$P_0 = 0$$

$$N = 0$$

$$0 < K_2 < K_1$$

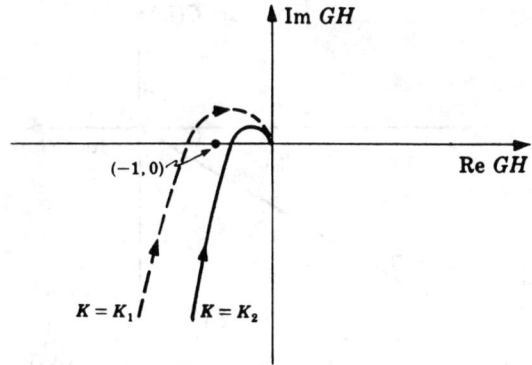

Further decrease of K does not alter stability.

Example 12.2.

The stable region for the $(-1, 0)$ point in the following plot is indicated by the portion of the real axis in the unshaded area:

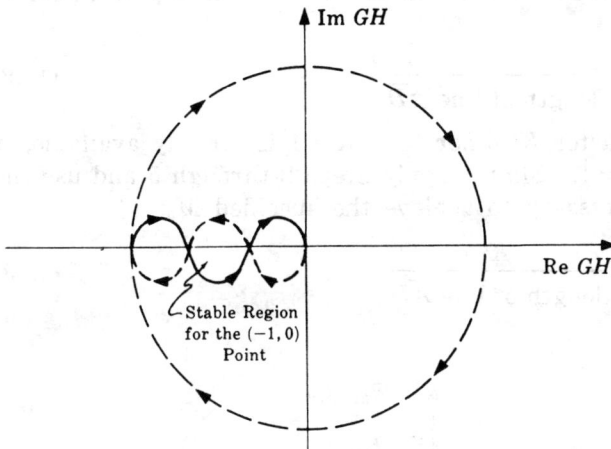

$$GH(s) = \frac{K(s + z_1)(s + z_2)}{s^2(s + p_1)(s + p_2)(s + p_3)}$$

$$P_0 = 0$$

$$z_1, z_2 > 0$$

$$p_i > 0$$

If the $(-1, 0)$ point falls in the stable region, an increase or decrease in K can cause enough shift in the GH contour to the left or the right to destabilize the system. This can happen because a shaded (unstable) region appears both to the left and the right of the unshaded (stable) region. This phenomenon is called **conditional stability.**

Although absolute stability can often be altered by adjustment of the gain factor alone, other performance criteria such as those concerned with relative stability usually require additional compensators.

12.3 GAIN-FACTOR COMPENSATION USING *M*-CIRCLES

The gain-factor K of $G(s)$ for a *unity feedback* system can be determined for a specific resonant peak M_p. The following procedure enables us to accomplish this task, drawing the Polar Plot only once.

Step 1: Draw the Polar Plot of $G(j\omega)$ for $K = 1$.

Step 2: Calculate Ψ_p, given by

$$\Psi_p = \sin^{-1}\left(\frac{1}{M_p}\right) \tag{12.1}$$

Step 3: Draw a radial line $\overline{AB}$ at an angle Ψ_p below the negative real axis, as shown in Fig. 12-1 below.

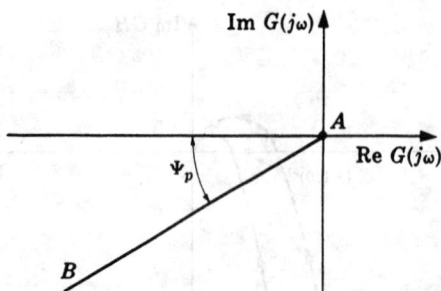

Fig. 12-1

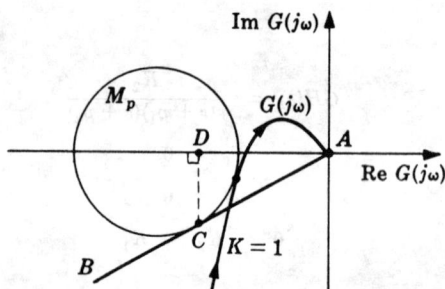

Fig. 12-2

Step 4: Draw the M_p circle tangent to both $G(j\omega)$ and line $\overline{AB}$ at C. Then draw a line $\overline{CD}$ perpendicular to the real axis as shown in the example Polar Plot shown in Fig. 12-2.

Step 5: Measure the length of line $\overline{AD}$ along the real axis. The required gain-factor K for the specified M_p is given by

$$K_{M_p} = \frac{1}{\text{length of line } \overline{AD}} \qquad (12.2)$$

If the Polar Plot of $G(s)$ for a gain-factor K' other than $K = 1$ is already available, it is not necessary to repeat this plot for $K = 1$. Simply apply Steps 2 through 5 and use the following formula for the gain-factor necessary to achieve the specified M_p:

$$K_{M_p} = \frac{K'}{\text{length of line } \overline{AD}} \qquad (12.3)$$

12.4 LEAD COMPENSATION

The transfer function for a lead network, presented in Chapter 6, Equation (6.2), Page 98, is

$$P_{\text{Lead}} = \frac{s + a}{s + b}$$

where $a < b$. The Polar Plot of P_{Lead} for $0 \leq \omega < \infty$ is shown in the adjoining diagram.

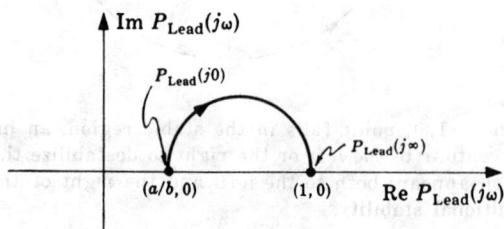

Fig. 12-3

For some systems in which lead compensation is applicable, appropriate choice of the zero at $-a$ and the pole at $-b$ permits an increase in the open-loop gain-factor K, providing greater accuracy (and sometimes stability), without adversely affecting transient performance. Conversely, for a given K, transient performance can be improved. In some cases, both steady-state and transient response can be favorably modified with lead compensation.

The lead network provides compensation by virtue of its phase-lead property in the low to medium frequency range and its negligible attenuation at high frequencies. The low to medium frequency range is defined as the vicinity of the resonant frequency ω_p. Several lead networks may be cascaded if a large phase-lead is required.

Lead compensation generally increases the bandwidth of a system.

Example 12.3.

The Polar Plot for $GH_1(s) = \dfrac{K_1}{s(s + p_1)(s + p_2)}$ $K_1, p_1, p_2 > 0$

is given below. The system is stable and the phase margin ϕ_{PM} is greater than 45 degrees. For a given

application, ϕ_{PM} is too large, causing a longer than desired delay time T_d in the system transient response.
The steady-state error is also too large. That is, the
velocity error constant K_v is too small by a factor of
$\lambda > 1$. We shall modify this system by a combination of
gain factor compensation, to meet the steady-state specification, and phase-lead compensation, to improve the
transient response. Assuming $H(s) = 1$, Equation (9.4),
Page 164, yields

$$K_{v1} = \lim_{s \to 0} [s\,GH_1(s)] = \frac{K_1}{p_1 p_2}$$

and hence

$$\lambda K_{v1} = \frac{\lambda K_1}{p_1 p_2}$$

Putting $K_2 \equiv \lambda K_1$, the open-loop transfer function becomes

$$GH_2 = \frac{K_2}{s(s + p_1)(s + p_2)}$$

The system represented by GH_2 has the desired velocity constant $K_{v2} = \lambda K_{v1}$. Let us consider what
would happen to K_{v2} of GH_2 if a lead network were to be introduced. The lead network acts like an
attenuator at low frequencies. That is,

$$\lim_{s \to 0} [s\,GH_2(s) \cdot P_{\text{Lead}}(s)] = \frac{K_2 a}{p_1 p_2 b} < \lambda K_{v1}$$

since $a/b < 1$. Therefore if a lead network is to be used to modify the transient response, the **gain-factor**
K_1 of GH_1 must be increased $\lambda(b/a)$ times in order to meet the steady-state requirement. The gain-factor
part of the total compensation should therefore be larger than that which would be called for if only the
steady-state specification has to be met. Hence we modify GH_2, yielding

$$GH_3 = \frac{\lambda K_1 (b/a)}{s(s + p_1)(s + p_2)}$$

As is often the case, increasing the gain-factor by an
amount as large as $\lambda(b/a)$ times destabilizes the system,
as shown in the adjoining Polar Plots of $GH_1(j\omega)$,
$GH_2(j\omega)$, and $GH_3(j\omega)$.

Now let us insert the lead network and determine
its effects. GH_3 becomes

$$GH_4 = \frac{\lambda K_1 (b/a)(s + a)}{s(s + p_1)(s + p_2)(s + b)}$$

First of all, $\lim_{s \to 0} [s\,GH_4(s)] = \lambda K_{v1}$ convinces us that
the steady-state specification has been met. In fact, in
the very low frequency region we have

$$GH_4(j\omega)\bigg|_{\omega \text{ very small}} \cong \frac{\lambda K_1}{j\omega(j\omega + p_1)(j\omega + p_2)}$$

$$= GH_2(j\omega)$$

Hence the $GH_4(j\omega)$ contour is almost coincident with the $GH_2(j\omega)$ contour in the very low **frequency range**.

In the very high frequency region,

$$GH_4(j\omega)\bigg|_{\omega \text{ very large}} \cong \frac{\lambda K_1(b/a)}{j\omega(j\omega + p_1)(j\omega + p_2)} = GH_3(j\omega)$$

Therefore $GH_4(j\omega)$ is almost coincident with $GH_3(j\omega)$ for very high frequencies.

In the mid-frequency range, where the phase lead property of the lead network substantially alters
the phase characteristic of $GH_4(j\omega)$, the $GH_4(j\omega)$ contour bends away from the $GH_2(j\omega)$ and **toward the**
$GH_4(j\omega)$ locus as ω is increased. This effect is better understood if we write $GH_4(j\omega)$ in the following form:

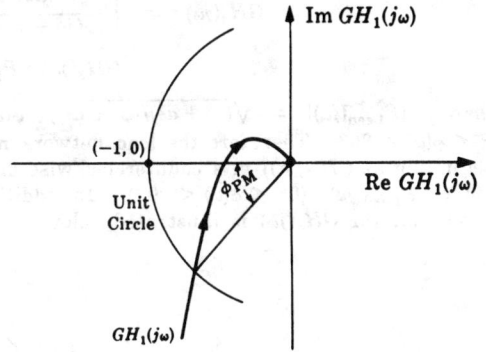

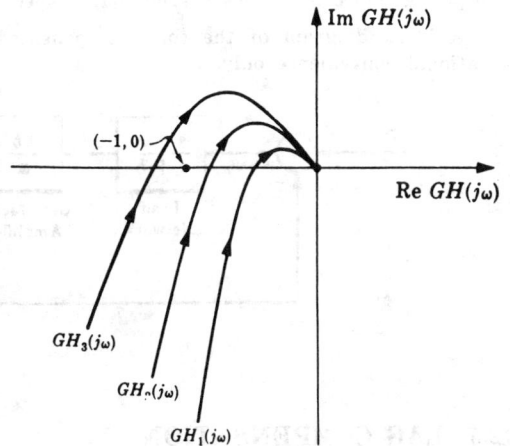

$$GH_4(j\omega) = \left[\frac{\lambda K_1(b/a)}{j\omega(j\omega + p_1)(j\omega + p_2)}\right] \cdot \left[\frac{j\omega + a}{j\omega + b}\right]$$

$$= GH_3(j\omega) \cdot P_{\text{Lead}}(j\omega) = GH_3(j\omega) \cdot |P_{\text{Lead}}(j\omega)| \underline{/\phi(\omega)}$$

where $|P_{\text{Lead}}(j\omega)| = \sqrt{(\omega^2 + a^2)/(\omega^2 + b^2)}$, $\phi(\omega) \equiv \tan^{-1}(\omega/a) - \tan^{-1}(\omega/b)$, $a/b < |P_{\text{Lead}}(j\omega)| < 1$, $0° < \phi(\omega) < 90°$. Therefore the lead network modifies $GH_3(j\omega)$ as follows. $GH_3(j\omega)$ is shifted downwards (beginning at $GH_3(j\infty)$) in a counterclockwise direction toward $GH_2(j\omega)$ due to the positive phase contribution of $P_{\text{Lead}}(j\omega)$ $(0° < \phi(\omega) < 90°)$. In addition, it is attenuated $(0 < |P_{\text{Lead}}(j\omega)| < 1)$. The resulting Polar Plot for $GH_4(j\omega)$ is illustrated below.

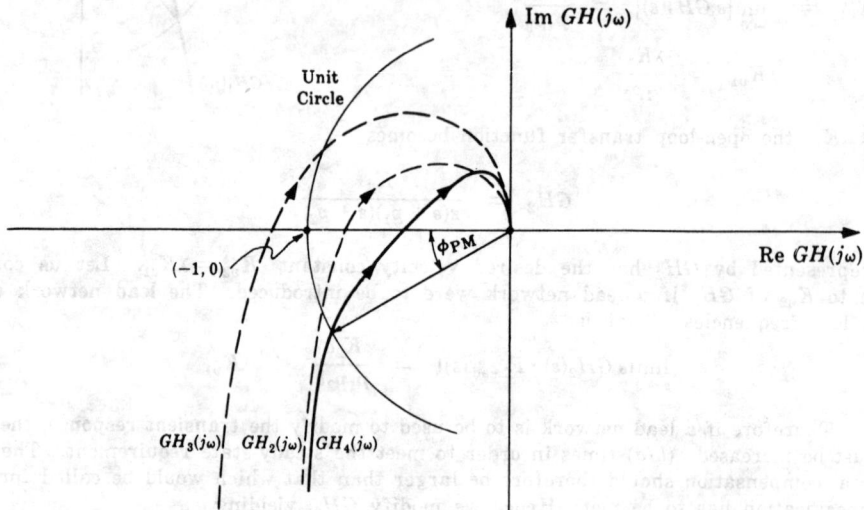

The system represented by GH_4 is clearly stable, and ϕ_{PM} is less than 45 degrees, reducing the delay time T_d of the original system represented by $GH_1(j\omega)$. By a trial-and-error procedure, the zero at $-a$ and the pole at $-b$ can be chosen such that a specific M_p can be achieved.

A block diagram of the fully compensated system is shown below. Unity-feedback is shown for notational convenience only.

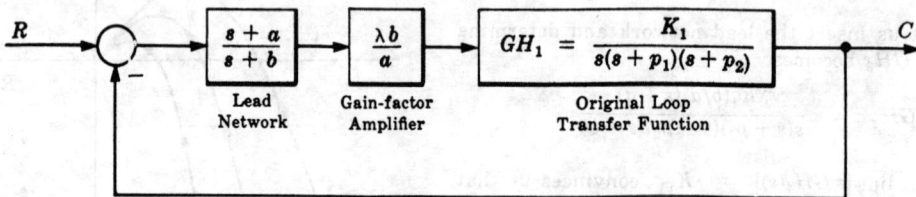

12.5 LAG COMPENSATION

The transfer function for a lag network, presented in Chapter 6, Equation (6.3), Page 98, is

$$P_{\text{Lag}} = \frac{a}{b}\left[\frac{s + b}{s + a}\right]$$

where $a < b$. The Polar Plot of $P_{\text{Lag}}(j\omega)$ for $0 \leqslant \omega < \infty$ is shown in Fig. 12-3 below.

The lag network usually provides compensation by virtue of its attenuation property in the high frequency portion of the Polar Plot, since $P_{\text{Lag}}(j0) = 1$ and $P_{\text{Lag}}(j\infty) = a/b < 1$. Several lag networks can be cascaded to provide even higher attenuation, if required. The phase-lag contribution of the lag network is often restricted by design to the very low frequency range.

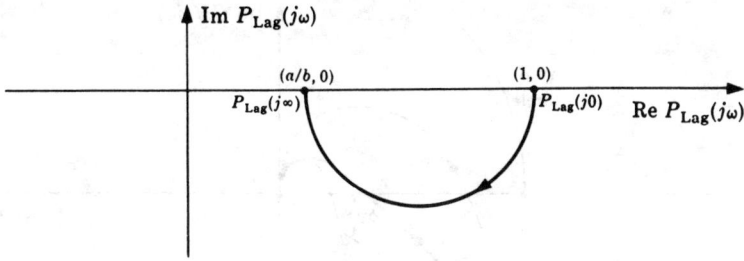

Fig. 12-3

Several general effects of lag compensation are:

1. The bandwidth of the system is usually decreased.

2. The predominant time constant τ of the system is usually increased, producing a more sluggish system.

3. For a given relative stability, the value of the error constant is increased.

4. For a given value of error constant, relative stability is improved.

The procedure for using lag compensation to improve the accuracy of some systems is essentially the same as that for lead compensation.

Example 12.4.

Let us redesign the system of Example 12.3 in the last section using gain-factor plus *lag* compensation. The original open-loop transfer function is

$$GH_1 \;=\; \frac{K_1}{s(s+p_1)(s+p_2)}$$

The gain-factor compensated transfer function is

$$GH_2 \;=\; \frac{\lambda K_1}{s(s+p_1)(s+p_2)}$$

Since $P_{\text{Lag}}(j0) = 1$, introduction of the lag network after the steady-state criterion has been met by gain-factor compensation does not require an additional increase in gain-factor.

Introducing the lag network, we get

$$GH_3' \;=\; \frac{\lambda K_1(a/b)(s+b)}{s(s+p_1)(s+p_2)(s+a)}$$

Now

$$\lim_{s \to 0} \big[s\, GH_3'(s)\big] \;=\; \lambda K_{v1}$$

where $K_{v1} = K_1/p_1 p_2$. Therefore the steady-state specification is met by GH_3'.

In the very low frequency region,

$$GH_3'(j\omega)\bigg|_{\omega \text{ very small}} \;\cong\; \frac{\lambda K_1}{j\omega(j\omega + p_1)(j\omega + p_2)} \;=\; GH_2(j\omega)$$

Hence $GH_3'(j\omega)$ is almost coincident with $GH_2(j\omega)$ at very low frequencies, with the lag property of this network manifesting itself in this range.

In the very high frequency region,

$$GH_3'(j\omega)\bigg|_{\omega \text{ very large}} \;\cong\; \frac{\lambda(a/b)K_1}{j\omega(j\omega + p_1)(j\omega + p_2)} \;=\; \lambda(a/b)GH_1(j\omega)$$

Therefore the $GH_3'(j\omega)$ contour lies above or below the $GH_1(j\omega)$ contour in this range if $\lambda > b/a$ or $\lambda < b/a$. If $\lambda = b/a$, the $GH_3'(j\omega)$ and $GH_1(j\omega)$ contours coincide.

In the mid-frequency range, the attenuation effect of $P_{\text{Lag}}(j\omega)$ increases as ω gets larger, and there is relatively small phase lag.

The resulting Polar Plot (with $\lambda = b/a$) and a block diagram of the fully compensated system are given by

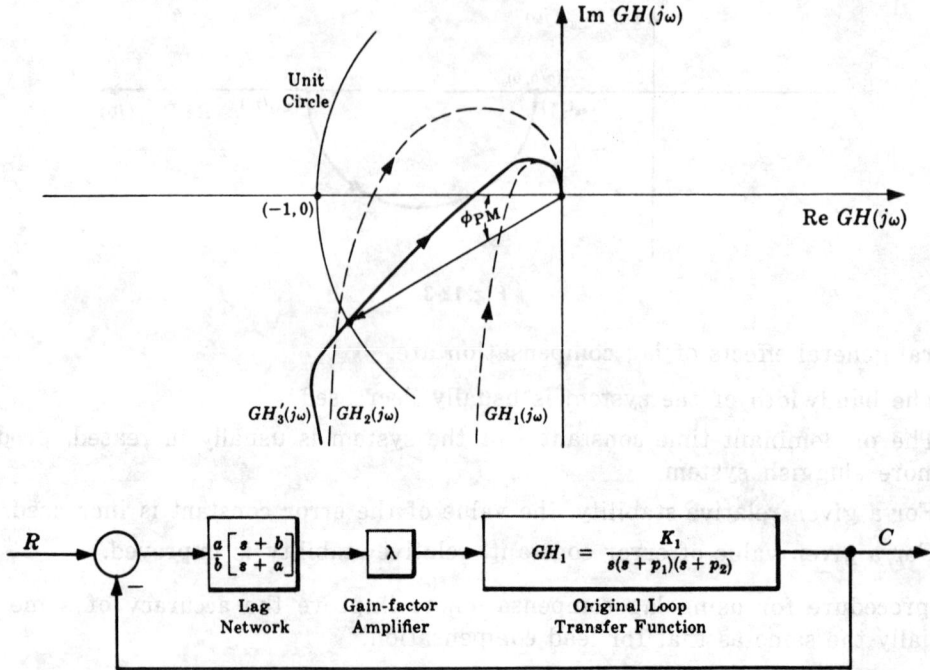

12.6 LAG-LEAD COMPENSATION

The transfer function for a lag-lead network, presented in Chapter 6, Equation (6.4), is

$$P_{LL} = \frac{(s + a_1)(s + b_2)}{(s + b_1)(s + a_2)}$$

where $\dfrac{a_1 b_2}{b_1 a_2} = 1$, $\dfrac{b_1}{a_1} = \dfrac{b_2}{a_2} > 1$, $a_i, b_i > 0$. The Polar Plot of $P_{LL}(j\omega)$ for $0 \leqq \omega \leqq \infty$ is shown in the adjoining figure.

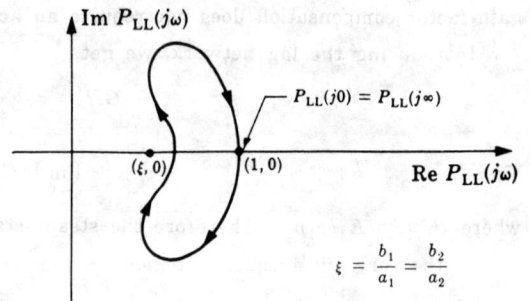

Lag-lead compensation has all of the advantages of both lag compensation and lead compensation, and only a minimum of their usually undesirable characteristics. Satisfaction of many system specifications is possible without the burden of excessive bandwidth and predominant time constants.

$$\xi = \frac{b_1}{a_1} = \frac{b_2}{a_2}$$

It is not easy to generalize about the application of lag-lead compensation or to prescribe a method for its employment, especially using Nyquist techniques. But, for illustrative purposes, we can describe how it alters the properties of the simple type 2 system in the following example.

Example 12.5.

The Nyquist Stability Plot for

$$GH = \frac{K}{s^2(s + p_1)} \qquad p_1, K > 0$$

is given in Fig. 12-4 below. Clearly the system is unstable, and no amount of gain-factor compensation can stabilize it because the contour for $0 < \omega < \infty$ always lies above the negative real axis. Lag compensation is also inapplicable for basically the same reason.

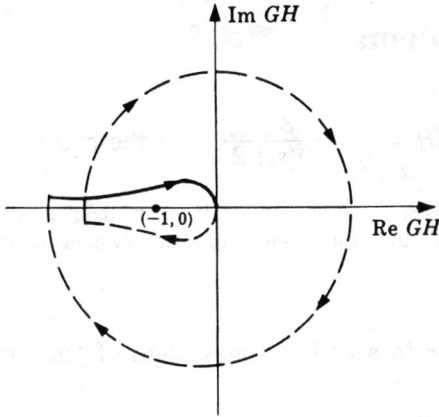

Fig. 12-4

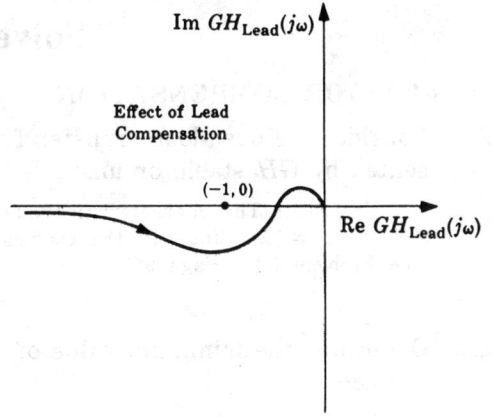

Fig. 12-5

Lead compensation may succeed in stabilizing the system, as shown in Fig. 12-5 above. But the desired application for the compensated system may call for a lower bandwidth than can be achieved with a lead network.

If a lag-lead network is used to compensate the system, the open-loop transfer function becomes

$$GH_{LL} = \frac{K(s + a_1)(s + b_2)}{s^2(s + p_1)(s + b_1)(s + a_2)}$$

and the Polar Plot is

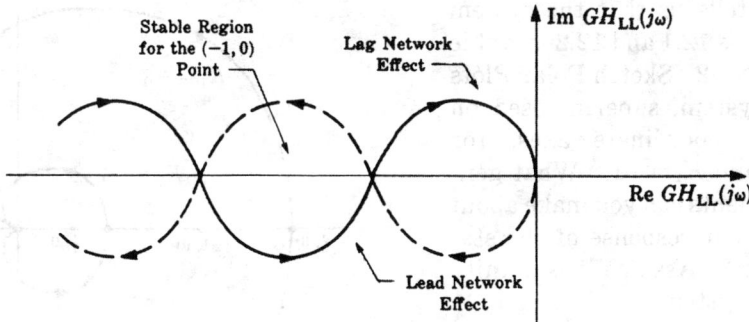

This system is conditionally stable if the $(-1, 0)$ point falls on the real axis in the unshaded region. By trial, error, and experience, the parameters of the lag-lead network can be chosen to yield good transient and steady-state performance for this previously unstable system. In addition, the bandwidth will be smaller than that of the lead compensated system.

12.7 OTHER COMPENSATION SCHEMES

There are many other practical physical networks which can be used to compensate feedback control systems. The Polar Plot is useful for illustrating how these networks alter the frequency response when the skill or equipment for generating Polar Plots is readily available.

Solved Problems

GAIN-FACTOR COMPENSATION

12.1. Consider the open-loop transfer function $GH = \dfrac{-3}{(s+1)(s+2)}$. Is the system represented by GH stable or unstable?

Unstable. The characteristic equation is determined from $1 + GH = 0$ and is given by $s^2 + 3s - 1 = 0$. Since all the coefficients do not have the same sign, the system is unstable (see Problem 5.24, Page 95).

12.2. Determine the minimum value of gain-factor to stabilize the system of the previous problem.

Let GH be written as $GH = \dfrac{K}{(s+1)(s+2)}$. Then the characteristic equation is $s^2 + 3s + 2 + K = 0$ and the Routh table (see Section 5.3, Page 87) is

s^2	1	$(2+K)$
s^1	3	0
s^0	$(2+K)$	

Hence the minimum gain-factor for stability is $K = -2 + \epsilon$, where ϵ is any small positive number.

12.3. The solution of the previous problem also tells us that the system of Problems 12.1 and 12.2 is stable for all $K > -2$. Sketch Polar Plots of this system, superimposed on the same coordinate axes, for $K_1 = -3$ and $K_2 = -1$. What general comments can you make about the transient response of the stable system? Assume it is a unity feedback system.

The required Polar Plots are shown in Fig. 12-6. The M-circle tangent to the plot for $K = -1$ has infinite radius; thus $M_p = 1$. This means that the peak overshoot is zero (no overshoot), and the system is either critically damped or overdamped.

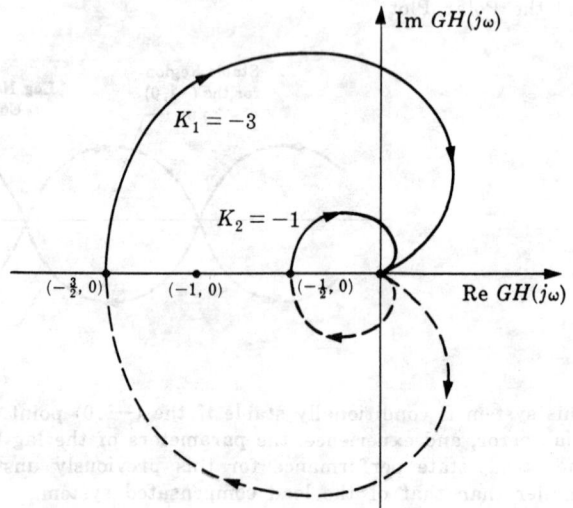

Fig. 12-6

12.4. Is the system represented by the characteristic equation $s^3 + 3s^2 + 3s + 1 + K = 0$ ever conditionally stable? Why?

Yes. The gain-factor range for stability of this system was determined in Example 5.3, Page 87, as $-1 < K < 8$. Since both limits are finite, an increase in the gain-factor above 8 or a decrease below -1 destabilizes the system.

12.5. Determine the gain-factor K of a unity feedback system whose open-loop transfer function is given by $G = \dfrac{K}{(s+1)(s+2)}$ for a resonant peak specified by $M_p = 2$.

From Equation (12.1) we have $\Psi_p = \sin^{-1}(\tfrac{1}{2}) = 30°$. The line $\overline{AB}$ drawn at an angle of $30°$ below the negative real axis is shown in Fig. 12-7, which is a replica of Fig. 12-6 for $K = -1$.

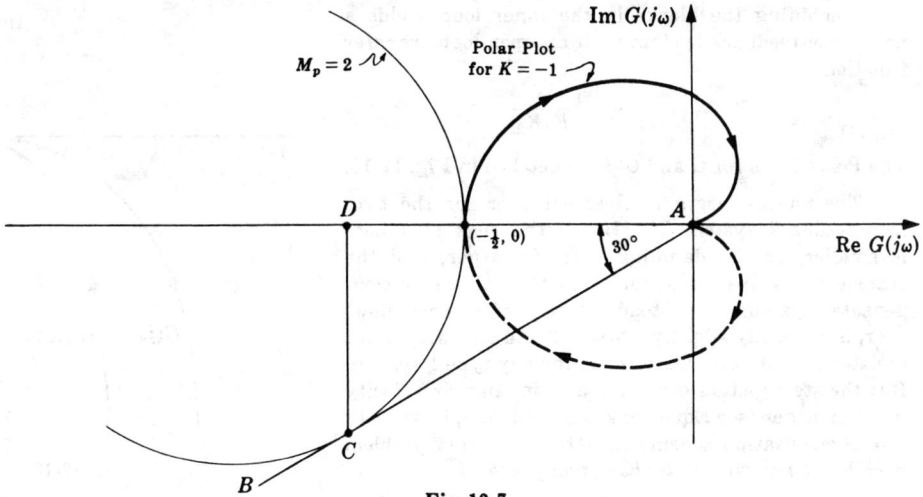

Fig. 12-7

The circle denoted by $M_p = 2$ has been drawn tangent to both $\overline{AB}$ and the Polar Plot for $K = -1$. Using the scale of this Polar Plot, line $\overline{AD}$ has a length equal to 0.76. Therefore Equation (12.3) yields

$$K_{M_p} = \frac{K'}{\text{length of } \overline{AD}} = \frac{-1}{0.76} = -1.32$$

It is also possible to compute a *positive* value of gain for $M_p = 2$ from a Polar Plot of $G(s)$ for any positive value of K. The Polar Plot for $K = 1$ is the same as that in Fig. 12-7, but rotated by 180°.

MISCELLANEOUS COMPENSATION

12.6. What kind of compensation is possible for a system whose Polar Plot is given by Fig. 12-8?

Lead, lag-lead, and simple gain-factor compensation are capable of stabilizing the system and improving the relative stability.

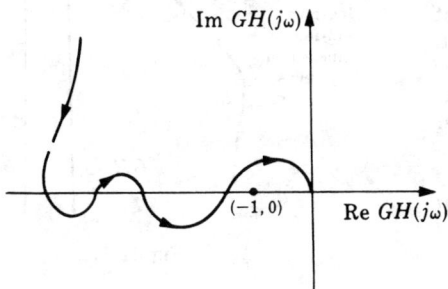

Fig. 12-8

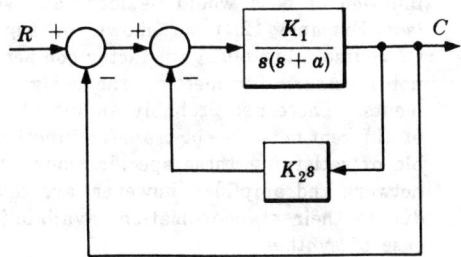

Fig. 12-9

12.7. Consider the unity-feedback system whose open-loop transfer function is given by

$$G = \frac{K_1}{s(s + a)} \qquad a, K_1 > 0$$

How would the inclusion of a minor feedback loop with a transfer function $K_2 s$ ($K_2 > 0$), as shown in the block diagram in Fig. 12-9, affect the transient and steady-state performance of the system?

Combining the blocks in the inner loop yields a new unity-feedback system with an open-loop transfer function

$$G' = \frac{K_1}{s(s + a + K_1 K_2)}$$

The Polar Plots for G and G' are sketched in Fig. 12-10.

The phase-margin is clearly larger for the two-loop feedback system G'. Hence the peak overshoot is smaller, or the damping ratio is larger, and the transient response is superior to that of the uncompensated system. The steady-state performance, however, is generally slightly worse. For a unit-step input the steady-state error is zero, as for any type 1 system. But the steady-state error for a unit-ramp or velocity input is larger (see Equations (9.4) and (9.5), Page 164). The compensation scheme illustrated by this problem is called *derivative or tachometric feedback*.

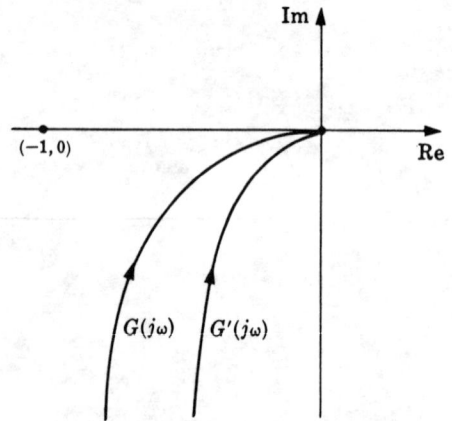

Fig. 12-10

12.8. Determine a type of compensator which will yield a phase-margin of approximately 45° when added to the fixed system components defined by

$$GH = \frac{4}{s(s^2 + 3.2s + 64)}$$

An additional requirement is that the high frequency response of the compensated system is to be approximately the same as that of the uncompensated system.

The Polar Plot for GH is sketched in Fig. 12-11. It is very close to the negative imaginary axis for almost all values of ω.

The phase-margin is almost 90°, and either an increase in gain-factor and/or a lag compensator is capable of satisfying the phase-margin requirement. But since the lag network may be designed to provide attenuation at high frequencies and lag in the low frequency range, a combination of both would be ideal and sufficient (see Example 12.4), as shown in Fig. 12-11. Of course a lag-plus-gain-factor compensator is not *necessary* for meeting the design requirements. There are probably an infinite number of different networks or transfer functions capable of satisfying these specifications. The lag network and amplifier, however, are *convenient* due to their standardization, availability, and ease of synthesis.

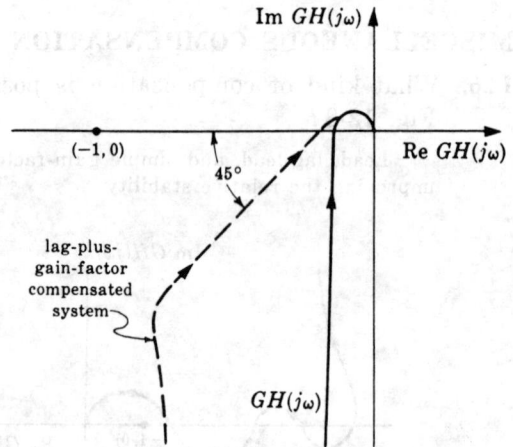

Fig. 12-11

12.9. Outline the design of a servomechanism capable of following a constant velocity input with zero steady-state error and approximately 25 percent maximum overshoot in the transient state. The fixed plant is given by $G_2 = \dfrac{50}{s^2(s + 5)}$.

Since the plant is type 2, it is capable of following a constant velocity input with zero steady-state error (see Chapter 9). However, the closed-loop system is unstable for any value of gain-factor (see Example 12.5). Since no demands on bandwidth have been made, lead-compensation should be sufficient (again see Example 12.5) to stabilize the system and meet the transient specification. But two lead networks in series are probably required because the phase-margin of the unstable system is negative, and 25 percent overshoot is equivalent to about +45° phase-margin. Most standard lead networks have a maximum phase lead of approximately 54° (see Fig. 16-2, Page 297). The details of the remainder of this design are very tedious by the methods of Nyquist analysis due to the fact that the Polar Plot must be drawn in some detail at least several times before converging to a satisfactory solution; unless, of course, the designer is very very lucky or talented, or has a machine to do the work for him.

This problem may be solved much more easily using the design methods introduced in subsequent chapters. For future reference, the two compensating lead networks can each have a transfer function of approximately $P_{Lead} = (s + 3)/(s + 20)$ in order to satisfy the specifications. If the maximum steady-state acceleration error is also specified, a preamplifier will be required with the lead networks. For example, if $K_a \equiv 50$, then a preamplifier of gain $5(20/3)^2$ is needed. This preamplifier should be placed *between* the two lead networks to prevent loading effects (see Section 8.7, Page 142).

12.10. Outline the design of a unity-feedback system with a plant given by

$$G_2 = \frac{2000}{s(s+5)(s+10)}$$

and the performance specifications:

(1) $\phi_{PM} \cong 45°$

(2) $K_v = 50$

(3)   The bandwidth BW of the compensated system must be approximately equal to or not much greater than that of the uncompensated system because high-frequency "noise" disturbances are present under normal operating conditions.

(4)   The compensated system should not respond sluggishly; that is, the predominant time constant τ of the system must be maintained at a value approximately the same as that of the uncompensated system.

A simple calculation clearly shows that the uncompensated system is unstable (e.g., Routh test). Therefore compensation is mandatory. But due to the stringent nature of the specifications, a detailed design for this system using Nyquist techniques requires too much effort. The techniques of the next few chapters provide a much simpler solution. A little intelligent analysis of the problem statement, however, will indicate the kind of compensation that is needed.

For G_2, $K_v = \lim_{s \to 0} sG_2(s) = 40$. Therefore satisfaction of (2) requires a gain-compensation of 5/4. But an increase in gain only makes the system more unstable. Therefore additional compensation is necessary. Lead compensation is probably not adequate due to (3), and lag compensation is not possible due to (4). Thus it appears that a lag-lead network and an amplifier will most likely satisfy all criteria. The lag portion of the lag-lead network will satisfy (3), and the lead portion (4) and (1).

12.11. What is the effect on the Polar Plot of the system

$$GH = \frac{\prod_{i=1}^{m}(s+z_i)}{\prod_{i=1}^{n}(s+p_i)}$$

where $m \leq n$, $0 < z_i < \infty$, $0 \leq p_i < \infty$, when k finite nonzero poles are included in GH in addition to the original n poles?

For low frequencies the Polar Plot is modified in magnitude only, since

$$\lim_{s \to 0} GH' = \lim_{s \to 0} \left[\frac{\prod_{i=1}^{m}(s+z_i)}{\prod_{i=1}^{n+k}(s+p_i)} \right] = \frac{\prod_{i=1}^{m} z_i}{\prod_{i=1}^{n+k} p_i} = \left(\frac{1}{\prod_{i=1}^{k} p_i} \right) \lim_{s \to 0} GH$$

For high frequencies the addition of k poles reduces the phase-angle of GH by $k\pi/2$ rad, since

$$\lim_{\omega \to \infty} \arg GH'(j\omega) = \lim_{\omega \to \infty} \left[\sum_{i=1}^{m} \tan^{-1}\left(\frac{\omega}{z_i}\right) - \sum_{i=1}^{n+k} \tan^{-1}\left(\frac{\omega}{p_i}\right) \right]$$

$$= \frac{m\pi}{2} - \frac{(n+k)\pi}{2} = \lim_{\omega \to \infty} \arg GH - \frac{k\pi}{2}$$

Therefore the portion of the Polar Plot near the origin is rotated clockwise by $k\pi/2$ rad when k poles are added.

Supplementary Problems

12.12. Determine a positive value of gain-factor K when $M_p = 2$ for the system of Problem 12.5.

12.13. Prove Equation (12.1), Page 225.

12.14. Prove Equations (12.2) and (12.3), Page 226.

12.15. Design a compensator which will yield a phase margin of approximately 45° for the system defined by $GH = \dfrac{84}{s(s+2)(s+6)}$.

12.16. Design a compensator which will yield a phase margin of about 40° and a velocity constant $K_v = 40$ for the system defined by $GH = \dfrac{4 \times 10^5}{s(s+20)(s+100)}$.

12.17. What kind of compensation can be used to yield a maximum overshoot of 20 percent for the system defined by $GH = \dfrac{4 \times 10^4}{s^2(s+100)}$?

12.18. Show that the addition of k non-numerically-zero finite zeros to the system of Problem 12.11 rotates the high-frequency portion of the Polar Plot by $k\pi/2$ rad in the counterclockwise direction.

Answers to Supplementary Problems

12.12. $K = 31.2$

12.15. $P_{\text{Lead}} = \dfrac{s+30}{s+120}$

12.16. $P_{\text{Lead}} = \dfrac{s+20}{s+100}$, no preamplifier required

12.17. Lag-lead, and possibly lead plus gain-factor compensation.

Chapter 13

Root-Locus Analysis

13.1 INTRODUCTION

It was shown in Chapters 4 and 6 that the poles of a transfer function can be displayed graphically in the s-plane by means of a pole-zero map. An analytical method is presented in this chapter for displaying the location of the poles of the closed-loop transfer function

$$\frac{G}{1 + GH}$$

as a function of the gain-factor K (see Section 6.2) of the open-loop transfer function GH. This method, called *root-locus analysis*, requires that only the location of the poles and zeros of GH be known, and does not require factorization of the characteristic polynomial.

Root-locus techniques permit accurate computation of the time-domain response in addition to yielding readily available frequency response information.

13.2 VARIATION OF CLOSED-LOOP SYSTEM POLES: THE ROOT-LOCUS

Consider the canonical feedback control system given by Fig. 13-1. The closed-loop transfer function of this system is

$$\frac{C}{R} = \frac{G}{1 + GH}$$

Let the open-loop transfer function GH be represented by

$$GH \equiv \frac{K N(s)}{D(s)} = \frac{K(s^m + a_{m-1}s^{m-1} + \cdots + a_0)}{s^n + b_{n-1}s^{n-1} + \cdots + b_0}$$

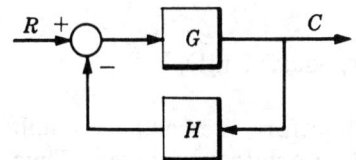

Fig. 13-1

where $N(s)$ and $D(s)$ are the finite polynomials in the complex variable s, $m \leqq n$, and K is the open-loop gain factor. The closed-loop transfer function then becomes

$$\frac{C}{R} = \frac{G}{1 + KN/D} = \frac{GD}{D + KN}$$

The closed-loop poles are roots of the characteristic equation

$$D(s) + K N(s) = 0 \qquad (13.1)$$

In general the location of these roots in the s-plane changes as the open-loop gain factor K is varied. A locus of these roots plotted in the s-plane as a function of K is called a **root-locus**.

For K equal to zero, the roots of (13.1) are the roots of the polynomial $D(s)$, which are the same as the poles of the open-loop transfer function GH. If K becomes very large, the roots approach those of the polynomial $N(s)$, which are the open-loop zeros. Thus as K is increased from zero to infinity, the loci of the closed-loop poles originate from the open-loop poles and proceed toward and terminate at the open-loop zeros.

Example 13.1.

Consider the open-loop transfer function

$$GH = \frac{K\,N(s)}{D(s)} = \frac{K(s+1)}{s^2 + 2s} = \frac{K(s+1)}{s(s+2)}$$

For $H = 1$, the closed-loop transfer function is

$$\frac{C}{R} = \frac{K(s+1)}{s^2 + 2s + K(s+1)}$$

The closed-loop poles of this system are easily determined by factoring the denominator polynomial:

$$s_1 = -\tfrac{1}{2}(2+K) + \sqrt{1 + \tfrac{1}{4}K^2}$$

$$s_2 = -\tfrac{1}{2}(2+K) - \sqrt{1 + \tfrac{1}{4}K^2}$$

The locus of these roots plotted as a function of K (for $K > 0$) is shown in the s-plane in Fig. 13-2. As observed in the figure, this root-locus has two *branches*: one for a closed-loop pole which moves from the open-loop pole at the origin to the open-loop zero at -1, and one from the open-loop pole at -2 to the open-loop zero at $-\infty$.

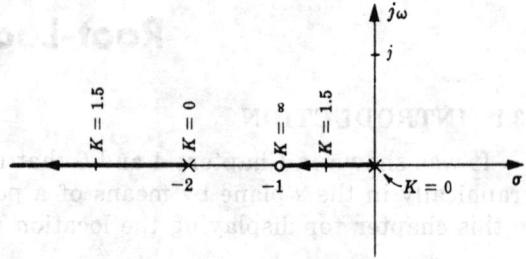

Fig. 13-2

In the example above, the root-locus is constructed by factoring the denominator polynomial of the system closed-loop transfer function. In the following sections, techniques are described which permit construction of root-loci without the need for factorization.

13.3 ANGLE AND MAGNITUDE CRITERIA

In order for a branch of a root-locus to pass through a particular point s_1 in the s-plane, it is necessary that s_1 be a root of the characteristic equation *(13.1)* for some real value of K. That is,

$$D(s_1) + K N(s_1) = 0 \tag{13.2}$$

or, equivalently,

$$GH(s_1) = \frac{K N(s_1)}{D(s_1)} = -1 \tag{13.3}$$

Therefore the complex number $GH(s_1)$ must have a phase angle of $180° + 360l°$, where l is an arbitrary integer. Thus we have the *angle criterion*

$$\arg GH(s_1) = 180° + 360l° = (2l+1)\pi \text{ radians}, \qquad l = 0, \pm 1, \pm 2, \ldots \tag{13.4a}$$

which can also be written as

$$\arg\left[\frac{N(s_1)}{D(s_1)}\right] = \begin{cases} (2l+1)\pi \text{ radians for } K > 0 \\ 2l\pi \text{ radians for } K < 0 \end{cases} \qquad l = 0, \pm 1, \pm 2, \ldots \tag{13.4b}$$

In order for s_1 to be a closed-loop pole of the system, that is, on the root-locus, it is necessary that Equation *(13.3)* be satisfied with regard to *magnitude* in addition to phase angle. That is, K must have the particular value that satisfies the *magnitude criterion*: $|GH(s_1)| = 1$, or

$$|K| = \left|\frac{D(s_1)}{N(s_1)}\right| \tag{13.5}$$

The angle and magnitude of $GH(s)$ at any point in the s-plane can be determined graphically as described in Sections 4.11 and 6.5. In this way, it is possible to construct the root-locus by a trial-and-error procedure of testing points in the s-plane. That is, the root-locus is drawn through all points which satisfy the angle criterion, Equation *(13.4b)*, and the magnitude criterion is used to determine the values of K at points along the loci. This construction procedure is considerably simplified by using certain shortcuts or construction rules as described in the following sections.

13.4   NUMBER OF LOCI

The number of loci, that is, the number of branches of the root-locus, is equal to the number of poles of the open-loop transfer function GH.

Example 13.2.

The open-loop transfer function $GH = \dfrac{K(s+2)}{s^2(s+4)}$ has three poles. Hence there are three separate loci in the root-locus plot.

13.5   REAL AXIS LOCI

Those sections of the root-locus on the real axis in the s-plane are determined by counting the total number of finite poles and zeros of GH to the right of the points in question. The following rule depends on whether the open-loop gain factor K is positive or negative.

Rule for $K > 0$

Points of the root-locus on the real axis lie to the left of an *odd* number of finite poles and zeros.

Rule for $K < 0$

Points of the root-locus on the real axis lie to the left of an *even* number of finite poles and zeros.

If no points on the real axis lie to the left of an odd number of finite poles and zeros, then no portion of the root-locus for $K > 0$ lies on the real axis. A similar statement is true for $K < 0$.

Example 13.3.

Consider the pole-zero map of an open-loop transfer function GH shown in Fig. 13-3. Since all the points on the real axis between 0 and −1 and between −1 and −2 lie to the left of an odd number of finite poles and zeros, these points are on the root-locus for $K > 0$. The portion of the real axis between −∞ and −4 lies to the left of an odd number of finite poles and zeros; hence these points are also on the root-locus for $K > 0$. All portions of the root-locus for $K > 0$ on the real axis are illustrated in Fig. 13-4. All remaining portions of the real axis, that is, between −2 and −4 and between 0 and ∞, lie on the root-locus for $K < 0$.

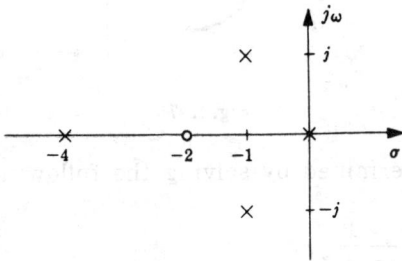

Fig. 13-3

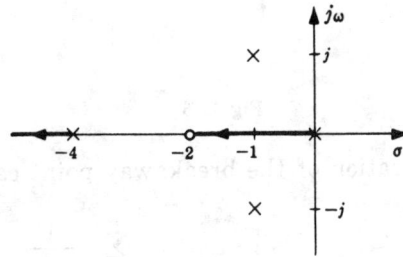

Fig. 13-4

13.6   ASYMPTOTES

For large distances from the origin in the s-plane, the branches of a root-locus approach a set of straight-line asymptotes. These asymptotes emanate from a point in the s-plane on the real axis called the **center of asymptotes** σ_c given by

$$\sigma_c = -\frac{\sum_{i=1}^{n} p_i - \sum_{i=1}^{m} z_i}{n - m} \tag{13.6}$$

where $-p_i$ are the poles, $-z_i$ are the zeros, n is the number of poles, and m the number of zeros of GH.

The angles between the asymptotes and the real axis are given by

$$\beta = \begin{cases} \dfrac{(2l+1)180}{n-m} \text{ degrees for } K > 0 \\[2mm] \dfrac{(2l)180}{n-m} \text{ degrees for } K < 0 \end{cases} \qquad (13.7)$$

for $l = 0, 1, 2, \ldots, n-m-1$. This results in a number of asymptotes equal to $n-m$.

Example 13.4.

The center of asymptotes for $GH = \dfrac{K(s+2)}{s^2(s+4)}$ is located at

$$\sigma_c = -\frac{4-2}{2} = -1$$

Since $n - m = 3 - 1 = 2$, there are two asymptotes. Their angles with the real axis are 90° and 270°, for $K > 0$, as shown in Fig. 13-5.

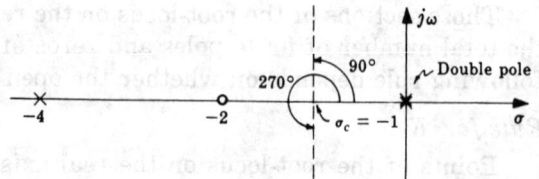

Fig. 13-5

13.7 BREAKAWAY POINTS

A **breakaway point** σ_b is a point on the real axis where two or more branches of the root-locus depart from or arrive at the real axis. Two branches leaving the real axis are illustrated in the root-locus plot in Fig. 13-6. Two branches coming onto the real axis are illustrated in Fig. 13-7.

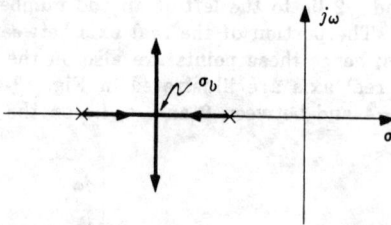

Fig. 13-6

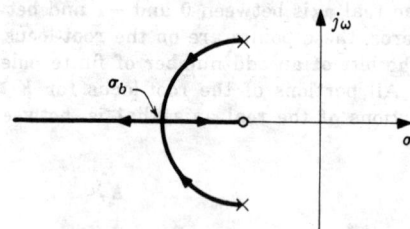

Fig. 13-7

The location of the breakaway point can be determined by solving the following equation for σ_b:

$$\sum_{i=1}^{n} \frac{1}{(\sigma_b + p_i)} = \sum_{i=1}^{m} \frac{1}{(\sigma_b + z_i)} \qquad (13.8)$$

where $-p_i$ and $-z_i$ are the poles and zeros of GH, respectively. The solution of this equation requires factorization of an $(n + m - 1)$ order polynomial in σ_b. Consequently the breakaway point can only be easily determined analytically for relatively simple GH. However, an approximate location can often be determined intuitively; then an iterative process can be used to solve the equation more exactly (see Problem 13.20).

Example 13.5.

To determine the breakaway points for $GH = \dfrac{K}{s(s+1)(s+2)}$, the following equation must be solved for σ_b:

$$\frac{1}{\sigma_b} + \frac{1}{\sigma_b + 1} + \frac{1}{\sigma_b + 2} = 0$$

$$(\sigma_b + 1)(\sigma_b + 2) + \sigma_b(\sigma_b + 2) + \sigma_b(\sigma_b + 1) = 0$$

which reduces to $3\sigma_b^2 + 6\sigma_b + 2 = 0$ whose roots are $\sigma_b = -0.423, -1.577$.

Applying the real axis rule of Section 13.5 for $K > 0$ indicates that there are branches of the root-locus between zero and -1 and between $-\infty$ and -2. Therefore the root at -0.423 is a breakaway point, as shown in Fig. 13-8. The value $\sigma_b = -1.577$ represents a breakaway point on the root-locus for negative values of K since the portion of the real axis between -1 and -2 is on the root-locus for $K < 0$.

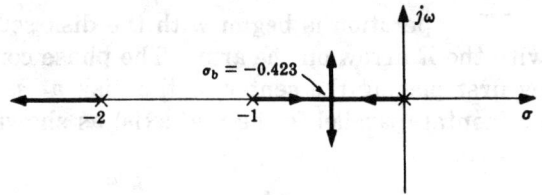

Fig. 13-8

13.8 DEPARTURE AND ARRIVAL ANGLES

The **departure angle** of the root-locus from a *complex pole* is given by

$$\theta_D = 180° + \arg GH' \qquad (13.9)$$

where $\arg GH'$ is the phase angle of GH computed at the complex pole, but ignoring the contribution of that particular pole. A simple technique for determining $\arg GH'$ is presented in Section 13.9.

Example 13.6.

Consider the open-loop transfer function

$$GH = \frac{K(s+2)}{(s+1+j)(s+1-j)}, \qquad K > 0$$

The departure angle of the root-locus from the complex pole at $s = -1 + j$ is determined as follows. The angle of GH for $s = -1 + j$, ignoring the contribution of the pole at $s = -1 + j$, is $-45°$. Therefore the departure angle is

$$\theta_D = 180° - 45° = 135°$$

and is illustrated in Fig. 13-9.

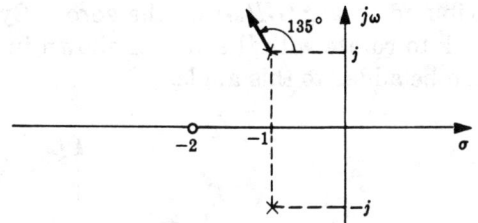

Fig. 13-9

The **angle of arrival** of the root-locus at a *complex zero* is given by

$$\theta_A = 180° - \arg GH'' \qquad (13.10)$$

where $\arg GH''$ is the phase angle of GH at the complex zero, ignoring the effect of that zero. A simple technique for determining $\arg GH''$ is given in Section 13.9.

Example 13.7.

Consider the open-loop transfer function

$$GH = \frac{K(s+j)(s-j)}{s(s+1)}, \qquad K > 0$$

The arrival angle of the root-locus for the complex zero at $s = j$ is $\theta_A = 180° - (-45°) = 225°$ as shown in Fig. 13-10.

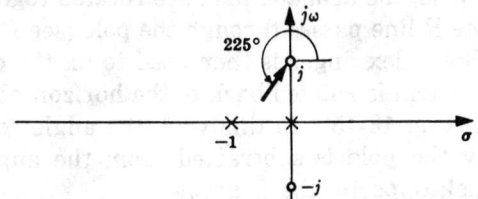

Fig. 13-10

13.9 THE SPIRULE

A mechanical device called the Spirule* was developed by Walter Evans to aid in the construction of root-locus plots. The Spirule can be used to determine the angle and magnitude of a transfer function at any point in the s-plane from a pole-zero map of the transfer function.

------

*The Spirule Co., 9728 El Venado, Whittier, Calif.

The *angle* of $GH(s_0)$, where s_0 is any specific point in the s-plane, is determined as follows.

The operation is begun with the disk set so that the arrow at $0°$ on the disk is aligned with the R arrow on the arm. The phase contribution to $GH(s_0)$ due to a zero is determined by first placing the center of the disk at s_0 with the arm orientated so that the R line is horizontal (parallel to the real axis) as shown in Fig. 13-11.

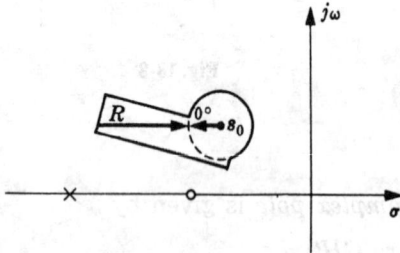

Fig. 13-11

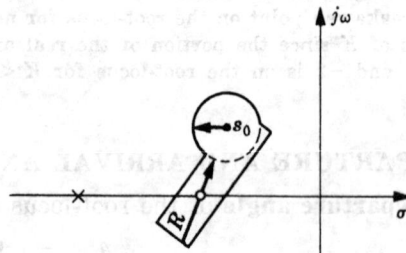

Fig. 13-12

The thumb of the right hand is used to fix the pivot point at s_0 by pressing on the eyelet at the center of the disk. With the index finger of the right hand fixing the disk so that it does not rotate, the arm is then rotated with the left hand until the R line passes through the zero (see Fig. 13-12). The number on the disk opposite the R arrow is the angle contributed to arg $GH(s_0)$ by the zero. By returning the arm to the horizontal, allowing the disk to rotate with the arm as shown in Fig. 13-13, the contribution of other zeros and poles can be added to this angle.

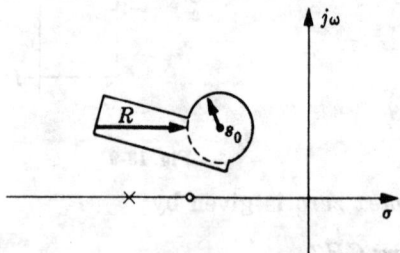

Fig. 13-13

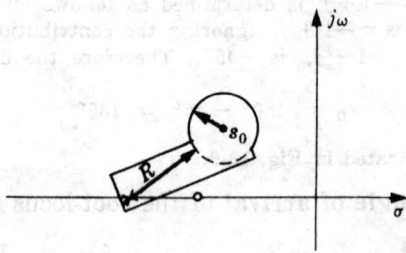

Fig. 13-14

To measure and add the angle contributed by a pole, the arm and disk are rotated together until the R line passes through the pole (see Fig. 13-14). The index finger is then used to fix the disk while the arm is rotated back to the horizontal as shown in Fig. 13-15. In this way, the angle contributed by the pole is subtracted from the angle on the disk opposite the R arrow.

By repeated rotations of the arm to all of the finite poles and zeros of GH, the total phase-angle of $GH(s_0)$ can be determined and read directly from the disk opposite the R arrow.

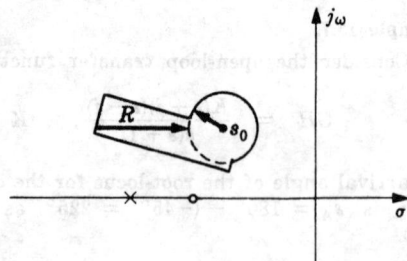

Fig. 13-15

If the total phase angle measured satisfies the angle criterion, Equation (13.4b), then $s_0 \equiv s_1$ and the test point s_0 is on the root-locus. If not, a new test point s_0' must be chosen and the Spirule manipulation repeated.

The Spirule can be used to determine the *magnitude* of $N(s_0)/D(s_0)$ in the following manner, where

$$GH = \frac{KN}{D} \quad \text{and} \quad \frac{N(s_0)}{D(s_0)} = \frac{(s_0 + z_1)(s_0 + z_2)\cdots(s_0 + z_m)}{(s_0 + p_1)(s_0 + p_2)\cdots(s_0 + p_n)}$$

The arm and disk are initially oriented with the R arrow opposite the $0°$ arrow on the disk and the pivot point at s_0. The arm and disk are first rotated together until the R line passes through a *zero* as in Fig. 13-16.

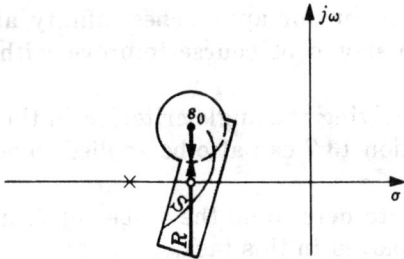

Fig. 13-16

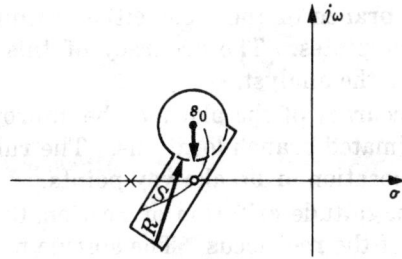

Fig. 13-17

Then the disk is held fixed while the arm is rotated until the S curve on the arm passes through the zero (see Fig. 13-17). The effects of other zeros are taken into account by repeating this procedure.

The contribution of a *pole* to $|N(s_0)/D(s_0)|$ is determined by rotating the arm and disk together until the S curve passes through the pole (see Fig. 13-18).

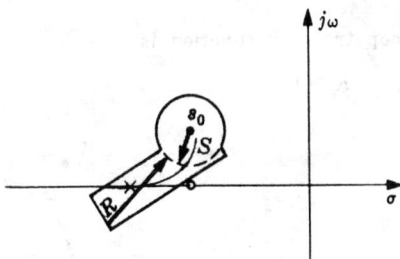

Fig. 13-18

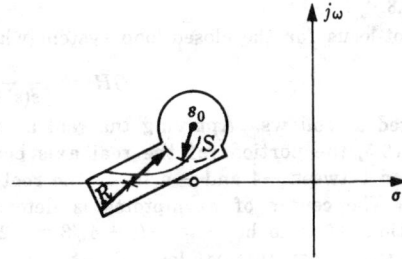

Fig. 13-19

The disk is then held fixed and the arm rotated until the R line passes through the pole as shown in Fig. 13-19. This procedure is repeated for each pole of GH.

When all zeros and poles have been accounted for, the magnitude of $N(s_0)/D(s_0)$ is read on the quarter-circle scale on the arm of the Spirule opposite whichever arrow on the disk falls in the range of the scale. The number beside the arrow: x1, x10, or x0.1, is a multiplying factor for the number read on the scale.

If the scale used for the s-plane is not the same as the scale on the top of the arm of the Spirule, the number read on the scale must be multiplied by an additional scale factor. The appropriate factor is

$$J^{m-n}$$

where J is the number of units of the s-plane scale corresponding to one unit of the scale on the top of the Spirule arm, m is the total number of finite zeros of GH, and n is the total number of finite poles of GH.

To determine the inverse of $|N(s_0)/D(s_0)|$, the procedure is the same as above, but the roles of the poles and zeros are reversed.

13.10 CONSTRUCTION OF THE ROOT-LOCUS

A root-locus plot may be easily and accurately sketched using the construction rules of Sections 13.4-13.8 and the Spirule as described in Section 13.9. An efficient procedure is the following: First, determine the portions of the root-locus on the real axis. Second, compute the center and angles of the asymptotes and draw the asymptotes on the plot. Then determine the departure and arrival angles at complex poles and zeros (if any) and indicate them on the plot. Next, make a rough sketch of the branches of the root-locus so that each branch of the locus either terminates at a zero or approaches infinity along one of the asymptotes. The accuracy of this last step should of course improve with the experience of the analyst.

The accuracy of the plot may be improved by applying the angle criterion in the vicinity of the estimated branch locations. The rule of Section 13.7 can also be applied to determine the exact location of breakaway points.

The magnitude criterion of Section 13.3 is used to determine the values of K along the branches of the root-locus. The spirule may be employed in this task.

Since complex poles of the system must occur in complex conjugate pairs (assuming real coefficients for the numerator and denominator polynomials of GH), the root-locus is symmetric about the real axis. Thus it is sufficient to plot only the upper half of the root-locus. However, it must be remembered that, in doing this, the lower halves of open-loop complex poles and zeros must be included when applying the magnitude and angle criteria.

Often, for analysis or design purposes, an accurate plot of the root-locus is required only in certain regions of the s-plane. In this case, the angle and magnitude criteria need only be applied in those regions of interest after a rough sketch has established the general shape of the plot.

Example 13.8.

The root-locus for the closed-loop system whose open-loop transfer function is

$$GH = \frac{K}{s(s+2)(s+4)}, \qquad K > 0$$

is constructed as follows. Applying the real axis rule of Section 13.5, the portions of the real axis between 0 and -2 and between -4 and $-\infty$ lie on the root-locus for $K > 0$. The center of asymptotes is determined from Equation (*13.6*) to be $\sigma_c = -(2+4)/3 = -2$, and there are three asymptotes located at angles of $\beta = 60°$, $180°$, and $300°$.

Since two branches of the root-locus for $K > 0$ come together on the real axis between 0 and -2, a breakaway point exists on that portion of the real axis. Hence the root-locus for $K > 0$ may be sketched by estimating the location of the breakaway point and continuing the branches of the root-locus to the asymptotes, as shown in Fig. 13-20. To improve the accuracy of this plot, the exact location of the breakaway point is determined from equation (*13.8*):

$$\frac{1}{\sigma_b} + \frac{1}{\sigma_b + 2} + \frac{1}{\sigma_b + 4} = 0$$

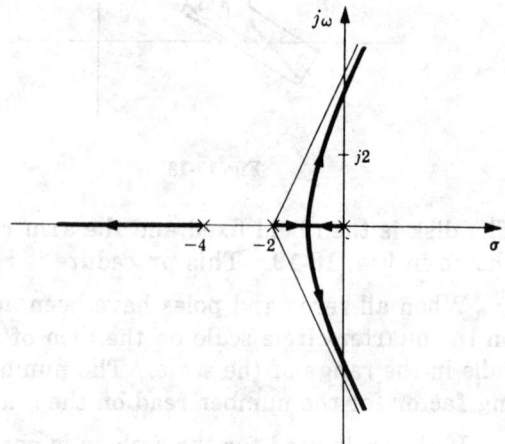

Fig. 13-20

which is simplified to $3\sigma_b^2 + 12\sigma_b + 8 = 0$. The appropriate solution of this equation is $\sigma_b = -0.845$.

The angle criterion is applied to points in the vicinity of the approximate root-locus to improve the accuracy of the location of the branches in the complex part of the s-plane; the magnitude criterion is used to determine the values of K along the root-locus. The resulting root-locus plot for $K > 0$ is shown in Fig. 13-21 below.

The root-locus for $K < 0$ is constructed in a similar manner. In this case, however, the portions of the real axis between 0 and ∞ and between -2 and -4 lie on the root-locus; the breakaway point is located at -3.155; and the asymptotes have angles of $0°$, $120°$, and $240°$. The root-locus for $K < 0$ is shown in Fig. 13-22 below.

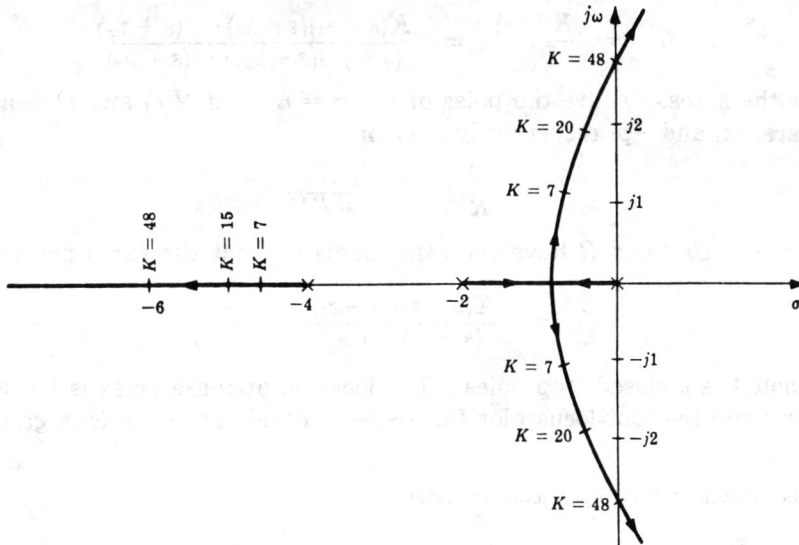

Fig. 13-21

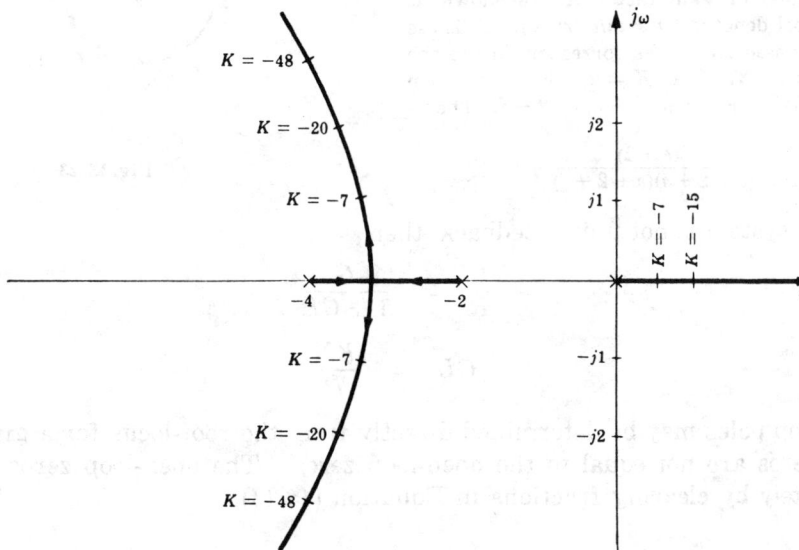

Fig. 13-22

13.11 THE CLOSED-LOOP TRANSFER FUNCTION AND THE TIME-DOMAIN RESPONSE

The closed-loop transfer function C/R is easily determined from the root-locus plot for a specified value of open-loop gain factor K. From this the time-domain response $c(t)$ may be determined, for a given input $r(t)$, by inversion of $C(s)$.

Consider the closed-loop transfer function C/R for the canonical *unity (negative) feedback* system

$$\frac{C}{R} \;=\; \frac{G}{1+G} \tag{13.11}$$

Open-loop transfer functions which are rational algebraic expressions can be written as

$$G = \frac{KN(s)}{D(s)} = \frac{K(s+z_1)(s+z_2)\cdots(s+z_m)}{(s+p_1)(s+p_2)\cdots(s+p_n)} \qquad (13.12)$$

where $-z_i$ are the zeros, $-p_i$ are the poles of G, $m \leqq n$, and $N(s)$ and $D(s)$ are polynomials whose roots are $-z_i$ and $-p_i$ respectively. Then

$$\frac{C}{R} = \frac{KN}{D+KN} \qquad (13.13)$$

and it is clear that C/R and G have the same zeros but not the same poles (unless $K = 0$). Hence

$$\frac{C}{R} = \frac{K(s+z_1)(s+z_2)\cdots(s+z_m)}{(s+\alpha_1)(s+\alpha_2)\cdots(s+\alpha_n)}$$

where $-\alpha_i$ denote the n closed-loop poles. The location of these poles is by definition determined directly from the root-locus plot for a specified value of open-loop gain K.

Example 13.9.

Consider the system whose open-loop transfer function is

$$G = \frac{K(s+2)}{(s+1)^2}, \qquad K > 0$$

The root-locus plot is given in Fig. 13-23.

Several values of gain factor K are shown at points on the loci denoted by *small triangles*. These points are the *closed-loop poles* corresponding to the specified values of K. For $K = 2$, the closed-loop poles are $-\alpha_1 = -2 + j$ and $-\alpha_2 = -2 - j$. Therefore

$$\frac{C}{R} = \frac{2(s+2)}{(s+2+j)(s+2-j)}$$

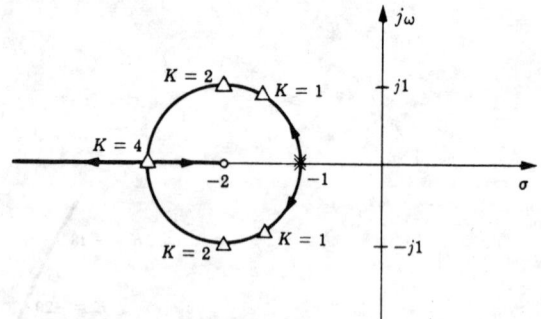

Fig. 13-23

When the system is not unity feedback, then

$$\frac{C}{R} = \frac{G}{1+GH} \qquad (13.14)$$

and

$$GH = \frac{KN}{D} \qquad (13.15)$$

The closed-loop poles may be determined directly from the root-locus for a given K, but the closed-loop zeros are not equal to the open-loop zeros. The open-loop zeros must be computed separately by clearing fractions in Equation (13.14).

Example 13.10.

Consider the system described by

$$G = \frac{K(s+2)}{s+1}, \qquad H = \frac{1}{s+1}, \qquad GH = \frac{K(s+2)}{(s+1)^2} \qquad K > 0$$

and

$$\frac{C}{R} = \frac{K(s+1)(s+2)}{(s+1)^2 + K(s+2)} = \frac{K(s+1)(s+2)}{(s+\alpha_1)(s+\alpha_2)}$$

The root-locus plot for this example is the same as that for Example 13.9. Hence for $K = 2$, $\alpha_1 = 2 + j$ and $\alpha_2 = 2 - j$. Thus

$$\frac{C}{R} = \frac{2(s+1)(s+2)}{(s+2+j)(s+2-j)}$$

13.12 GAIN AND PHASE MARGINS FROM THE ROOT-LOCUS

The *gain margin* is the factor by which the design value of the gain factor K can be multiplied before the closed-loop system becomes unstable. It can be determined from the root-locus using the following formula:

$$\text{gain margin} = \frac{\text{value of } K \text{ at imaginary-axis crossover}}{\text{design value of } K} \qquad (13.16)$$

If the root-locus does not cross the $j\omega$ axis, the gain margin is infinite.

Example 13.11.

Consider the system in Fig. 13-24. The design value for the gain factor is 8, producing the closed-loop poles (denoted by small triangles) shown in the root-locus of Fig. 13-25. The gain factor at the $j\omega$-axis crossing is 64; hence the gain margin for this system is $64/8 = 8$.

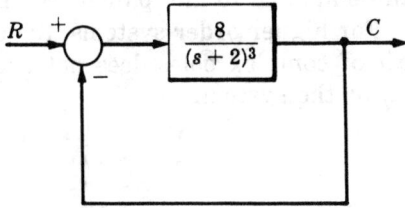

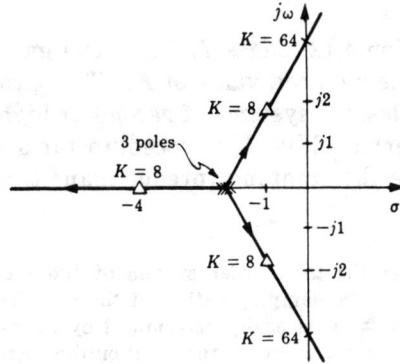

Fig. 13-24   **Fig. 13-25**

The *phase margin* for a feedback control system can also be determined from the root-locus. In this case it is necessary to find the point $j\omega_1$ on the $j\omega$ axis for which $|GH(j\omega_1)| = 1$ for the design value of K; that is,

$$|D(j\omega_1)/N(j\omega_1)| = K_{\text{design}}$$

It is usually necessary to use a trial-and-error procedure to locate $j\omega_1$. The phase margin is then computed from $\arg GH(j\omega_1)$ as

$$\phi_{PM} = 180° + \arg GH(j\omega_1) \qquad (13.17)$$

Example 13.12.

For the system of Example 13.11, $|GH(j\omega_1)| = |8/(j\omega_1 + 2)^3| \equiv 1$ when $\omega_1 = 0$; the phase angle of $GH(j0)$ is $0°$. The phase margin is therefore $180°$.

Example 13.13.

Consider the system of Fig. 13-26. The root-locus for this system is shown in Fig. 13-27. The point

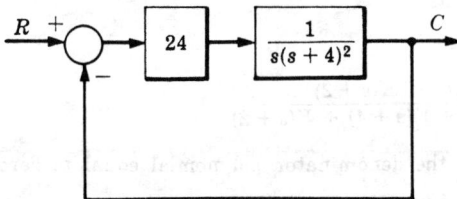

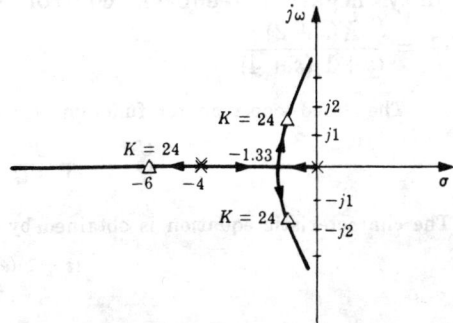

Fig. 13-26   **Fig. 13-27**

on the $j\omega$ axis for which $|GH(j\omega_1)| = \left|\dfrac{24}{j\omega_1(j\omega_1 + 4)^2}\right| \equiv 1$ is at $\omega_1 = 1.35$; the angle of $GH(j1.35)$ is $-129.6°$. Therefore the phase margin is $\phi_{PM} = 180° - 129.6° = 50.4°$.

13.13 DAMPING RATIO FROM THE ROOT-LOCUS

The gain-factor K required to give a specified damping ratio ζ (or vice-versa) for the second-order system

$$GH = \frac{K}{(s+p_1)(s+p_2)} \qquad K, p_1, p_2 > 0$$

is easily determined from the root-locus. Simply draw a line from the origin at an angle of plus or minus θ with the negative real axis, where

$$\theta = \cos^{-1}\zeta \qquad\qquad (13.18)$$

(See Section 4.12, pages 71-72). The gain factor at the point of intersection with the root-locus is the required value of K. This procedure can be applied to any pair of complex conjugate poles, for systems of second or higher order. For higher order systems, the damping ratio determined by this procedure for a *specific pair* of complex poles does not necessarily determine the damping (predominant time constant) of the system.

Example 13.14.

Consider the third-order system of the Example 13.13. The damping ratio ζ of the *complex poles* for $K = 24$ is easily determined by drawing a line from the origin to the point on the root-locus where $K = 24$, as shown in Fig. 13-28. The angle θ is measured as 60°; hence

$$\zeta = \cos\theta = 0.5$$

This value of ζ is a good approximation for the damping of the third-order system with $K = 24$ because the complex poles dominate the response.

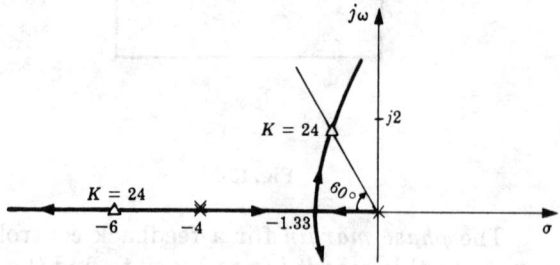

Fig. 13-28

Solved Problems

VARIATION OF SYSTEM CLOSED-LOOP POLES

13.1. Determine the closed-loop transfer function and the characteristic equation of the unity negative feedback control system whose open-loop transfer function is
$$G = \frac{K(s+2)}{(s+1)(s+4)}.$$

The closed-loop transfer function is
$$\frac{C}{R} = \frac{G}{1+G} = \frac{K(s+2)}{(s+1)(s+4)+K(s+2)}$$

The characteristic equation is obtained by setting the denominator polynomial equal to zero:
$$(s+1)(s+4) + K(s+2) = 0$$

13.2. How would the closed-loop poles of the system of Problem 13.1 be determined for $K = 2$ from its root-locus plot?

The root-locus is a plot of the closed-loop poles of the feedback system as a function of K. Therefore the closed-loop poles for $K = 2$ are determined by the points on the root-locus which correspond to $K = 2$ (one point on each branch of the locus).

13.3. How can a root-locus be employed to factor the polynomial $s^2 + 6s + 18$?

Since the root-locus is a plot of the roots of the characteristic equation of a system, Equation (13.1), as a function of its open-loop gain-factor, the roots of the above polynomial can be determined from the root-locus of any system whose characteristic polynomial is equivalent to it for some value of K. For example, the root-locus for $GH = \dfrac{K}{s(s+6)}$ factors the characteristic polynomial $s^2 + 6s + K$. For $K = 18$ this polynomial is equivalent to the one we desire to factor. Thus the desired roots are located on this root-locus at the points corresponding to $K = 18$.

Note that other forms for GH could be chosen, such as $GH = \dfrac{K}{(s+2)(s+4)}$ whose closed-loop characteristic polynomial corresponds to the one we wish to factor, but now for $K = 10$.

ANGLE AND MAGNITUDE CRITERIA

13.4. Show that the point $s_1 = -0.5$ satisfies the angle criterion, Equation (13.4), and the magnitude criterion, Equation (13.5), when $K = 1.5$ in the open-loop transfer function of Example 13.1.

$$\arg GH(s_1) = \arg \frac{K(s_1+1)}{s_1(s_1+2)} = \arg \frac{1.5(0.5)}{-0.5(1.5)} = 180°, \quad |GH(s_1)| = \left| \frac{1.5(0.5)}{-0.5(1.5)} \right| = 1$$

or $\qquad \left| \dfrac{D(s_1)}{N(s_1)} \right| = \left| \dfrac{-0.5(1.5)}{0.5} \right| = 1.5 = K$

Thus as illustrated on the root-locus plot of Example 13.1, the point $s_1 = -0.5$ is on the root-locus and is a closed-loop pole for $K = 1.5$.

13.5. Determine the angle and magnitude of $GH(j2)$ for $GH = \dfrac{K}{s(s+2)^2}$. What value of K satisfies $|GH(j2)| = 1$?

$$GH(j2) = \frac{K}{j2(j2+2)^2}, \quad \arg GH(j2) = \begin{cases} -180° & \text{for } K > 0 \\ 0° & \text{for } K < 0 \end{cases}, \quad |GH(j2)| = \frac{|K|}{2(8)} = \frac{|K|}{16}$$

and for $|GH(j2)| = 1$ it is necessary that $|K| = 16$.

13.6. Illustrate the graphical computation of $\arg GH(j2)$ and $|GH(j2)|$ in Problem 13.5.

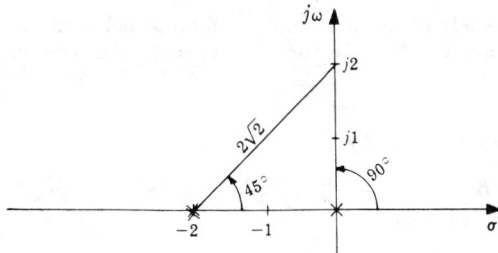

Fig. 13-29

$$\arg GH(j2) = -90° - 45° - 45° = -180°$$

$$|GH(j2)| = \frac{|K|}{2(2\sqrt{2})^2} = \frac{|K|}{16}$$

13.7. Show that the point $s_1 = -1 + j\sqrt{3}$ is on the root-locus for

$$GH(s) = \frac{K}{(s+1)(s+2)(s+4)}, \quad K > 0$$

and determine K at this point.

$$\arg \frac{N(s_1)}{D(s_1)} = \arg \frac{1}{j\sqrt{3}\,(1 + j\sqrt{3}\,)(3 + j\sqrt{3}\,)} = -90° - 60° - 30° = -180°$$

The angle criterion, Equation (*13.4b*), is thus satisfied for $K > 0$ and the point $s_1 = -1 + j\sqrt{3}$ is on the root-locus. From Equation (*13.5*)

$$K = \left| \frac{j\sqrt{3}\,(1 + j\sqrt{3}\,)(3 + j\sqrt{3}\,)}{1} \right| = \sqrt{3(4)12} = 12$$

NUMBER OF LOCI

13.8. Why must the number of loci equal the number of open-loop poles for $m \leq n$?

Each branch of the root-locus represents the locus of one closed-loop pole. Consequently there must be as many branches or loci as there are closed-loop poles. Since the number of closed-loop poles is equal to the number of open-loop poles for $m \leq n$, the number of loci must equal the number of open-loop poles.

13.9. How many loci are in the root-locus for $GH(s) = \dfrac{K(s+1)(s+2)}{s(s+1+j)(s+1-j)}$?

Since the number of open-loop poles is three, there are three loci in the root-locus plot.

REAL AXIS LOCI

13.10. Prove the real axis loci rules.

For any point on the real axis, the angle contributed to arg GH by any real axis pole or zero is either $0°$ or $180°$, depending on whether or not the point is to the right or to the left of the pole or zero. The total angle contributed to arg $GH(s)$ by a pair of complex poles or zeros is zero because

$$\arg (s + \sigma_1 + j\omega_1) + \arg (s + \sigma_1 - j\omega_1) = 0$$

for all real values of s. Thus arg $GH(s)$ for real values of s $(s = \sigma)$ may be written as

$$\arg GH(\sigma) = 180 n_r + \arg K$$

where n_r = total number of finite poles and zeros to the right of σ. In order to satisfy the angle criterion, n_r must be odd for positive K and even for negative K. Thus for $K > 0$, points of the root-locus on the real axis lie to the left of an odd number of finite poles and zeros; and for $K < 0$, points of the root-locus on the real axis lie to left of an even number of finite poles and zeros.

13.11. Determine which parts of the real axis are on the root-locus for

$$GH = \frac{K(s+2)}{(s+1)(s+3+j)(s+3-j)}, \qquad K > 0$$

The points on the real axis which lie to the left of an odd number of finite poles and zeros are only those points between -1 and -2. Therefore by the rule for $K > 0$, only the portion of real axis between -1 and -2 lies on the root-locus.

13.12. Which parts of the real axis are on the root-locus for

$$GH = \frac{K}{s(s+1)^2(s+2)}, \qquad K > 0$$

Points on the real axis between 0 and -1 and between -1 and -2 lie to the left of an odd number of poles and zeros and therefore are on the root-locus for $K > 0$.

ASYMPTOTES

13.13. Prove that the angles of the asymptotes are given by

$$\beta = \begin{cases} \dfrac{(2l+1)180}{n-m} & \text{degrees for } K > 0 \\[2mm] \dfrac{(2l)180}{n-m} & \text{degrees for } K < 0 \end{cases} \qquad (13.7)$$

For points s far from the origin in the s-plane, the angle contributed to arg GH by each of m zeros is

$$\arg (s + z_i)\big|_{|s| \gg |z_i|} \cong \arg (s)$$

Similarly, the angle contributed to arg GH by each of n poles is approximately equal to $-\arg (s)$. Therefore

$$\arg [N(s)/D(s)] \cong -(n - m) \cdot \arg (s) = -(n - m)\beta$$

where $\beta \equiv \arg (s)$. In order for s to be on the root-locus the angle criterion, **Equation** $(13.4b)$, must be satisfied. Thus

$$\arg \left[\frac{N(s_1)}{D(s_1)}\right] = -(n - m)\beta = \begin{cases} (2l + 1)\pi & \text{for } K > 0 \\ (2l)\pi & \text{for } K < 0 \end{cases}$$

and, since $\pm\pi$ radians ($\pm 180°$) are the same angle in the s-plane, then

$$\beta = \begin{cases} \dfrac{(2l + 1)180}{n - m} \text{ degrees} & \text{for } K > 0 \\[4mm] \dfrac{(2l)180}{n - m} \text{ degrees} & \text{for } K < 0 \end{cases}$$

13.14. Show that the center of asymptotes is given by

$$\sigma_c = -\frac{\displaystyle\sum_{i=1}^{n} p_i - \sum_{i=1}^{m} z_i}{n - m} \qquad (13.6)$$

The points on the root-locus satisfy the characteristic equation $D(s) + KN(s) = 0$, or

$$s^n + b_{n-1}s^{n-1} + \cdots + b_0 + K(s^m + a_{m-1}s^{m-1} + \cdots + a_0) = 0$$

Dividing by the numerator polynomial $N(s)$, this becomes

$$s^{n-m} + (b_{n-1} - a_{m-1})s^{n-m-1} + \cdots + K = 0$$

When the first coefficient of a polynomial is unity, the second coefficient is equal to minus the sum of the roots [8]. Thus from $D(s) = 0$, $b_{n-1} = \displaystyle\sum_{i=1}^{n} p_i$. From $N(s) = 0$, $a_{m-1} = \displaystyle\sum_{i=1}^{m} z_i$; and $-(b_{n-1} - a_{m-1})$ is equal to the sum of $n - m$ roots of the characteristic equation.

Now for large values of K and correspondingly large distances from the origin these $n - m$ roots approach the straight-line asymptotes and, along the asymptotes, the sum of the $n - m$ roots is equal to $-(b_{n-1} - a_{m-1})$. Since $b_{n-1} - a_{m-1}$ is a real number, the asymptotes must intersect at a point on the real axis. The center of asymptotes is therefore given by the point on the real axis where $n - m$ equal roots add up to $-(b_{n-1} - a_{m-1})$. Thus

$$\sigma_c = -\frac{b_{n-1} - a_{m-1}}{n - m} = -\frac{\displaystyle\sum_{i=1}^{n} p_i - \sum_{i=1}^{m} z_i}{n - m}$$

For a more detailed proof, see [9].

13.15. Find the angles and center of, and sketch the asymptotes for

$$GH = \frac{K(s + 2)}{(s + 1)(s + 3 + j)(s + 3 - j)(s + 4)}, \quad K > 0$$

The center of asymptotes is

$$\sigma_c = -\frac{1 + 3 + j + 3 - j + 4 - 2}{4 - 1} = -3$$

There are three asymptotes located at angles of $\beta = 60°, 180°, 300°$ as shown in Fig. 13-30.

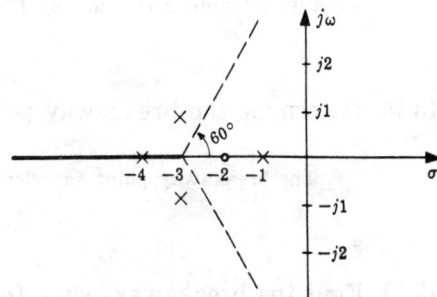

Fig. 13-30

13.16. Sketch the asymptotes for $K > 0$ and $K < 0$ for

$$GH = \frac{K}{s(s+2)(s+1+j)(s+1-j)}$$

The center of asymptotes is $\sigma_c = -\dfrac{0+2+1+j+1-j}{4} = -1.$

For $K > 0$, the angles of the asymptotes are $\beta = 45°, 135°, 225°, 315°$ as shown in Fig. 13-31.

For $K < 0$, the angles of the asymptotes are $\beta = 0°, 90°, 180°, 270°$ as shown in Fig. 13-32.

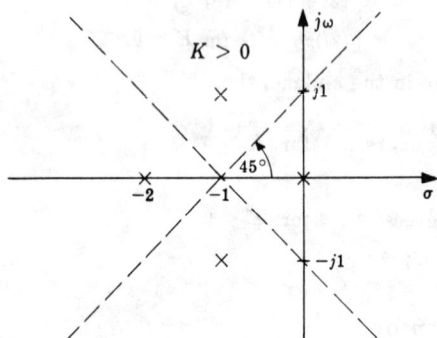

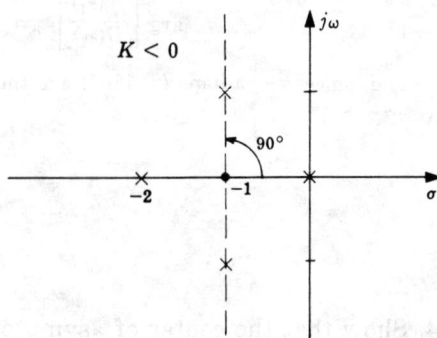

Fig. 13-31   Fig. 13-32

BREAKAWAY POINTS

13.17. Show that a breakaway point σ_b satisfies

$$\sum_{i=1}^{n} \frac{1}{(\sigma_b + p_i)} = \sum_{i=1}^{m} \frac{1}{(\sigma_b + z_i)} \tag{13.8}$$

A breakaway point is a point on the real axis where the gain factor K along the real axis portion of the root-locus is a maximum for poles leaving the real axis, or a minimum for poles coming onto the real axis, (see Section 13.2). The gain factor along the root-locus is given by

$$|K| = |D/N| \tag{13.5}$$

On the real axis, $s = \sigma$ and the magnitude signs may be dropped because $D(\sigma)$ and $N(\sigma)$ are both real. Then

$$K = D(\sigma)/N(\sigma)$$

To find the value of σ for which K is a maximum or minimum, the derivative of K with respect to σ is set equal to zero:

$$\frac{dK}{d\sigma} = \frac{d}{d\sigma}\left[\frac{(\sigma+p_1)\cdots(\sigma+p_n)}{(\sigma+z_1)\cdots(\sigma+z_m)}\right] = 0$$

By repeated differentiation and factorization, this can be written as

$$\frac{dK}{d\sigma} = \sum_{i=1}^{n} \frac{1}{(\sigma+p_i)}\left[\frac{D(\sigma)}{N(\sigma)}\right] - \sum_{i=1}^{m} \frac{1}{(\sigma+z_i)}\left[\frac{D(\sigma)}{N(\sigma)}\right] = 0$$

Finally, dividing both sides by $D(\sigma)/N(\sigma)$ yields the required result.

13.18. Determine the breakaway point for $GH = \dfrac{K}{s(s+3)^2}.$

The breakaway point satisfies $\dfrac{1}{\sigma_b} + \dfrac{1}{\sigma_b + 3} + \dfrac{1}{\sigma_b + 3} = 0,$ from which $\sigma_b = -1.$

13.19. Find the breakaway point for $GH = \dfrac{K(s+2)}{(s+1+j\sqrt{3})(s+1-j\sqrt{3})}.$

From Equation (13.8), $\dfrac{1}{\sigma_b + 1 + j\sqrt{3}} + \dfrac{1}{\sigma_b + 1 - j\sqrt{3}} = \dfrac{1}{\sigma_b + 2}$ which gives $\sigma_b^2 + 4\sigma_b = 0$.

This equation has the solutions $\sigma_b = 0$ and $\sigma_b = -4$; $\sigma_b = -4$ is a breakaway point for $K > 0$ and $\sigma_b = 0$ is a breakaway point for $K < 0$, as shown in Fig. 13-33.

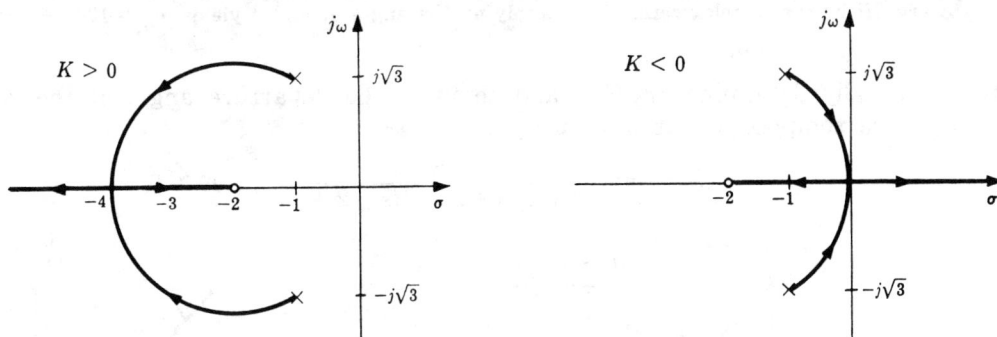

Fig. 13-33

13.20. Find the breakaway point between 0 and -1 for $GH = \dfrac{K}{s(s+1)(s+3)(s+4)}$.

The breakaway point must satisfy

$$\frac{1}{\sigma_b} + \frac{1}{(\sigma_b+1)} + \frac{1}{(\sigma_b+3)} + \frac{1}{(\sigma_b+4)} = 0$$

If this equation were simplified, a third order polynomial would be obtained. To avoid solving a third order polynomial, the following procedure may be used. As a first guess, assume $\sigma_b = -0.5$ and use this value in the two terms for the poles furthest from the breakaway point. Then

$$\frac{1}{\sigma_b} + \frac{1}{\sigma_b + 1} + \frac{1}{2.5} + \frac{1}{3.5} = 0$$

which simplifies to $\sigma_b^2 + 3.92\sigma_b + 1.46 = 0$ and has the root $\sigma_b = -0.43$ between 0 and -1. This value is used to obtain a better approximation as follows:

$$\frac{1}{\sigma_b} + \frac{1}{\sigma_b + 1} + \frac{1}{2.57} + \frac{1}{3.57} = 0, \qquad \sigma_b^2 + 3.99\sigma_b + 1.496 = 0, \qquad \sigma_b = -0.424$$

The second computation did not result in a value much different from the first. A reasonable first guess can often result in a fairly accurate approximation with only one computation.

DEPARTURE AND ARRIVAL ANGLES

13.21. Show that the departure angle of the root-locus from a complex pole is given by

$$\theta_D = 180° + \arg GH' \qquad\qquad (13.9)$$

Consider a circle of infinitesimally small radius around the complex pole. Clearly the phase angle $\arg GH'$ of GH neglecting the contribution of the complex pole, is constant around this circle. If θ_D represents the departure angle, the total phase angle of GH at the point on the circle where the root-locus crosses it is

$$\arg GH = \arg GH' - \theta_D$$

since $-\theta_D$ is the phase angle contributed to $\arg GH$ by the complex pole. In order to satisfy the angle criteria, $\arg GH = \arg GH' - \theta_D = 180°$ or $\theta_D = 180° + \arg GH'$ since $+180°$ and $-180°$ are equivalent.

13.22. Determine the relationship between the departure angle from a complex pole for $K > 0$ with that for $K < 0$.

Since $\arg GH'$ changes by $180°$ if K changes from a positive number to a negative one, the departure angle for $K < 0$ is $180°$ different from the departure angle for $K > 0$.

13.23. Show that the arrival angle at a complex zero satisfies

$$\theta_A = 180° - \arg GH''$$

(13.10)

In the same manner as in the solution to Problem 13.21, the phase angle of GH in the vicinity of the complex zero is given by $\arg GH = \arg GH'' + \theta_A$ since θ_A is the phase angle contributed to $\arg GH$ by the complex zero. Then applying the angle criterion yields $\theta_A = 180° - \arg GH''$.

13.24. Graphically determine $\arg GH'$ and compute the departure angle of the root-locus from the complex pole at $s = -2 + j$ for

$$GH = \frac{K}{(s+1)(s+2-j)(s+2+j)}, \qquad K > 0$$

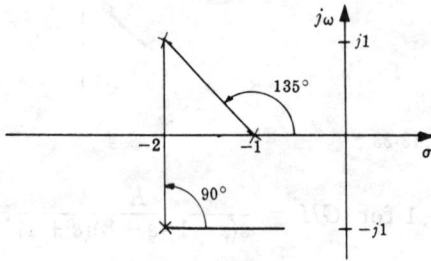

Fig. 13-34

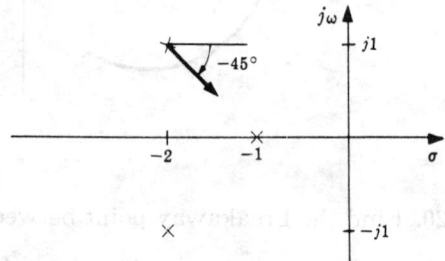

Fig. 13-35

From Fig. 13-34, $\arg GH' = -135° - 90° = -225°$; and $\theta_D = 180° - 225° = -45°$ as shown in Fig. 13-35.

13.25. Determine the departure angles from the complex poles and the arrival angles at the complex zeros for the open-loop transfer function

$$GH = \frac{K(s+1+j)(s+1-j)}{s(s+2j)(s-2j)}, \qquad K > 0$$

For the complex pole at $s = 2j$,

$$\arg GH' = 45° + 71.6° - 90° - 90° = -63.4° \quad \text{and} \quad \theta_D = 180° - 63.4° = 116.6°$$

Since the root-locus is symmetric about the real axis, the departure angle from the pole at $s = -2j$ is $-116.6°$. For the complex zero at $s = -1 + j$,

$$\arg GH'' = 90° - 108.4° - 135° - 225° = -18.4° \quad \text{and} \quad \theta_A = 180° - (-18.4°) = 198.4°$$

Thus the arrival angle at the complex zero $s = -1 - j$ is $\theta_A = -198.4°$.

THE SPIRULE

13.26. The technique presented in Section 13.9 describes the use of the Spirule for measuring the phase angle of $GH(s)$ at s_0. Show that the angles measured by the Spirule are equivalent to the angles measured using the graphical procedures described in Sections 4.11 and 6.5.

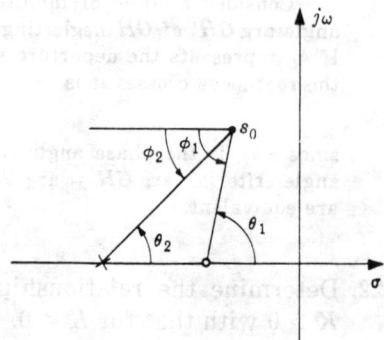

Fig. 13-36

Consider the s-plane diagram in Fig. 13-36. Using the techniques described in Sections 4.11 and 6.5, the angle contributed by the zero and the pole to $\arg GH(s_0)$ are determined by measuring θ_1 and θ_2 respectively. By simple geometry, $\phi_1 = \theta_1$ and $\phi_2 = \theta_2$. Thus $\arg GH(s_0)$ can be computed by measuring ϕ_1 and ϕ_2. These are the angles measured by the Spirule.

13.27. What is the significance of the S curve when using the Spirule to compute $N(s_0)/D(s_0)$?

The S curve is a logarithmic spiral used to convert the logarithm of a distance into an angle. The multiplications and divisions necessary to compute $N(s_0)/D(s_0)$ can then be performed by adding and subtracting angles. The resulting number is read from the quarter-circle scale, which effectively computes the antilogarithm.

CONSTRUCTION OF THE ROOT-LOCUS

13.28. Construct the root-locus for

$$GH = \frac{K}{(s+1)(s+2-j)(s+2+j)}, \qquad K > 0$$

The real axis from -1 to $-\infty$ is on the root-locus. The center of asymptotes is at

$$\sigma_c = \frac{-1-2+j-2-j}{3} = -1.67$$

There are three asymptotes $(n-m=3)$, located at angles of $60°$, $180°$ and $300°$. The departure angle from the complex pole at $s=-2+j$, which was computed in Problem 13.24, is $-45°$. A sketch of the resulting root-locus is shown in Fig. 13-37. An accurate root-locus plot is obtained by checking the angle criterion at points along the sketched branches, adjusting the location of the branches if necessary, and then applying the magnitude criterion to determine the values of K at selected points along the branches. The completed root-locus is shown in Fig. 13-38.

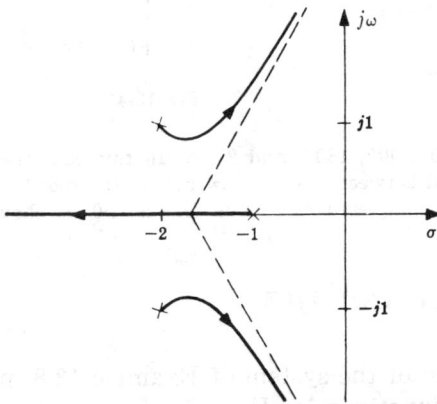

Fig. 13-37

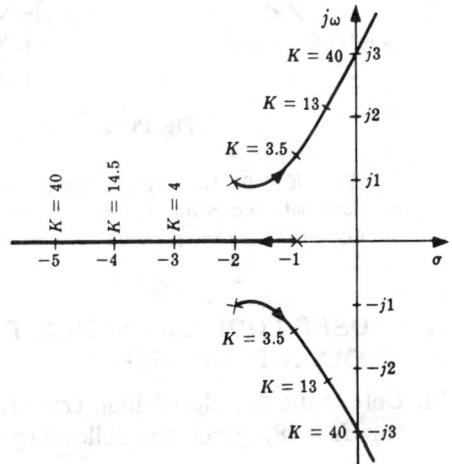

Fig. 13-38

13.29. Sketch the branches of the root-locus for the transfer function

$$GH = \frac{K(s+2)}{(s+1)(s+3+j)(s+3-j)}$$

$K > 0$.

The real axis between -1 and -2 is on the root-locus (Problem 13.11). There are two asymptotes with angles of $90°$ and $270°$. The center of asymptotes is easily computed as $\sigma_c = -2.5$ and the departure angle from the complex pole at $s=-3+j$ as $72°$. By symmetry, the departure angle from the pole at $-3-j$ is $-72°$. The branches of the root-locus may therefore be sketched as shown in Fig. 13-39.

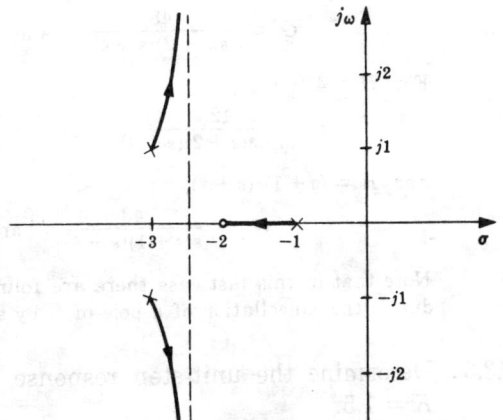

Fig. 13-39

13.30. Construct the root-locus for $K > 0$ and $K < 0$ for the transfer function

$$GH = \frac{K}{s(s+1)(s+3)(s+4)}$$

For this transfer function the center of asymptotes is simply $\sigma_c = -2$; and $n - m = 4$. Therefore for $K > 0$ the asymptotes have angles of $45°$, $135°$, $225°$, and $315°$. The real axis sections between 0 and -1 and between -3 and -4 lie on the root-locus for $K > 0$ and it was determined in Problem 13.20 that a breakaway point is located at $\sigma_b = -0.424$. From the symmetry of the pole locations, another breakaway point is located at -3.576. This can be verified by substituting this value into the relation for the breakaway point, Equation (13.8). The completed root-locus for $K > 0$ is shown in Fig. 13-40.

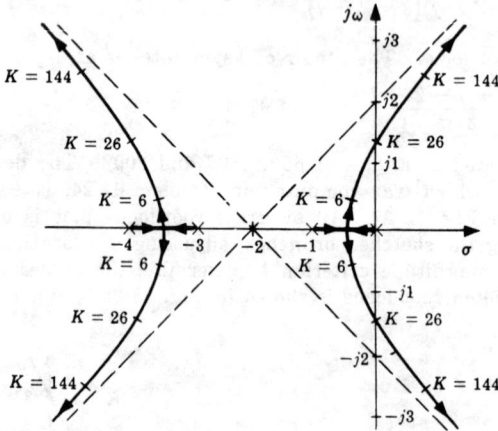

Fig. 13-40

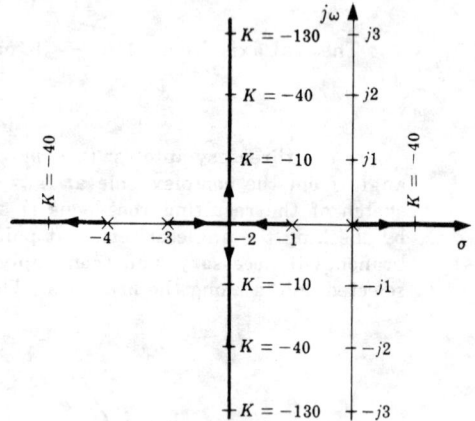

Fig. 13-41

For $K < 0$, the asymptotes have angles of $0°$, $90°$, $180°$, and $270°$. In this case the real axis portions between ∞ and 0, between -1 and -3 and between -4 and $-\infty$ are on the root-locus. There is only one breakaway point, located at -2. The completed root-locus for $K < 0$ is shown in Fig. 13-41.

THE CLOSED-LOOP TRANSFER FUNCTION AND THE TIME-DOMAIN RESPONSE

13.31. Determine the closed-loop transfer function of the system of Example 13.8, page 244, for $K = 48$, given the following transfer functions for H:

(a) $H = 1$,   (b) $H = 4/(s+1)$,   (c) $H = (s+1)/(s+2)$.

From the root-locus plot of Example 13.8, the closed-loop poles for $K = 48$ are located at $s = -6$, $j2.83$, and $-j2.83$. For $H = 1$,

$$G = \frac{48}{s(s+2)(s+4)} \quad \text{and} \quad \frac{C}{R} = \frac{GH}{1+GH} = \frac{48}{(s+6)(s-j2.83)(s+j2.83)}$$

For $H = 4/(s+1)$

$$G = \frac{12(s+1)}{s(s+2)(s+4)} \quad \text{and} \quad \frac{C}{R} = \frac{1}{H}\left(\frac{GH}{1+GH}\right) = \frac{12(s+1)}{(s+6)(s-j2.83)(s+j2.83)}$$

For $H = (s+1)/(s+2)$,

$$G = \frac{48}{s(s+1)(s+4)} \quad \text{and} \quad \frac{C}{R} = \frac{48(s+2)}{(s+1)(s+6)(s-j2.83)(s+j2.83)}$$

Note that in this last case there are four closed-loop poles, while GH has only three poles. This is due to the cancellation of a pole of G by a zero of H.

13.32. Determine the unit-step response of the system of Example 13.1, page 238, with $K = 1.5$.

The closed-loop transfer function of this system is $\dfrac{C}{R} = \dfrac{1.5(s+1)}{(s+0.5)(s+3)}$. For $R = 1/s$,

$$C = \frac{1.5(s+1)}{s(s+0.5)(s+3)} = \frac{1}{s} + \frac{-0.6}{(s+0.5)} + \frac{-0.4}{(s+3)}$$

and the unit-step response is $\mathcal{L}^{-1}[C(s)] = c(t) = 1 - 0.6e^{-0.5t} - 0.4e^{-3t}$.

13.33. Determine the relationship between the closed-loop zeros and the poles and zeros of G and H, assuming there are no cancellations.

Let $G = N_1/D_1$ and $H = N_2/D_2$, where N_1 and D_1 are numerator (zeros) and denominator (poles) polynomials of G, and N_2 and D_2 are the numerator and denominator polynomials of H. Then

$$\frac{C}{R} = \frac{G}{1+GH} = \frac{N_1 D_2}{D_1 D_2 + N_1 N_2}$$

Thus the closed-loop zeros are equal to the zeros of G and the poles of H.

GAIN AND PHASE MARGINS

13.34. Find the gain margin of the system of Example 13.8, page 244, if the design value of K is 6.

The gain factor at the $j\omega$-axis crossover is 48, as shown. Hence the gain margin is $48/6 = 8$.

13.35. Show how a Routh Table (Section 5.3, page 87) can be used to determine the frequency and the gain at the $j\omega$-axis crossover.

In Section 5.3 it was pointed out that a row of zeros in the s^1 row of the Routh table indicates that the polynomial has a pair of roots which satisfy the auxiliary equation $As^2 + B = 0$ where A and B are the first and second elements of the s^2 row. If A and B have the same sign, the roots of the auxiliary equation are imaginary (on the $j\omega$ axis). Thus if a Routh table is constructed for the characteristic equation of a system, the values of K and ω corresponding to $j\omega$-axis crossovers can be determined. For example, consider the system with the open-loop transfer function

$$GH = \frac{K}{s(s+2)^2}$$

The characteristic equation for this system is

$$s^3 + 4s^2 + 4s + K = 0$$

The Routh table for the characteristic polynomial is

s^3	1	4
s^2	4	K
s^1	$(16-K)/4$	
s^0	K	

The s^1 row is zero for $K = 16$. The auxiliary equation then becomes

$$4s^2 + 16 = 0$$

Thus for $K = 16$ the characteristic equation has solutions (closed-loop poles) at $s = \pm j2$, and the root-locus crosses the $j\omega$-axis at $j2$.

13.36. Determine the phase margin for the system of Example 13.8 with K set at 6.

First, the point on the $j\omega$ axis for which $|GH(j\omega)| = 1$ is found by trial-and-error to be $j0.7$. Then $\arg GH(j0.7)$ is computed as $-120°$. Hence the phase margin is $180° - 120° = 60°$.

13.37. Is it necessary to construct the entire root-locus in order to determine the gain and phase margins of a system in the s-plane?

No. Only one point on the root-locus is required to determine the gain margin. This point, $j\omega_\pi$, where the root-locus crosses the $j\omega$-axis, can be determined by trial-and-error or by the use of a Routh table as described in Problem 13.35. To determine the phase margin, it is only necessary to determine the point $j\omega_1$ on the $j\omega$ axis where $|GH(j\omega)| = 1$. Although the entire root-locus plot is not necessary, it can often be helpful, especially in the case of multiple $j\omega$-axis crossings.

DAMPING RATIO FROM THE ROOT-LOCUS

13.38. Prove Equation (13.18), page 248.

The roots of $s^2 + 2\zeta\omega_n s + \omega_n^2$ are (see Page 39) $s_{1,2} = -\zeta\omega_n \pm j\omega_n\sqrt{1 - \zeta^2}$. Then

$$|s_1| = |s_2| = \sqrt{\zeta^2\omega_n^2 + \omega_n^2(1 - \zeta^2)} = \omega_n$$

and

$$\arg s_{1,2} = \mp\tan^{-1}(\sqrt{1 - \zeta^2}/\zeta) \equiv 180° \pm \theta$$

or $s_{1,2} = \omega_n/180° \pm \theta$. Thus $\cos\theta = \zeta\omega_n/\omega_n = \zeta$.

13.39. Determine the positive value of gain which results in a damping ratio of 0.55 for the complex poles on the root-locus shown in Fig. 13-21.

The angle of the desired poles is $\theta = \cos^{-1} 0.55 = 56.6°$. A line drawn from the origin at an angle of 55.6° with the negative real axis intersects the root-locus of Fig. 13-21 at $K = 7$.

13.40. Find the damping ratio of the complex poles of Problem 13.28 for $K = 3.5$.

A line drawn from the root-locus at $K = 3.5$ to the origin makes an angle of 53° with the negative real axis. Hence the damping ratio of the complex poles is $\zeta = \cos 53° = 0.6$.

Supplementary Problems

13.41. Determine the angle and magnitude of $GH = \dfrac{16(s + 1)}{s(s + 2)(s + 4)}$ at the following points in the s-plane: (a) $s = j2$, (b) $s = -2 + j2$, (c) $s = -4 + j2$, (d) $s = -6$, (e) $s = -3$.

13.42. Determine the angle and magnitude of $GH = \dfrac{20(s + 10 + j10)(s + 10 - j10)}{(s + 10)(s + 15)(s + 25)}$ at the following points in the s-plane: (a) $s = j10$, (b) $s = j20$, (c) $s = -10 + j20$, (d) $s = -20 + j20$, (e) $s = -15 + j5$.

13.43. For each transfer function, find the breakaway points on the root-locus: (a) $GH = \dfrac{K}{s(s + 6)(s + 8)}$, (b) $GH = \dfrac{K(s + 5)}{(s + 2)(s + 4)}$, (c) $GH = \dfrac{K(s + 1)}{s^2(s + 9)}$.

13.44. Find the departure angle of the root-locus from the pole at $s = -10 + j10$ for

$$GH = \frac{K(s + 8)}{(s + 14)(s + 10 + j10)(s + 10 - j10)}, \qquad K > 0$$

13.45. Find the departure angle of the root-locus from the pole at $s = -15 + j9$ for

$$GH = \frac{K}{(s+5)(s+10)(s+15+j9)(s+15-j9)}, \qquad K > 0$$

13.46. Find the arrival angle of the root-locus to the zero at $s = -7 + j5$ for

$$GH = \frac{K(s+7+j5)(s+7-j5)}{(s+3)(s+5)(s+10)}, \qquad K > 0$$

13.47. Construct the root-locus for $K > 0$ for the transfer function of Problem 13.43(a).

13.48. Construct the root-locus for $K > 0$ for the transfer function of Problem 13.43(c).

13.49. Construct the root-locus for $K > 0$ for the transfer function of Problem 13.44.

13.50. Construct the root-locus for $K > 0$ for the transfer function of Problem 13.45.

13.51. Determine the gain and phase margins for the system with the open-loop transfer function of Problem 13.45 if the gain factor K is set equal to 20,000.

Answers to Supplementary Problems

13.41. (a) arg $GH = -99°$, $|GH| = 1.5$; (b) arg $GH = -153°$, $|GH| = 2.3$; (c) arg $GH = -232°$, $|GH| = 1.8$; (d) arg $GH = 0°$, $|GH| = 1.7$; (e) arg $GH = -180°$, $|GH| = 10.7$

13.42. (a) arg $GH = -38°$, $|GH| = 0.68$; (b) arg $GH = -40°$, $|GH| = 0.37$; (c) arg $GH = -41°$, $|GH| = 0.60$; (d) arg $GH = -56°$, $|GH| = 0.95$; (e) arg $GH = +80°$, $|GH| = 6.3$

13.43. (a) $\sigma_b = -2.25, -7.07$; (b) $\sigma_b = -3.27, -6.73$; (c) $\sigma_b = 0, -3$

13.44. $\theta_D = 124°$

13.45. $\theta_D = 193°$

13.46. $\theta_A = 28°$

13.51. Gain margin = 3.7; phase margin = $102°$

Chapter 14

Root-locus Design

14.1 THE DESIGN PROBLEM

The root-locus method can be used very effectively in the design of feedback control systems because it graphically illustrates the variation of a system's closed-loop poles as a function of the gain factor K. In its simplest form, the design is accomplished by choosing a value of K which results in a satisfactory closed-loop behavior. This is called *gain factor compensation* (see Section 12.2). If it is not possible to meet system specifications in this way, another form of compensation can be added to the system to alter the root-locus in the required manner.

In order to accomplish system design in the s-plane using root-locus techniques, it is necessary to interpret the system specifications in terms of allowable s-plane configurations.

The closed-loop system transient and frequency responses are determined by the location of the closed-loop poles. If the system has only two closed-loop poles and no finite zeros, performance parameters such as percentage overshoot, rise time, and 3 db bandwidth are directly related to the damping ratio ζ and undamped natural frequency ω_n of the poles. For higher order systems, the relationship of these parameters to the closed-loop poles and zeros becomes more complicated and it is necessary to resort to standard curves or tables for this information [10]. However, in many instances a higher order system may be approximated by a second or third order system by using a *dominant pole-zero approximation*, which is discussed in detail in Section 14.5.

Specifications on allowable steady-state errors usually take the form of a minimum open-loop gain factor, expressed in terms of the error constants (Chapter 9).

Example 14.1.

Consider the design of a unity feedback system with the plant $G = \dfrac{K}{(s+1)(s+3)}$ and the following specifications: (1) Overshoot less than 20%, (2) $K_p \geqq 4$, (3) 10 to 90% rise-time less 1 second.

The root-locus for this system is shown in Fig. 14-1. The system closed-loop transfer function may be written

$$\frac{C}{R} = \frac{K}{s^2 + 2\zeta\omega_n s + \omega_n^2}$$

where ζ and ω_n can be determined from the root-locus for a given value of K. In order to satisfy the first specification, ζ must be greater than 0.45 (see Fig. 3-8, Page 40). Then from the root-locus we see that K must be less than 16 (see Section 13.13). For this system, K_p is given by $K/3$. Thus in order to satisfy the second specification, K must be greater than 12. The rise time is a function of both ζ and ω_n. Suppose a trial value of $K = 13$ is chosen. In this case, $\zeta = 0.5$, $\omega_n = 4$ and the rise-time is 0.5 seconds. Hence all the specifications can be met by setting $K = 13$. Note that if the specification on K_p was greater than 5.33, or the specification on rise time was less than 0.34 seconds, all the specifications could not be met by simply adjusting the open-loop gain factor.

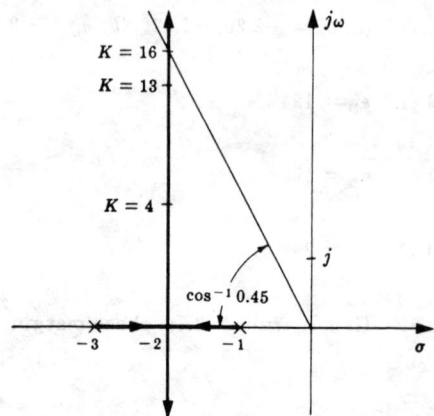

Fig. 14-1

14.2 CANCELLATION COMPENSATION

If the pole-zero configuration of the plant is such that the system specifications cannot be met by an adjustment of the open-loop gain factor, a cascade compensator, as shown in the adjoining figure, can be added to the system to cancel some or all of the poles and zeros of the plant. Due to realiza-

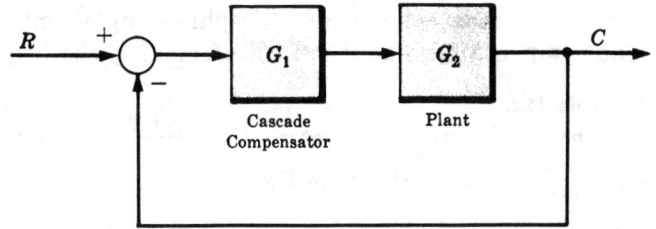

bility considerations (Pages 63 and 64), the compensator must have no more zeros than poles. Consequently, when poles of the plant are cancelled by zeros of the compensator, the compensator also adds new poles to the forward loop transfer function. The philosophy of this compensation technique is then to replace undesirable with desirable poles.

The difficulty encountered in applying this scheme is that it is not always apparent what open-loop pole-zero configuration is desirable from the standpoint of meeting specifications on the closed-loop system performance.

Some situations where cancellation compensation can be used to advantage are the following:

1. If the specifications on system rise time or bandwidth cannot be met without compensation, cancellation of low frequency poles and replacement with high frequency poles is helpful.

2. If the specifications on allowable steady-state errors cannot be met, a low frequency pole can be cancelled and replaced with a lower frequency pole, yielding a larger forward-loop gain at low frequencies.

3. If poles with small damping ratios are present in the plant transfer function, they may be cancelled and replaced with poles which have larger damping ratios.

14.3 PHASE COMPENSATION: LEAD AND LAG NETWORKS

A cascade compensator can be added to a system to alter the phase characteristics of the open-loop transfer function in a manner which favorably affects system performance. These effects were illustrated in the frequency domain for lead, lag, and lag-lead networks using Polar Plots in Chapter 12. Sections 12.4-12.6 summarize the general effects of these networks.

The pole-zero maps of a lead and a lag network are shown in Figs. 14-2 and 14-3. Note that a lead network makes a positive, and a lag network a negative phase contribution. A lag-lead network may be obtained by appropriately combining a lag and a lead network in series, or from the mechanization described in Problem 6.14, Page 103.

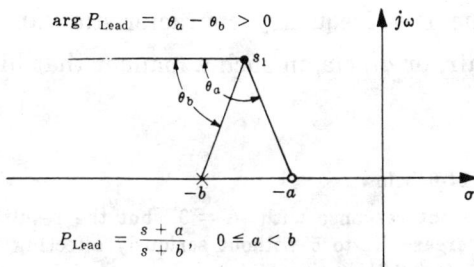

$$\arg P_{\text{Lead}} = \theta_a - \theta_b > 0$$

$$P_{\text{Lead}} = \frac{s+a}{s+b}, \quad 0 \le a < b$$

Fig. 14-2

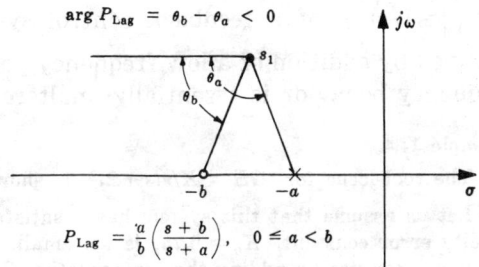

$$\arg P_{\text{Lag}} = \theta_b - \theta_a < 0$$

$$P_{\text{Lag}} = \frac{a}{b}\left(\frac{s+b}{s+a}\right), \quad 0 \le a < b$$

Fig. 14-3

Since the compensated system root-locus is determined by the points in the s-plane for which the phase-angle of $G = G_1G_2$ is equal to $-180°$, the branches of the locus can be moved by proper selection of the phase angle contributed by the compensator. In general, lead compensation has the effect of moving the loci to the left.

Example 14.2.

The phase lead compensator $G_1 = \dfrac{s + 2}{s + 8}$ alters the root-locus of the system with the plant $G_2 = \dfrac{K}{(s + 1)^2}$, as illustrated in Fig. 14-4.

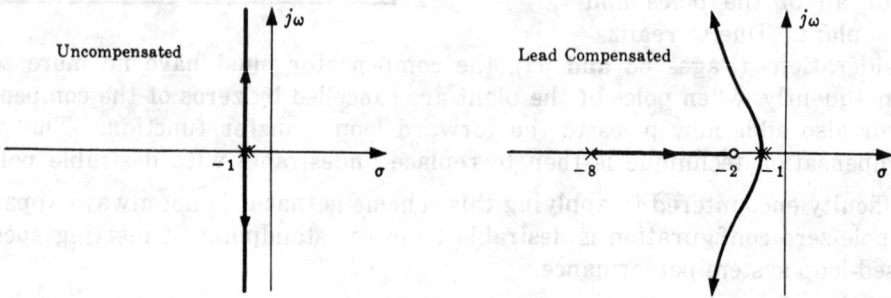

Fig. 14-4

Example 14.3.

The use of simple lag compensation (one pole at -1, no zero) to alter the breakaway angle of a root-locus from a pair of complex poles is illustrated in Fig. 14-5.

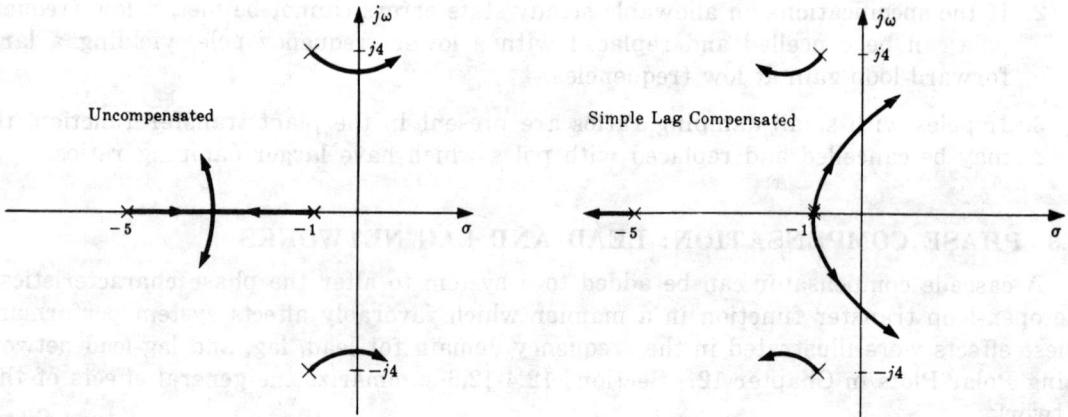

Fig. 14-5

14.4 MAGNITUDE COMPENSATION

Compensation networks may be employed to alter the magnitude, or gain, characteristic $\left(\left| \dfrac{C}{R}(j\omega) \right| \right)$ of a feedback control system. The low frequency characteristic can be modified by addition of a low frequency pole-zero pair, or *dipole*, in such a manner that high frequency behavior is essentially unaltered.

Example 14.4.

The root-locus for $GH = K/s(s + 2)^2$ is shown in Fig. 14-6 below.

Let us assume that this system has a satisfactory transient response with $K = 3$, but the resulting velocity error constant, $K_v = 0.75$, is too small. We can increase K_v to 5 without seriously affecting the transient response by adding the compensator $G_1 = (s + 0.1)/(s + 0.015)$ since

$$K_v' = K_v G_1(0) = 0.75(0.1)/0.015 = 5$$

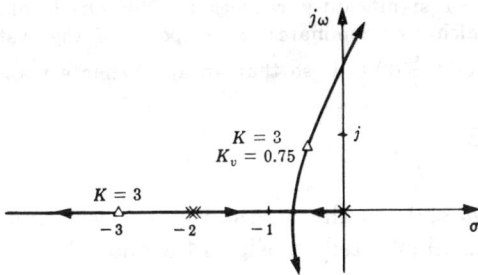

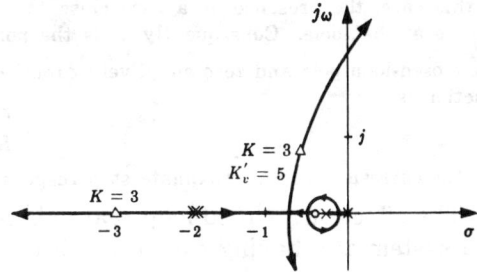

Fig. 14-6   Fig. 14-7

The resulting root-locus is shown in Fig. 14-7. The high frequency portion of the root-locus and the transient response are essentially unaffected because the closed-loop transfer function has a low frequency pole-zero pair which approximately cancel each other.

A low frequency dipole for magnitude compensation, with the pole at the origin, can be synthesized using a proportional plus integral compensator whose transfer function is

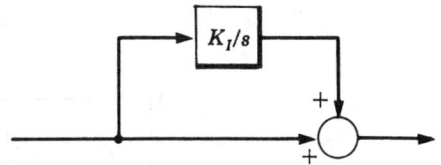

$$G_1 \;=\; \frac{s + K_I}{s}$$

14.5 DOMINANT POLE-ZERO APPROXIMATIONS

The root-locus method offers the advantage of a graphical display of the systems closed-loop poles and zeros. Theoretically, the designer can determine the system's response characteristics from the closed-loop pole-zero map. Practically, however, this task becomes increasingly difficult for systems with four or more poles and zeros. In some cases the problem can be considerably simplified if the response is dominated by two or three poles and zeros.

The influence of a particular pole (or pair of complex poles) on the response is mainly determined by two factors: the real part σ of the pole $p = \sigma + j\omega$, and the relative magnitude of the residue at the pole. The real part σ of the pole p determines the rate at which the transient term due to that pole decays; the larger σ, the faster the rate of decay. The relative magnitude of the residue determines the percentage of the total response due to that particular pole.

Example 14.5.

Consider a system with the closed-loop transfer function $\dfrac{C}{R} = \dfrac{5}{(s+1)(s+5)}$. The step response of this system is

$$c \;=\; 1 - 1.25e^{-t} + 0.25e^{-5t}$$

The term in the response due to the pole at $s_1 = \sigma_1 = -5$ decays five times as fast as the term due to the pole at $s_2 = \sigma_2 = -1$. Furthermore, the residue at the pole at $s_1 = -5$ is only $\frac{1}{5}$ that of the one at $s_2 = -1$. Therefore for most practical purposes the effect of the pole at $s_1 = -5$ can be ignored and the system approximated by

$$\frac{C}{R} \;\cong\; \frac{1}{s+1}$$

The pole at $s_1 = -5$ has been removed from the transfer function and the numerator has been adjusted to maintain the same steady-state gain $\left(\dfrac{C}{R}(0) = 1\right)$. The response of the approximated system is $c = 1 - e^{-t}$.

Example 14.6.

The system with the closed-loop transfer function $\dfrac{C}{R} = \dfrac{5.5(s + 0.91)}{(s+1)(s+5)}$ has the step response

$$c \;=\; 1 + 0.125e^{-t} - 1.125e^{-5t}$$

In this case, the presence of a zero close to the pole at -1 significantly reduces the magnitude of the residue at that pole. Consequently, it is the pole at -5 which now dominates the response of the system. The closed-loop pole and zero effectively cancel each other and $\frac{C}{R}(0) = 1$ so that an approximate transfer function is

$$\frac{C}{R} \cong \frac{5}{s+5}$$

and the corresponding approximate step response is $c \cong 1 - e^{-5t}$.

The effect of a closed-loop real-axis *pole* at $-p_r < 0$ on the overshoot and rise time T_r of a system also having complex poles $-p_c, -p_c^*$ is illustrated in Fig. 14-8 and 14-9. For

$$\frac{p_r}{\zeta \omega_n} > 5 \tag{14.1}$$

the overshoot and rise time approach that of a second-order system containing only complex poles (see Fig. 3-8, Page 40). Therefore p_r can be neglected in determining overshoot and rise time if $\zeta > 0.5$ and

$$p_r > 5 |\mathrm{Re}\, p_c| = 5\zeta\omega_n \tag{14.2}$$

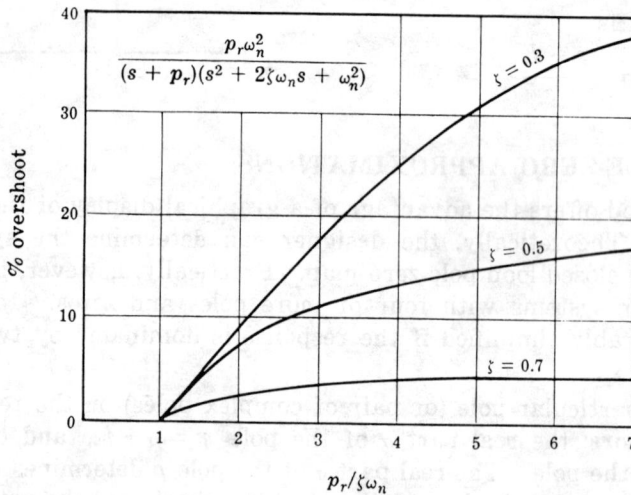

Fig. 14-8

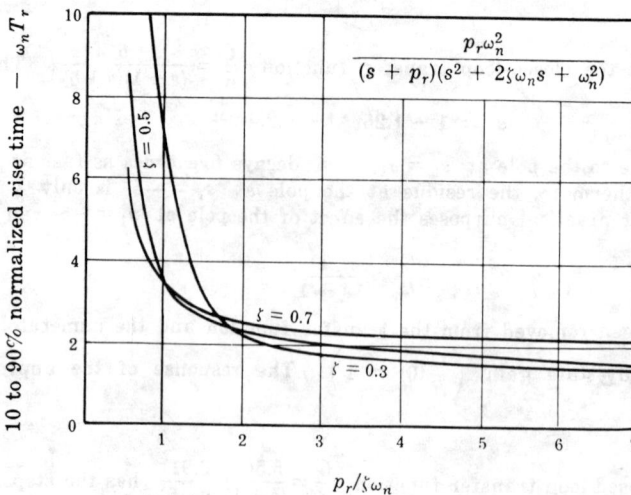

Fig. 14-9

There is no overshoot if

$$p_r \leq |\mathrm{Re}\ p_c| = \zeta\omega_n \qquad (14.3)$$

and the rise time approaches that of a first-order system containing only the real-axis pole.

The effect of a closed-loop real-axis *zero* at $-z_r < 0$ on the overshoot and rise time T_r of a system also having complex poles $-p_c, -p_c^*$ is illustrated in Fig. 14-10 and 14-11. These graphs show that z_r can be neglected in determining overshoot and rise time if $\zeta > 0.5$ and

$$z_r > 5\,|\mathrm{Re}\ p_c| = 5\zeta\omega_n \qquad (14.4)$$

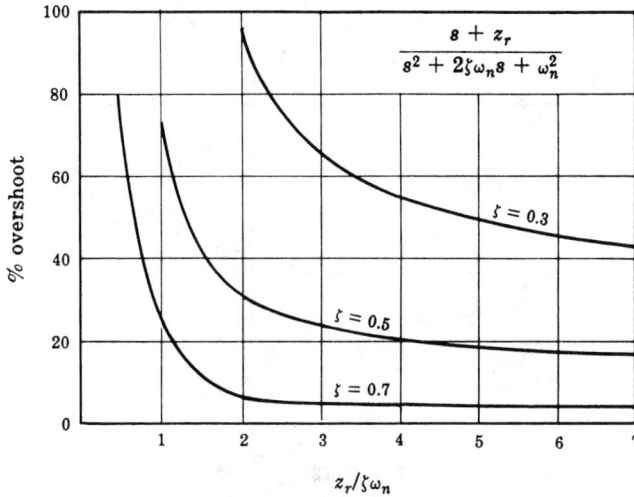

Fig. 14-10

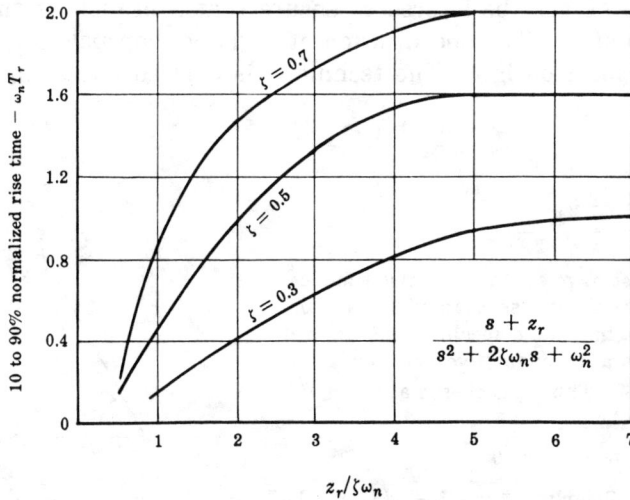

Fig. 14-11

Example 14.7.

The closed-loop transfer function of a particular system is represented by the pole-zero map shown in Fig. 14-12 below. Given that the steady-state gain $\dfrac{C}{R}(j0) = 1$, a dominant pole-zero approximation is

$$\frac{C}{R} \cong \frac{4}{s^2 + 2s + 4}$$

This is a reasonable approximation because the pole and zero near $s = -2$ effectively cancel each other and all other poles and zeros satisfy Equations (14.2) and (14.4) with $-p_c = -1 + j\sqrt{3}$ and $\zeta = 0.5$.

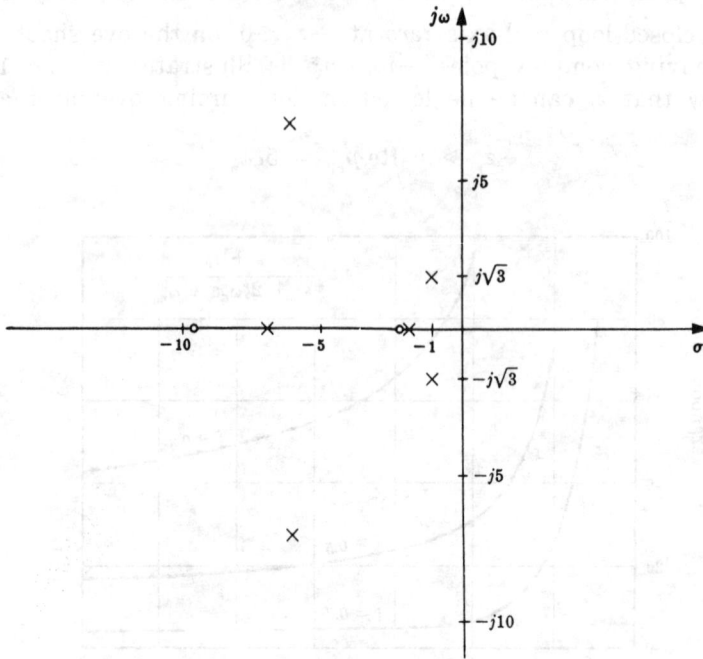

Fig. 14-12

14.6 POINT DESIGN

If an allowable closed-loop pole position s_1 can be determined from the system specifications, the system root-locus may be altered to ensure that a branch of the locus will pass through the required point s_1. The specification of a closed-loop pole at a particular point in the s-plane is called **point design**. The technique is carried out using phase and magnitude compensation.

Example 14.8.

Consider the plant

$$G_2 = \frac{K}{s(s+2)^2}$$

The closed-loop response must have a 10 to 90% rise time less than one second, and an overshoot less than 20%. We observe from Fig. 3-8, Page 40, that these specifications are met if the closed-loop system has a dominant two pole configuration with $\zeta = 0.5$ and $\omega_n = 2$. Thus s_1 is chosen at $-1 + j\sqrt{3}$, which is a solution of

$$s_1^2 + 2\zeta\omega_n s_1 + \omega_n^2 = 0$$

for $\zeta = 0.5$ and $\omega_n = 2$. Clearly, $s_1^* = -1 - j\sqrt{3}$ is the remaining solution of this quadratic equation. The orientation of s_1 with respect to the poles of G_2 is shown in Fig. 14-13.

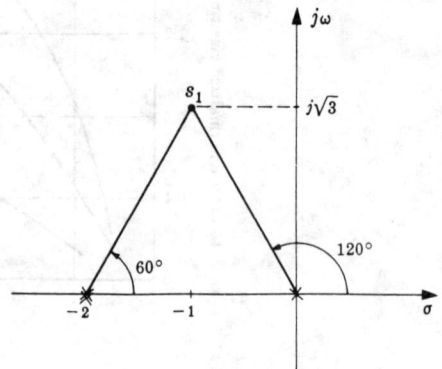

Fig. 14-13

The phase angle of G_2 is $-240°$ at s_1. In order for a branch of the root-locus to pass through s_1, the system must be modified so that the phase angle of the compensated system is $-180°$ at s_1. This can be accomplished by adding a cascade lead network having a phase angle of $240° - 180° = 60°$ at s_1, which is satisfied by

$$G_1 = P_{\text{Lead}} = (s+1)/(s+4)$$

as shown in the pole-zero map of the compensated open-loop transfer function G_1G_2 in Fig. 14-14 below.

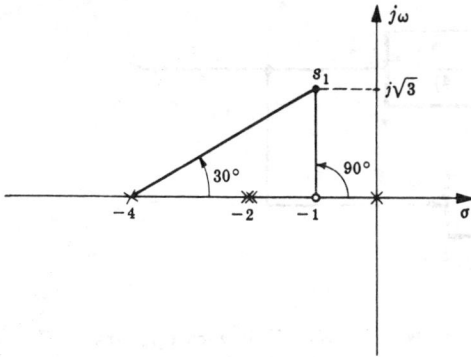

Fig. 14-14

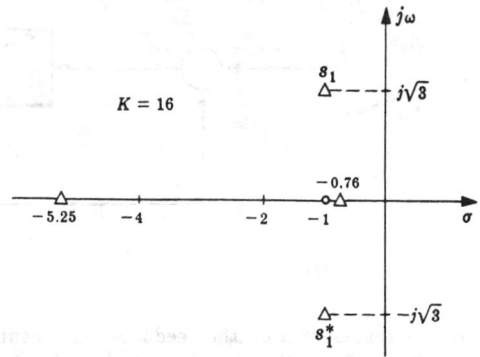

Fig. 14-15

The closed-loop pole can now be located at s_1 by choosing a value for K which satisfies the root-locus magnitude criterion. Solution of Equation (13.5) yields $K = 16$. The root-locus or closed-loop pole-zero map of the compensated system should be sketched to check the validity of the dominant two pole assumption. Figure 14-15 illustrates that the poles at s_1 and s_1^* dominate the response.

14.7 FEEDBACK COMPENSATION

The addition of compensation elements to a feedback path of a control system can be employed in root-locus design in a manner similar to that discussed in the preceding sections. The compensation elements affect the root-locus of the system's open-loop transfer function in the same manner. But although the root-locus is the same when the compensator is in either the forward or feedback path, the closed-loop transfer function may be significantly different. It was shown in Problem 13.33 that feedback *zeros* do not appear in the closed-loop transfer function, while feedback *poles* become zeros of the closed-loop transfer function (assuming no cancellations).

Example 14.9.

Suppose a feedback compensator were added to a system with the forward transfer function

$$G = \frac{K}{(s+1)(s+4)(s+5)}$$

in an attempt to cancel the pole at -1 and replace it with a pole at -6. Then the compensator would be $H = (s+1)/(s+6)$, GH would be given by $GH = K/(s+4)(s+5)(s+6)$ and the closed-loop transfer function would become

$$\frac{C}{R} = \frac{K(s+6)}{(s+1)[(s+4)(s+5)(s+6)+K]}$$

Although the pole at -1 is cancelled from GH, it reappears as a *closed-loop* pole. Furthermore, the feedback pole at -6 becomes a closed-loop zero. Consequently, *the cancellation technique does not work with a compensator in the feedback path.*

Example 14.10.

The block diagram in Fig. 14-16 contains two feedback paths.

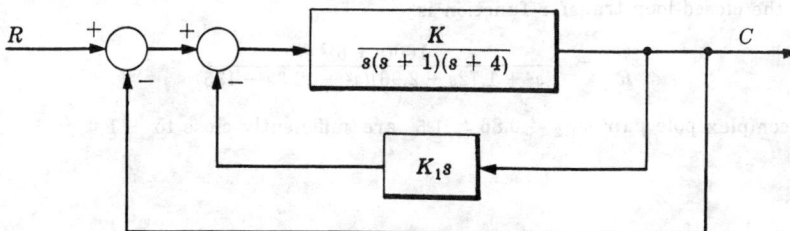

Fig. 14-16

These two paths may be combined, as shown in Fig. 14-17.

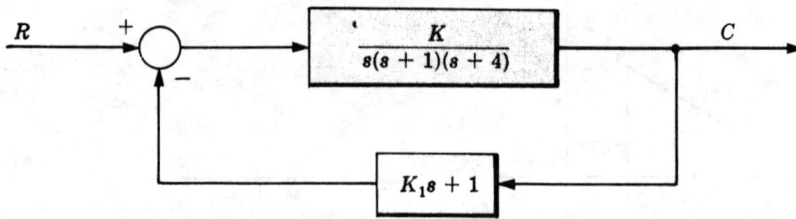

Fig. 14-17

In this representation the feedback path contains a zero at $s = -1/K_1$. This zero appears in GH and consequently affects the root-locus. However, it does not appear in the closed-loop transfer function, which contains three poles no matter where the zero is located.

The fact that feedback zeros do not appear in the closed-loop transfer function may be used to advantage in the following manner. If closed-loop poles are desired at certain locations in the s-plane, feedback zeros can be placed at these points. Since branches of the root-locus will terminate on these zeros, the desired closed-loop pole locations can be obtained by setting the open-loop gain factor sufficiently high.

Example 14.11.

The feedback compensator
$$H = \frac{s^2 + 2s + 4}{(s + 6)^2}$$

is added to the system with the forward-loop transfer function
$$G = \frac{K}{s(s + 2)}$$

in order to guarantee that the dominant closed-loop poles will be near $s = -1 \pm j\sqrt{3}$. The resulting root-locus is shown in Fig. 14-18.

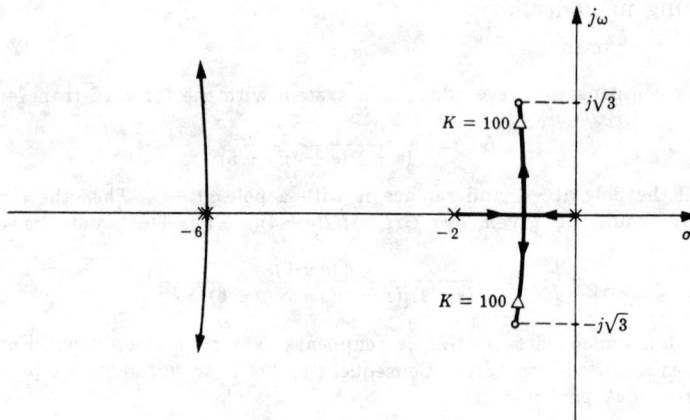

Fig. 14-18

If K is set at 100, the closed-loop transfer function is
$$\frac{C}{R} = \frac{100(s + 6)^2}{(s^2 + 1.72s + 2.96)(s^2 + 12.3s + 135)}$$

and the dominant complex pole pair $s_{1,2} = 0.86 \pm j1.5$ are sufficiently close to $-1 \pm j\sqrt{3}$.

Solved Problems

THE DESIGN PROBLEM

14.1. Determine the value of the gain factor K for which the system with the open-loop transfer function

$$GH = \frac{K}{s(s+2)(s+4)}$$

has closed-loop poles with a damping ratio $\zeta = 0.5$.

The closed-loop poles will have a damping ratio of 0.5 when they make an angle of 60° with the negative real axis [Equation *(13.18)*]. The desired value of K is determined at the point where the root-locus crosses the $\zeta = 0.5$ line in the s-plane. A sketch of the root-locus is shown in Fig. 14-19. The desired value of K is 8.3.

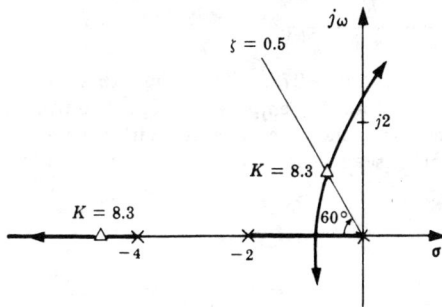

Fig. 14-19

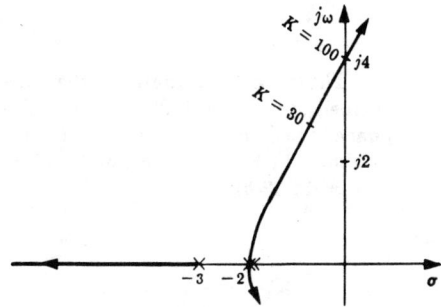

Fig. 14-20

14.2. Determine a value of K for which the system with the open-loop transfer function

$$GH = \frac{K}{(s+2)^2(s+3)}$$

satisfies the following specifications: (a) $K_p \geq 2$, (b) gain margin ≥ 3.

For this system, K_p is equal to $K/12$. Hence, in order to satisfy the first specification, K must be greater than 24. The value of K at the $j\omega$-axis crossover of the root-locus is equal to 100, as shown in Fig. 14-20. Then, in order to satisfy the second specification, K must be less than $100/3 = 33.3$. A value of K that will satisfy both specifications is 30.

CANCELLATION COMPENSATION

14.3. Can right-half s-plane poles of a plant be effectively cancelled by a compensator with a right-half s-plane zero?

No. For example, suppose a particular plant has the transfer function

$$G_2 = \frac{K}{s-1}, \qquad K > 0$$

and a cascade compensator is added with the transfer function $G_1 = (s-1+\epsilon)/(s+1)$. The ϵ term in the transfer function represents any small error between the desired zero location at $+1$ and the actual location. The closed-loop transfer function is then

$$\frac{C}{R} = \frac{K(s-1+\epsilon)}{s^2 + Ks + K\epsilon - K - 1}$$

By applying the Hurwitz or Routh Stability Criterion (Chapter 5) to the denominator of this transfer function, it can be seen that the system is unstable for any value of K if ϵ is less than $(1+K)/K$, which is usually the case because ϵ represents the error in the desired zero location.

PHASE COMPENSATION

14.4. It is desired to add to a system a compensator with a *zero* at $s = -1$ to produce $60°$ phase lead at $s = -2 + j3$. How can the proper location of the *pole* be determined?

With reference to Fig. 14-2, we want the phase contribution of the network to be $\theta_a - \theta_b = 60°$. From Fig. 14-21, $\theta_a = 108°$. Hence $\theta_b = \theta_a - 60° = 48°$ and the pole should be located at $s = -4.7$, as shown in Fig. 14-21.

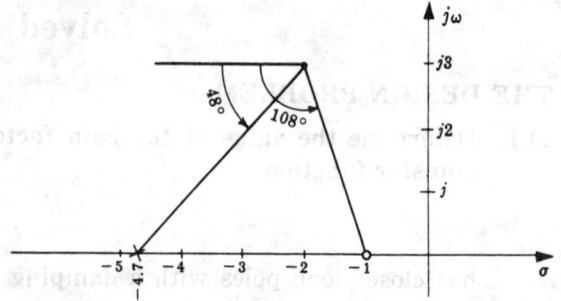

Fig. 14-21

14.5. Determine a compensator which will change the departure angle of the root-locus from the pole at $s = -0.5 + j$ to $-135°$ for the plant transfer function

$$G_2 = \frac{K}{s(s^2 + s + 1.25)}$$

The departure angle of the uncompensated system is $-27°$. To change this to $-135°$, a lag compensator with $108°$ phase lag at $s = -0.5 + j$ can be employed. The required amount of phase lag could be supplied by a *simple* lag compensator (one pole, no zero) with a pole at $s = -0.18$, as shown in Fig. 14-22(a), or by two simple lags in cascade with two poles at $s = -1.22$, as shown in Fig. 14-22(b).

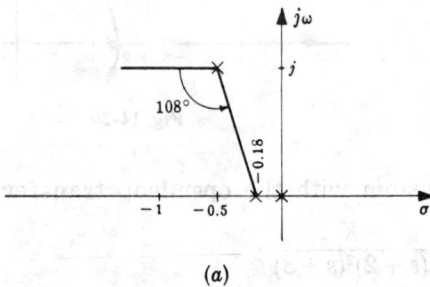

(a)

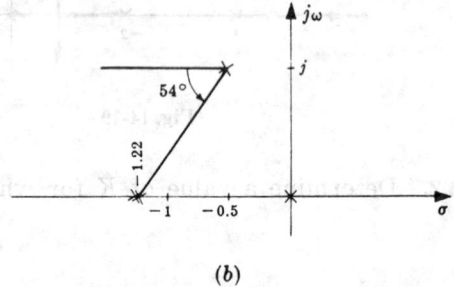

(b)

Fig. 14-22

MAGNITUDE COMPENSATION

14.6. In Example 14.4, Page 262, the velocity error constant K_v was increased by a factor of $6\frac{2}{3}$ without increasing the gain factor. How was this accomplished?

It was assumed that the compensator G_1 had a high frequency gain of one and a low frequency (d.c.) gain of $6\frac{2}{3}$. This compensator cannot be mechanized passively because a passive lag compensator has a d.c. gain of one. Consequently, G_1 must include an amplifier. An alternative method would be to let G_1 be the passive lag compensator $G_1' = \frac{0.015}{0.1}\left(\frac{s + 0.1}{s + 0.015}\right)$ and then amplify the gain factor by $6\frac{2}{3}$. However, when root-locus techniques are employed it is usually more convenient to assume the compensator just adds a pole and zero, as was done in Example 14.4. Appropriate adjustments can be made in the final stages of design to achieve the simplest and/or least expensive compensator mechanization.

DOMINANT POLE-ZERO APPROXIMATIONS

14.7. Determine the overshoot and rise time of the system with the transfer function

$$\frac{C}{R} = \frac{1}{(s + 1)(s^2 + s + 1)}$$

For this system, $\omega_n = 1$, $\zeta = 0.5$, $p_r = 1$ and $p_r/\zeta\omega_n = 2$. From Fig. 14-8, Page 264, the percentage overshoot is about 8%. The rise time from Fig. 14-9 is 2.4 seconds. The corresponding numbers for a system with the complex poles only are 18% and 1.6 seconds. Thus the real axis pole reduces the overshoot and slows down the response.

14.8. Determine the overshoot and rise time of the system with the transfer function

$$\frac{C}{R} = \frac{(s+1)}{(s^2+s+1)}$$

For this system $\omega_n = 1$, $\zeta = 0.5$, $z_r = 1$ and $z_r/\zeta\omega_n = 2$. From Fig. 14-10, Page 265, the percentage overshoot is 31%. From Fig. 14-11, the 10 to 90% rise time is 1.0 second. The corresponding numbers for a system without the zero are 18% and 1.6 seconds. The real axis zero thus increases the overshoot and decreases the rise time, i.e. speeds up the response.

14.9. What is a suitable dominant pole-zero approximation for the following system?

$$\frac{C}{R} = \frac{2(s+8)}{(s+1)(s^2+2s+3)(s+6)}$$

The real axis pole at $s = -6$ and the real axis zero at $s = -8$ satisfy Equations (14.2), Page 264, and (14.4), Page 265, respectively, with regard to the complex poles ($\zeta\omega_n = 1$ and $\zeta > 0.5$) and therefore may be neglected. The real axis pole at $s = -1$ and the complex poles cannot be neglected. Hence a suitable approximation (with the same d.c. gain) is $\frac{C}{R} = \frac{8}{3(s+1)(s^2+2s+3)}$.

POINT DESIGN

14.10. Determine K, a, and b so that the system with the open-loop transfer function

$$GH = \frac{K(s+a)}{(s+b)(s+2)^2(s+4)}$$

has a closed-loop pole at $s_1 = -2 + j3$.

The angle contributed to $\arg GH(s_1)$ by the poles at $s = -2$ and $s = -4$ is $-237°$. This calculation is easily performed with the Spirule. To satisfy the angle criterion, the angle contributions of the zero at $s = -a$ and the pole $s = -b$ must total $-180° - (-237°) = 57°$. Since this is a positive angle, the zero must be farther to the right than the pole ($b > a$). Either a or b may be chosen arbitrarily as long as the remaining one can be fixed in the finite left-half s-plane to give a total contribution of 57°. Let a be set equal to 2, resulting in a 90° phase contribution. Then b must be placed where the contribution of the pole is $-33°$. A line drawn from s_1 at 33° intercepts the real axis at $6.6 = b$, as shown in Fig. 14-23.

The necessary value of K required to satisfy the magnitude criterion at s_1 can now be computed using the chosen values of a and b. From the Spirule or the following calculation, the required value of K is

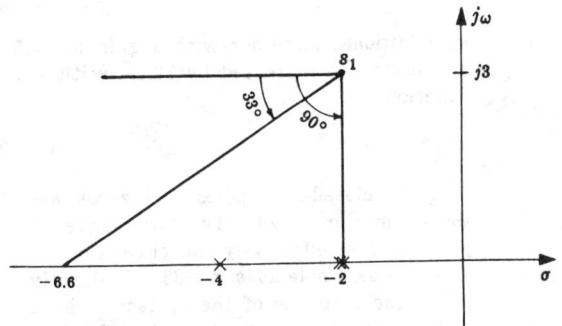

Fig. 14-23

$$\left| \frac{(s_1+6.6)(s_1+2)^2(s_1+4)}{(s_1+2)} \right|_{s_1=-2+j3} = 60$$

14.11. Determine the required compensation for a system with the plant transfer function

$$G_2 = \frac{K}{(s+8)(s+14)(s+20)}$$

to satisfy the following specifications: (a) overshoot $\leq 5\%$, (b) 10 to 90% rise time $T_r \leq 150$ milliseconds, (c) $K_p > 6$.

The first specification may be satisfied with a closed-loop transfer function whose response is dominated by two complex poles with $\zeta \geq 0.7$, as seen from Fig. 3-8, Page 40. A wide variety of dominant pole-zero configurations can satisfy the overshoot specification; but the two pole configuration is usually the simplest obtainable form. We also see from Fig. 3-8 that, if $\zeta = 0.7$, the normalized 10 to 90% rise time is about $\omega_n T_r = 2.2$. Thus in order to satisfy the second specification with $\zeta = 0.7$, we have $T_r = 2.2/\omega_n \leq 0.15$ s or $\omega_n \geq 14.7$ rad/s.

But let us choose $\omega_n = 17$ so as to achieve some margin with respect to the rise time specification. Other closed-loop poles, which may appear in the final design, may slow down the response. Thus in order to satisfy the first two specifications, we shall design the system to have a dominant two pole response with $\zeta = 0.7$ and $\omega_n = 17$. An s-plane evaluation of $\arg G_2(s_1)$, where $s_1 = -12 + j12$ (corresponding to $\zeta = 0.7$, $\omega_n = 17$), yields $\arg G_2(s_1) = -245°$. Then, to satisfy the angle criterion at s_1, we must compensate the system with phase lead so that the total angle becomes $-180°$. Hence we add a cascade lead compensator with $245° - 180° = 65°$ phase lead at s_1. Arbitrarily placing the zero of the lead compensator at $s = -8$ results in $\theta_a = 108°$ (see Fig. 14-2). Then, since we want $\theta_a - \theta_b = 65°$, $\theta_b = 108° - 65° = 43°$. Drawing a line from s_1 to the real axis at the required θ_b determines the pole location at $s = -25$. Addition of the lead compensator with $a = 8$ and $b = 25$ yields an open-loop transfer function

$$G_2 G_{\text{Lead}} = \frac{K}{(s+14)(s+20)(s+25)}$$

The value of K necessary to satisfy the magnitude criterion at s_1 is $K = 3100$ (as determined with the Spirule). The resulting positional error constant for this design is $K_p = 3100/(14)(20)(25) = 0.444$, which is substantially less than the specified value of 6 or more. K_p could be increased slightly by trying other design points (higher ω_n); but the required K_p cannot be achieved without some form of low frequency magnitude compensation. The required increase is $6/0.444 = 13.5$ and may be obtained with a low frequency lag compensator with $b/a = 13.5$. The only other requirement is that a and b for the lag compensator must be small enough so as not to affect the high frequency design accomplished with the lead network. That is,

$$\arg P_{\text{Lag}}(s_1) \cong 0$$

Let $b = 1$ and $a = 0.074$. Then the required compensator is

$$G_{\text{Lag}} = (s+1)/(s+0.074)$$

To synthesize this compensator using a conventional lag network with the transfer function

$$P_{\text{Lag}} = 0.074(s+1)/(s+0.074)$$

an additional amplifier with a gain of 13.5 is required; equivalently, the design value of K chosen above may be increased by 13.5. With either practical mechanization, the total open-loop transfer function is

$$GH = \frac{3100(s+1)}{(s+0.075)(s+14)(s+20)(s+25)}$$

The closed-loop poles and zeros are shown in Fig. 14-24. The low frequency pole and zero effectively cancel each other. The real-axis pole at $s = -35$ will slightly affect the response of the system because $p_r/\zeta\omega_n$ for this pole is only about 3 [Equation (14.2)]. However, reference to Fig. 14-8 and 14-9 verify that the overshoot and rise time are still well within the specifications. If the system had been designed to barely meet the required rise time specification with the dominant two pole approximation, the presence of the additional pole in the closed-loop transfer function may have slowed the response enough to dissatisfy the specification.

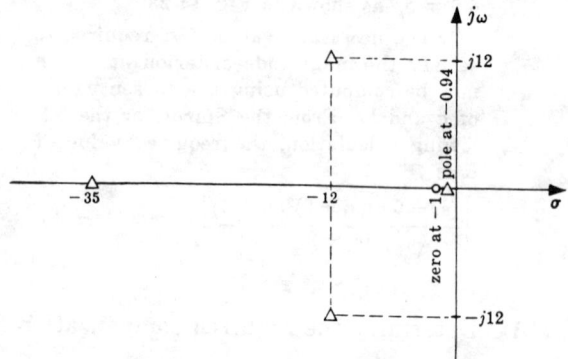

Fig. 14-24

FEEDBACK COMPENSATION

14.12. A positional control system with a tachometer feedback path has the block diagram shown in Fig. 14-25 below. Determine values of K_1 and K_2 which result in a system design which yields a 10 to 90% rise time of less than one second and an overshoot of less than 20%.

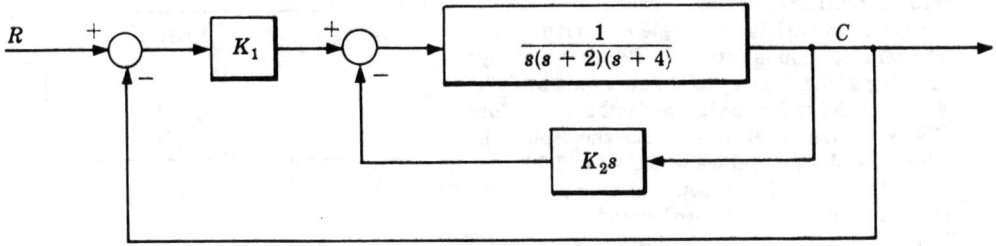

Fig. 14-25

A straightforward way to accomplish this design is to determine a suitable design point in the s-plane and use the point design technique. If the two feedback paths are combined, the block diagram shown in Fig. 14-26 is obtained.

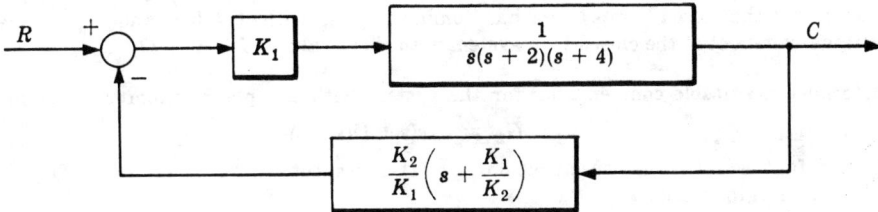

Fig. 14-26

For this configuration

$$GH = \frac{K_2(s + K_1/K_2)}{s(s + 2)(s + 4)}$$

The zero location at $s = -K_1/K_2$ appears in the feedback path and the gain factor is K_2. Thus for a fixed zero location (ratio of K_1/K_2), a root-locus for the system may be constructed as a function of K_2. The closed-loop transfer function will then contain three poles, but no zeros. Rough sketches of the root-locus (Fig. 14-27) reveal that if the ratio K_1/K_2 is set anywhere between 0 and 4, the closed-loop transfer function will probably contain two complex poles (if K_2 is large enough) and a real axis pole near the value of $-K_1/K_2$.

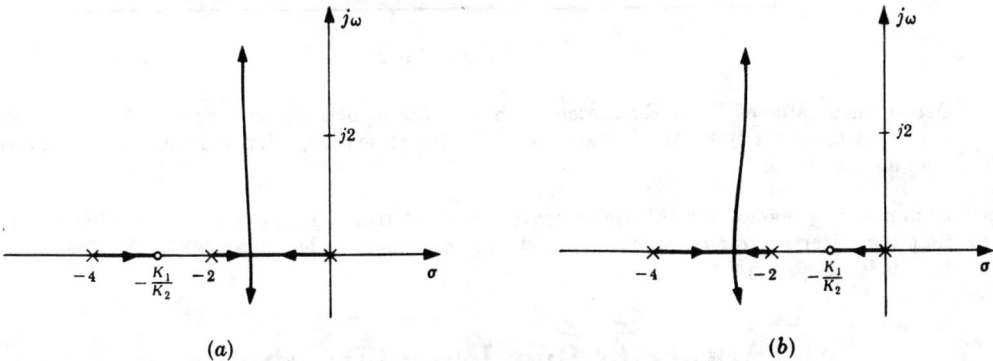

Fig. 14-27

A three pole dominant configuration may then be appropriate for the design. A value of $\zeta = 0.5$ for the complex poles will satisfy the overshoot requirement. For $\zeta = 0.5$ and $p_r/\zeta\omega_n = 2$, Fig. 14-9, Page 264, shows a normalized rise time $\omega_n T_r = 2.3$. Thus $T_r = 2.3/\omega_n < 1$ s or $\omega_n > 2.3$ rad/s. If $p_r/\zeta\omega_n$ turns out to be greater than 2, the rise time will be faster, and vice versa. In order to have a little margin in case $p_r/\zeta\omega_n$ is smaller than 2, let us choose $\omega_n = 2.6$. The design point in the s-plane is therefore $s_1 = -1.3 + j2.3$, corresponding to $\zeta = 0.5$ and $\omega_n = 2.6$.

Using the Spirule on Fig. 14-28, the contribution of the poles at $s = 0$, -2 and -4 to $\arg GH(s_1)$ is $-233°$. The contribution of the zero must therefore be $-180° - (-233°) = 53°$ at s_1 to satisfy the angle criterion at s_1. The zero location is determined at $s = -3$ by drawing a line from s_1 to the real axis at $53°$. With $K_1/K_2 = 3$, the gain factor at s_1 for GH is 7.5 (as determined with the Spirule). Thus the design values are $K_2 = 7.5$ and $K_1 = 22.5$. The closed-loop real-axis pole is to the left of, but near the zero located at $s = -3$. Therefore $p_r/\zeta\omega_n$ for this design is at least $3/1.3 = 2.3$.

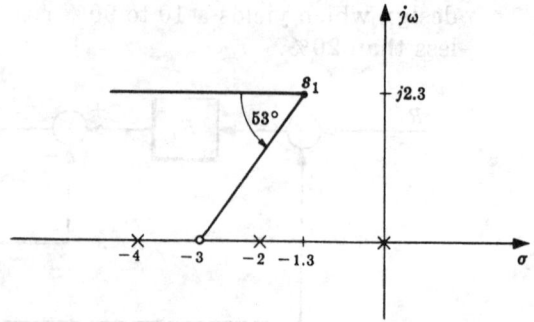

Fig. 14-28

Supplementary Problems

14.13. For the system with the open-loop transfer function $GH = K(s + a)/(s^2 - 1)(s + 5)$ determine K and a such that the closed-loop system has dominant poles with $\zeta = 0.5$ and $\omega_n = 2$. What is the percentage overshoot of the closed-loop system with these values of K and a?

14.14. Determine a suitable compensator for the system with the plant transfer function

$$G_2 = 1/s(s + 1)(s + 4)$$

to satisfy the following specifications: (1) overshoot $< 20\%$,   (2) 10 to 90% rise time $\leqslant 1$ s, (3) gain margin $\geqslant 5$.

14.15. Determine suitable compensation for the system with the plant transfer function $G_2 = 1/s(s + 4)^2$ to satisfy the following specifications: (1) overshoot $< 20\%$,   (2) velocity error constant $K_v \geqslant 10$.

14.16. For the system shown in the block diagram of Fig. 14-29, determine K_1 and K_2 such that the system has closed-loop poles at $s = -2 \pm j2$.

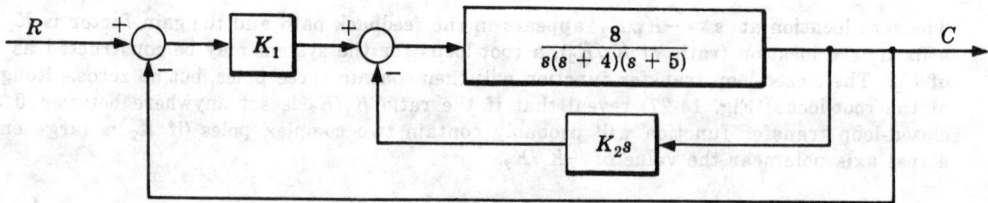

Fig. 14-29

14.17. Determine a value of K for the system with the open-loop transfer function $GH = K/s(s^2 + 6s + 25)$ such that the velocity error constant $K_v > 1$, the closed-loop step response has no overshoot, and the gain margin > 5.

14.18. Design a compensator for the system with the plant transfer function $G_2 = 63/s(s + 7)(s + 9)$ such that the velocity error constant $K_v > 30$, the overshoot is less than 20%, and the 10 to 90% rise time is less than 0.5 s.

Answers to Supplementary Problems

14.13. $K = 11.25$, $a = 1.6$, overshoot $= 38\%$; note that the system has a closed-loop zero at $s = -a = -1.6$.

14.14. $G_1 = 24(s + 1)/(s + 4)$

14.15. $G_1 = 24(s + 0.2)/(s + 0.03)$ **14.17.** $K = 28$

14.16. $K_2 = 1$, $K_1 = 5$ **14.18.** $G_1 = 3(s + 0.5)/(s + 0.05)$

Chapter 15

Bode Analysis

15.1 INTRODUCTION

The analysis of feedback control systems using the Bode method is equivalent to Nyquist analysis in that both techniques employ graphical representations of $GH(j\omega)$. However, **Bode plots** consist of two graphs: the magnitude of $GH(j\omega)$, and the phase angle of $GH(j\omega)$, both plotted as a function of frequency ω. Logarithmic scales are usually used for the frequency axes and for $|GH(j\omega)|$.

Bode plots clearly illustrate the relative stability of a system. In fact, gain and phase margins are often defined in terms of Bode plots (see Example 10.1, page 181). These measures of relative stability can be determined for a particular system with a minimum of computational effort using Bode plots, especially for those cases where experimental frequency response data is available.

15.2 LOGARITHMIC SCALES AND BODE PLOTS

Logarithmic scales are used for Bode plots because they considerably simplify their construction, manipulation and interpretation.

A logarithmic scale is used for the ω axes (abscissas) because the magnitude and phase angle may be graphed over a greater range of frequencies than with linear frequency axes, all frequencies being equally emphasized, and such graphs often result in straight lines (Section 15.4).

The magnitude $|P(j\omega)|$ of any transfer function $P(j\omega)$ for any value of ω is plotted on a logarithmic scale in decibel (db) units, where

$$\text{db} \equiv 20 \log_{10} |P(j\omega)| \qquad (15.1)$$

[Also see Equation (10.4), page 181.]

Example 15.1.

If $|P(j2)| \equiv |GH(j2)| = 10$, the magnitude is $20 \log_{10} 10 = 20$ db.

Since the decibel is a logarithmic unit, the *db magnitude* of a frequency response function composed of a *product* of terms is equal to the *sum* of the db magnitudes of the individual terms. Thus when a logarithmic scale is employed, the magnitude plot of a frequency response function expressible as a product of more than one term can be obtained by adding the individual db magnitude plots for each product term.

The *db magnitude versus log ω* plot is called the **Bode magnitude plot,** and the *phase angle versus log ω* plot is the **Bode phase angle plot.** The Bode magnitude plot is sometimes called the *Log-modulus plot* in the literature.

Example 15.2.

The Bode magnitude plot for $P(j\omega) = \dfrac{100[1 + j(\omega/10)]}{1 + j\omega}$ may be obtained by adding the Bode magnitude plots for: 100, $1 + j(\omega/10)$, and $1/(1 + j\omega)$.

15.3 THE BODE FORM AND THE BODE GAIN

It is convenient to use the so-called *Bode form* of a frequency response function for constructing Bode plots.

The **Bode form** for the function

$$\frac{K(j\omega + z_1)(j\omega + z_2)\cdots(j\omega + z_m)}{(j\omega)^l(j\omega + p_1)(j\omega + p_2)\cdots(j\omega + p_n)}$$

where l is a nonnegative integer, is obtained by factoring out all z_i and p_i and rearranging it in the form

$$\frac{\left[K\prod_{i=1}^{m} z_i \Big/ \prod_{i=1}^{n} p_i\right](1 + j\omega/z_1)(1 + j\omega/z_2)\cdots(1 + j\omega/z_m)}{(j\omega)^l(1 + j\omega/p_1)(1 + j\omega/p_2)\cdots(1 + j\omega/p_n)} \qquad (15.2)$$

The **Bode gain** K_B is defined as the coefficient of the numerator in (15.2):

$$K_B \equiv \frac{K\prod_{i=1}^{m} z_i}{\prod_{i=1}^{n} p_i} \qquad (15.3)$$

15.4 BODE PLOTS OF SIMPLE FREQUENCY RESPONSE FUNCTIONS AND THEIR ASYMPTOTIC APPROXIMATIONS

The constant K_B has a magnitude $|K_B|$, a phase angle of $0°$ if K_B is positive, and $-180°$ if K_B is negative. Therefore the Bode plots for K_B are simply horizontal straight lines as shown in Fig. 15-1 and 15-2.

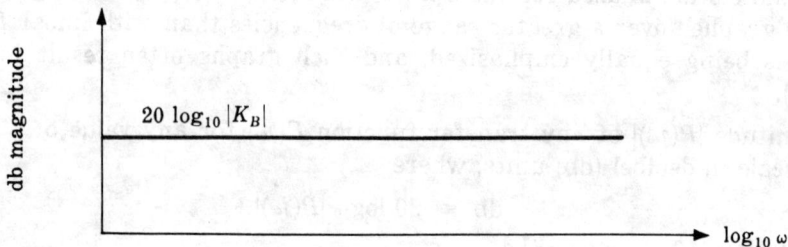

Fig. 15-1

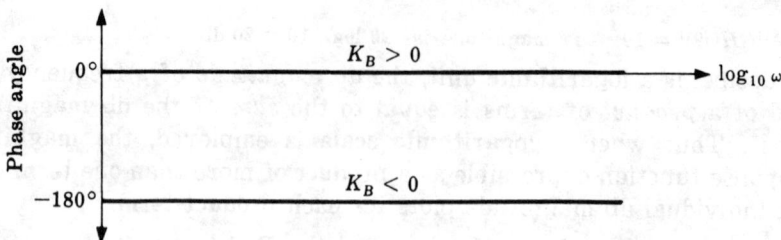

Fig. 15-2

The frequency response function (or sinusoidal transfer function) for a *pole of order l at the origin* is

$$\frac{1}{(j\omega)^l} \qquad (15.4)$$

The Bode plots for this function are straight lines, as shown in Fig. 15-3 and 15-4.

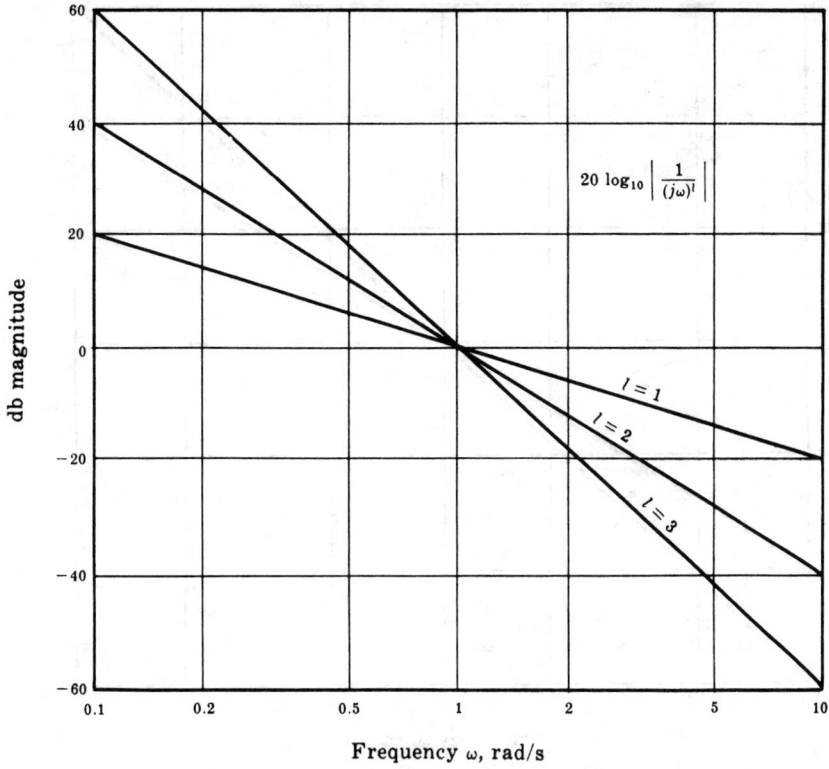

Fig. 15-3

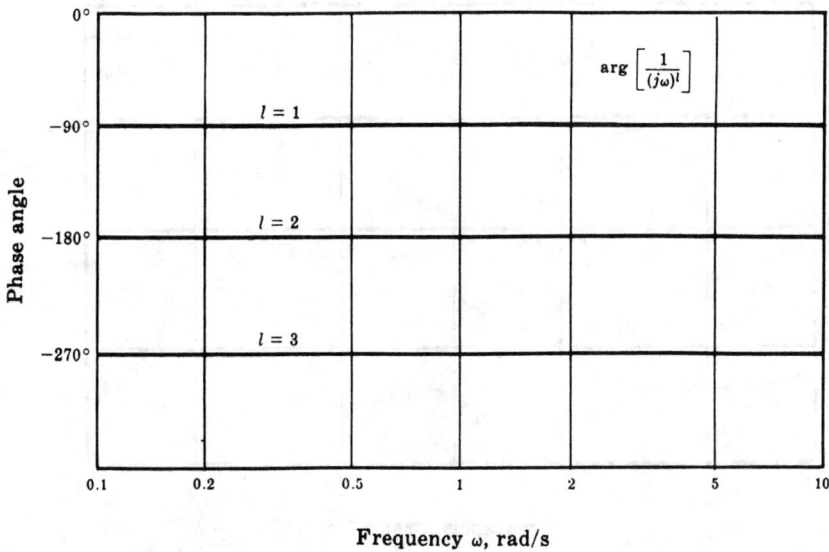

Fig. 15-4

For a *zero of order l at the origin,*

$$(j\omega)^l \qquad\qquad (15.5)$$

the Bode plots are the reflections about the 0 db and 0° lines of Fig. 15-3 and 15-4, as shown in Fig. 15-5 and 15-6.

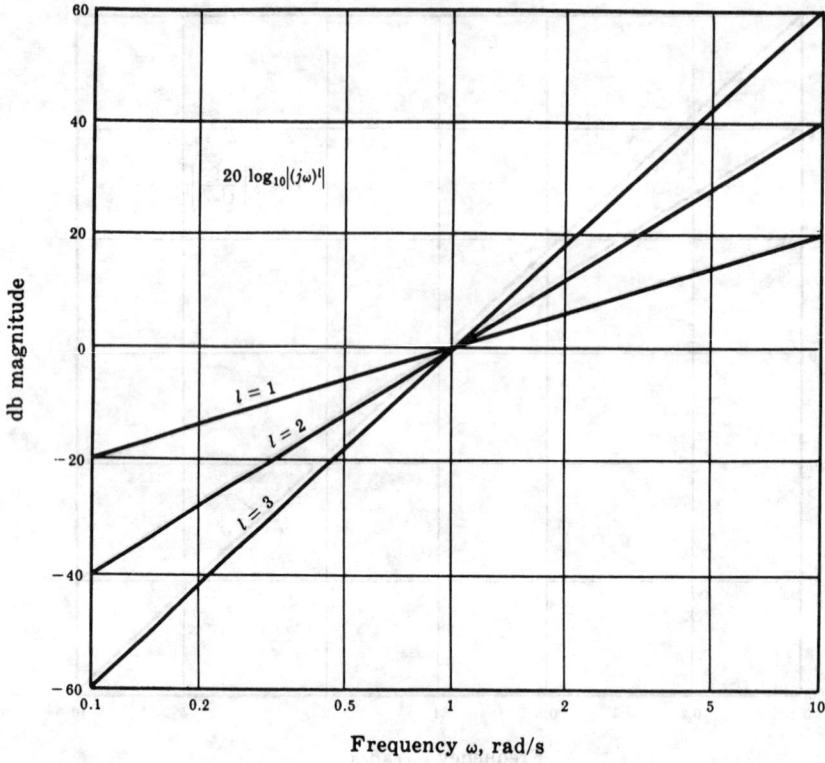

Fig. 15-5

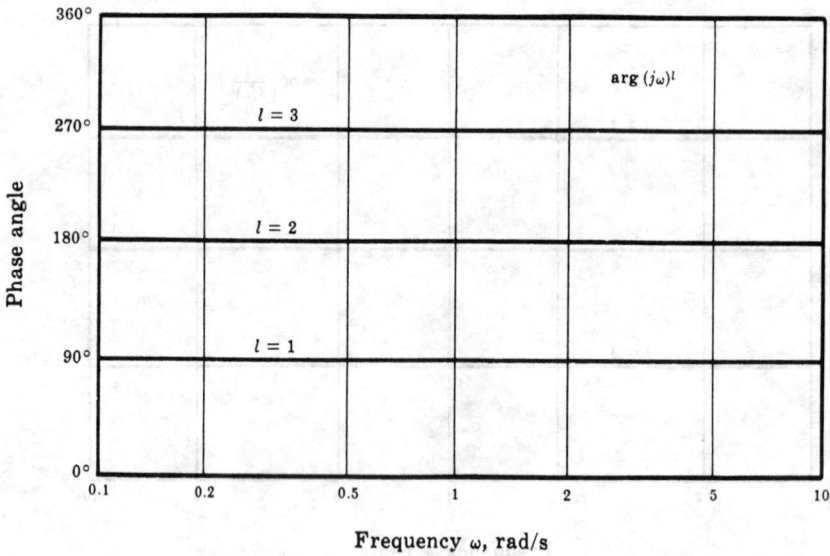

Fig. 15-6

Consider the *single pole* transfer function $\dfrac{p}{s+p}$, $p > 0$. The Bode plots for its frequency response function

$$\frac{1}{1 + j\omega/p} \tag{15.6}$$

are given in Fig. 15-7 and 15-8. Note that the logarithmic frequency scale is normalized in terms of p. To determine the *asymptotic approximations* for these Bode plots, we see that for $\omega/p \ll 1$, or $\omega \ll p$,

$$20 \log_{10} \left| \frac{1}{1 + j\omega/p} \right| \cong 20 \log_{10} 1 = 0 \text{ db}$$

and for $\omega/p \gg 1$, or $\omega \gg p$,

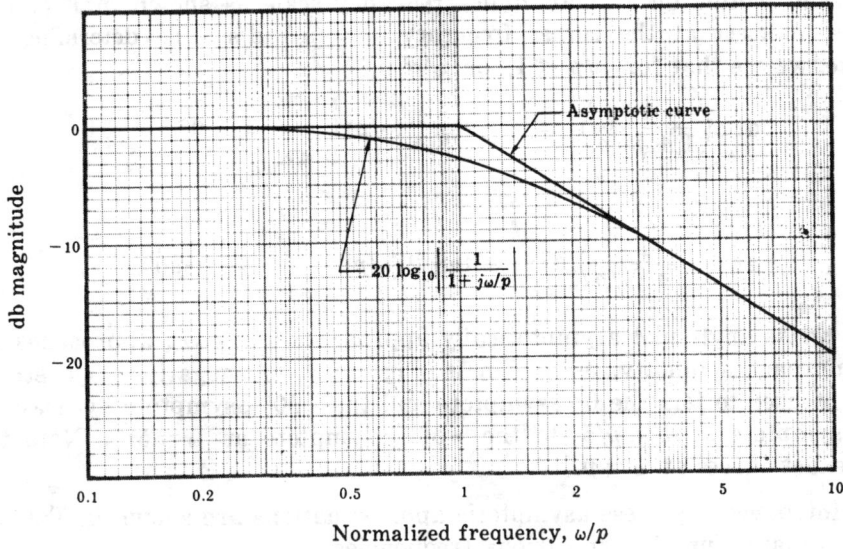

Fig. 15-7

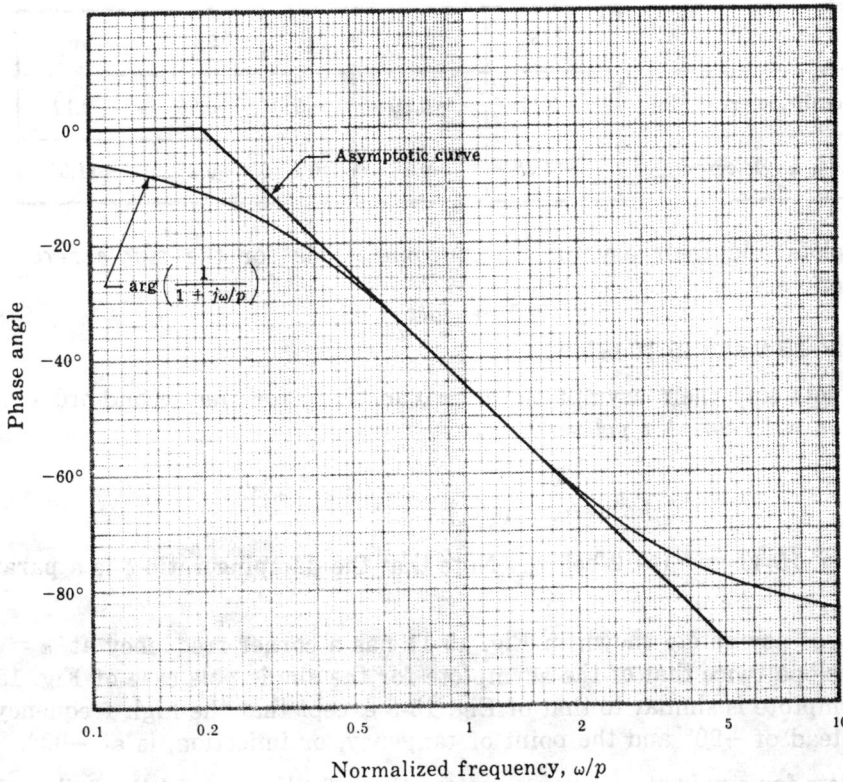

Fig. 15-8

$$20 \log_{10} \left| \frac{1}{1 + j\omega/p} \right| \quad \cong \quad 20 \log_{10} \left| \frac{1}{j\omega/p} \right| \quad = \quad -20 \log_{10}(\omega/p)$$

Therefore the Bode magnitude plot asymptotically approaches a horizontal straight line at 0 db as ω/p approaches zero and $-20 \log_{10}(\omega/p)$ as ω/p approaches infinity (Fig. 15-7). Note that this high frequency asymptote is a straight line with a slope of -20 db/decade, or -6 db/octave when plotted on a logarithmic frequency scale as shown in Fig. 15-7. The two asymptotes intersect at the **corner frequency** $\omega = p$ rad/s. To determine the phase angle asymptote, we see that for $\omega/p \ll 1$, or $\omega \ll p$,

$$\arg\left(\frac{1}{1 + j\omega/p}\right) \quad = \quad -\tan^{-1}(\omega/p)\bigg|_{\omega \ll p} \quad \cong \quad 0°$$

and for $\omega/p \gg 1$, or $\omega \gg p$,

$$\arg\left(\frac{1}{1 + j\omega/p}\right) \quad = \quad -\tan^{-1}(\omega/p)\bigg|_{\omega \gg p} \quad \cong \quad -90°$$

Thus the Bode phase angle plot asymptotically approaches $0°$ as ω/p approaches zero, and $-90°$ as ω/p approaches infinity, as shown in Fig. 15-8. A negative-slope straight line asymptote can be used to join the $0°$ asymptote and the $-90°$ asymptote by drawing a line from the $0°$ asymptote at $\omega = p/5$ to the $-90°$ asymptote at $\omega = 5p$. Note that it is tangent to the exact curves at $\omega = p$.

The *errors* introduced by these asymptotic approximations are shown in Table 15-1 for the single-pole transfer function at various frequencies.

Table 15-1. Asymptotic Errors for $\dfrac{1}{1 + j\omega/p}$

ω	$p/5$	$p/2$	p	$2p$	$5p$
Magnitude error (db)	-0.17	-0.96	-3	-0.96	-0.17
Phase angle error	$-11.3°$	$-0.8°$	$0°$	$+0.8°$	$+11.3°$

The Bode plots and their asymptotic approximations for the *single zero* frequency response function

$$1 + j\omega/z \tag{15.7}$$

are shown in Fig. 15-9 and 15-10 below.

The Bode plots and their asymptotic approximations for the second-order frequency response function with *complex poles*,

$$\frac{1}{1 + j2\zeta\omega/\omega_n - (\omega/\omega_n)^2} \qquad 0 \le \zeta \le 1 \tag{15.8}$$

are shown in Fig. 15-11 and 15-12 below. Note that the damping ratio ζ is a parameter on these graphs.

The magnitude asymptote shown in Fig. 15-11 has a corner frequency at $\omega = \omega_n$ and a high frequency slope twice that of the asymptote for the single pole case of Fig. 15-7. The phase angle asymptote is similar to that of Fig. 15-8 except that the high frequency portion is at $-180°$ instead of $-90°$ and the point of tangency, or inflection, is at $-90°$.

The Bode plots for a pair of *complex zeros* are the reflections about the 0 db and $0°$ lines of those for the complex poles.

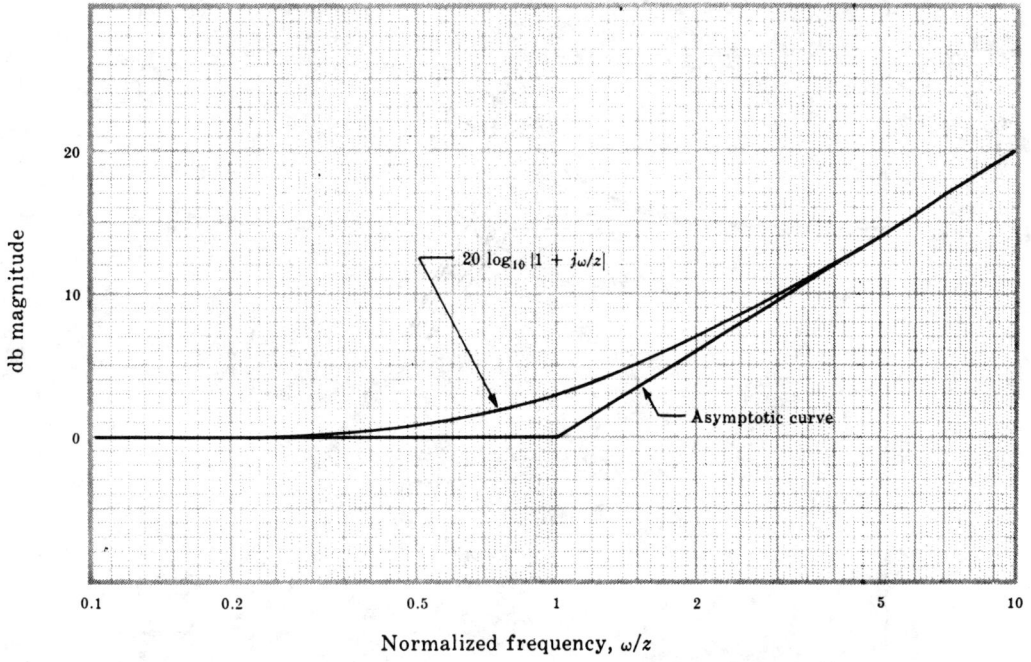

Fig. 15-9

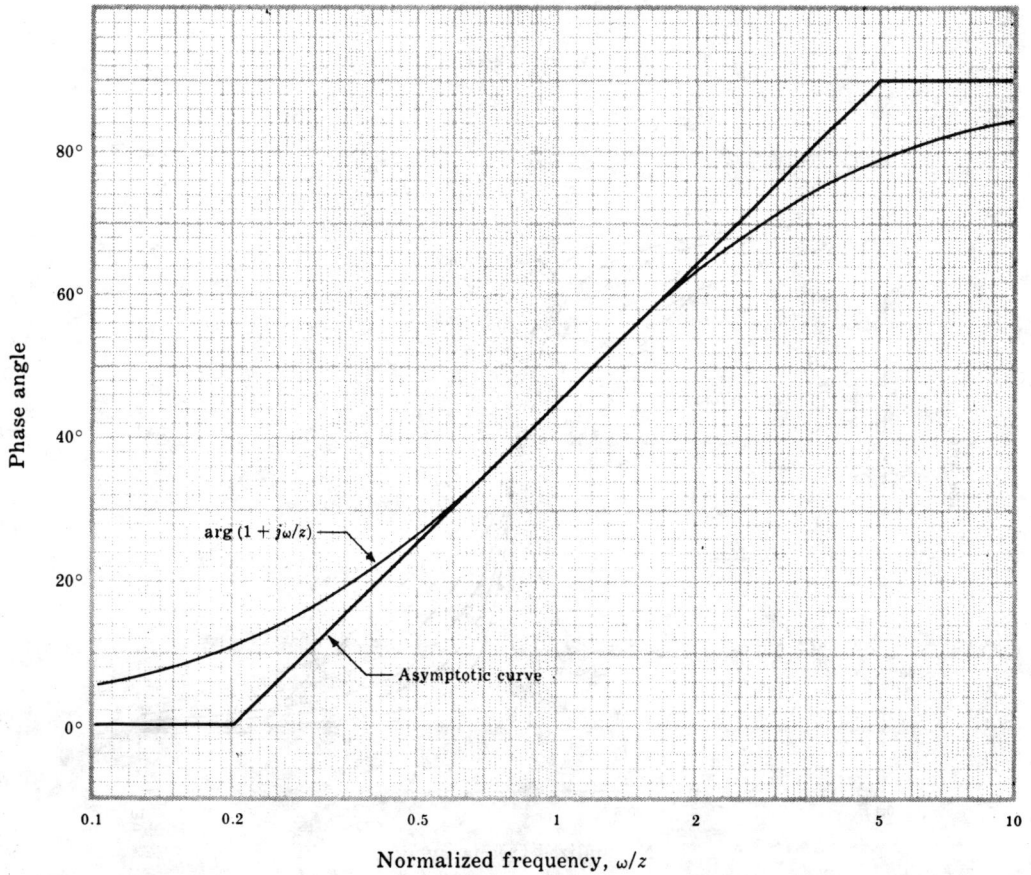

Fig. 15-10

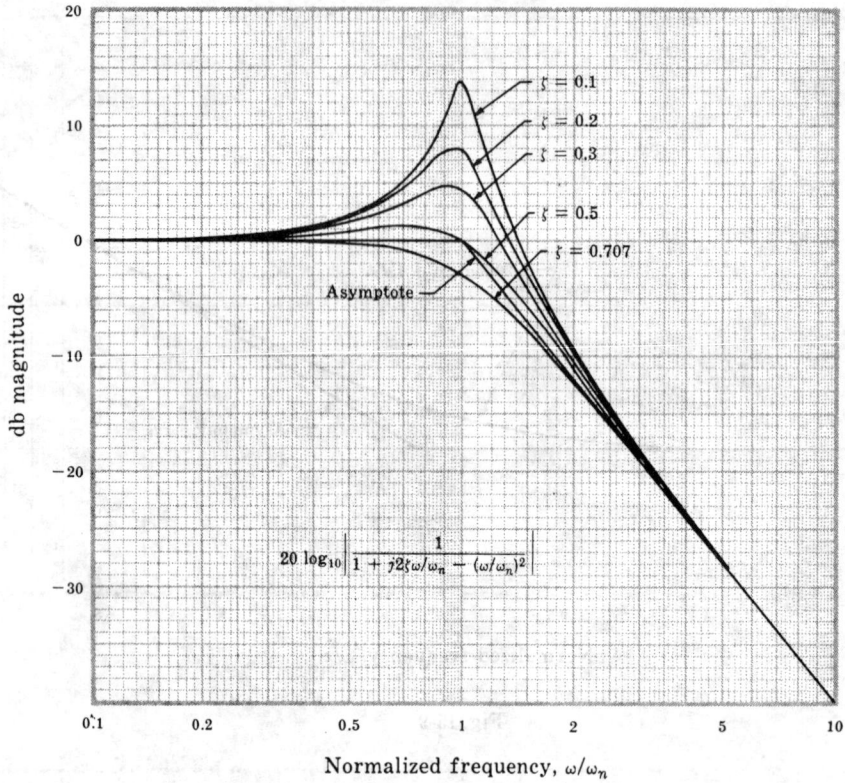

$$20 \log_{10} \left| \frac{1}{1 + j2\zeta\omega/\omega_n - (\omega/\omega_n)^2} \right|$$

Normalized frequency, ω/ω_n

Fig. 15-11

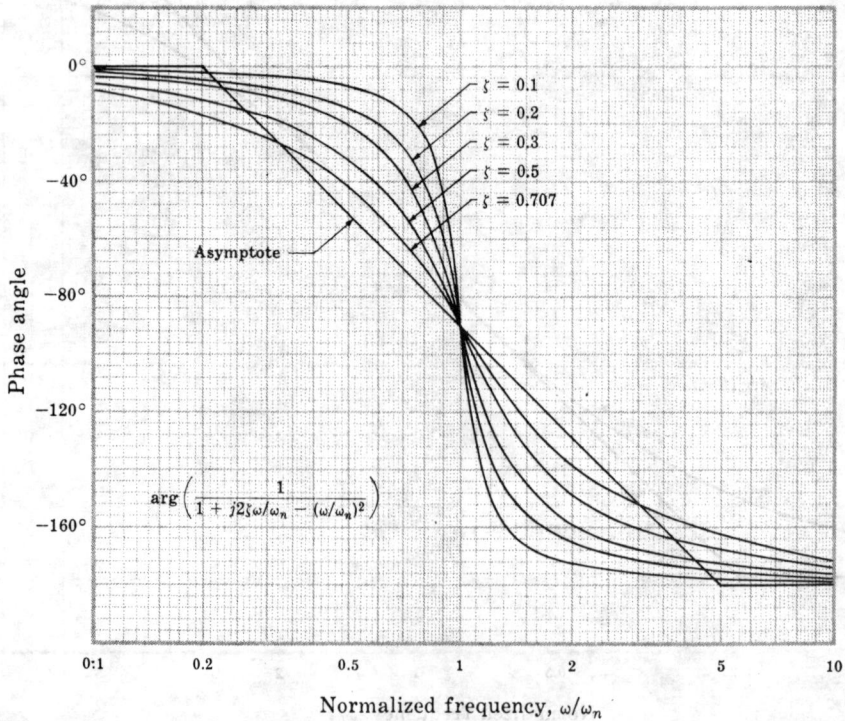

$$\arg\left(\frac{1}{1 + j2\zeta\omega/\omega_n - (\omega/\omega_n)^2} \right)$$

Normalized frequency, ω/ω_n

Fig. 15-12

15.5 CONSTRUCTION OF BODE PLOTS

Bode plots of complex frequency response functions are constructed by summing the magnitude and phase angle contributions of each pole and zero (or pairs of complex poles and zeros). The asymptotic approximations of these plots are often sufficient. If more accurate plots are desired, the errors introduced by the approximations may be determined at selected frequencies and then added to the asymptotic values. Table 15-1 and Fig. 15-1 to 15-12 are helpful in this regard.

For the general open-loop frequency response function

$$GH(j_\varphi) \;=\; \frac{K_B(1 + j\omega/z_1)(1 + j\omega/z_2)\cdots(1 + j\omega/z_m)}{(j\omega)^l(1 + j\omega/p_1)(1 + j\omega/p_2)\cdots(1 + j\omega/p_n)} \qquad (15.9)$$

where l is a positive integer or zero, the magnitude and phase angle are given by

$$20\log_{10}|GH(j\omega)| \;=\; 20\log_{10}|K_B| \;+\; 20\log_{10}|1 + j\omega/z_1| \;+\; \cdots \;+\; 20\log_{10}|1 + j\omega/z_m|$$
$$+\; 20\log_{10}\frac{1}{|(j\omega)^l|} \;+\; 20\log_{10}\frac{1}{|1 + j\omega/p_1|} \;+\; \cdots \;+\; 20\log_{10}\frac{1}{|1 + j\omega/p_n|}$$
$$(15.10)$$

and

$$\arg GH(j\omega) \;=\; \arg K_B \;+\; \arg(1 + j\omega/z_1) \;+\; \cdots \;+\; \arg(1 + j\omega/z_m)$$
$$+\; \arg\left(\frac{1}{(j\omega)^l}\right) \;+\; \arg\left(\frac{1}{1 + j\omega/p_1}\right) \;+\; \cdots \;+\; \arg\left(\frac{1}{1 + j\omega/p_n}\right)$$
$$(15.11)$$

The Bode plots for each of the terms in (15.10) and (15.11) were given in Fig. 15-1 to 15-12. If $GH(j\omega)$ has complex poles or zeros, terms having a form similar to (15.8) are simply added to (15.10) and (15.11). The construction procedure is best illustrated by an example.

Example 15.3.

The asymptotic Bode plots for the frequency response function

$$GH(j\omega) \;=\; \frac{10(1 + j\omega)}{(j\omega)^2[1 + j\omega/4 - (\omega/4)^2]}$$

are constructed using Equations (15.10) and (15.11):

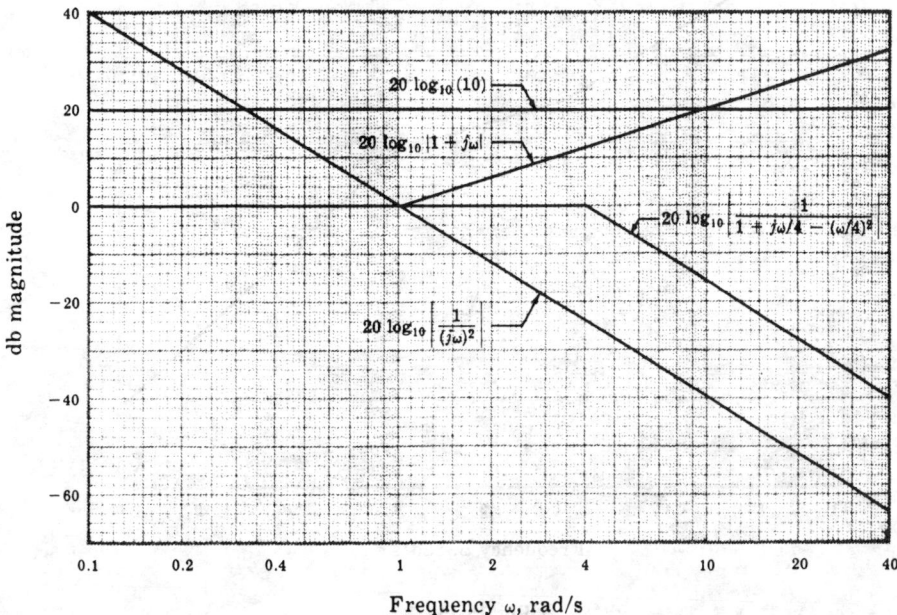

Fig. 15-13

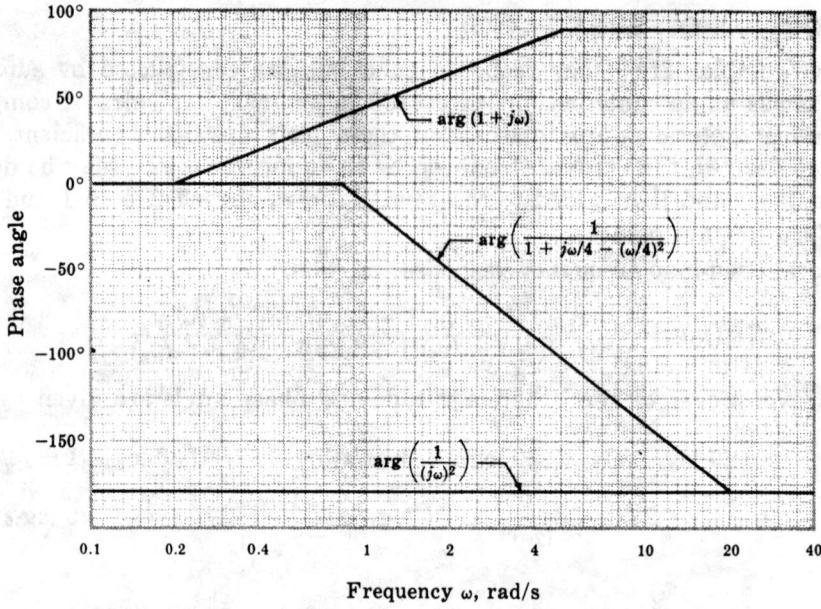

Fig. 15-14

$$20 \log_{10} |GH(j\omega)| \;=\; 20 \log_{10} 10 \;+\; 20 \log_{10} |1+j\omega| \;+\; 20 \log_{10} \left| \frac{1}{(j\omega)^2} \right| \;+\; 20 \log_{10} \left| \frac{1}{1+j\omega/4 - (\omega/4)^2} \right|$$

$$\arg GH(j\omega) \;=\; \arg (1+j\omega) \;+\; \arg (1/j\omega)^2 \;+\; \arg \left(\frac{1}{1+j\omega/4 - (\omega/4)^2} \right)$$

The graphs for each of the terms in these equations are obtained from Fig. 15-1 to 15-12 and are shown in Fig. 15-13 and 15-14 above. The asymptotic Bode plots for $GH(j\omega)$ are obtained by adding these curves, resulting in Fig. 15-15 and 15-16 below.

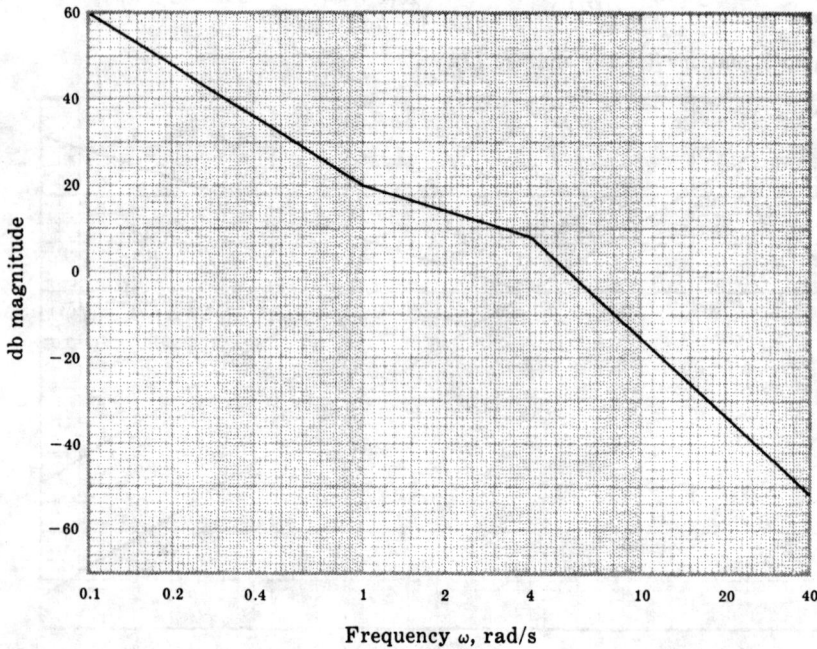

Fig. 15-15

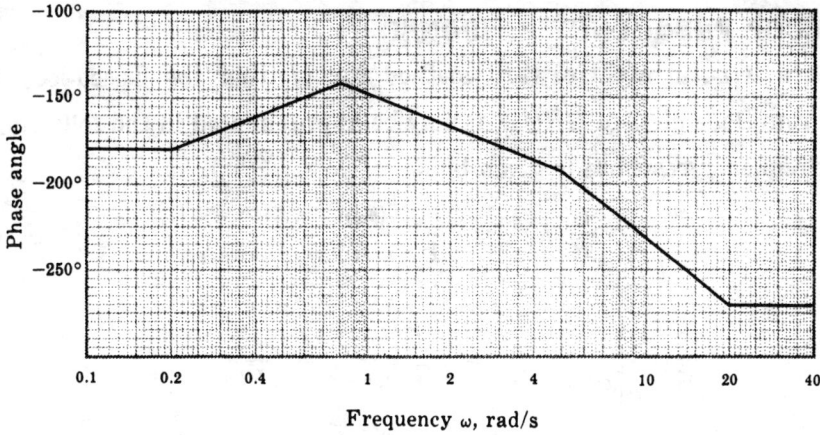

Fig. 15-16

15.6 RELATIVE STABILITY

The relative stability indicators "gain margin" and "phase margin" are defined in terms of a system's open-loop frequency response, as described in Section 10.4. Consequently these parameters are easily determined from the Bode plots of $GH(j\omega)$ as illustrated in Example 10.1, page 181. Since 0 db corresponds to a magnitude of 1, the *gain margin* is the number of db that $|GH(j\omega)|$ is *below* 0 db at the phase crossover frequency ω_π ($\arg GH(j\omega_\pi) = 180°$). The *phase margin* is the number of degrees $\arg GH(j\omega)$ is above $-180°$ at the gain crossover frequency ω_1 ($|GH(j\omega_1)| = 1$). If asymptotic curves are used for the Bode plots, they must be corrected at a sufficient number of points to accurately determine ω_π, ω_1 and the gain and phase margins.

In most cases positive gain and phase margins, as defined above, will ensure stability of the closed-loop system. However, a Nyquist Stability Plot (Chapter 11) may be sketched, or one of the methods of Chapter 5 employed to verify the absolute stability of the system.

Example 15.4.

The system whose Bode plots are shown in Fig. 15-17 has a gain margin of 8 db and a phase margin of 40°.

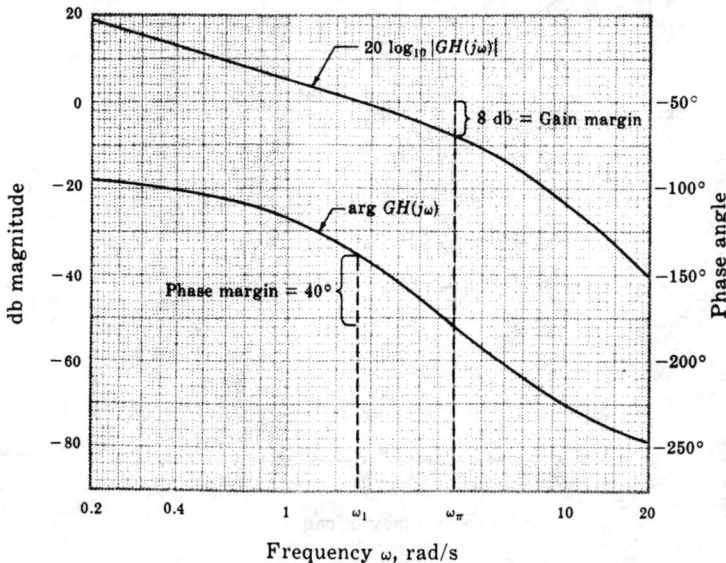

Fig. 15-17

15.7 CLOSED-LOOP FREQUENCY RESPONSE

Although there is no straightforward method for plotting the closed-loop frequency response $\frac{C}{R}(j\omega)$ from Bode plots of $GH(j\omega)$, it may be approximated in the following manner. The closed-loop frequency response is given by

$$\frac{C}{R}(j\omega) = \frac{G(j\omega)}{1 + GH(j\omega)}$$

If $|GH(j\omega)| \gg 1$,

$$\frac{C}{R}(j\omega)\bigg|_{|GH(j\omega)| \gg 1} \cong \frac{G(j\omega)}{GH(j\omega)} = \frac{1}{H(j\omega)}$$

If $|GH(j\omega)| \ll 1$,

$$\frac{C}{R}(j\omega)\bigg|_{|GH(j\omega)| \ll 1} \cong G(j\omega)$$

The open-loop frequency response of most systems is characterized by high gain for low frequencies and decreasing gain for higher frequencies, due to the usual excess of poles over zeros. Thus the closed-loop frequency response *for a unity feedback system* $(H = 1)$ is approximated by a magnitude of 1 (0 db) and phase angle of 0° for frequencies below the gain crossover frequency ω_1. For frequencies above ω_1, the closed-loop frequency response may be approximated by the magnitude and phase angle of $G(j\omega)$. *An approximate closed-loop bandwidth for many systems is the gain crossover frequency ω_1.*

Example 15.5.

The open-loop Bode magnitude plot and approximate closed-loop Bode magnitude plot for the unity feedback system represented by $G(j\omega) = \frac{10}{j\omega(1 + j\omega)}$ are shown in Fig. 15-18.

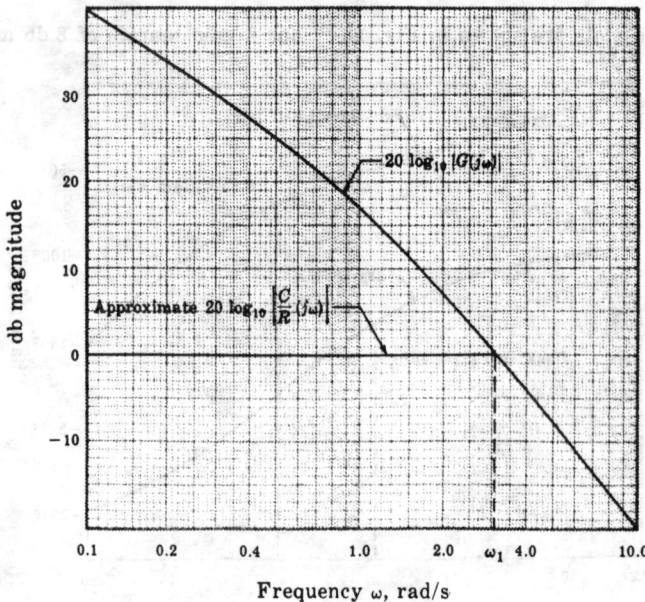

Fig. 15-18

Solved Problems

LOGARITHMIC SCALES

15.1. Express the following quantities in decibel (db) units: (a) 2, (b) 4, (c) 8, (d) 20, (e) 25, (f) 140.

From equation (15.1), page 275,

$$db_a = 20 \log_{10} 2 = 20(0.301) = 6.02 \qquad db_d = 20 \log_{10} 20 = 20(1.301) = 26.02$$
$$db_b = 20 \log_{10} 4 = 20(0.602) = 12.04 \qquad db_e = 20 \log_{10} 25 = 20(1.398) = 27.96$$
$$db_c = 20 \log_{10} 8 = 20(0.903) = 18.06 \qquad db_f = 20 \log_{10} 140 = 20(2.146) = 42.92$$

Note that since $4 = 2 \times 2$, then for part (b) we have

$$20 \log_{10} 4 = 20 \log_{10} 2 + 20 \log_{10} 2 = 12.04$$

and since $8 = 2 \times 4$, then for part (c) we have

$$20 \log_{10} 8 = 20 \log_{10} 2 + 20 \log_{10} 4 = 6.02 + 12.04 = 18.06$$

THE BODE FORM AND THE BODE GAIN

15.2. Determine the Bode form and the Bode gain for the transfer function

$$GH = \frac{K(s+2)}{s^2(s+4)(s+6)}$$

Factoring 2 from the numerator, 4 and 6 from the denominator and putting $s = j\omega$ results in the Bode form

$$GH(j\omega) = \frac{(K/12)(1 + j\omega/2)}{(j\omega)^2(1 + j\omega/4)(1 + j\omega/6)}$$

The Bode gain is $K_B = K/12$.

15.3. When is the Bode gain equal to the d.c. gain (zero frequency magnitude) of a transfer function?

The Bode gain is equal to the d.c. gain of any transfer function with no poles or zeros at the origin ($l = 0$ in equation (15.2), page 276).

BODE PLOTS OF SIMPLE FREQUENCY-RESPONSE FUNCTIONS

15.4. Prove that the Bode magnitude plot for $(j\omega)^l$ is a straight line.

The Bode magnitude plot for $(j\omega)^l$ is a plot of $20 \log_{10} \omega^l$ versus $\log_{10} \omega$. Thus,

$$\text{slope} = \frac{d(20 \log_{10} \omega^l)}{d(\log_{10} \omega)} = \frac{20l \, d(\log_{10} \omega)}{d(\log_{10} \omega)} = 20l$$

Since the slope is constant for any l, the Bode magnitude plot is a straight line.

15.5. Determine: (1) the conditions under which the Bode magnitude plot for a pair of complex poles has a peak at a nonzero, finite value of ω; and (2) the frequency at which the peak occurs.

The Bode magnitude is given by

$$20 \log_{10} \left| \frac{1}{1 + j2\zeta\omega/\omega_n - (\omega/\omega_n)^2} \right|$$

Since the logarithm is a monotonically increasing function, the magnitude in db has a peak (maximum) if and only if the magnitude itself is maximum. The magnitude squared, which is maximum when the magnitude is maximum, is

$$\frac{1}{[1 - (\omega/\omega_n)^2]^2 + 4(\zeta\omega/\omega_n)^2}$$

Taking the derivative of this function and setting it equal to zero yields

$$\frac{(4\omega/\omega_n^2)[1 - (\omega/\omega_n)^2] - 8\zeta^2\omega/\omega_n^2}{\{[1 - (\omega/\omega_n)^2]^2 + 4(\zeta\omega/\omega_n)^2\}^2} = 0$$

or $1 - (\omega/\omega_n)^2 - 2\zeta^2 = 0$

and the frequency at the peak is $\omega = \omega_n\sqrt{1 - 2\zeta^2}$. Since ω must be real, by definition, the magnitude has a peak at a nonzero value ω only if $1 - 2\zeta^2 > 0$ or $\zeta < 1/\sqrt{2} = 0.707$. For $\zeta \geq 0.707$, the Bode magnitude is monotonically decreasing.

CONSTRUCTION OF BODE PLOTS

15.6. Construct the asymptotic Bode plots for the frequency response function

$$GH(j\omega) = \frac{1 + j\omega/2 - (\omega/2)^2}{j\omega(1 + j\omega/0.5)(1 + j\omega/4)}$$

The asymptotic Bode plots are determined by summing the graphs of the asymptotic representations for each of the terms of $GH(j\omega)$, as in equations (15.10) and (15.11), page 283. The asymptotes for each of these terms are shown in Fig. 15-19 and 15-20 and the asymptotic Bode plots for $GH(j\omega)$ in Fig. 15-21 and 15-22.

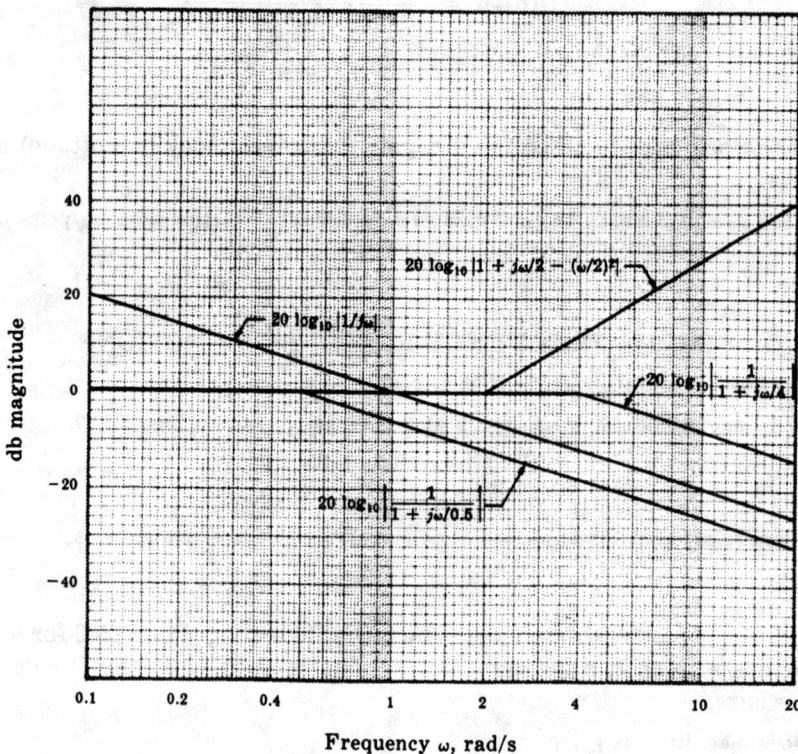

Fig. 15-19

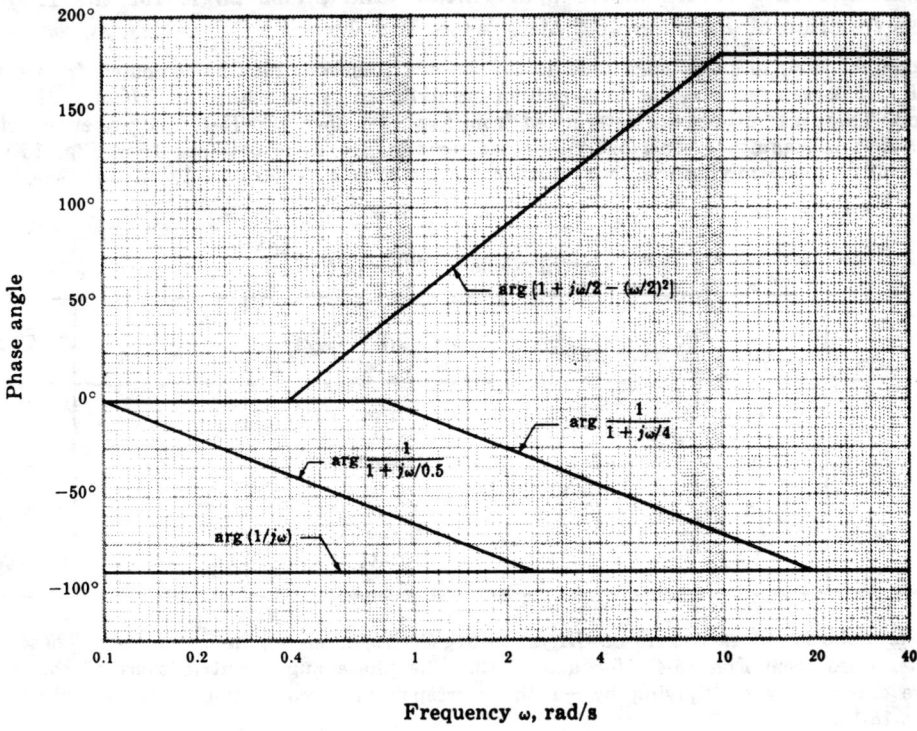

Fig. 15-20

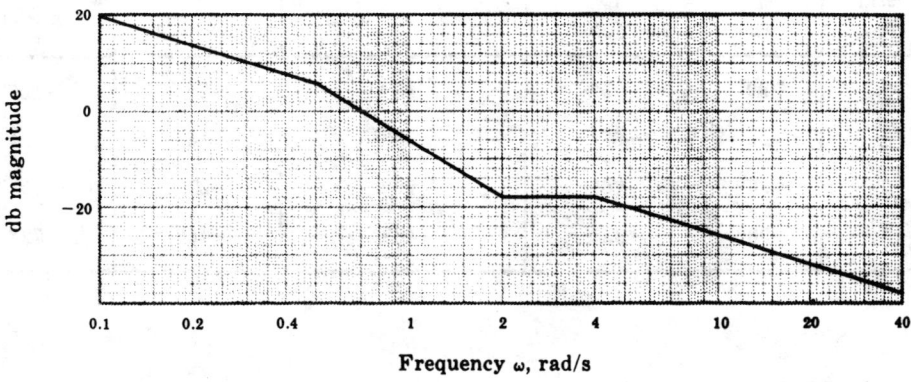

Fig. 15-21

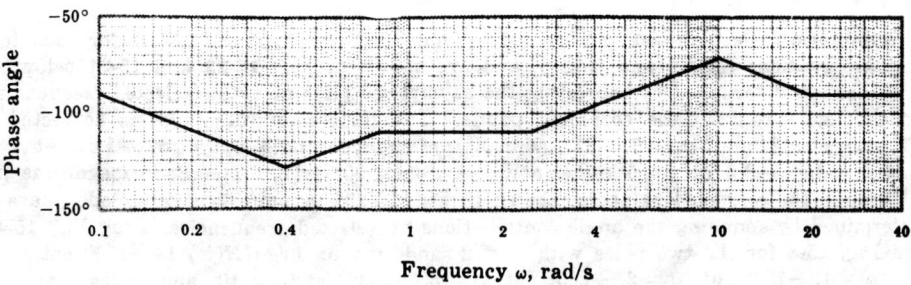

Fig. 15-22

15.7. Determine accurate values of Bode magnitude and phase angle for the frequency response function of Problem 15.6 at $\omega = 0.5, 1, 2, 4$.

The actual value of Bode magnitude at any specific frequency may be computed from equation (15.1). But it is easier to sum the db magnitudes contributed by each term of $GH(j\omega)$ at the desired frequency. These magnitudes are determined from Fig. 15-3, 15-7, and 15-11 and are summarized in Table 15-2. The magnitude contributions of the complex zeros are obtained from Fig. 15-11 (for complex poles) by multiplying the db magnitude for the complex poles by -1. In this case, $\zeta = 0.5$ and $\omega_n = 2$.

Table 15-2.   Contributions to $20 \log_{10} |GH(j\omega)|$, in db

Term ω	$\dfrac{1}{j\omega}$	$\dfrac{1}{1+j\omega/0.5}$	$1 + j\omega/2 - (\omega/2)^2$	$\dfrac{1}{1+j\omega/4}$	Total
0.5	6	−3	0	0	3
1.0	0	−7	−0.8	0	−7.8
2.0	−6	−12	0	−1	−19
4.0	−12	−18	−11	−3	−44

The contribution of each term of $GH(j\omega)$ to arg $GH(j\omega)$ is shown in Table 15-3. These values were determined from Fig. 15-4, 15-8 and 15-12. The phase angle contributions of the complex zeros are obtained by multiplying by -1 the corresponding contributions due to complex poles from Fig. 15-12.

Table 15-3.   Contributions to Arg $GH(j\omega)$

Term ω	$\dfrac{1}{j\omega}$	$\dfrac{1}{1+j\omega/0.5}$	$1 + j\omega/2 - (\omega/2)^2$	$\dfrac{1}{1+j\omega/4}$	arg $GH(j\omega)$
0.5	−90°	−45°	15°	−7°	−127°
1.0	−90°	−63°	33°	−14°	−134°
2.0	−90°	−76°	90°	−27°	−103°
4.0	−90°	−83°	146°	−45°	−72°

15.8. Construct Bode plots for the frequency response function

$$GH(j\omega) = \frac{2}{j\omega(1+j\omega/2)(1+j\omega/5)}$$

The asymptotic Bode plots are constructed by summing the asymptotic plots for each term of $GH(j\omega)$, as in equations (15.10) and (15.11), and are shown in Fig. 15-23 and 15-24 below. More accurate curves are determined by correcting the asymptotic plots at a few selected frequencies and sketching the loci through these corrected points. The Bode magnitude plot is corrected using Fig. 15-7 with $p = 2$ and $p = 5$. The sum of the errors for both poles are -1 db at 1 rad/s, -3 db at 2 rad/s, -3.6 db at 5 rad/s, and -1 db at 10 rad/s. The Bode magnitude plot is sketched through the corrected points as shown in Fig. 15-23. A more accurate Bode phase angle plot is determined by summing the angle contributions at selected frequencies, using Fig. 15-4 with $l = 1$ and Fig. 15-8 for the two poles with $p = 2$ and $p = 5$; arg $GH(j\omega)$ is $-106°$ at $\omega = 0.4$, $-128°$ at $\omega = 1$, $-157°$ at $\omega = 2$, $-203°$ at $\omega = 5$, $-232°$ at $\omega = 10$ and $-255°$ at $\omega = 25$. The asymptotic phase angle plot is modified by sketching a curve through these points, resulting in the Bode phase angle plot shown in Fig. 15-24.

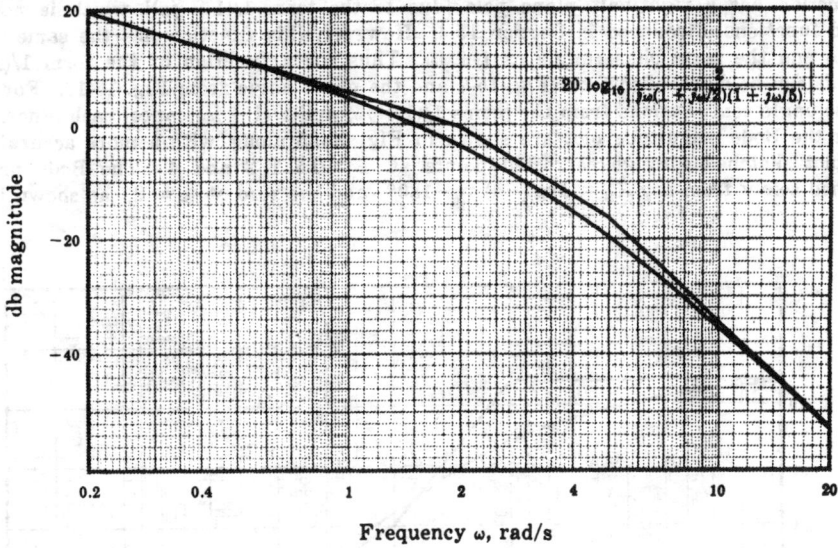

Fig. 15-23

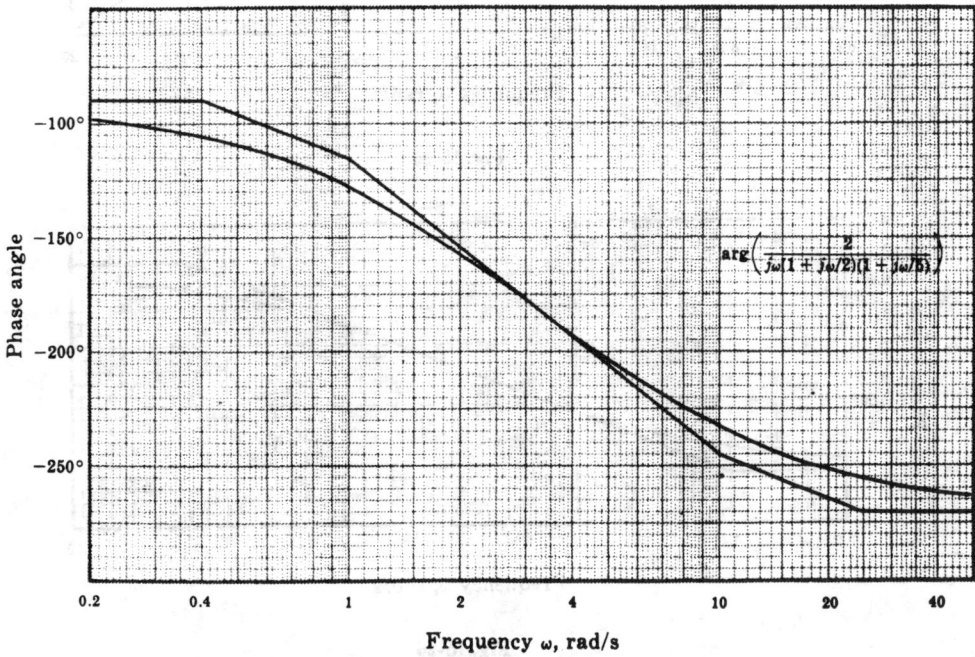

Fig. 15-24

15.9. Construct the Bode plots for the open-loop transfer function $GH = \dfrac{2(s+2)}{(s^2-1)}$.

With $s = j\omega$, the Bode form for this transfer function is

$$GH(j\omega) = \frac{-4(1+j\omega/2)}{(1+j\omega)(1-j\omega)}$$

This function has a right-half plane pole [due to the term $1/(1 - j\omega)$] which is not one of the standard functions introduced in Section 15.4. However, this function has the same magnitude as $1/(1 + j\omega)$ and the same phase angle as $1 + j\omega$. Thus for a function of the form $1/(1 - j\omega/p)$, the magnitude can be determined from Fig. 15-7 and the phase angle from Fig. 15-10. For this problem the phase angle contributions from the terms $1/(1 + j\omega)$ and $1/(1 - j\omega)$ cancel each other. The asymptotes for the Bode magnitude plot are shown in Fig. 15-25 along with a more accurate Bode magnitude plot sketched through corrected points at $\omega = 0.5, 1, 2$ and 4. The Bode phase angle is determined solely from $\arg K_B = \arg(-4) = -180°$ and the zero at $\omega = 2$, as shown in Fig. 15-26.

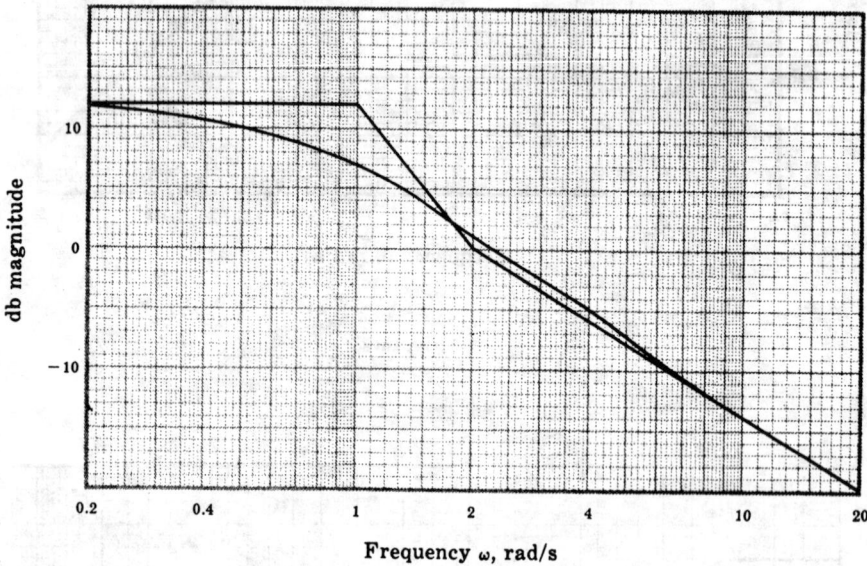

Fig. 15-25

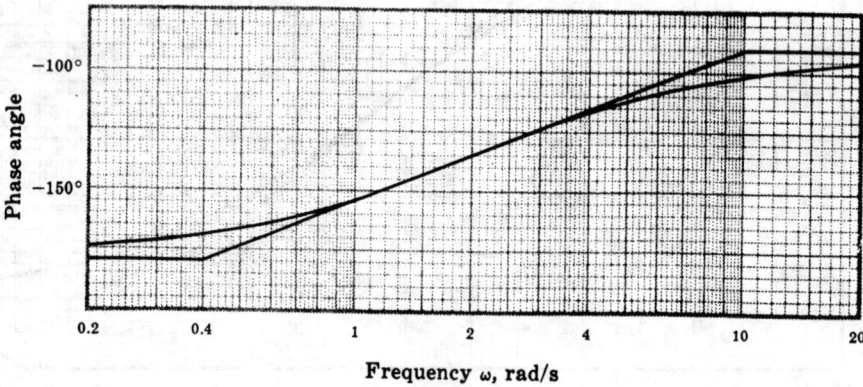

Fig. 15-26

RELATIVE STABILITY

15.10. For the system with the open-loop transfer function of Problem 15.6, find ω_1, ω_π, the gain margin and the phase margin.

Using the actual values for the Bode magnitude determined in Problem 15.7 for $\omega = 0.5$ and 1.0, the actual magnitude curve is sketched as shown in Fig. 15-27 and the gain crossover frequency is $\omega_1 = 0.62$. The phase crossover frequency ω_π is indeterminate because $\arg GH(j\omega)$ never crosses

$-180°$. Using linear extrapolation of the phase angle values at $\omega = 0.5$ and $\omega = 1$ in Table 15-3, $\arg GH(j\omega_1) = \arg GH(j0.62)$ is estimated as $-129°$. Hence the phase margin is $-129° + 180° = 51°$. Since ω_π is indeterminate, the gain margin is also indeterminate.

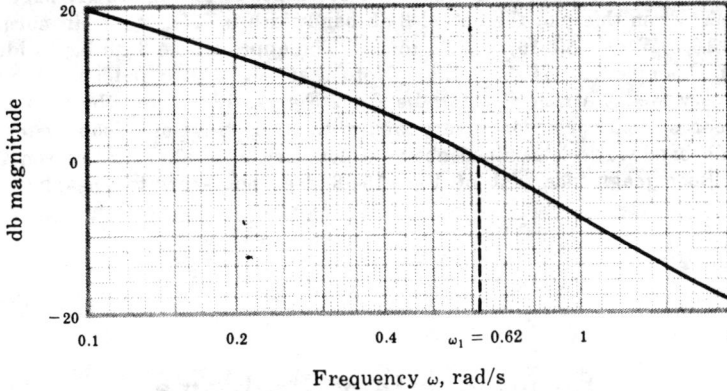

Fig. 15-27

15.11. Determine the gain and phase margins for the system with the open-loop frequency response function of Problem 15.8.

From Fig. 15-23, $\omega_1 = 1.5$; and from Fig. 15-24, $\arg GH(j\omega_1) = -144°$. Therefore the phase margin is $180° - 144° = 36°$. From Fig. 15-24, $\omega_\pi = 3.2$; and the gain margin is read from Fig. 15-23 as $-20 \log_{10} |GH(j\omega_\pi)| = 11$ db.

15.12. Determine the gain and phase margins for the system with the open-loop transfer function of Problem 15.9.

From Fig. 15-25, $\omega_1 = 2.3$ rad/s. From Fig. 15-26, $\arg GH(j\omega_1) = -127°$. Hence the phase margin is $180° - 127° = 53°$. As shown in Fig. 15-26, $\arg GH(j\omega)$ approaches $-180°$ as ω decreases. Since $\arg GH(j\omega) = -180°$ only at $\omega = 0$, then $\omega_\pi = 0$. Therefore the gain margin is $-20 \log_{10} |GH(j\omega_\pi)| = -12$ db using the normal procedure. Although a negative gain margin indicates instability for most systems, this system is stable as verified by sketching the Nyquist Stability Plot as shown in Fig. 15-28. Remember that the system has an open-loop right-half plane pole; but the zero of GH at -2 acts to stabilize the system for $K = 2$.

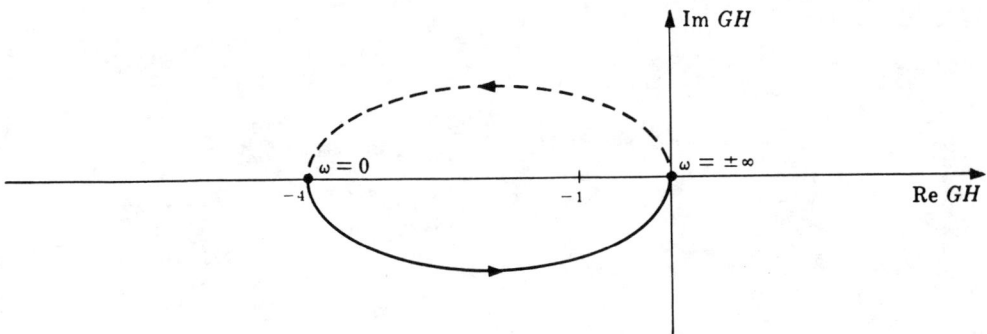

Fig. 15-28

CLOSED-LOOP FREQUENCY RESPONSE

15.13. For the system of Example 15.5 (page 286) with $H = 1$, determine the closed-loop frequency response function and compare the actual closed-loop Bode magnitude plot with the approximate one of Example 15.5.

For this system, $GH = \dfrac{10}{s(s+1)}$. Then $\dfrac{C}{R} = \dfrac{10}{s^2 + s + 10}$ and

$$\frac{C}{R}(j\omega) \;=\; \frac{1}{1 + j\omega/10 - \omega^2/10}$$

Therefore the closed-loop Bode magnitude plot corresponds to Fig. 15-11, page 282, with $\zeta = 0.18$ and $\omega_n = 3.16$. From this plot the actual 3 db bandwidth is $\omega/\omega_n = 1.5$ in normalized form; hence, since $\omega_n = 3.16$, BW $= 1.5(3.16) = 4.74$ rad/s. The approximate 3 db bandwidth determined from Fig. 15-18 of Example 15.5 is 3.7 rad/s. Note that $\omega_n = 3.16$ rad/s for the closed-loop system corresponds very well with $\omega_1 = 3.1$ rad/s from Fig. 15-18. Thus the gain crossover frequency of the open-loop system corresponds very well with ω_n of the closed-loop system, although the approximate 3 db bandwidth determined above is not very accurate. The reason for this is that the approximate Bode magnitude plot of Fig. 15-18 does not show the peaking that occurs in the exact curve.

Supplementary Problems

15.14. Construct the Bode plots for the open-loop frequency response function

$$GH(j\omega) \;=\; \frac{4(1 + j\omega/2)}{(j\omega)^2(1 + j\omega/8)(1 + j\omega/10)}$$

15.15. Construct the Bode plots and determine the gain and phase margins for the system with the open-loop frequency response function

$$GH(j\omega) \;=\; \frac{4}{(1 + j\omega)(1 + j\omega/3)^2}$$

15.16. Solve Problems 13.34 and 13.36, page 257, by constructing the Bode plots.

15.17. Work Problem 13.51, page 259, using Bode plots.

15.18. Work Problem 11.59, page 220, using Bode plots.

Chapter 16

Bode Design

16.1 DESIGN PHILOSOPHY

Design of a feedback control system using Bode techniques entails shaping and reshaping the Bode magnitude and phase angle plots until the system specifications are satisfied. These specifications are most conveniently expressed in terms of frequency-domain figures of merit such as gain and phase margin for the transient performance and the error constants (Chapter 9) for the steady-state time-domain response.

Shaping the asymptotic Bode plots by adding cascade or feedback compensation is a relatively simple procedure. Bode plots for several common compensation networks are presented in Sections 16.3, 16.4, and 16.5. With these graphs, the magnitude and phase angle contributions of a particular compensator can be added directly to the uncompensated system Bode plots.

It is usually necessary to correct the asymptotic Bode plots in the final stages of design to accurately verify satisfaction of the performance specifications.

16.2 GAIN-FACTOR COMPENSATION

It is possible in some cases to satisfy all system specifications by simply adjusting the open-loop gain-factor K. But, when working with Bode plots, it is more convenient to use the Bode gain:

$$ K_B = \frac{K \prod_{i=1}^{m} z_i}{\prod_{i=1}^{n} p_i} $$

where $-p_i$ and $-z_i$ are the finite poles and zeros of GH.

Adjustment of the Bode gain does not affect the phase angle plot. It only shifts the magnitude plot up or down to correspond to the increase or decrease in K_B. The simplest procedure is to alter the db scale of the magnitude plot in accordance with the change in K_B instead of replotting the curve. For example, if K_B is doubled, the db scale should be shifted down by $20 \log_{10} 2 = 6.02$ db.

Example 16.1.

The Bode plots for

$$ GH(j\omega) = \frac{K_B}{j\omega(1 + j\omega/2)} $$

are shown in Fig. 16-1 for $K_B = 1$.

The maximum amount K_B may be increased to improve the system steady-state performance without decreasing the phase margin below 45° is determined as follows. In Fig. 16-1 below, the phase margin is 45° if the gain crossover frequency ω_1 is 2 rad/s and the magnitude plot can be raised by as much as 9 db before ω_1 becomes 2 rad/s. Thus K_B can be increased by 9 db without decreasing the phase margin below 45°.

295

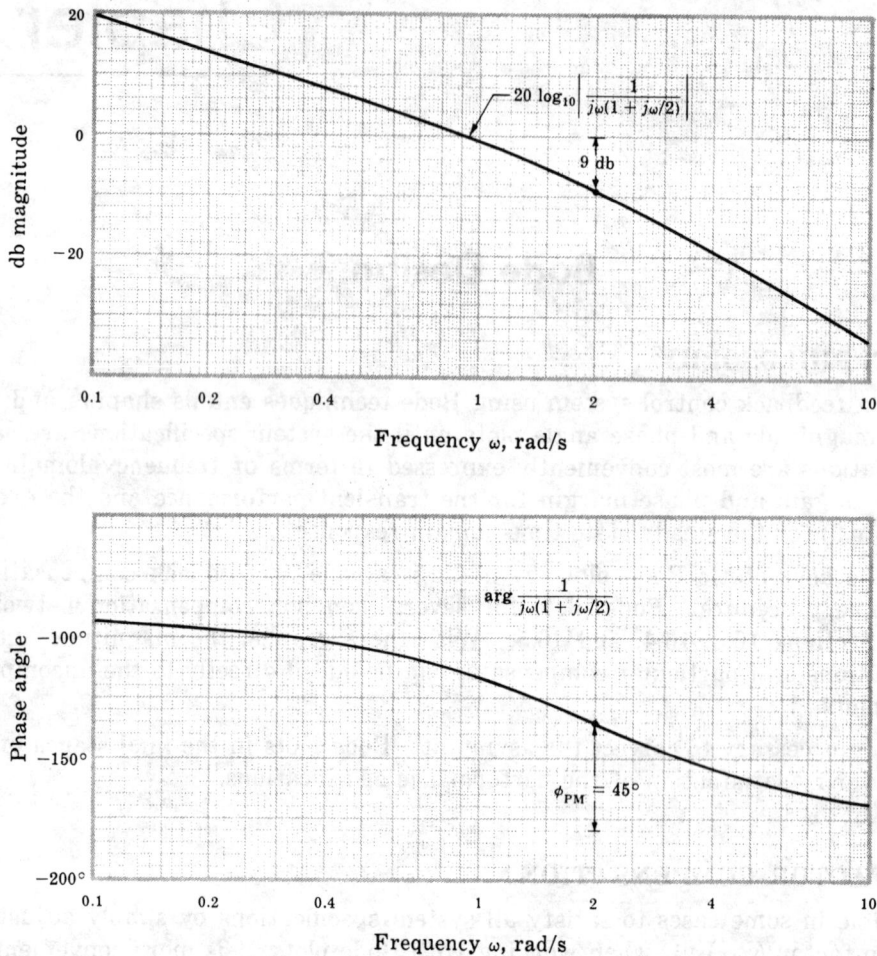

Fig. 16-1

16.3   LEAD COMPENSATION

The lead compensator, presented in Sections 6.3 and 12.4, has the following Bode form frequency response function:

$$P_{\text{Lead}}(j\omega) = \frac{\frac{a}{b}(1 + j\omega/a)}{1 + j\omega/b} \tag{16.1}$$

The Bode plots for this compensator, for various *lead ratios* a/b, are shown in Fig. 16-2 below. These graphs illustrate that addition of a cascade lead compensator to a system lowers the overall magnitude curve in the low frequency region and raises the overall phase angle curve in the low to mid-frequency region. Other properties of the lead compensator are discussed in Section 12.4, page 226.

The amount of low frequency attenuation and phase lead produced by the lead compensator depends on the lead ratio a/b. Maximum phase lead occurs at the frequency $\omega_m = \sqrt{ab}$ and is equal to

$$\phi_{\text{max}} = (90 - 2\tan^{-1}\sqrt{a/b}) \text{ degrees} \tag{16.2}$$

Lead compensation is normally used to increase the gain and/or phase margins of a system or increase its bandwidth. Additional modification of the Bode gain K_B is usually required with lead networks, as described in Section 12.4.

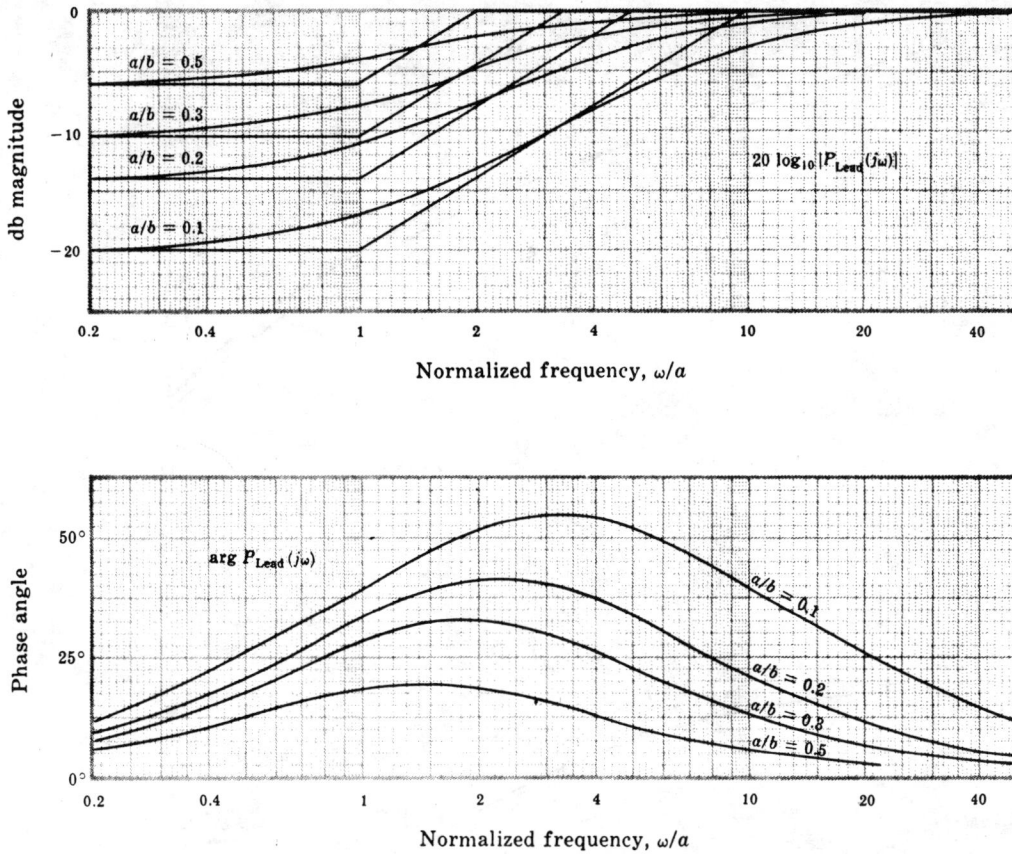

$$20 \log_{10} |P_{\text{Lead}}(j\omega)|$$

Normalized frequency, ω/a

$$\arg P_{\text{Lead}}(j\omega)$$

Normalized frequency, ω/a

Fig. 16-2

Example 16.2.

An uncompensated system whose open-loop transfer function is

$$GH = \frac{24}{s(s+2)(s+6)}, \qquad H = 1$$

is to be designed to meet the following performance specifications:

(1) when the input is a ramp with slope (velocity) equal 2π rad/s, the steady-state error in position must be less than or equal to $\pi/10$ rad.

(2) $\phi_{\text{PM}} = 45° \pm 5°$

(3) gain crossover frequency $\omega_1 \geqq 1$ rad/s.*

Lead compensation is appropriate, as previously outlined in detail in Example 12.3, page 226. Transforming $GH(j\omega)$ into Bode form,

$$GH(j\omega) = \frac{2}{j\omega(1 + j\omega/2)(1 + j\omega/6)}$$

we note that the Bode gain K_B is equal to the velocity error constant $K_{v1} = 2$. The Bode plots for this system are shown in Fig. 16-3 below.

The steady-state error $e(\infty)$ is given by equation (9.5), page 164, as $1/K_v$ for a unit-ramp function input. Therefore if $e(\infty) \leqq \pi/10$ radians and the ramp has a slope of 2π instead of one, then the required velocity error constant is

$$K_{v2} \geqq 2\pi/(\pi/10) = 20 \text{ s}^{-1}$$

---

*When using Bode techniques, closed-loop system *bandwidth* specifications are often interpreted in terms of the gain crossover frequency ω_1, which is easily determined from the Bode magnitude plot. The bandwidth and ω_1 are not generally equivalent; but, when one increases or decreases, the other usually follows. As described in Section 15.7, page 286, ω_1 is often a reasonable approximation for the bandwidth.

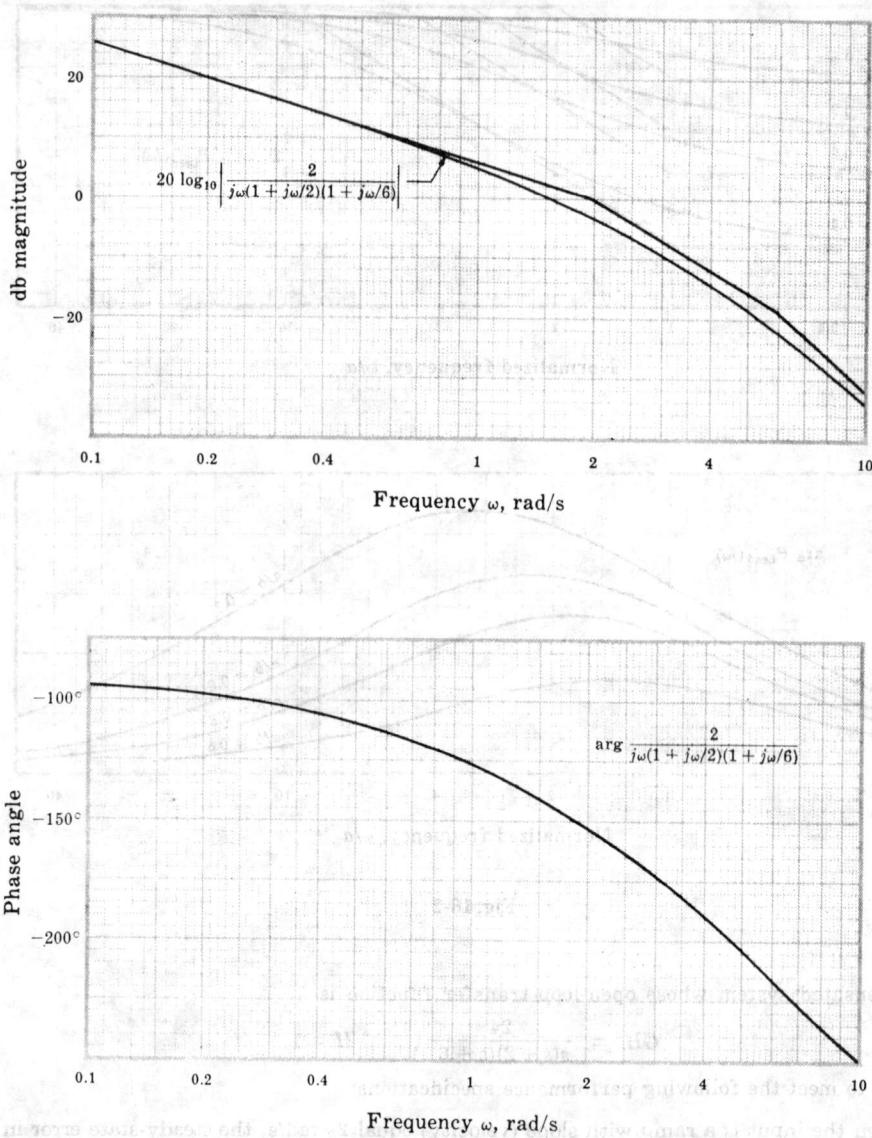

Fig. 16-3

Thus a cascade amplifier with a gain of $\lambda = 10$, or 20 db, will satisfy the steady-state specification. But this gain must be further increased after the lead network parameters are chosen, as described in Example 12.3, page 226. When the Bode gain is increased by 20 db, the gain margin is -8 db and the phase margin $-28°$, as read directly from the plots of Fig. 16-3. Therefore the lead compensator must be chosen to bring the phase margin up to $45°$. This requires a large amount of phase lead. Furthermore, since addition of the lead compensator must be accompanied by an increase in gain of b/a, the net effect is to increase the gain at mid and high frequencies, thus raising the gain crossover frequency. Hence a phase margin of $45°$ has to be established at a higher frequency, requiring even more phase lead. For these reasons we add two cascaded lead networks (with the necessary isolation to reduce loading effects, if required).

To determine the parameters of the lead compensator, assume that the Bode gain has been increased by 20 db so that the 0 db line is effectively lowered by 20 db. If we choose $b/a = 10$, then the lead compensator plus an additional Bode gain increase of $(b/a)^2$ for the two networks has the following combined form:

$$[10\,P_{\text{Lead}}\,(j\omega)]^2 \;=\; G_c(j\omega) \;=\; \frac{(1 + j\omega/a)^2}{(1 + j\omega/10a)^2}$$

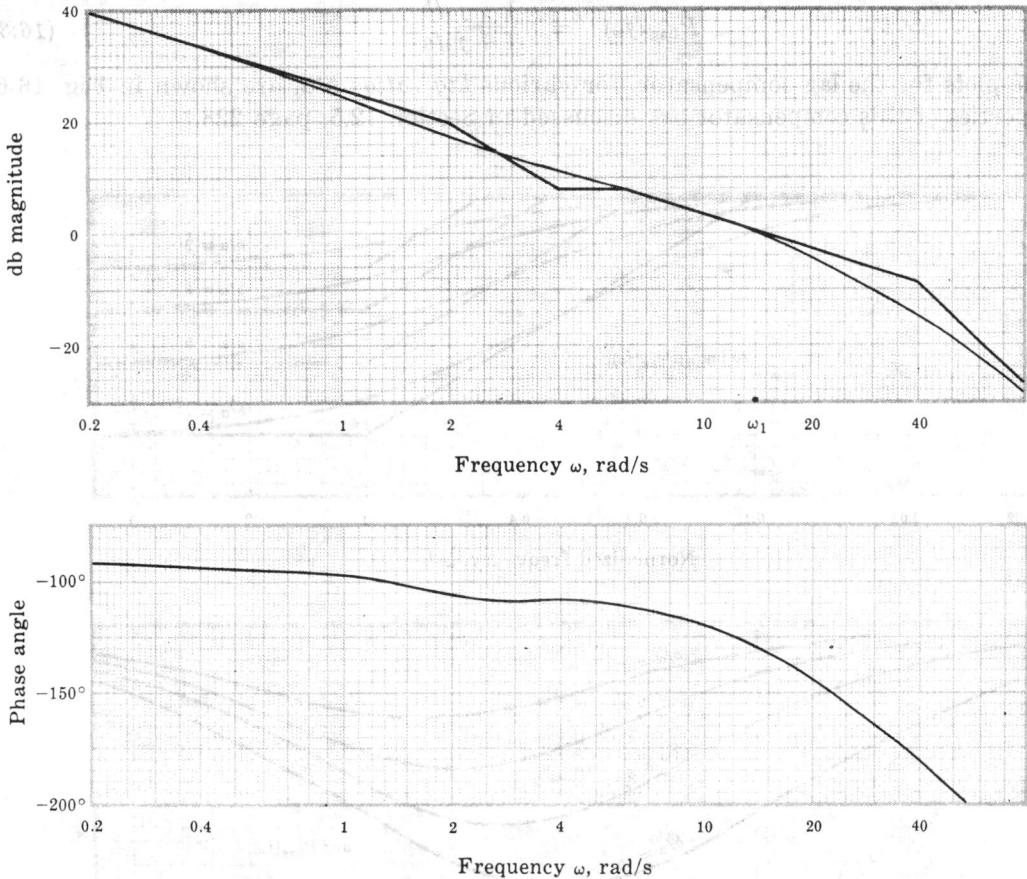

Fig. 16-4

Now we must choose an appropriate value for a. A useful method for improving system stability is to try to cross the 0 db line at a slope of -6 db/octave. Crossing at a slope of -12 db/octave usually results in too low a value for the phase margin. If a is set equal to 2, a sketch of the asymptotes reveals that the 0 db line is crossed at -12 db/octave. If $a = 4$, the 0 db line is crossed at a slope of -6 db/octave. The Bode magnitude and phase angle plots for the system with $a = 4$ rad/s are shown in Fig. 16-4. The gain margin is 14 db and the phase margin is 50°. Thus the second specification is satisfied. The gain crossover frequency $\omega_1 = 14$ rad/s is substantially higher than the value specified, indicating that the system will respond a good deal faster than required by the third specification. The compensated system block diagram is shown in Fig. 16-5. A properly designed amplifier may additionally serve the purpose of load-effect isolation if it is placed *between* the two lead networks.

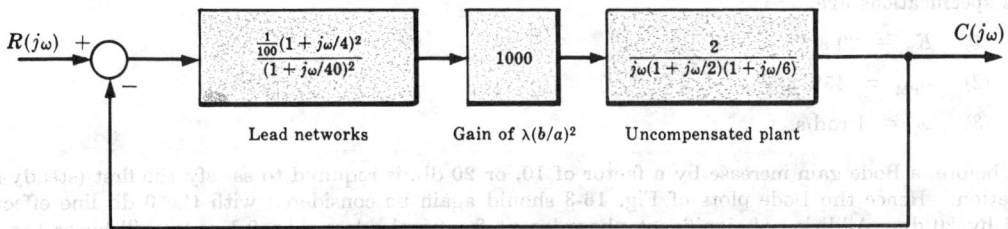

Fig. 16-5

16.4 LAG COMPENSATION

The lag compensator, presented in Sections 6.3 and 12.5, has the following Bode form frequency response function:

$$P_{\text{Lag}}(j\omega) \;=\; \frac{1 + j\omega/b}{1 + j\omega/a} \qquad\qquad (16.3)$$

The Bode plots for the lag compensator, for various *lag ratios b/a*, are shown in Fig. 16-6. The properties of this compensator are discussed in Section 12.5, page 228.

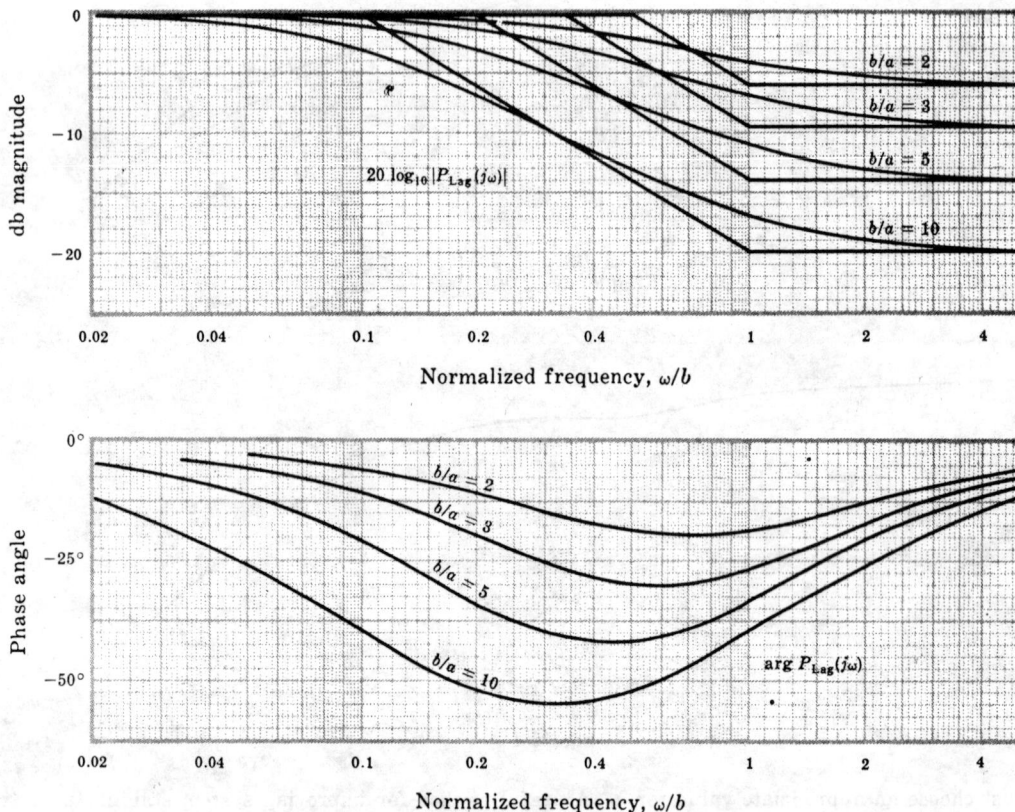

Fig. 16-6

Example 16.3.

Let us redesign the system of Example 16.2 using gain factor plus *lag* compensation, as previously outlined in detail in Example 12.4, page 229. The uncompensated system is, again, represented by

$$GH(j\omega) \;=\; \frac{2}{j\omega(1 + j\omega/2)(1 + j\omega/6)}$$

and the specifications are

    (1) $K_v \;\geq\; 20 \text{ s}^{-1}$

    (2) $\phi_{\text{PM}} \;=\; 45° \pm 5°$

    (3) $\omega_1 \;\geq\; 1 \text{ rad/s}.$

As before, a Bode gain increase by a factor of 10, or 20 db, is required to satisfy the first (steady-state) specification. Hence the Bode plots of Fig. 16-3 should again be considered with the 0 db line effectively lowered by 20 db. Addition of significant phase lag at frequencies less than 0.1 rad/s will lower the curve or effectively raise the 0 db line by an amount corresponding to b/a. Thus the ratio b/a must be chosen so that the resulting phase margin is 45°. From the Bode phase angle plot (Fig. 16-3) we see that a 45° phase margin is obtained if the gain crossover frequency is $\omega_1 = 1.3$ rad/s. From the Bode magnitude plot, this requires that the magnitude curve be lowered by $2 + 20 = 22$ db. Thus a gain decrease of 22 db, or a factor of 14, is needed. This can be obtained using a lag compensator with $b/a = 14$. The actual location of the compensator is arbitrary as long as the phase shift produced at ω_1 is negligible. Values of $a = 0.01$ and $b = 0.14$ rad/s are adequate. The compensated system block diagram is shown in Fig. 16-7 below.

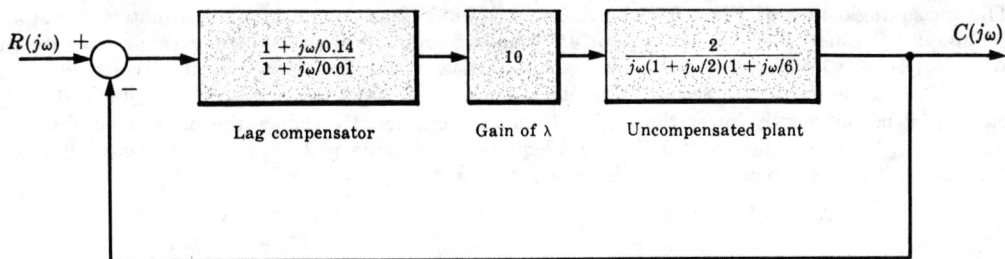

Fig. 16-7

16.5 LAG-LEAD COMPENSATION

It is sometimes desirable, as discussed in Section 12.6, page 230, to simultaneously employ both lead and lag compensation. Although one each of these two networks can be connected in series to achieve the desired effect, it is usually more convenient to mechanize the combined lag-lead compensator described in Example 6.6, page 99. This compensator can be constructed with a single RC network, as shown in Problem 6.14, page 103.

The Bode form of the frequency response function for the lag-lead compensator is

$$P_{LL}(j\omega) \;=\; \frac{(1 + j\omega/a_1)(1 + j\omega/b_2)}{(1 + j\omega/b_1)(1 + j\omega/a_2)}$$

with $b_1 > a_1$, $b_2 > a_2$ and $a_1 b_2 = b_1 a_2$. A typical Bode magnitude plot in which $a_1 > b_2$ is shown in Fig. 16-8. The Bode plots for a specific lag-lead compensator can be determined by combining the Bode plots for the lag portion from Fig. 16-6 with those for the lead portion from Fig. 16-2. Additional properties of the lag-lead compensator are discussed in Section 12.6.

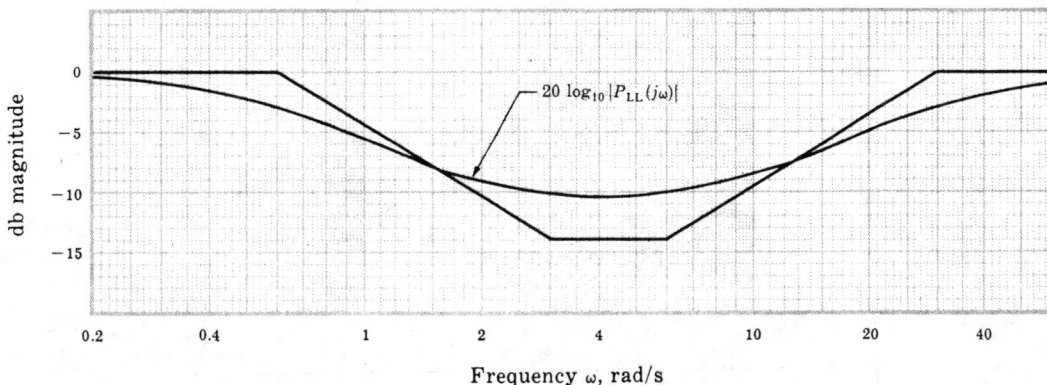

Fig. 16-8

Example 16.4.

Let us redesign the system of Example 16.2 using lag-lead compensation. Suppose, for example, that we want the gain crossover frequency ω_1 (approximate closed-loop bandwidth) to be greater than 2 rad/s but less than 5 rad/s, with all the other specifications the same as Example 16.2. For this application, we shall see that the lag-lead compensator has advantages over either lag or lead compensation. The uncompensated system is, again, represented by

$$GH(j\omega) \;=\; \frac{2}{j\omega(1 + j\omega/2)(1 + j\omega/6)}$$

The Bode plots are shown in Fig. 16-3. As in Example 16.2, a Bode gain increase of 20 db is required to satisfy the specification on steady-state performance. Once again referring to Fig. 16-3 with the 0 db line shifted down by 20 db to correspond to the Bode gain increase, the parameters of the lag-lead compensator must be chosen to result in a gain crossover frequency between 2 and 5 rad/s, with a phase margin of about

$45°$. The phase angle plot of Fig. 16-3 shows about $-188°$ phase angle at approximately 4 rad/s. Thus we need about $53°$ phase lead to establish a $45°$ phase margin in that frequency range. Let us choose a lead ratio of $a_1/b_1 = 0.1$ to make sure we have enough phase lead. To place it in about the right frequency range, let $a_1 = 0.8$ and $b_1 = 8$ rad/s. The lag portion must have the same ratio $a_2/b_2 = 0.1$; but the lag portion must be sufficiently lower than a_1 so as not to significantly reduce the phase lead obtained from the lead portion; $b_2 = 0.2$ and $a_2 = 0.02$ are adequate. The Bode plots for the compensated system are shown in Fig. 16-9; and the block diagram is shown in Fig. 16-10.

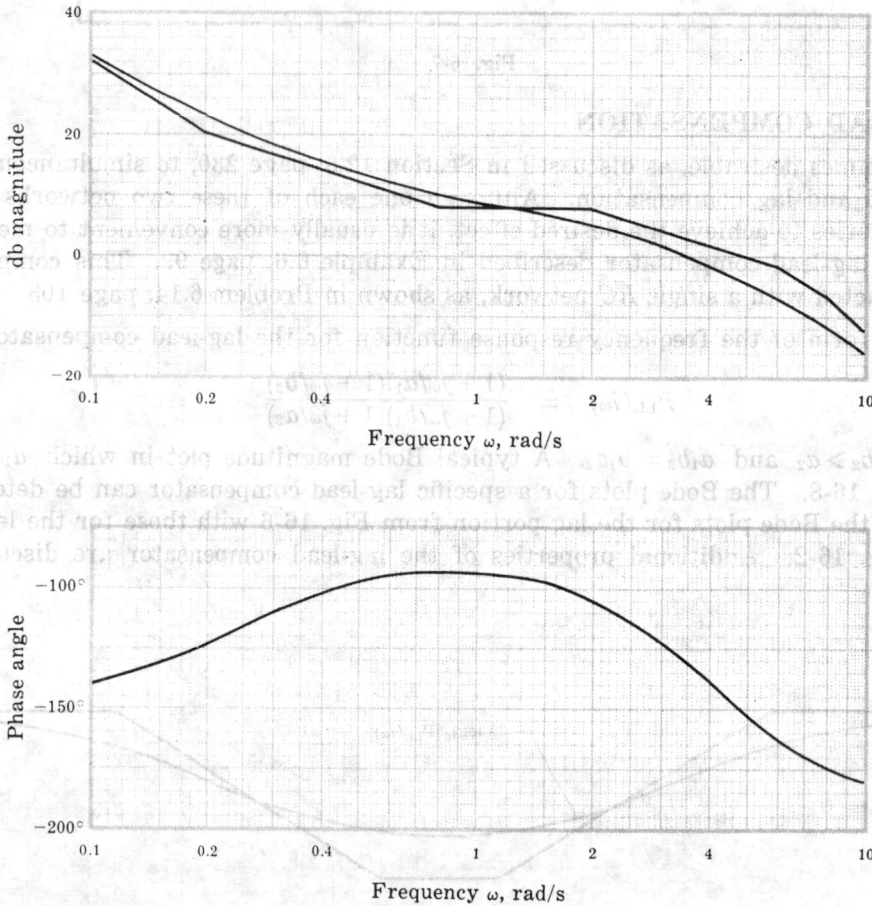

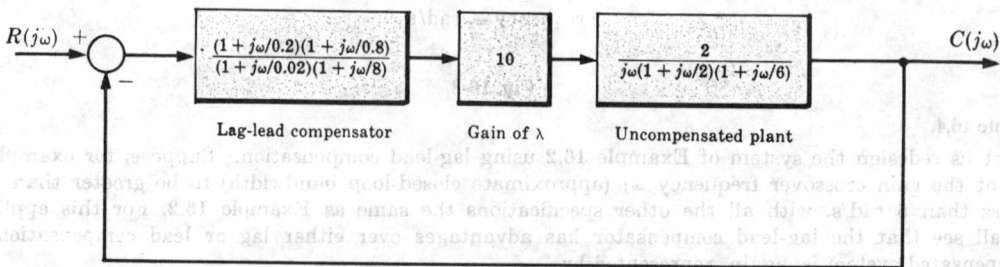

Fig. 16-9

Fig. 16-10

We note that the lag-lead compensator produces no magnitude attenuation at either high or low frequencies. Therefore a smaller gain factor adjustment (as obtained with lag compensation in Example 16.3) and a larger bandwidth and gain crossover frequency (as that resulting from lead compensation in Example 16.2) are obtained using lag-lead compensation.

Solved Problems

GAIN-FACTOR COMPENSATION

16.1. Determine the maximum value for the Bode gain K_B which will result in a gain margin of 6 db or more and a phase margin of 45° or more for the system with the open-loop frequency response function

$$GH(j\omega) \;=\; \frac{K_B}{j\omega(1 + j\omega/5)^2}$$

The Bode plots for this system with $K_B = 1$ are shown in Fig. 16-11.

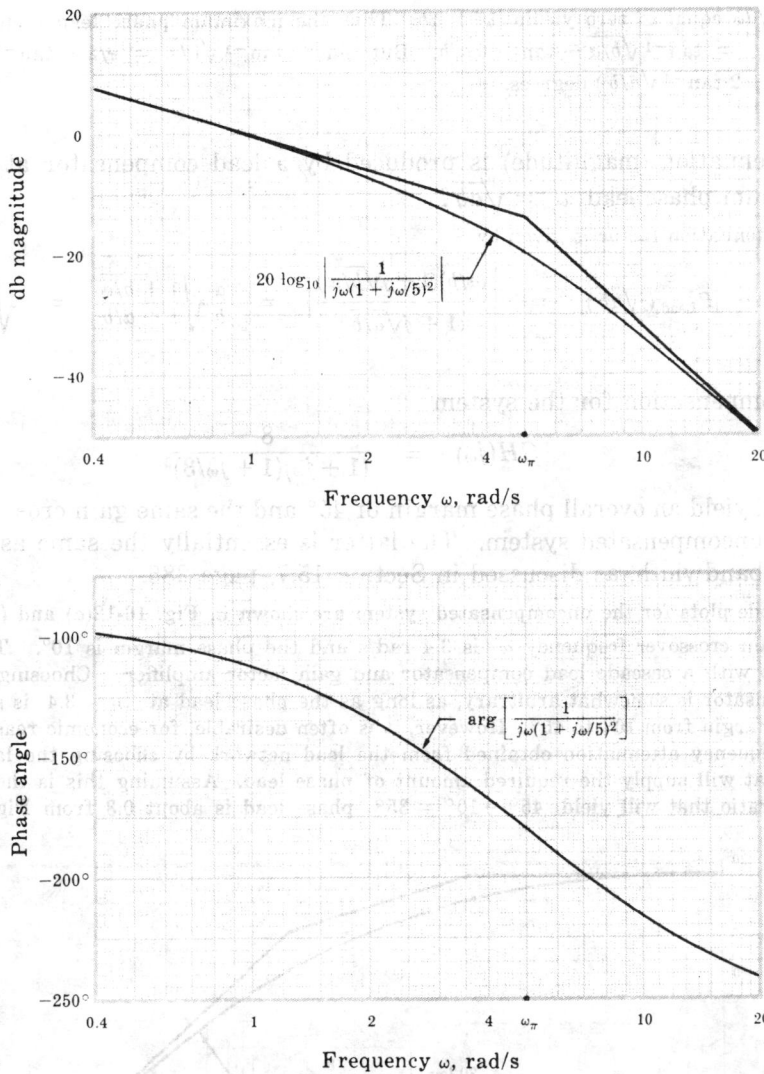

Fig. 16-11

The gain margin, measured at $\omega_\pi = 5$ rad/s, is 20 db. Thus the Bode gain can be raised by as much as $20 - 6 = 14$ db and still satisfy the gain margin requirement. However, the Bode phase angle plot indicates that, for $\phi_{PM} \geq 45°$, the gain crossover frequency ω_1 must be less than about 2 rad/s. The magnitude curve can be raised by as much as 7.5 db before ω_1 exceeds 2 rad/s. Thus the maximum value of K_B satisfying *both* specifications is 7.5 db, or 2.37.

16.2. Design the system of Problem 15.8, page 290, to have à phase margin of 55°.

The Bode phase angle plot in Fig. 15-24 indicates that the gain crossover frequency ω_1 must be 0.9 rad/s for 55° phase margin. From the Bode magnitude plot of Fig. 15-23, K_B must be reduced by 6 db, or a factor of 2, to achieve $\omega_1 = 0.9$ rad/s and hence $\phi_{PM} = 55°$.

LEAD COMPENSATION

16.3. Show that the maximum phase lead of the lead compensator [equation(16.1)] occurs at $\omega_m = \sqrt{ab}$ and prove equation (16.2).

The phase angle of the lead compensator is $\phi = \arg P_{Lead}(j\omega) = \tan^{-1}\omega/a - \tan^{-1}\omega/b$. Then

$$\frac{d\phi}{d\omega} = \frac{1}{a[1 + (\omega/a)^2]} - \frac{1}{b[1 + (\omega/b)^2]}$$

Setting $d\phi/d\omega$ equal to zero yields $\omega^2 = ab$. Thus the maximum phase lead occurs at $\omega_m = \sqrt{ab}$. Hence $\phi_{max} = \tan^{-1}\sqrt{b/a} - \tan^{-1}\sqrt{a/b}$. But since $\tan^{-1}\sqrt{b/a} = \pi/2 - \tan^{-1}\sqrt{a/b}$, we have $\phi_{max} = (90 - 2\tan^{-1}\sqrt{a/b})$ degrees.

16.4. What attenuation (magnitude) is produced by a lead compensator at the frequency of maximum phase lead $\omega_m = \sqrt{ab}$?

The attenuation factor is given by

$$|P_{Lead}(j\sqrt{ab})| = \left|\frac{(a/b)(1 + j\sqrt{b/a})}{(1 + j\sqrt{a/b})}\right| = \frac{a}{b}\sqrt{\frac{1 + b/a}{1 + a/b}} = \sqrt{\frac{a}{b}}$$

16.5. Design compensation for the system

$$GH(j\omega) = \frac{8}{(1 + j\omega)(1 + j\omega/3)^2}$$

which will yield an overall phase margin of 45° and the same gain crossover frequency ω_1 as the uncompensated system. The latter is essentially the same as designing for the same bandwidth, as discussed in Section 15.7, page 286.

The Bode plots for the uncompensated system are shown in Fig. 16-12(a) and (b).

The gain crossover frequency ω_1 is 3.4 rad/s and the phase margin is 10°. The specifications can be met with a cascade lead compensator and gain factor amplifier. Choosing a and b for the lead compensator is somewhat arbitrary, as long as the phase lead at $\omega_1 = 3.4$ is sufficient to raise the phase margin from 10° to 45°. However, it is often desirable, for economic reasons, to minimize the low frequency attenuation obtained from the lead network by choosing the largest lead ratio $a/b < 1$ that will supply the required amount of phase lead. Assuming this is the case, the maximum lead ratio that will yield $45° - 10° = 35°$ phase lead is about 0.3 from Fig. 16-2. Solution

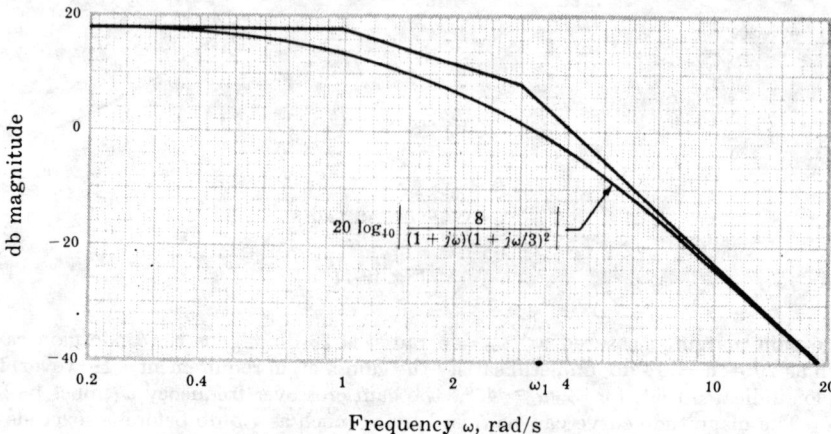

Fig. 16-12(a)

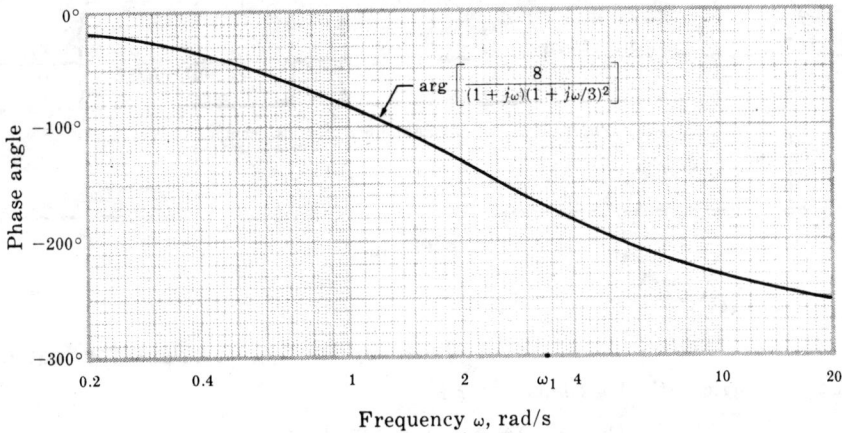

$$\text{arg}\left[\frac{8}{(1+j\omega)(1+j\omega/3)^2}\right]$$

Fig. 16-12(b)

of equation (16.2) yields a value of $a/b = 0.27$. But we shall use $a/b = 0.3$ because we have the curves available for this value in Fig. 16-2. We want to choose a and b such that the maximum phase lead, which occurs at $\omega_m = \sqrt{ab}$, is obtained at $\omega_1 = 3.4$ rad/s. Thus, $\sqrt{ab} = 3.4$. Substituting $a = 0.3b$ into this equation and solving for b, we find $b = 6.2$ and $a = 1.86$. But this compensator produces $20 \log_{10} \sqrt{6.2/1.86} = 5.2$ db attenuation at $\omega_1 = 3.4$ rad/s (see Problem 16.4). Thus an amplifier with a gain of 5.2 db, or 1.82, is required, in addition to the lead compensator, to maintain ω_1 at 3.4 rad/s. The Bode plots for the compensated system are shown in Fig. 16-13 and the block diagram in Fig. 16-14 below.

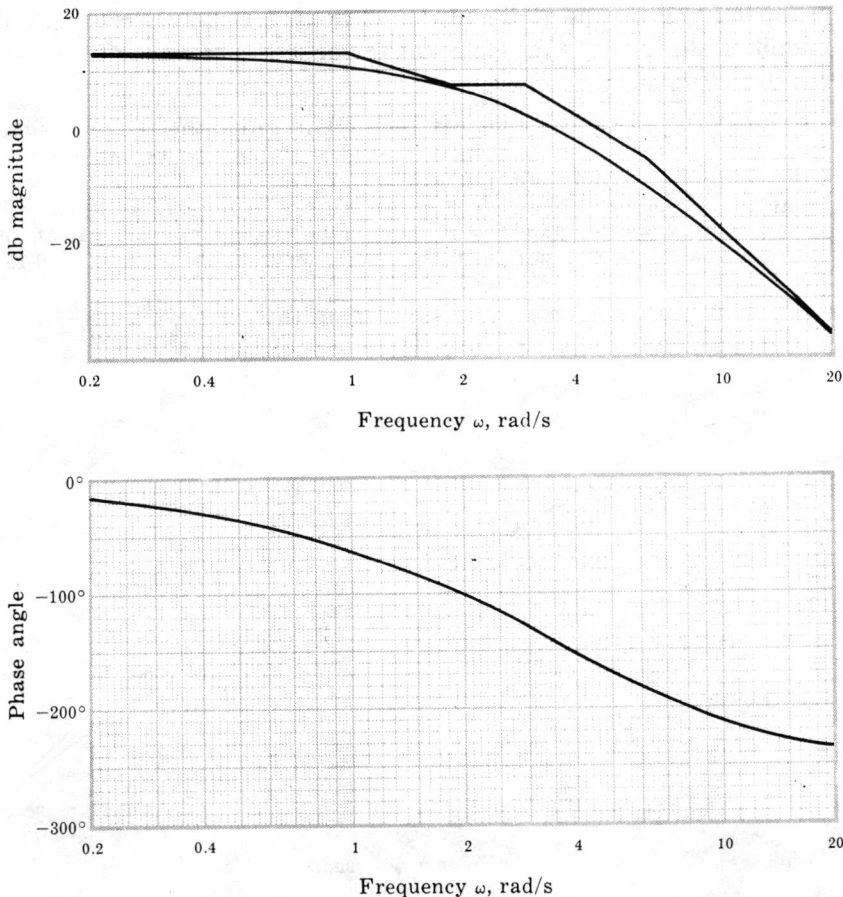

Fig. 16-13

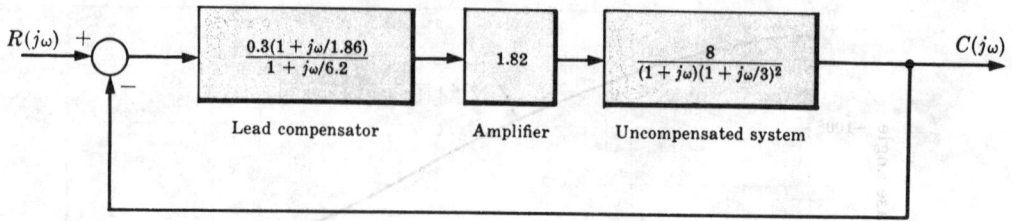

Fig. 16-14

LAG COMPENSATION

16.6. What is the maximum phase lag produced by the lag compensator [equation *(16.3)*]?

The phase angle of the lag compensator is

$$\arg P_{\text{Lag}}(j\omega) = \tan^{-1}\omega/b - \tan^{-1}\omega/a = -\arg P_{\text{Lead}}(j\omega)$$

Thus the maximum phase lag (negative phase angle) of the lag compensator is the same as the maximum phase lead of the lead compensator with the same values of a and b. Hence the maximum also occurs at $\omega_m = \sqrt{ab}$ and, from equation *(16.2)*, we get

$$\phi_{\text{max}} = (90 - 2\tan^{-1}\sqrt{a/b}) \text{ degrees}$$

Expressed in terms of the lag ratio b/a, this equation becomes

$$\phi_{\text{max}} = (2\tan^{-1}\sqrt{b/a} - 90) \text{ degrees}$$

16.7. Design compensation for the system of Problem 16.1 to satisfy the same specifications and, in addition, to have a gain crossover frequency ω_1 less than or equal to 1.0 rad/s and a velocity error constant $K_v > 5$.

The Bode plots for this sytem, shown in Fig. 16-11, indicate that $\omega_1 = 1$ rad/s for $K_B = 1$. Hence $K_v = K_B = 1$ for $\omega_1 = 1$. The gain and phase margin requirements are easily met with any $K_B < 2.37$; but the steady-state specification requires $K_v = K_B > 5$. Therefore a low frequency cascade lag compensator with $b/a = 5$ can be used to increase K_v to 5, while maintaining the crossover frequency and the gain and phase margins at their previous values. A lag compensator with $b = 0.5$ and $a = 0.1$ satisfies these requirements, as shown in Fig. 16-15(a) and (b).

The compensated open-loop frequency response function is $\dfrac{5(1 + j\omega/0.5)}{j\omega(1 + j\omega/0.1)(1 + j\omega/5)^2}$.

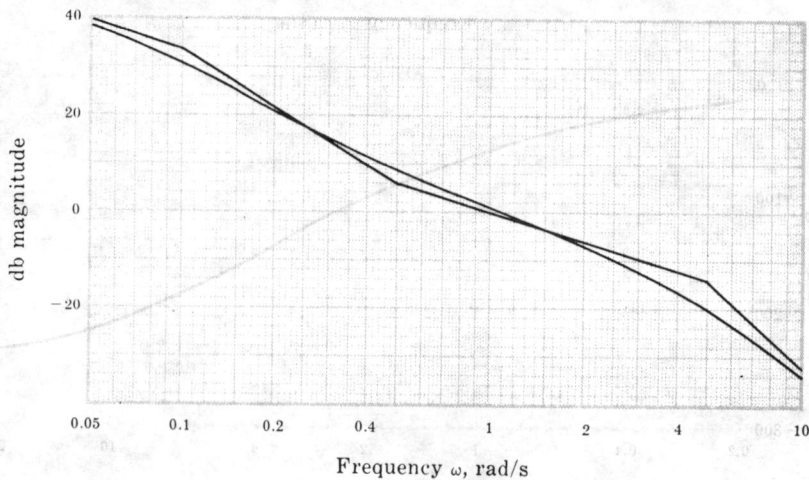

Fig. 16-15(a)

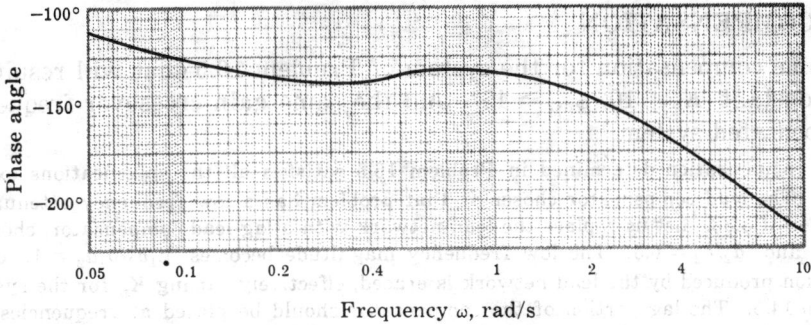

Fig. 16-15(b)

16.8. Design a unity-feedback system, with the fixed plant

$$G_2(j\omega) = \frac{1}{(1 + j\omega/3)^3}$$

satisfying the specifications: (1) $K_p \geq 4$, (2) gain margin ≥ 12 db, (3) phase margin $\geq 45°$.

The specification on the position error constant K_p requires a Bode gain increase by a factor of 4. The Bode plots for this system, with the gain increased by $20 \log_{10} 4 = 12$ db, are shown in Fig. 16-16.

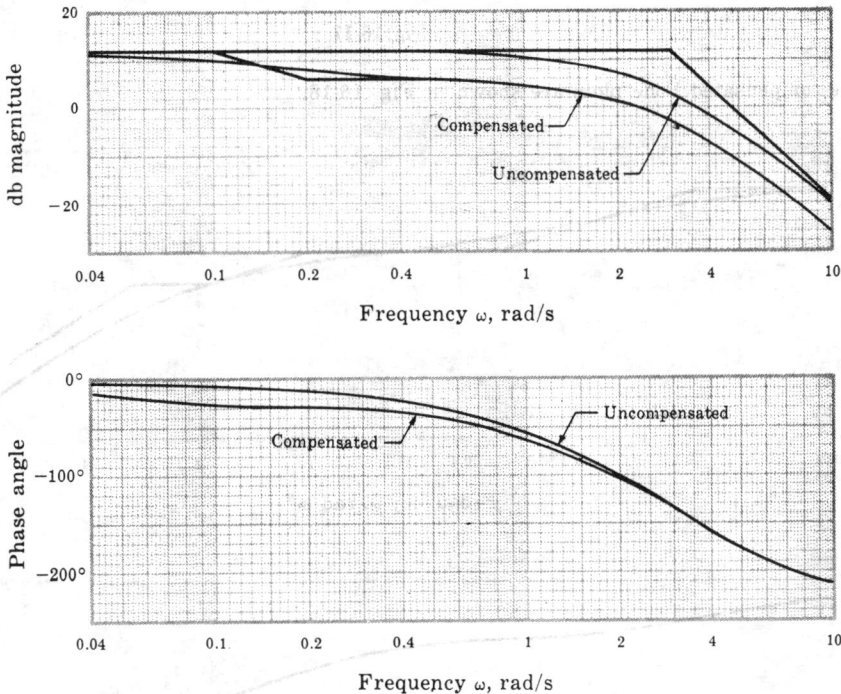

Fig. 16-16

The gain margin is 6 db and the phase margin is 30°. These margins can be increased by adding a lag compensator. To get the gain margin up to 12 db, the high frequency magnitude must be reduced by 6 db. To raise the phase margin to 45°, ω_1 must be lowered to 3.0 rad/s or less. This requires a magnitude attenuation of 3 db at that frequency. Therefore let us choose a lag ratio $b/a = 2$ to yield a high frequency attenuation of $20 \log_{10} 2 = 6$ db. For $a = 0.1$ and $b = 0.2$ the phase margin is 65° and the gain margin is 12 db, as shown in the compensated Bode plots of Fig. 16-16.

The compensated open-loop frequency response function is $\dfrac{4(1 + j\omega/0.2)}{(1 + j\omega/0.1)(1 + j\omega/3)^3}$.

LAG-LEAD COMPENSATION

16.9. Determine compensation for the system of Problem 16.5 that will result in a position error constant $K_p \geqq 10$, $\phi_{PM} \geqq 45°$ and the same gain crossover frequency ω_1 as the uncompensated system.

   The compensation determined in Problem 16.5 satisfies all the specifications except that K_p is only 4.4. The lead compensator chosen in that problem has a low frequency attenuation of 10.4 db, or a factor of 3.33. Let us replace the lead network with a lag-lead compensator, choosing $a_1 = 1.86$, $b_1 = 6.2$ and $a_2/b_2 = 0.3$. The low frequency magnitude becomes $a_1 b_2/b_1 a_2 = 1$, or 0 db, and the attenuation produced by the lead network is erased, effectively raising K_p for the system by a factor of 3.33 to 14.5. The lag portion of the compensator should be placed at frequencies sufficiently low so that the phase margin is not reduced below the specified value of 45°. This can be accomplished with $a_2 = 0.09$ and $b_2 = 0.3$. The compensated system block diagram is shown in Fig. 16-17. Note that an amplifier with a gain of 1.82 is included, as in Problem 16.5, to maintain $\omega_1 = 3.4$.

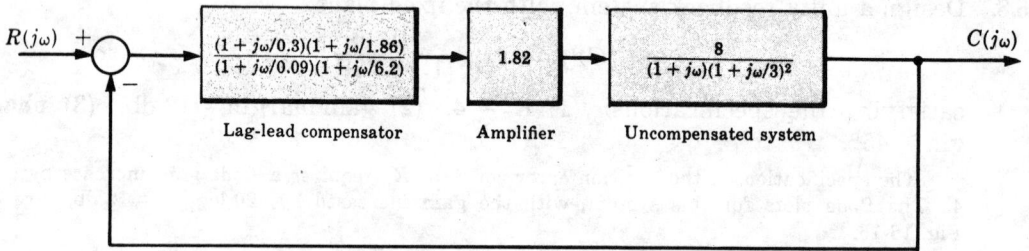

Fig. 16-17

The compensated Bode plots are shown in Fig. 16-18.

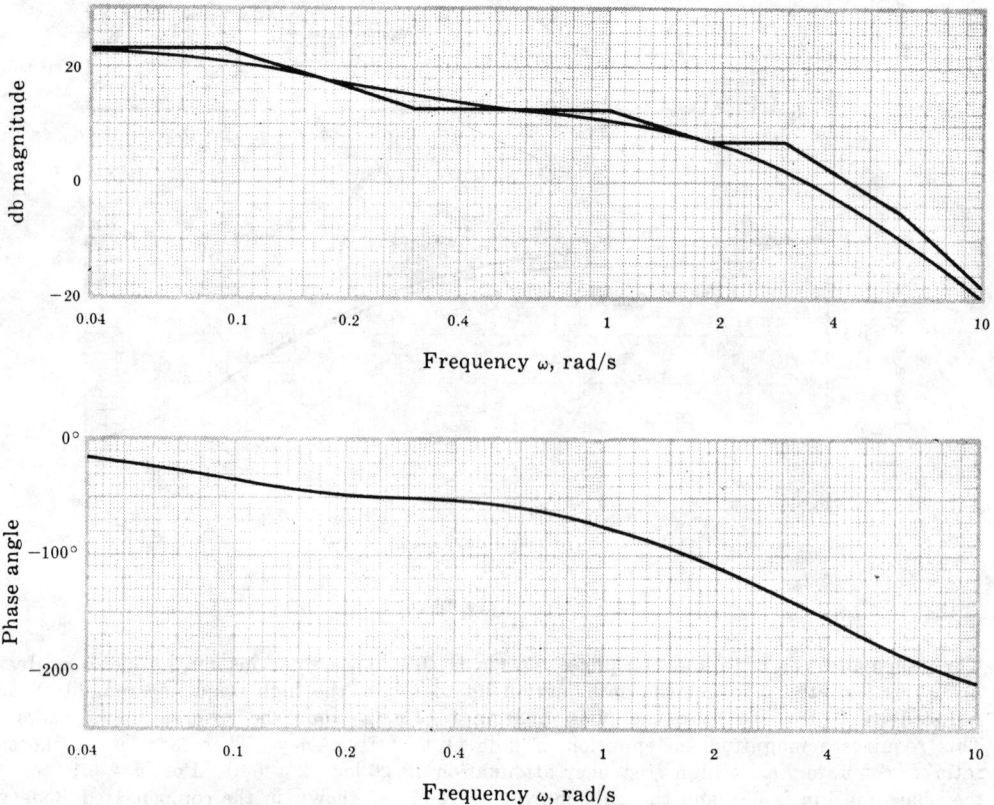

Fig. 16-18

16.10. Design cascade compensation for a unity feedback control system, with the plant

$$G_2(j\omega) \;=\; \frac{1}{j\omega(1 + j\omega/8)(1 + j\omega/20)}$$

to meet the following specifications:

    (1) $K_v \geq 100$ (3) gain margin ≥ 10 db

    (2) $\omega_1 \geq 10$ rad/s   (4) phase margin $\phi_{PM} \geq 45°$

To satisfy the first specification, a Bode gain increase by a factor of 100 is required since the uncompensated $K_v = 1$. The Bode plots for this system, with the gain increased to 100, are shown in Fig. 16-19.

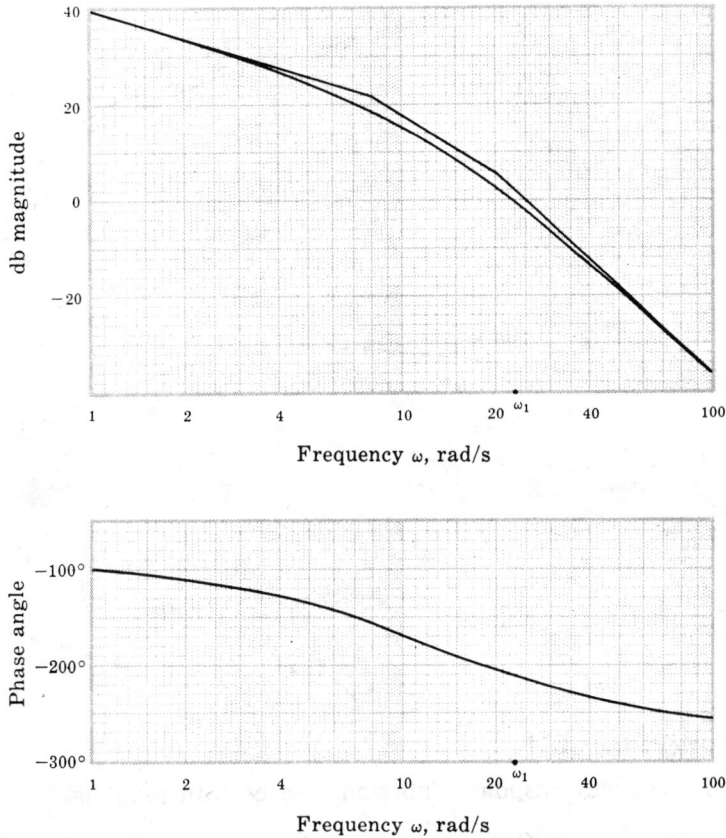

Fig. 16-19

The gain crossover frequency $\omega_1 = 23$ rad/s, the phase margin is $-30°$, and the gain margin is -12 db. Lag compensation could be used to increase the gain and phase margins by reducing ω_1. However, ω_1 would have to be lowered to less than 8 rad/s to achieve a $45°$ phase margin and to less than 6 rad/s for a 10 db gain margin. Consequently, we would not satisfy the second specification. With lead compensation, an additional Bode gain increase by a factor of b/a would be required and ω_1 would be increased, thus requiring substantially more than the $75°$ phase lead for $\omega_1 = 23$ rad/s. These disadvantages can be overcome using lag-lead compensation. The lead portion produces attenuation and phase lead. The frequencies at which these effects occur must be positioned near ω_1 so that ω_1 is slightly reduced and the phase margin is increased. Note that, although *pure* lead compensation increases ω_1, the lead portion of a lag-lead compensator decreases ω_1 because the gain factor increase of b/a is unnecessary, thereby lowering the magnitude characteristic. The lead portion can be determined independently using the curves of Fig. 16-2; but it must be kept in mind that, when the lag portion is included, the attenuation and phase lead may be somewhat reduced. Let us try a lead ratio of $a_1/b_1 = 0.1$, with $a_1 = 5$ and $b_1 = 50$. The maximum phase lead then occurs at 15.8 rad/s. This enables the magnitude asymptote to cross the 0 db line with a slope of -6 db/octave (see Example 16.2). The compensated Bode plots are shown in Fig. 16-20 with a_2 and b_2 chosen as 0.1 and 1.0 rad/s, respectively. The resulting parameters are $\omega_1 = 12$ rad/s, gain margin $= 14$ db and $\phi_{PM} = 52°$, as shown on the graphs. The compensated open-loop frequency response function is

$$\frac{100(1 + j\omega)(1 + j\omega/5)}{j\omega(1 + j\omega/0.1)(1 + j\omega/8)(1 + j\omega/20)(1 + j\omega/50)}$$

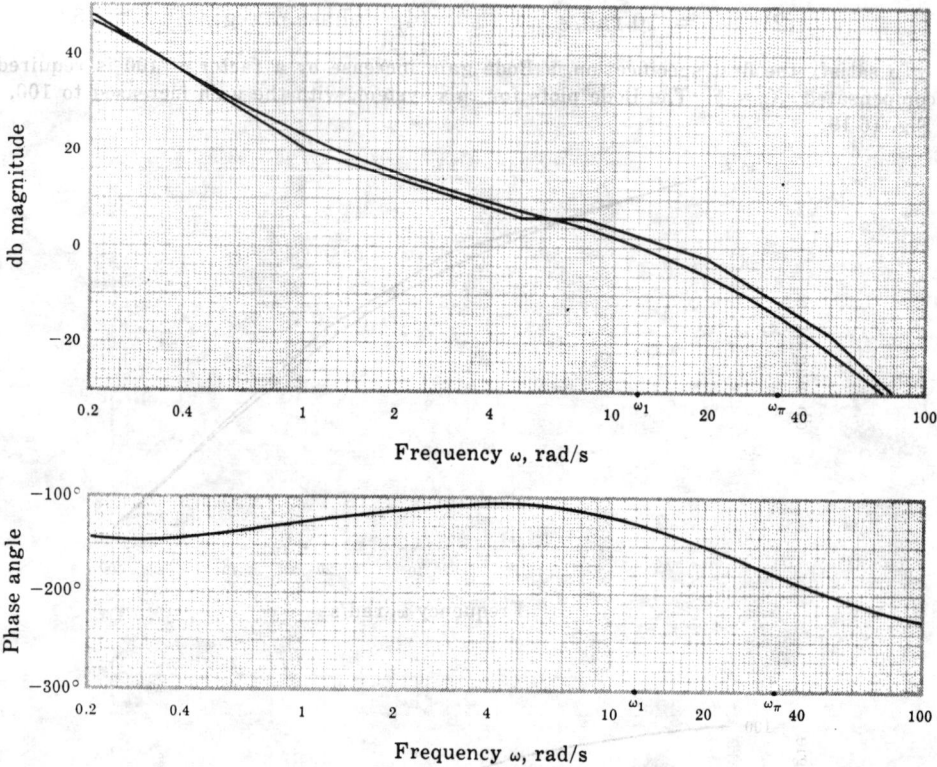

Fig. 16-20

MISCELLANEOUS PROBLEM

16.11. The nominal frequency response function of a certain plant is

$$G_2(j\omega) = \frac{1}{j\omega(1 + j\omega/8)(1 + j\omega/20)}$$

A feedback control system must be designed to control the output of this plant for a certain application and it must satisfy the following frequency domain specifications:

(1)   gain margin ≥ 6 db

(2)   phase margin $(\phi_{PM}) \geq 30°$

In addition, it is known that the "fixed" parameters of the plant may vary slightly during operation of the system. The effects of this variation on the system response must be minimized over the frequency range of interest, which is $0 \leq \omega \leq 8$ rad/s, and the actual requirement can be interpreted as a specification on the sensitivity of $(C/R)(j\omega)$ with respect to $|G_2(j\omega)|$, i.e.,

(3) $20 \left| \log_{10} S^{(C/R)(j\omega)}_{|G_2(j\omega)|} \right| \leq -10$ db   for $0 \leq \omega \leq 8$ rad/s

It is also known that the plant will be subjected to an uncontrollable, additive disturbance input, represented in the frequency domain by $U(j\omega)$. For this application, the system response to this disturbance input must be suppressed in the frequency range $0 \leq \omega \leq 8$ rad/s. Therefore the design problem includes the additional constraint on the magnitude ratio of the output to the disturbance input given by

$$(4) \qquad 20 \log_{10} \left| \frac{C}{U}(j\omega) \right| \leq -20 \text{ db} \qquad \text{for} \qquad 0 \leq \omega \leq 8 \text{ rad/s}$$

Design a system which satisfies these four specifications.

The general system configuration, which includes the possibility of either or both cascade and feedback compensators, is shown in Fig. 16-21.

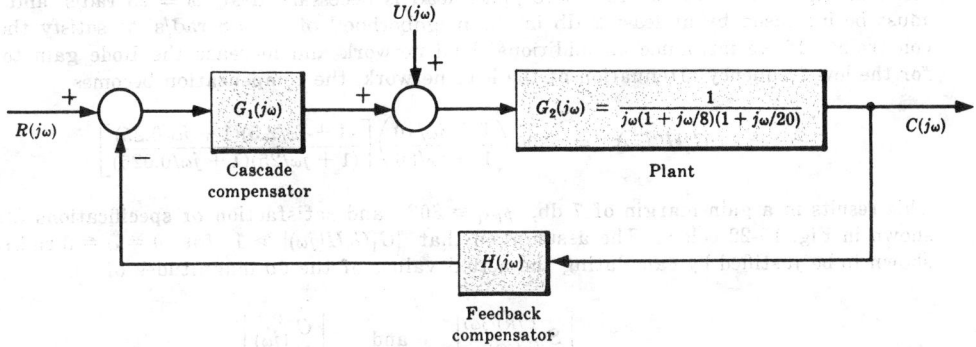

Fig. 16-21

From Fig. 16-21, we see that

$$\frac{C}{U}(j\omega) = \frac{G_2(j\omega)}{1 + G_1 G_2 H(j\omega)} \qquad \text{and} \qquad \frac{C}{R}(j\omega) = \frac{G_1 G_2(j\omega)}{1 + G_1 G_2 H(j\omega)}$$

In a manner similar to that of Example 9.17, page 169, it is easily shown that

$$S_{|G_2(j\omega)|}^{(C/R)(j\omega)} = \frac{1}{1 + G_1 G_2 H(j\omega)}$$

If we assume that $|G_1 G_2 H(j\omega)| \gg 1$ in the frequency range $0 \leq \omega \leq 8$ rad/s (this inequality must be checked upon completion of the design and, if it is not satisfied, the compensation may have to be recomputed), then specification (3) may be approximated by

$$20 \log_{10} \left| S_{|G_2(j\omega)|}^{(C/R)(j\omega)} \right| \cong 20 \log_{10} \left| \frac{1}{G_1 G_2 H(j\omega)} \right|$$

$$= -20 \log_{10} |G_1 G_2 H(j\omega)| \leq -10 \text{ db}$$

or $$20 \log_{10} |G_1 G_2 H(j\omega)| \geq 10 \text{ db}$$

Similarly, specification (4) can be approximated by

$$20 \log_{10} \left| \frac{C}{U}(j\omega) \right| \cong 20 \log_{10} \frac{|G_2(j\omega)|}{|G_1 G_2 H(j\omega)|}$$

$$= 20 \log_{10} |G_2(j\omega)| - 20 \log_{10} |G_1 G_2 H(j\omega)| \leq -20 \text{ db}$$

or $$20 \log_{10} |G_1 G_2 H(j\omega)| \geq [20 + 20 \log_{10} |G_2(j\omega)|] \text{ db}$$

Specifications (3) and (4) can therefore be translated into the following combined form. We require that the open-loop frequency response, $G_1 G_2 H(j\omega)$, lie in a region on a Bode magnitude plot which simultaneously satisfies the two inequalities:

$$20 \log_{10} |G_1 G_2 H(j\omega)| \geq 10 \text{ db}$$

$$20 \log_{10} |G_1 G_2 H(j\omega)| \geq [20 + 20 \log_{10} |G_2(j\omega)|] \text{ db}$$

This region lies above the broken line shown in the Bode magnitude plot in Fig. 16-22, which also includes the Bode plots of $G_2(j\omega)$. The design may be completed by determining compensation which satisfies the gain and phase margin requirements, (1) and (2), subject to this magnitude constraint.

A 32 db increase in Bode gain, which is necessary at $\omega = 8$ rad/s, would satisfy specifications (3) and (4), but not (1) and (2). Therefore a more complicated compensation is required. For a second trial, we find that the lag-lead compensation:

$$G_1H'(j\omega) \;=\; \frac{100(1 + j\omega/2.5)(1 + j\omega/0.25)}{(1 + j\omega/25)(1 + j\omega/0.025)}$$

results in a system with a gain margin of 6 db and $\phi_{PM} \cong 26°$, as shown in Fig. 16-22. We see from the figure that 10° to 15° more phase lead is necessary near $\omega = 25$ rad/s and $|G_1H'(j\omega)|$ must be increased by at least 2 db in the neighborhood of $\omega = 8$ rad/s to satisfy the magnitude constraint. If we introduce an additional lead network and increase the Bode gain to compensate for the low frequency attenuation of the lead network, the compensation becomes

$$G_1H''(j\omega) \;=\; 300\left(\frac{1 + j\omega/10}{1 + j\omega/30}\right)\left[\frac{(1 + j\omega/2.5)(1 + j\omega/0.25)}{(1 + j\omega/25)(1 + j\omega/0.025)}\right]$$

This results in a gain margin of 7 db, $\phi_{PM} \cong 30°$, and satisfaction of specifications (3) and (4), as shown in Fig. 16-22 below. The assumption that $|G_1G_2H(j\omega)| \gg 1$ for $0 \leq \omega \leq 8$ rad/sec is easily shown to be justified by calculating the actual values of the db magnitudes of

$$\left|S_{|G_2(j\omega)|}^{(C/R)(j\omega)}\right| \quad \text{and} \quad \left|\frac{C}{U}(j\omega)\right|$$

The compensator $G_1H''(j\omega)$ can be divided between the forward and feedback paths, or put all in one path, depending on the form desired for $(C/R)(j\omega)$ if such a form is specified by the application.

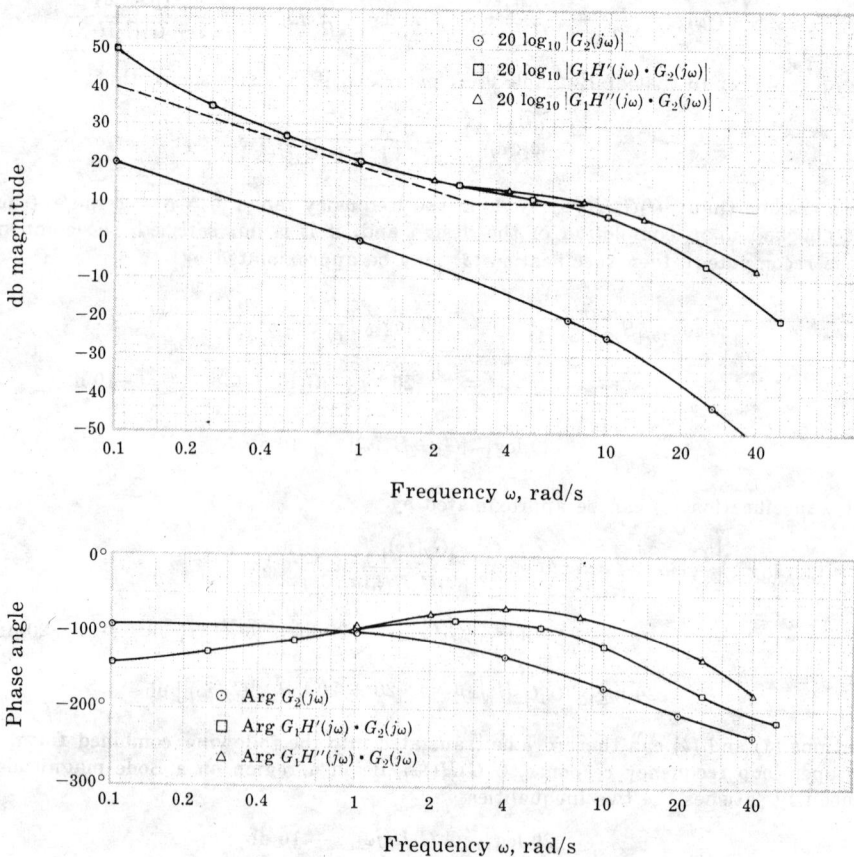

Fig. 16-22

Supplementary Problems

16.12. Design a compensator for the system with the open-loop frequency response function

$$GH(j\omega) = \frac{20}{j\omega(1 + j\omega/10)(1 + j\omega/25)(1 + j\omega/40)}$$

to result in a closed-loop system with a gain margin of at least 10 db and a phase margin of at least 45°.

16.13. Determine a compensator for the system of Problem 16.1 which will result in the same gain and phase margins but with a crossover frequency ω_1 of at least 4 rad/s.

16.14. Design a compensator for the system with the open-loop frequency response function

$$GH(j\omega) = \frac{2}{(1 + j\omega)[1 + j\omega/10 - (\omega/4)^2]}$$

which will result in a closed-loop system with a gain margin of at least 6 db and a phase margin of at least 40°.

16.15. Work Problem 12.9, page 234, using Bode plots. Assume a maximum of 25 percent overshoot will be ensured if the system has a phase margin of at least 45°.

16.16. Work Problem 12.10, page 235, using Bode plots.

16.17. Work Problem 12.15, page 236, using Bode plots.

16.18. Work Problem 12.16, page 236, using Bode plots.

Chapter 17

Nichols Chart Analysis

17.1 INTRODUCTION

Nichols chart analysis, a frequency response method, is a modification of the Nyquist and Bode methods. The *Nichols chart* is essentially a transformation of the M and N circles on the Polar Plot (Section 11.12) into noncircular M and N contours on a db magnitude versus phase angle plot in rectangular coordinates. The open-loop frequency response function $GH(j\omega)$ plotted on a Nichols chart is called a *Nichols chart plot* of $GH(j\omega)$. The relative stability of the closed-loop system is easily obtained from this graph. The determination of absolute stability, however, is generally impractical with this method and either the techniques of Chapter 5 or the Nyquist Stability Criterion (Section 11.10) are preferred.

The reasons for using Nichols chart analysis are the same as those for the other frequency response methods, the Nyquist and Bode techniques, and are discussed in Chapters 11 and 15. The Nichols chart plot has at least two advantages over the polar plot: (1) a much wider range of magnitudes can be graphed because $|GH(j\omega)|$ is plotted on a logarithmic scale; and (2) the graph of $GH(j\omega)$ is obtained by algebraic summation of the individual magnitude and phase angle contributions of its poles and zeros. While both of these properties are also shared by Bode plots, $|GH(j\omega)|$ and $\arg GH(j\omega)$ are included on a single Nichols chart plot rather than on two Bode plots.

Nichols chart techniques are useful for directly plotting $\frac{C}{R}(j\omega)$ and are especially applicable in system design, as shown in the next chapter.

17.2 db MAGNITUDE—PHASE ANGLE PLOTS

The polar form of the open-loop frequency response function is

$$GH(j\omega) = |GH(j\omega)| \underline{/\arg GH(j\omega)} \tag{17.1}$$

Definition 17.1: The **db magnitude-phase angle plot** of $GH(j\omega)$ is a graph of $|GH(j\omega)|$ in db versus $\arg GH(j\omega)$, in degrees, on rectangular coordinates with ω as a parameter.

Example 17.1.

The db magnitude-phase angle plot of

$$GH(j\omega) = 1 + j\omega = \sqrt{1 + \omega^2} \underline{/\tan^{-1}\omega}$$

is shown in Fig. 17-1 below.

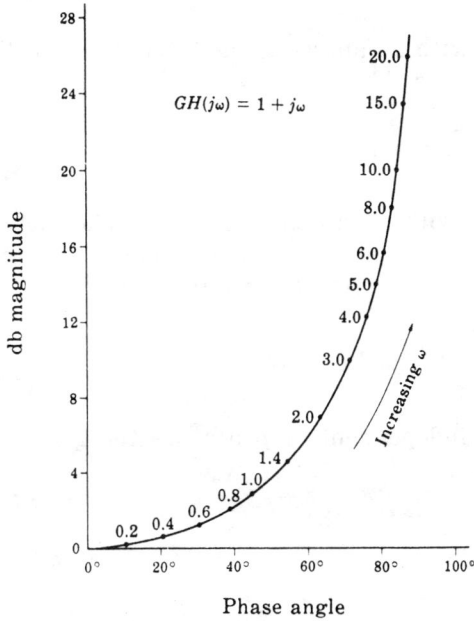

Fig. 17-1

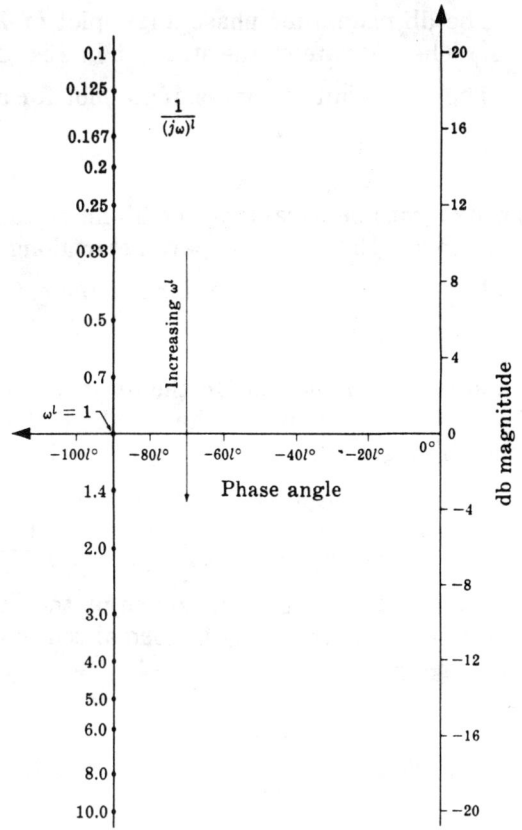

Fig. 17-2

17.3   CONSTRUCTION OF db MAGNITUDE—PHASE ANGLE PLOTS

The db magnitude-phase angle plot can be constructed directly by evaluating $20 \log_{10} |GH(j\omega)|$ and $\arg GH(j\omega)$, for a sufficient number of values of ω, and plotting the results. This is tedious, however, and the following method is generally simpler. First write $GH(j\omega)$ in the *Bode form* (Section 15.3):

$$GH(j\omega) \;=\; \frac{K_B(1 + j\omega/z_1)\cdots(1 + j\omega/z_m)}{(j\omega)^l(1 + j\omega/p_1)\cdots(1 + j\omega/p_n)}$$

where l is a nonnegative integer. For $K_B > 0$ (if $K_B < 0$, add 180° to $\arg GH(j\omega)$),

$$20 \log_{10} |GH(j\omega)| \;=\; 20 \log_{10} K_B \;+\; 20 \log_{10} |1 + j\omega/z_1|$$

$$+ \;\cdots\; + \; 20 \log_{10} |1 + j\omega/z_m| \;+\; 20 \log_{10} |1/(j\omega)^l| \qquad (17.2)$$

$$+ \; 20 \log_{10} \left|\frac{1}{1 + j\omega/p_1}\right| \;+\; \cdots \;+\; 20 \log_{10} \left|\frac{1}{1 + j\omega/p_n}\right|$$

$$\arg GH(j\omega) \;=\; \arg (1 + j\omega/z_1) \;+\; \cdots \;+\; \arg (1 + j\omega/z_m) \;+\; \arg [1/(j\omega)^l]$$

$$+ \; \arg \frac{1}{1 + j\omega/p_1} \;+\; \cdots \;+\; \arg \frac{1}{1 + j\omega/p_n} \qquad (17.3)$$

Using Equations (*17.2*) and (*17.3*), the db magnitude-phase angle plot of $GH(j\omega)$ is generated by summing the db magnitudes and phase angles of the poles and zeros, or pairs of poles and zeros when they are complex conjugates.

The db magnitude-phase angle plot of K_B is a straight line parallel to the phase angle axis. The ordinate of the straight line is $20 \log_{10} K_B$.

The db magnitude-phase angle plot for a *pole of order l at the origin*,

$$\frac{1}{(j\omega)^l} \tag{17.4}$$

is a straight line parallel to the db magnitude axis with an abscissa $-90l°$ as shown in Fig. 17-2 above. Note that the parameter along the curve is ω^l.

The plot for a *zero of order l at the origin*,

$$(j\omega)^l \tag{17.5}$$

is a straight line parallel to the db magnitude axis with an abscissa of $90l°$. The plot for $(j\omega)^l$ is the diagonal mirror image about the origin of the plot for $1/(j\omega)^l$. That is, for fixed ω the db magnitude and phase angle of $1/(j\omega)^l$ are the negatives of those for $(j\omega)^l$.

The db magnitude-phase angle plot for a *real pole*,

$$\frac{1}{1 + j\omega/p}, \quad p > 0 \tag{17.6}$$

is shown in Fig. 17-3. The shape of the graph is independent of p because the frequency parameter along the curve is normalized to ω/p.

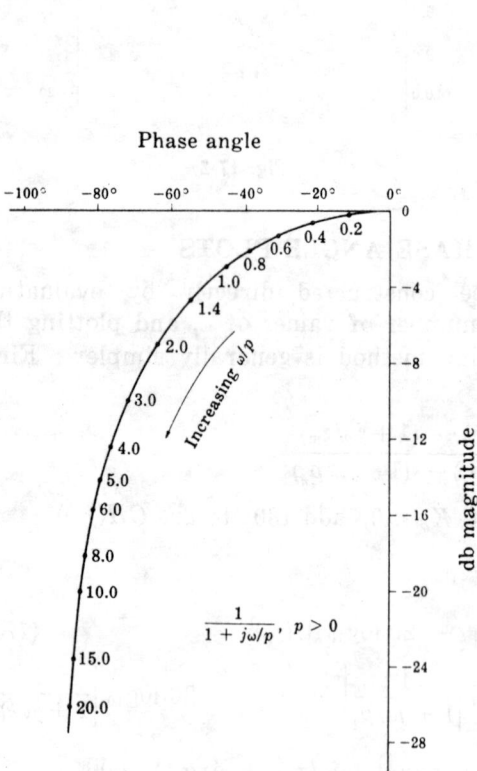

Fig. 17-3

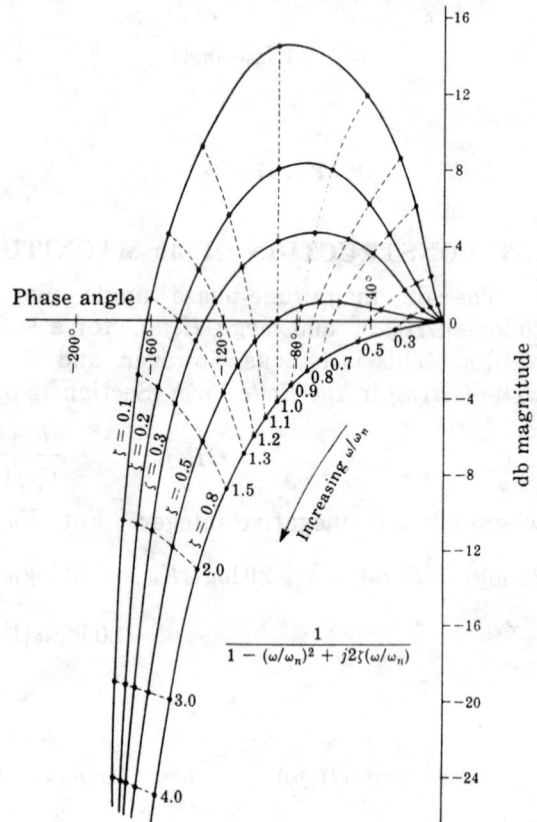

Fig. 17-4

The plot for a *real zero*,

$$1 + j\omega/z, \quad z > 0 \qquad (17.7)$$

is the diagonal mirror image about the origin of Fig. 17-3.

A set of db magnitude-phase angle plots of several pairs of *complex conjugate poles*,

$$\frac{1}{1 - (\omega/\omega_n)^2 + j2\zeta(\omega/\omega_n)}, \quad 0 < \zeta < 1 \qquad (17.8)$$

are shown in Fig. 17-4 above. For fixed ζ, the graphs are independent of ω_n because the frequency parameter is normalized to ω/ω_n.

The plots for *complex conjugate zeros*,

$$1 - (\omega/\omega_n)^2 + j2\zeta(\omega/\omega_n), \quad 0 < \zeta < 1 \qquad (17.9)$$

are diagonal mirror images about the origin of Fig. 17-4.

Example 17.2.

The db magnitude-phase angle plot of

$$GH(j\omega) = \frac{10(1 + j\omega/2)}{(1 + j\omega)[1 - (\omega/2)^2 + j\omega/2]}$$

is constructed by adding the db magnitudes and phase angles of the individual factors:

$$10, \quad 1 + j\omega/2, \quad \frac{1}{1 + j\omega}, \quad \frac{1}{1 - (\omega/2)^2 + j\omega/2}$$

Tabulation of these factors is helpful, as shown in Table 17.1. The first row contains the db magnitude and phase angle of the Bode gain $K_B = 10$ for several frequency values. The db magnitude is 20 db and the phase angle is 0° for all ω. The second row contains the db magnitude and phase angle of the term $(1 + j\omega/2)$ for the same values of ω. These were obtained from Fig. 17-3 by letting $p = 2$ and taking the negatives of the values on the curve for the frequencies in the table. The third row corresponds to the term $\frac{1}{1 + j\omega}$ and was also obtained from Fig. 17-3. The fourth row was taken from the $\zeta = 0.5$ curve of Fig. 17-4 by letting $\omega_n = 2$. The sum of the db magnitudes and phase angles of the individual terms for the frequencies in the table is given in the last row. These values are plotted in Fig. 17-5 below, the db magnitude-phase angle plot of $GH(j\omega)$.

Table 17.1

Term \ Frequency ω	0	0.4	0.8	1.2	1.6	2	2.8	4	6	8
10	20 db	20	20	20	20	20	20	20	20	20
	0°	0°	0°	0°	0°	0°	0°	0°	0°	0°
$1 + j\omega/2$	0 db	0.2	0.6	1.3	2.2	3.0	4.7	7	10	12.3
	0°	11°	21°	31°	39°	45°	54°	63°	71°	76°
$\frac{1}{1 + j\omega}$	0 db	−0.6	−2.2	−3.8	−5.4	−7.0	−9.4	−12.3	−15.7	−18.1
	0°	−21°	−39°	−50°	−57°	−63°	−70°	−76°	−81°	−83°
$\frac{1}{1 - (\omega/2)^2 + j\omega/2}$	0 db	0.3	0.6	0.9	1.0	0	−4.8	−12	−19.5	−24.5
	0°	−12°	−26°	−46°	−68°	−90°	−126°	−148°	−160°	−166°
Sum = $GH(j\omega)$	20 db	19.9	19.0	18.4	17.8	16	10.5	2.7	−5.2	−10.3
	0°	−22°	−44°	−65°	−86°	−108°	−142°	−161°	−170°	−173°

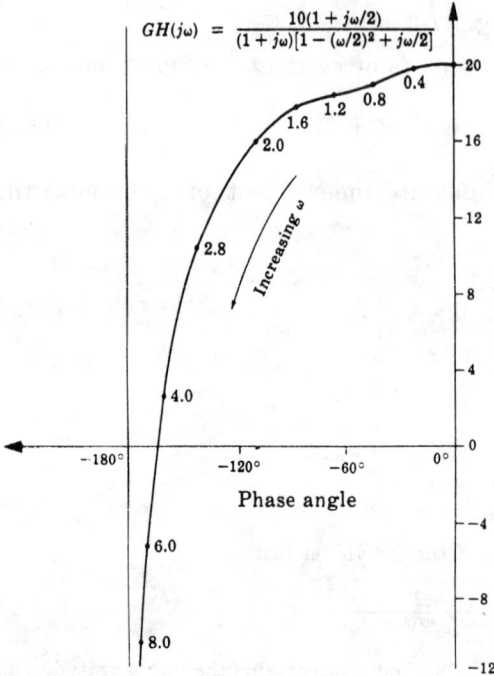

Fig. 17-5

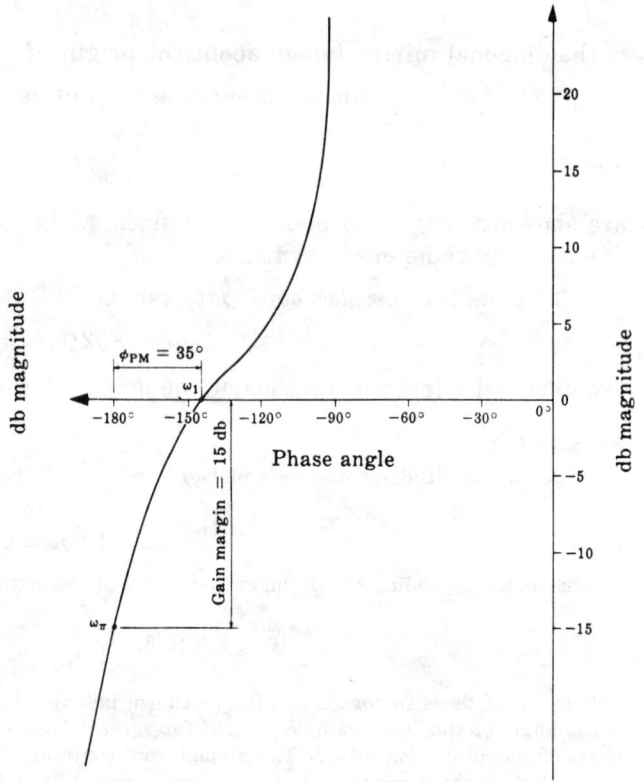

Fig. 17-6

17.4 RELATIVE STABILITY

The gain and phase margins are readily determined from the db magnitude-phase angle plot of $GH(j\omega)$.

The *phase crossover* frequency ω_π is the frequency at which the graph of $GH(j\omega)$ intersects the $-180°$ line on the db magnitude-phase angle plot. The *gain margin in db* is given by

$$\text{gain margin} = -20 \log_{10} |GH(j\omega_\pi)| \quad \text{db} \qquad (17.10)$$

and is read directly from the db magnitude-phase angle plot.

The *gain crossover* frequency ω_1 is the frequency at which the graph of $GH(j\omega)$ intersects the 0 db line on the db magnitude-phase angle plot. The phase margin is given by

$$\text{phase margin} = 180° + \arg GH(j\omega_1)$$

and can be read directly from the db magnitude-phase angle plot.

In most cases, positive gain and phase margins will ensure stability of the closed-loop system; however, absolute stability should be established by some other means (for example, see Chapters 5 and 11) to guarantee that this is true.

Example 17.3.

For a stable system, the db magnitude-phase angle plot of $GH(j\omega)$ is shown in Fig. 17-6. The gain margin is 15 db and the phase margin is 35°, as indicated.

17.5 THE NICHOLS CHART

The remaining discussion is restricted to unity feedback systems. The results are easily generalized to non-unity feedback systems, as illustrated in **Example 17.8.**

The closed-loop frequency response function of a unity feedback system may be written in polar form as

$$\frac{C}{R}(j\omega) \;=\; \left|\frac{C}{R}(j\omega)\right| \underline{/\arg \frac{C}{R}(j\omega)} \;=\; \frac{G(j\omega)}{1 + G(j\omega)} \;=\; \frac{|G(j\omega)|\,\underline{/\phi_G}}{1 + |G(j\omega)|\,\underline{/\phi_G}} \qquad (17.11)$$

where $\phi_G \equiv \arg G(j\omega)$.

The locus of points on a db magnitude-phase angle plot for which

$$\left|\frac{C}{R}(j\omega)\right| \;=\; M \;=\; \text{constant}$$

is defined by the equation

$$|G(j\omega)|^2 \;+\; \frac{2M^2}{M^2 - 1}|G(j\omega)|\cos\phi_G \;+\; \frac{M^2}{M^2 - 1} \;=\; 0 \qquad (17.12)$$

For a fixed value of M, this locus can be plotted in three steps: (1) choose numerical values for $|G(j\omega)|$; (2) solve the resultant equations for ϕ_G, excluding values of $|G(j\omega)|$ for which $|\cos\phi_G| > 1$; and (3) plot the points obtained on a db magnitude-phase angle plot. Note that for fixed values of M and $|G(j\omega)|$, ϕ_G is multiple-valued because it appears in the equation as $\cos\phi_G$.

Example 17.4.

The locus of points for which $\left|\dfrac{C}{R}(j\omega)\right| = \sqrt{2}$ or, equivalently,

$$20\log_{10}\left|\frac{C}{R}(j\omega)\right| \;=\; 3\ \text{db}$$

is graphed in Fig. 17-7. A similar curve appears at all odd multiples of 180° along the arg $G(j\omega)$ axis.

The locus of points on a db magnitude-phase angle plot for which arg $\dfrac{C}{R}(j\omega)$ is constant or, equivalently,

$$\tan\left[\arg\frac{C}{R}(j\omega)\right] \;=\; N \;=\; \text{constant}$$

is defined by the equation

$$|G(j\omega)| \;+\; \cos\phi_G \;-\; \frac{1}{N}\sin\phi_G \;=\; 0 \qquad (17.13)$$

For a fixed value of N, this locus of points can be plotted in three steps: (1) choose values for ϕ_G; (2) solve the resultant equations for $G(j\omega)$; and (3) plot the points obtained on a db magnitude-phase angle plot.

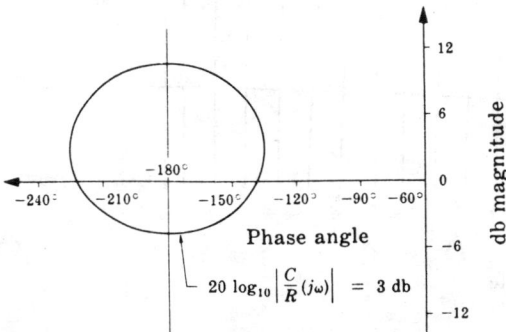

Fig. 17-7

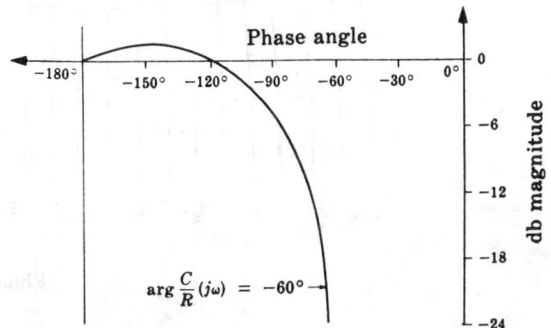

Fig. 17-8

Example 17.5.

The locus of points for which arg $\dfrac{C}{R}(j\omega) = -60°$ or, equivalently,

$$\tan\left[\arg \frac{C}{R}(j\omega)\right] = -\sqrt{3}$$

is graphed in Fig. 17-8 above. A similar curve appears at all multiples of 180° along the arg $G(j\omega)$ axis.

Definition 17.2: A **Nichols chart** is a db magnitude-phase angle plot of the loci of constant db magnitude and phase angle of $\frac{C}{R}(j\omega)$, graphed as $|G(j\omega)|$ versus arg $G(j\omega)$.

Example 17.6.

A Nichols chart is shown in Fig. 17-9. The range of arg $G(j\omega)$ on this chart is well suited to control system analysis.

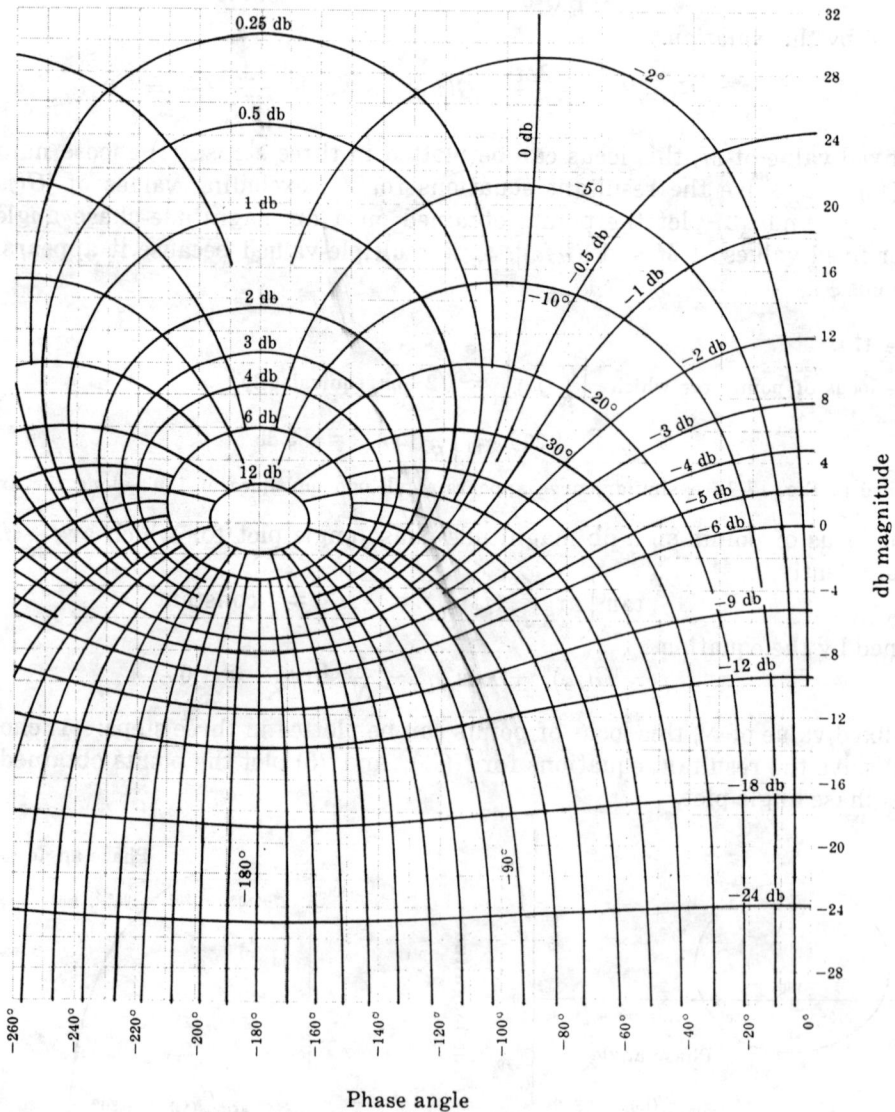

Fig. 17-9

Definition 17.3: A **Nichols chart plot** is a db magnitude-phase angle plot of a frequency response function $P(j\omega)$ superimposed on a Nichols chart.

17.6 CLOSED-LOOP FREQUENCY RESPONSE FUNCTIONS

The frequency response function $\frac{C}{R}(j\omega)$ of a unity feedback system is easily determined from the Nichols chart plot of $G(j\omega)$. Values of $\left|\frac{C}{R}(j\omega)\right|$ in db and arg $G(j\omega)$ are determined directly from the plot as the points where the graph of $G(j\omega)$ intersects the graphs of loci of constant $\left|\frac{C}{R}(j\omega)\right|$ and arg $\frac{C}{R}(j\omega)$.

Example 17.7.

The Nichols chart plot of $GH(j\omega)$ for the system of Example 17.2 is shown in Fig. 17-10. Assuming that it is a unity feedback system $(H = 1)$, values for $\frac{C}{R}(j\omega)$ and arg $\frac{C}{R}(j\omega)$ are obtained from this graph and plotted as a db magnitude-phase angle plot of $\frac{C}{R}(j\omega)$ in Fig. 17-11.

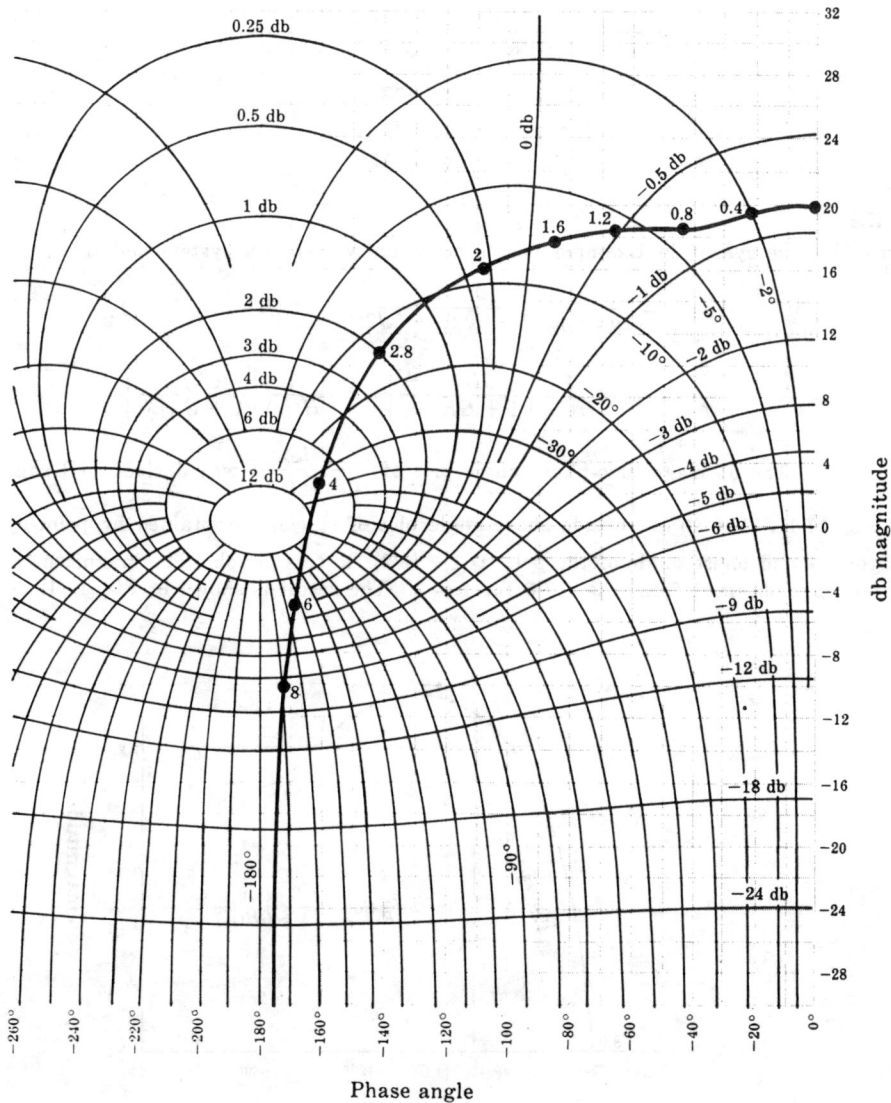

Fig. 17-10

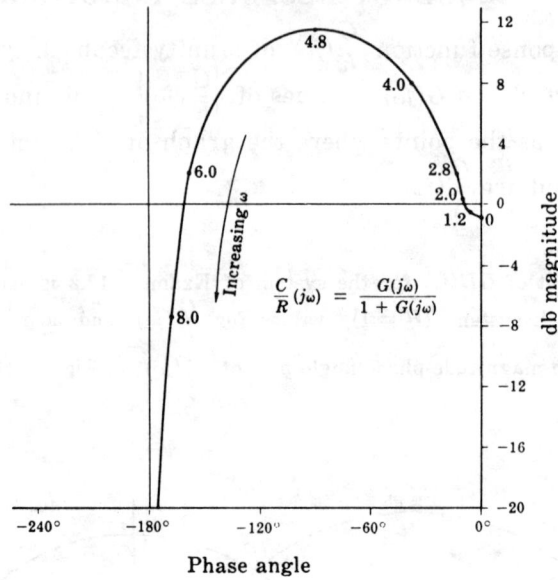

Fig. 17-11

Example 17.8.

Assume that the system in Example 17.2 is not a unity feedback system and that

$$G(j\omega) = \frac{10}{(1 + j\omega)[1 - (\omega/2)^2 + j\omega/2]}; \qquad H(j\omega) = 1 + j\frac{\omega}{2}$$

Then

$$\frac{C}{R}(j\omega) = \frac{1}{H(j\omega)}\left[\frac{GH(j\omega)}{1 + GH(j\omega)}\right] = \frac{1}{H(j\omega)}\left[\frac{G'(j\omega)}{1 + G'(j\omega)}\right]$$

where $G' \equiv GH$. The db magnitude-phase angle plot of $\dfrac{G'(j\omega)}{1 + G'(j\omega)}$ was derived in Example 17.7 and is

shown in Fig. 17-11. The db magnitude-phase angle plot of $\dfrac{C}{R}(j\omega)$ is obtained by point-by-point addition

of the magnitude and phase angle of the pole $1/(1 + j\omega/2)$ to this graph. The magnitude and phase angle

of $1/(1 + j\omega/2)$ are obtained from Fig. 17-3 for $p = 2$. The result is shown in Fig. 17-12.

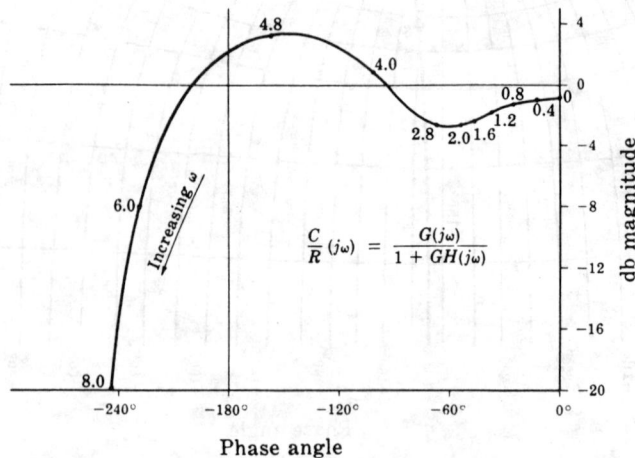

Fig. 17-12

Solved Problems

db MAGNITUDE—PHASE ANGLE PLOTS

17.1. Show that the db magnitude-phase angle plot for a pole of order l at the origin, $1/(j\omega)^l$, is a straight line parallel to the db magnitude axis with an abscissa of $-90l°$ for $\omega \geq 0$.

In polar form, $j\omega = \omega\underline{/90°}$, $\omega \geq 0$. Therefore

$$\frac{1}{(j\omega)^l} = \frac{1}{\omega^l}\underline{/-90l°}, \qquad \omega \geq 0$$

$$20\log_{10}\left|\frac{1}{(j\omega)^l}\right| = 20\log_{10}\frac{1}{\omega^l} = -20\log_{10}\omega^l$$

and $\arg 1/(j\omega)^l = -90l°$. We see that $\arg 1/(j\omega)^l$ is independent of ω; hence the abscissa of the plot is a constant $-90l°$. In addition, for the region $0 \leq \omega \leq +\infty$, the db magnitude ranges from $+\infty$ to $-\infty$. Thus the abscissa is fixed and the ordinate takes on all values. The result is a straight line as shown in Fig. 17-2.

17.2. Construct the db magnitude-phase angle plot for the open-loop transfer function

$$GH = \frac{2}{s(1+s)(1+s/3)}$$

The db magnitude of $GH(j\omega)$ is

$$20\log_{10}|GH(j\omega)| = 20\log_{10}\frac{2}{|j\omega|\,|1+j\omega|\,|1+j\omega/3|}$$

$$= 20\log_{10}2 - 20\log_{10}\left[\omega\sqrt{1+\omega^2}\,\sqrt{1+\omega^2/9}\,\right]$$

$$= 6.02 - 10\log_{10}\left[\omega^2(1+\omega^2)(1+\omega^2/9)\right]$$

The phase angle of $GH(j\omega)$ is

$$\arg[GH(j\omega)] = -\arg[j\omega] - \arg[1+j\omega] - \arg[1+j\omega/3]$$

$$= -90° - \tan^{-1}\omega - \tan^{-1}(\omega/3)$$

We now evaluate $20\log_{10}|GH(j\omega)|$ and $\arg GH(j\omega)$ for several values of ω, as shown in Table 17.2. The plot of the values in the table is given in Fig. 17-13.

Table 17.2

ω	$20\log_{10}|GH(j\omega)|$	$\arg GH(j\omega)$
0	∞	$-90°$
0.1	25.97 db	$-97.4°$
0.2	19.80	$-105.0°$
0.5	10.95	$-126.1°$
1.0	2.55	$-153.5°$
1.5	-3.59	$-172.9°$
2.0	-8.59	$-187.1°$
3.0	-16.54	$-206.6°$

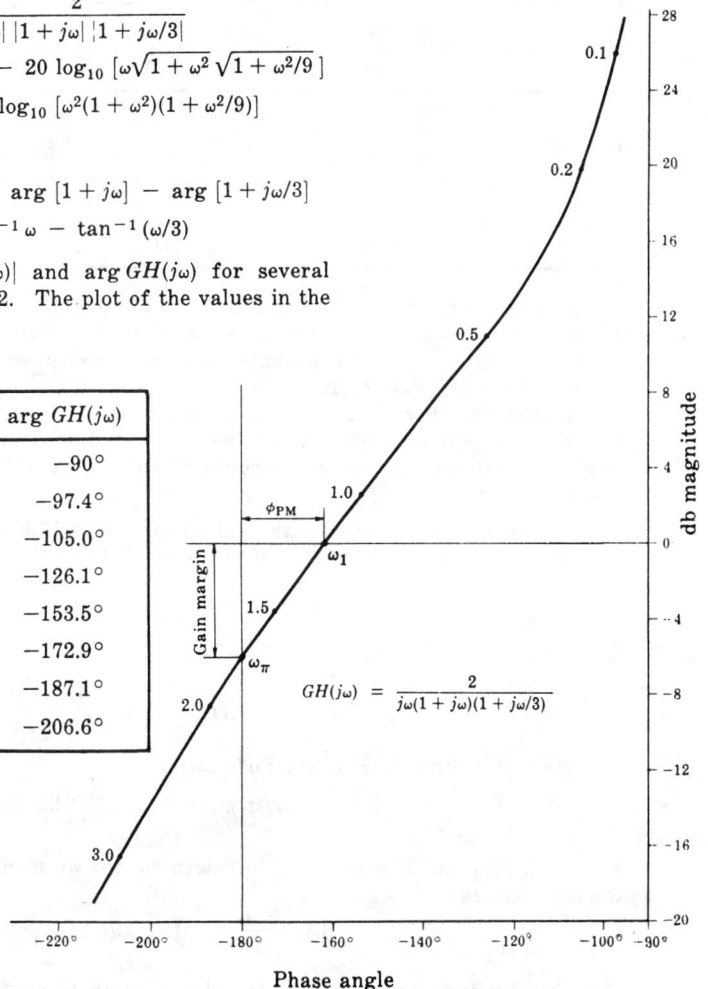

Fig. 17-13

17.3. Using the plots in Fig. 17-2 and Fig. 17-3, reconstruct Table 17.2.

We rewrite $GH(j\omega)$ as $GH(j\omega) = (2)\left(\dfrac{1}{j\omega}\right)\left(\dfrac{1}{1+j\omega}\right)\left(\dfrac{1}{1+j\omega/3}\right)$.   The db magnitude of $GH(j\omega)$ is

$$20 \log_{10} |GH(j\omega)| = 20 \log_{10} 2 + 20 \log_{10}\left|\frac{1}{j\omega}\right| + 20 \log_{10}\left|\frac{1}{1+j\omega}\right| + 20 \log_{10}\left|\frac{1}{1+j\omega/3}\right|$$

The phase angle is

$$\arg GH(j\omega) = \arg (2) + \arg\left(\frac{1}{j\omega}\right) + \arg\left(\frac{1}{1+j\omega}\right) + \arg\left(\frac{1}{1+j\omega/3}\right)$$

We now construct Table 17.3.

Table 17.3

Term \ Frequency ω	0	0.1	0.2	0.5	1.0	1.5	2.0	3.0
2	6 db 0°	6 0°	6 0°	6 0°	6 0°	6 0°	6 0°	6 0°
$\dfrac{1}{j\omega}$	∞ −90°	20 −90°	14 −90°	6 −90°	0 −90°	−3.6 −90°	−6 −90°	−9.5 −90°
$\dfrac{1}{1+j\omega}$	0 0°	−0.1 −5.5°	−0.3 −11°	−1.0 −26°	−3.0 −45°	−5.2 −57°	−7.0 −63°	−10 −72°
$\dfrac{1}{1+j\omega/3}$	0 0°	0 −2°	−0.1 −4°	−0.2 −9°	−0.5 −17.5°	−1.0 −26°	−1.6 −33°	−3.0 −45°
Sum $= GH(j\omega)$	∞ −90°	25.9 −97.5°	19.6 −105°	10.8 −125°	2.5 −152.5°	−3.8 −173°	−8.6 −186°	−16.5 −207°

The first row contains the db magnitude and phase angle of the Bode gain $K_B = 2$. The second row contains the db magnitude and phase angle of the term $1/j\omega$ for several values of ω. These are obtained from Fig. 17-2 by letting $l = 1$ and taking values from the curve for the frequencies given. The third row corresponds to the term $1/(1+j\omega)$ and is obtained from Fig. 17-3 for $p = 1$. The fourth row corresponds to the term $1/(1+j\omega/3)$ and is obtained from Fig. 17-3 for $p = 3$. Each pair of entries in the final row is obtained by summing the db magnitudes and phase angles in each column and corresponds to the db magnitude and phase angle of $GH(j\omega)$ for the given value of ω.

The discrepancies between the values in Tables 17.2 and 17.3 are due to the graphical inter-polation used to determine values of ω in Fig. 17-2 and 17-3.

17.4. Construct the db magnitude-phase angle plot for the open-loop transfer function

$$GH = \frac{4(s + 0.5)}{s^2(s^2 + 2s + 4)}$$

We first write $GH(j\omega)$ in the Bode form,

$$GH(j\omega) = \frac{0.5(1 + j\omega/0.5)}{(j\omega)^2[1 - (\omega/2)^2 + j\omega/2]}$$

Fig. 17-2, 17-3 and 17-4 are used to determine the db magnitude and phase angle contributions of the four factors:

$$0.5, \quad \frac{1}{(j\omega)^2}, \quad 1 + j\omega/0.5, \quad \frac{1}{1 - (\omega/2)^2 + j\omega/2}$$

The db magnitude and phase angle value of each term, for several values of ω, are summarized in Table 17.4.

Table 17.4

Term \ Frequency ω	0	0.2	0.5	1.0	1.5	2.0	3.0	5.0
0.5	−6 db 0°	−6 0°	−6 0°	−6 0°	−6 0°	−6 0°	−6 0°	−6 0°
$\dfrac{1}{(j\omega)^2}$	∞ −180°	28 −180°	12 −180°	0 −180°	−6.8 −180°	−12 −180°	−19 −180°	−28 −180°
$1 + j\omega/0.5$	0 0	0.6 21°	3.0 45°	7.0 63°	10.0 71°	12.3 76°	15.7 81°	20 84°
$\dfrac{1}{1 - (\omega/2)^2 + j\omega/2}$	0 0°	0.2 −6°	0.4 −15°	0.7 −34°	1.0 −63°	0 −90°	−6 −130°	−15 −152°
Sum = $GH(j\omega)$	∞ −180°	22.8 −165°	9.4 −150°	1.7 −151°	−1.8 −172°	−5.7 −194°	−15.3 −229°	−29 −248°

The first row contains db magnitude and phase angle values of the term $1/(j\omega)^2$. These were obtained from Fig. 17-2 by letting $l = 2$ and taking values from the curve for the frequencies in the table. The second row is obtained from Fig. 17-3 by letting $p = 0.5$, and the third row from Fig. 17-4 letting $\xi = 0.5$ and $\omega = 2$. The final row is obtained by summing the db magnitudes and phase angles in each column and contains the values of db magnitude and phase angle for $GH(j\omega)$. These are plotted in Fig. 17-14.

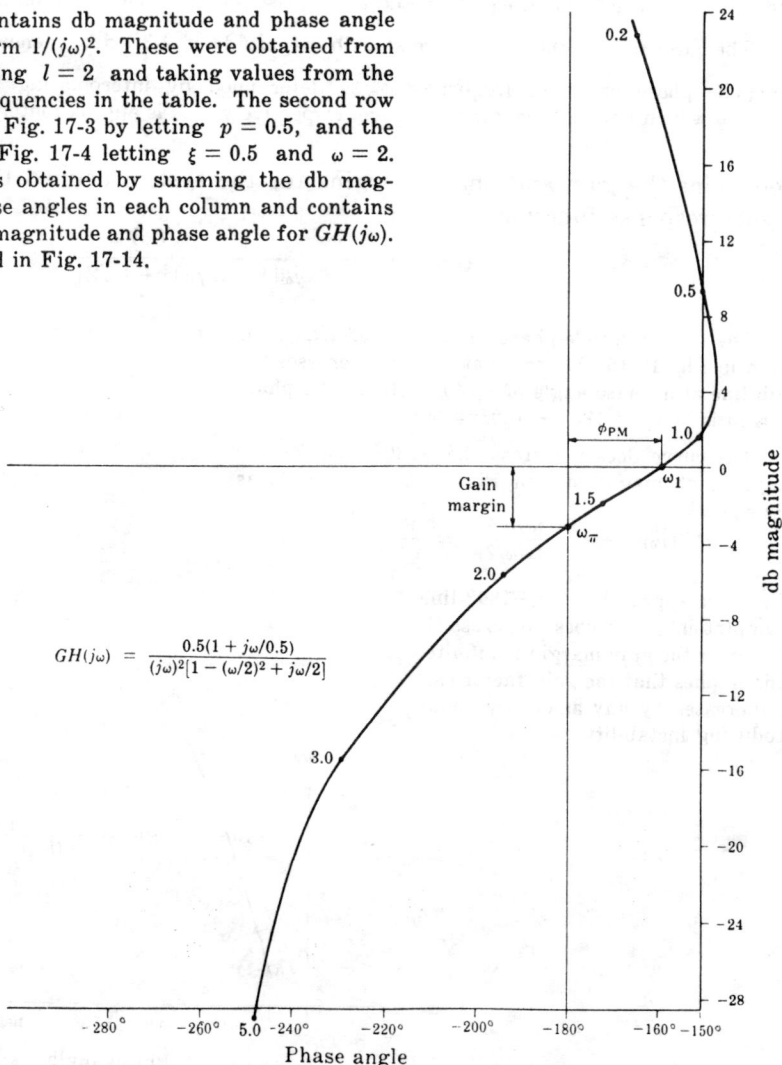

$$GH(j\omega) = \frac{0.5(1 + j\omega/0.5)}{(j\omega)^2[1 - (\omega/2)^2 + j\omega/2]}$$

Fig. 17-14

GAIN AND PHASE MARGINS

17.5. Determine the gain and phase margins for the system of Problem 17.2.

The db magnitude-phase angle plot for the open-loop transfer function of this system is given in Fig. 17-13 (Problem 17.2). We see that the curve crosses the 0 db line at a phase angle of $-162°$. Therefore the phase margin is $\phi_{PM} = 180° - 162° = 18°$.

(The gain crossover frequency ω_1 is determined by interpolating along the curve between $\omega = 1.0$ and $\omega = 1.5$ which bound ω_1 below and above, respectively. ω_1 is approximately 1.2 rad/s.)

The curve crosses the $-180°$ line at a db magnitude of -6 db. Hence, gain margin $= -(-6) = 6$ db.

(The phase crossover frequency ω_π is determined by interpolating along the curve between $\omega = 1.5$ and $\omega = 2.0$ which bound ω_π below and above. ω_π is approximately 1.75 rad/s.)

17.6. Determine the gain and phase margins for the system of Problem 17.4.

The db magnitude-phase angle plot for the open-loop transfer function of this system is given in Fig. 17-14 (Problem 17.4). We see that the curve crosses the 0 db line at a phase angle of $-159°$. Therefore the phase margin is $\phi_{PM} = 180° - 159° = 21°$.

(The gain crossover frequency ω_1 is found by interpolating along the curve between $\omega = 1.0$ and $\omega = 1.5$ which bound ω_1 below and above, respectively. ω_1 is approximately 1.2 rad/s.)

The curve crosses the $-180°$ line at a db magnitude of -3.1 db. Hence, gain margin $= 3.1$ db.

(The phase crossover frequency ω_π is determined by interpolating between $\omega = 1.5$ and $\omega = 2.0$ which bound ω_π below and above, respectively. ω_π is approximately 1.7 rad/s.)

17.7. Determine the gain and phase margins for the system defined by the open-loop frequency response function

$$GH(j\omega) = \frac{1 + j\omega/0.5}{j\omega[1 - (\omega/2)^2 + j\omega/2]}$$

The db magnitude-phase angle plot of $GH(j\omega)$ is given in Fig. 17-15. We see that the curve crosses the 0 db line at a phase angle of $-140°$. Hence the phase margin is $\phi_{PM} = 180° - 140° = 40°$.

The curve does not cross the $-180°$ line for the range of db magnitudes in Fig. 17-15. However, as $\omega \to \infty$,

$$GH(j\omega) \longrightarrow \frac{j\omega/0.5}{-j\omega(\omega/2)^2} = \frac{8}{\omega^2}\underline{/-180°}$$

The curve approaches the $-180°$ line asymptotically but does not cross it. Therefore the gain margin is infinite. This implies that the gain factor can be increased by any amount without producing instability.

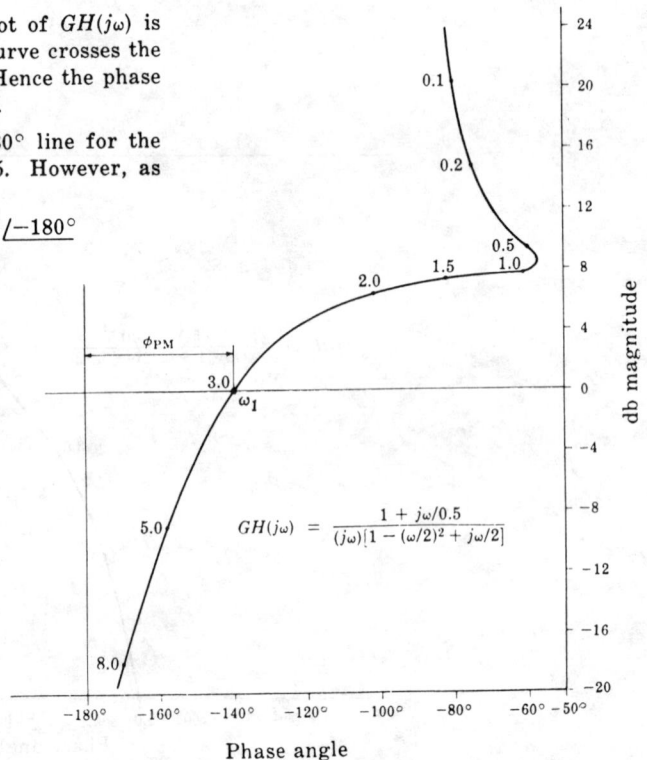

$$GH(j\omega) = \frac{1 + j\omega/0.5}{(j\omega)[1 - (\omega/2)^2 + j\omega/2]}$$

Phase angle

Fig. 17-15

NICHOLS CHART

17.8. Show that the locus of points on a db magnitude-phase angle plot for which the magnitude of the closed-loop frequency response function $\frac{C}{R}(j\omega)$ of a unity feedback system equals a constant M is defined by Equation (*17.12*), page 319.

Using Equation (*17.11*), $\left|\dfrac{C}{R}(j\omega)\right|$ can be written as

$$\left|\frac{C}{R}(j\omega)\right| = \left|\frac{|G(j\omega)|\ \underline{/\phi_G}}{1 + |G(j\omega)|\ \underline{/\phi_G}}\right|$$

Since $|G(j\omega)|\ \underline{/\phi_G} = |G(j\omega)|\cos\phi_G + j|G(j\omega)|\sin\phi_G$, this can be rewritten as

$$\left|\frac{C}{R}(j\omega)\right| = \left|\frac{|G(j\omega)|\cos\phi_G + j|G(j\omega)|\sin\phi_G}{1 + |G(j\omega)|\cos\phi_G + j|G(j\omega)|\sin\phi_G}\right|$$

$$= \sqrt{\frac{|G(j\omega)|^2\cos^2\phi_G + |G(j\omega)|^2\sin^2\phi_G}{[1 + |G(j\omega)|\cos\phi_G]^2 + |G(j\omega)|^2\sin^2\phi_G}} = \sqrt{\frac{|G(j\omega)|^2}{1 + 2|G(j\omega)|\cos\phi_G + |G(j\omega)|^2}}$$

If we set the last expression equal to M, square both sides and clear the fraction, we obtain

$$M^2[\,|G(j\omega)|^2 + 2|G(j\omega)|\cos\phi_G + 1\,] = |G(j\omega)|^2$$

which can be written as

$$(M^2 - 1)\,|G(j\omega)|^2 + 2M^2\,|G(j\omega)|\cos\phi_G + M^2 = 0$$

Dividing by $(M^2 - 1)$ we obtain Equation (*17.12*), as required.

17.9. Show that the locus of points on a db magnitude-phase angle plot for which the tangent of the argument of the closed-loop frequency response function $\frac{C}{R}(j\omega)$ of a unity feedback system equals a constant N is defined by Equation (*17.13*), page 319.

Using Equation (*17.11*), $\arg\dfrac{C}{R}(j\omega)$ can be written as

$$\arg\left[\frac{C}{R}(j\omega)\right] = \arg\left[\frac{|G(j\omega)|\ \underline{/\phi_G}}{1 + |G(j\omega)|\ \underline{/\phi_G}}\right]$$

Since $|G(j\omega)|\ \underline{/\phi_G} = |G(j\omega)|\cos\phi_G + j|G(j\omega)|\sin\phi_G$,

$$\arg\left[\frac{C}{R}(j\omega)\right] = \arg\left[\frac{|G(j\omega)|\cos\phi_G + j|G(j\omega)|\sin\phi_G}{1 + |G(j\omega)|\cos\phi_G + j|G(j\omega)|\sin\phi_G}\right]$$

Multiplying numerator and denominator of the term in brackets by the complex conjugate of the denominator yields

$$\arg\left[\frac{C}{R}(j\omega)\right] = \arg\left[\frac{(|G(j\omega)|\cos\phi_G + j|G(j\omega)|\sin\phi_G)(1 + |G(j\omega)|\cos\phi_G - j|G(j\omega)|\sin\phi_G)}{(1 + |G(j\omega)|\cos\phi_G)^2 + |G(j\omega)|^2\sin^2\phi_G}\right]$$

Since the denominator of the term in the last brackets is real, $\arg\left[\dfrac{C}{R}(j\omega)\right]$ is determined by the numerator only. That is,

$$\arg\left[\frac{C}{R}(j\omega)\right] = \arg\,[(|G(j\omega)|\cos\phi_G + j|G(j\omega)|\sin\phi_G)(1 + |G(j\omega)|\cos\phi_G - j|G(j\omega)|\sin\phi_G)]$$

$$= \arg\,[\,|G(j\omega)|\cos\phi_G + |G(j\omega)|^2 + j|G(j\omega)|\sin\phi_G]$$

using $\cos^2\phi_G + \sin^2\phi_G = 1$. Therefore

$$\tan\left[\arg\frac{C}{R}(j\omega)\right] = \frac{|G(j\omega)|\sin\phi_G}{|G(j\omega)|\cos\phi_G + |G(j\omega)|^2}$$

Equating this to N, cancelling the common $|G(j\omega)|$ term and clearing the fraction, we obtain

$$N[\cos\phi_G + |G(j\omega)|] = \sin\phi_G$$

which can be rewritten in the form of Equation (*17.13*), as required.

17.10. Construct the db magnitude-phase angle plot of the locus defined by Equation (*17.12*) for db magnitude of $\dfrac{C}{R}(j\omega)$ equal to 6 db.

$$20 \log_{10} \left| \frac{C}{R}(j\omega) \right| = 6 \text{ db} \quad \text{implies that} \quad \left| \frac{C}{R}(j\omega) \right| = 2. \quad \text{Therefore we let} \quad M = 2 \quad \text{in Equation}$$

(*17.12*) and obtain

$$|G(j\omega)|^2 + \frac{8}{3}|G(j\omega)| \cos \phi_G + \frac{4}{3} = 0$$

as the equation defining the locus. Since $|\cos \phi_G| \leqq 1$, $|G(j\omega)|$ may take on only those values for which this constraint is satisfied. To determine bounds of $|G(j\omega)|$, we let $\cos \phi_G$ take on its two extreme values of plus and minus unity. For $\cos \phi_G = 1$, the locus equation becomes

$$|G(j\omega)|^2 + \frac{8}{3}|G(j\omega)| + \frac{4}{3} = 0$$

with solutions $|G(j\omega)| = -2$ and $|GH(j\omega)| = -\frac{2}{3}$. Since an absolute value cannot be negative, these solutions are discarded. This implies that the locus does not exist on the 0° line (in general, any line which is a multiple of 360°), which corresponds to $\cos \phi_G = 1$.

For $\cos \phi_G = -1$, the locus equation becomes

$$|G(j\omega)|^2 - \frac{8}{3}|G(j\omega)| + \frac{4}{3} = 0$$

with solutions $|G(j\omega)| = 2$ and $|G(j\omega)| = \frac{2}{3}$. These are valid solutions for $|G(j\omega)|$ and are the extreme values which $|G(j\omega)|$ can assume.

Solving the locus equation for $\cos \phi_G$, we obtain

$$\cos \phi_G = \frac{-[\frac{4}{3} + |G(j\omega)|^2]}{\frac{8}{3}|G(j\omega)|}$$

The curves obtained from this relationship are periodic with period 360°. The plot is restricted to a single cycle in the vicinity of the −180° line and is obtained by solving for ϕ_G at several values of $|G(j\omega)|$ between the bounds 2 and $\frac{2}{3}$. The results are given in Table 17.5.

Table 17.5

$|G(j\omega)|$	$20 \log_{10} |G(j\omega)|$	$\cos \phi_G$	ϕ_G	
2.0	6 db	−1	−180°	——
1.59	4	−0.910	−204.5°	−155.5°
1.26	2	−0.867	−209.9°	−150.1°
1.0	0	−0.873	−209.2°	−150.8°
0.79	−2	−0.928	−201.9°	−158.1°
0.67	−3.5	−1	−180°	——

Note that there are two values of ϕ_G whenever $|\cos \phi_G| < 1$. The resulting plot is shown in Fig. 17-16.

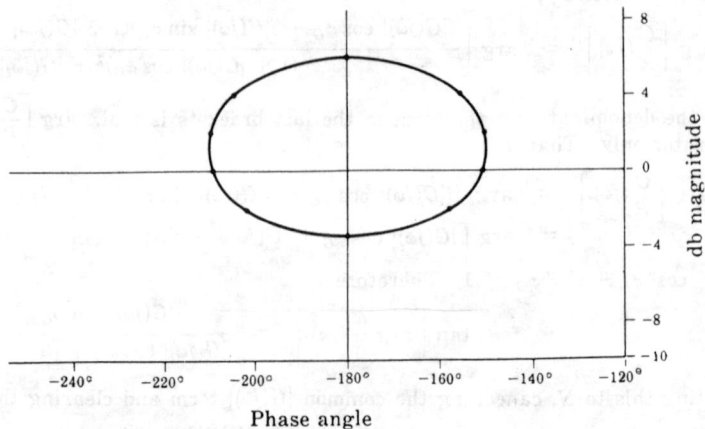

Fig. 17-16

17.11. Construct the db magnitude-phase angle plot of the locus defined by Equation (*17.13*), page 319, for $\tan\left[\arg\dfrac{C}{R}(j\omega)\right] = N = -\infty$.

$\tan\left[\arg\dfrac{C}{R}(j\omega)\right] = -\infty$ implies that $\arg\dfrac{C}{R}(j\omega) = -90° + k360°$, $k = 0, \pm1, \pm2, \ldots$, or $\arg\dfrac{C}{R}(j\omega) = -270° + k360°$, $k = 0, \pm1, \pm2, \ldots$. We will plot only the cycle between $-360°$ and $0°$, which corresponds to $k = 0$. Setting $N = -\infty$ in Equation (*17.13*), we obtain the locus equation

$$|G(j\omega)| + \cos\phi_G = 0 \quad \text{or} \quad \cos\phi_G = -|G(j\omega)|$$

Since $|\cos\phi_G| \leq 1$, the locus exists only for $0 \leq |G(j\omega)| \leq 1$ or, equivalently,

$$-\infty \leq 20\log_{10}|G(j\omega)| \leq 0$$

To obtain the plot we use the locus equation to calculate values of db magnitude of $G(j\omega)$ corresponding to several values of ϕ_G. The results of these calculations are given in Table 17.6. The desired plot is shown in Fig. 17-17.

Table 17.6

ϕ_G		$\cos\phi_G$	$|G(j\omega)|$	$20\log_{10}|G(j\omega)|$
$-180°$	—	-1	1	0 db
$-153°$	$-207°$	-0.893	0.893	-1.0
$-135°$	$-222.5°$	-0.707	0.707	-3
$-120°$	$-240°$	-0.5	0.5	-6
$-110.7°$	$-249.3°$	-0.354	0.354	-9
$-104.5°$	$-255.5°$	-0.25	0.25	-12
$-100.3°$	$-259.8°$	-0.178	0.178	-15

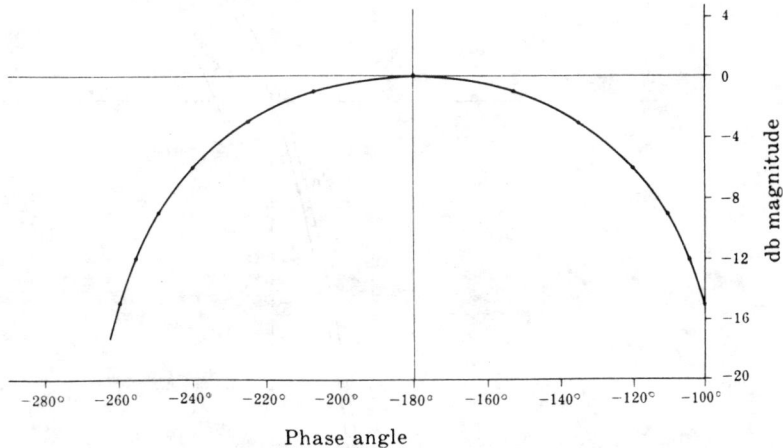

Fig. 17-17

CLOSED-LOOP FREQUENCY RESPONSE FUNCTIONS

17.12. Construct the db magnitude-phase angle plot of the closed-loop frequency response function $\dfrac{C}{R}(j\omega)$ of the unity feedback system whose open-loop transfer function is

$$G = \frac{2}{s(1+s)(1+s/3)}$$

$$\frac{C}{R}(j\omega) = \frac{G(j\omega)}{1 + G(j\omega)} = \frac{6}{(j\omega)^3 + 4(j\omega)^2 + 3j\omega + 6} = \frac{6}{(6 - 4\omega^2) + j(3\omega - \omega^3)}$$

Therefore $20 \log_{10} \left| \dfrac{C}{R}(j\omega) \right| = 10 \log_{10} \left| \dfrac{C}{R}(j\omega) \right|^2 = 10 \log_{10} \dfrac{36}{(6 - 4\omega^2)^2 + (3\omega - \omega^3)^2}$

and $\arg \left[\dfrac{C}{R}(j\omega) \right] = -\tan^{-1} \dfrac{3\omega - \omega^3}{6 - 4\omega^2}$

Now we calculate $20 \log_{10} \left| \dfrac{C}{R}(j\omega) \right|$ and $\arg \left[\dfrac{C}{R}(j\omega) \right]$ for several values of ω. The results are tabulated in Table 17.7.

Table 17.7

ω	$20 \log_{10} \left\| \dfrac{C}{R}(j\omega) \right\|$	$\arg \left[\dfrac{C}{R}(j\omega) \right]$
0	0 db	$0°$
0.2	0.20	$-5.8°$
0.5	1.27	$-15.4°$
1.0	6.54	$-45°$
1.25	10.38	$-97.9°$
1.5	5.46	$-159.5°$
2.0	-4.62	$-191.3°$
3.0	-15.4	$-210.5°$

The db magnitude-phase angle plot of $\dfrac{C}{R}(j\omega)$ is graphed using the values in the table, as shown by the solid line in Fig. 17-18.

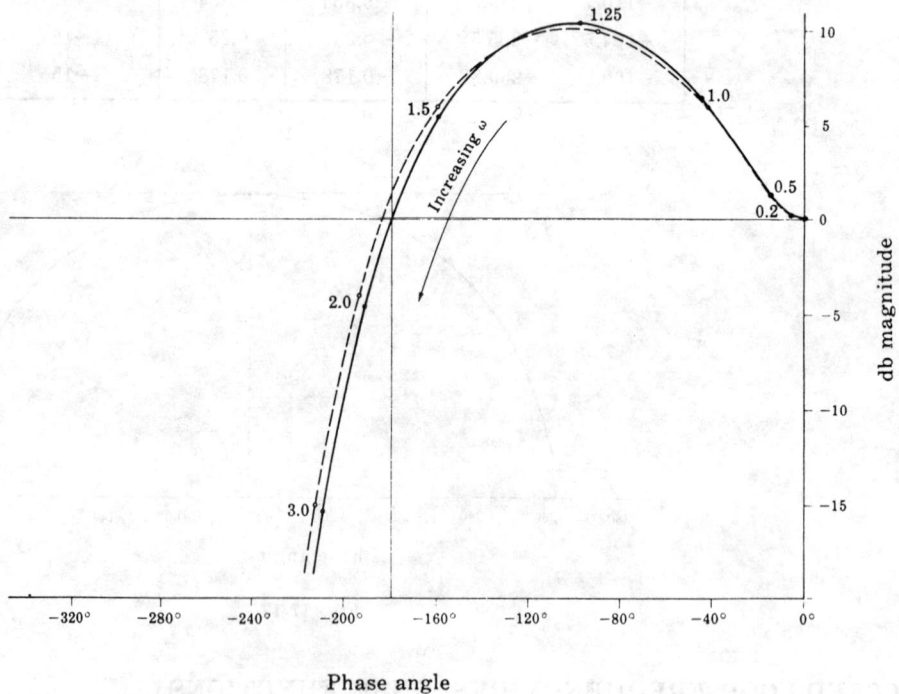

Fig. 17-18

17.13. Using the technique discussed in Section 17.6, page 321, solve Problem 17.12 again.

The Nichols chart plot of $G(j\omega)$ is shown in Fig. 17-19 below. We determine values for the db magnitude of $\left| \dfrac{C}{R}(j\omega) \right|$ and $\arg \left[\dfrac{C}{R}(j\omega) \right]$ by interpolating values of db magnitude and phase angle on the Nichols chart plot for $\omega = 0, 0.2, 0.5, 1.0, 1.25, 1.5, 2.0, 3.0$. These values are given in Table 17.8.

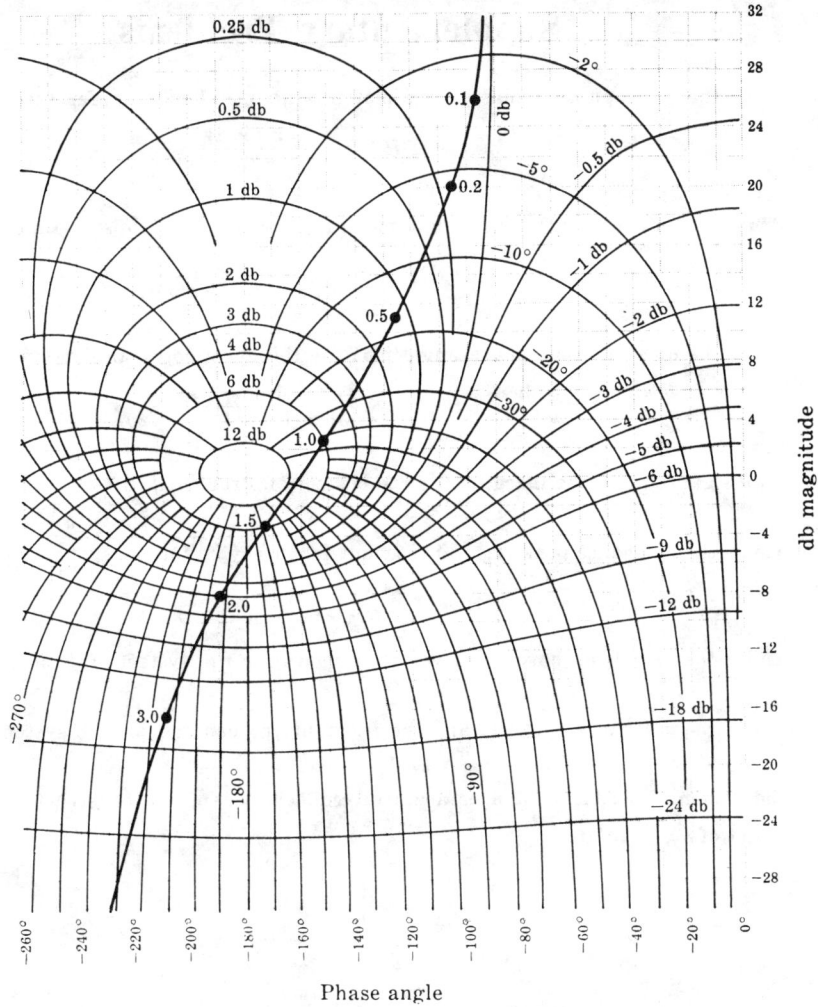

Fig. 17-19

Table 17.8

ω	$20 \log_{10} \left| \dfrac{C}{R}(j\omega) \right|$	$\arg \left[\dfrac{C}{R}(j\omega) \right]$
0	0 db	0°
0.2	0.2	−6°
0.5	1.2	−15°
1.0	6.0	−42°
1.25	10.0	−90°
1.5	6.0	−155°
2.0	−4.0	−194°
3.0	−15.0	−212°

The discrepancies between the values in Tables 17.7 and 17.8 are due to the fact that the values in Table 17.8 are produced by interpolating along the curve to obtain values of db magnitude and phase angle. The db magnitude-phase angle plot of $\dfrac{C}{R}(j\omega)$, graphed using the values in the table, is illustrated by the broken line in Fig. 17-18.

Supplementary Problems

17.14. Construct the db magnitude-phase angle plot for the open-loop transfer function

$$GH = \frac{5(s+2)}{s(s+3)(s+5)}$$

17.15. Construct the db magnitude-phase angle plot for the open-loop transfer function

$$GH = \frac{10}{s(1+s/5)(1+s/50)}$$

17.16. Construct the db magnitude-phase angle plot for the open-loop transfer function

$$GH = \frac{1+s/2}{s(1+s)(1+s/4)(1+s/20)}$$

17.17. Determine gain and phase margins for the system of Problem 17.15.

17.18. Determine the resonance peak M_p and resonant frequency ω_p for the system whose open-loop transfer function is $GH = \dfrac{1}{s(1+s)(1+s/4)}$.

17.19. Determine the gain and phase crossover frequencies for the system of Problem 17.15.

17.20. Determine the resonance peak M_p and the resonant frequency ω_p of the system in Problem 17.15.

17.21. Let the system of Problem 17.15 be a unity feedback system and construct the db magnitude-phase angle plot of $\dfrac{C}{R}(j\omega)$.

Answers to Supplementary Problems

17.17. Gain margin $= 9.5$ db, $\phi_{PM} = 25°$.

17.18. $M_p = 1.3$ db, $\omega_p = 0.9$ rad/s.

17.19. $\omega_1 = 7$ rad/sec., $\omega_\pi = 14.5$ rad/s.

17.20. $M_p = 8$ db, $\omega_p = 7.2$ rad/s.

Chapter 18

Nichols Chart Design

18.1 DESIGN PHILOSOPHY

Design by analysis in the frequency domain using Nichols chart techniques is performed in the same general manner as the design methods described in previous chapters: appropriate compensation networks are introduced in the forward and/or feedback paths and the behavior of the resulting system is critically analyzed. In this manner, the Nichols chart plot is shaped and reshaped until the performance specifications are met. These specifications are most conveniently expressed in terms of frequency domain figures of merit such as gain and phase margin for transient performance and the error constants (Chapter 9) for the steady-state time-domain response.

Since the Nichols chart plot is a graph of the open-loop frequency response function $GH(j\omega)$, compensation can be introduced in the forward and/or feedback paths, thus changing $G(j\omega)$, $H(j\omega)$, or both.

We emphasize that no single compensation scheme is universally applicable.

18.2 GAIN FACTOR COMPENSATION

We have seen in several previous chapters (5, 12, 13, 16) that an unstable feedback system can sometimes be stabilized, or a stable system destabilized, by adjustment of the gain factor K of $GH(s)$. Nichols chart plots are particularly well suited for determining gain factor adjustments. However, when using Nichols techniques, it is more convenient to use the Bode gain K_B (Section 15.4), expressed in decibels (db), than the gain factor K. Changes in K_B and K, when given in db, are equal.

Example 18.1

The db magnitude-phase angle plot for an unstable system, represented by $GH(j\omega)$ with the Bode gain $K_B = 5$, is shown in Fig. 18-1. The instability of this system can be verified by a sketch of the Nyquist plot, or application of the Routh criterion. The Nyquist plot in Example 12.1, page 224, illustrates the general shape for all Nyquist plots of systems with one pole at the origin and two real poles in the left half plane. This graph indicates that positive phase and gain margins guarantee stability and negative phase and gain margins guarantee instability for such a system, which implies that a sufficient decrease in the Bode gain stabilizes the system. If the Bode gain is decreased from $20 \log_{10} 5$ db to $20 \log_{10} 2$ db, the system is stabilized. The db magnitude-phase angle plot for the compensated system is shown in Fig. 18-2. Further decrease in gain does not alter stability.

Note that the curves for $K_B = 5$ and $K_B = 2$ have identical shapes, the only difference being that the ordinates on the $K_B = 5$ curve exceed those on the $K_B = 2$ curve by $20 \log_{10}(5/2)$ db. Therefore changing the gain on a db magnitude-phase angle plot is accomplished by simply shifting the locus of $GH(j\omega)$ up or down by an appropriate number of db.

Even though absolute stability can often be altered by gain factor adjustment, this form of compensation is inadequate for most designs because other performance criteria such as those concerned with relative stability cannot usually be met without the inclusion of other types of compensators.

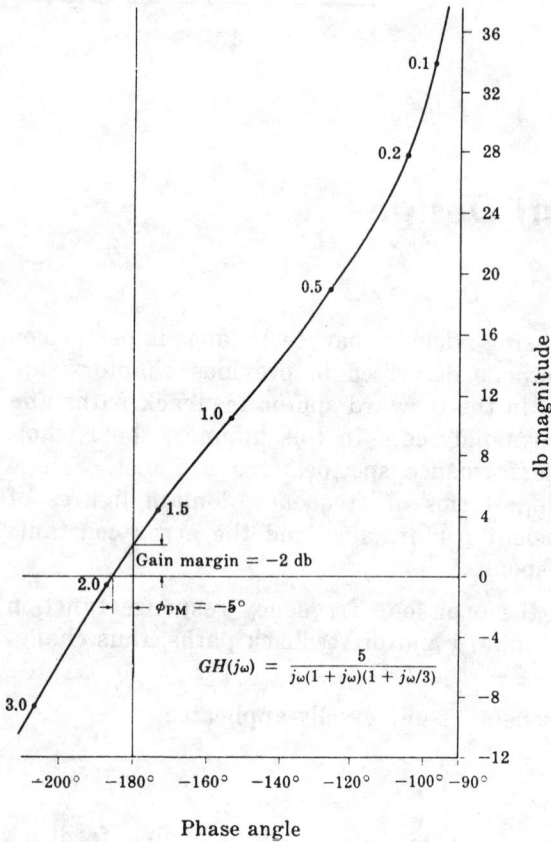

Fig. 18-1

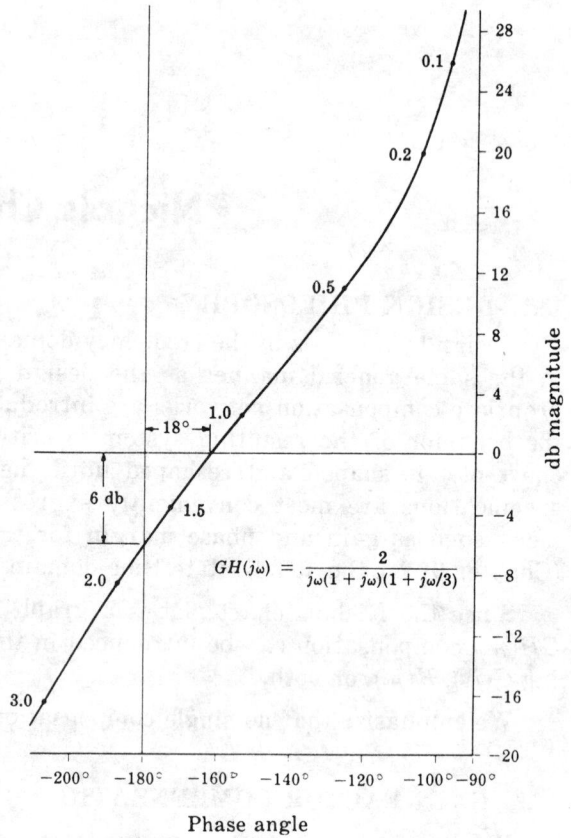

Fig. 18-2

18.3 GAIN FACTOR COMPENSATION USING CONSTANT AMPLITUDE CURVES

The Nichols chart may be used to determine K_B (for a *unity feedback* system) for a specified resonant peak M_p (in db). The following procedure requires drawing the db magnitude-phase angle plot only once.

Step 1: Draw the db magnitude-phase angle plot of $G(j\omega)$ for $K_B = 1$ on tracing paper. The scale of the graph must be the same as that on the Nichols chart.

Step 2: Overlay this plot on the Nichols chart so that the magnitude and phase angle scales of each sheet are aligned.

Step 3: Fix the Nichols chart and slide the plot up or down until it is just tangent to the constant amplitude curve of M_p db. The amount of shift in db is the required value of K_B.

Example 18.2

In Fig. 18-3(a) below, the db magnitude-phase angle plot of the open-loop transfer function of a particular unity feedback system with $K_B = 1$ is shown superimposed on a Nichols chart. The desired M_p is 4 db. We see in Fig. 18-3(b) below that, if the overlay is shifted upward by 4 db, then the resonant peak M_p of the system is 4 db. Thus the desired K_B is 4 db.

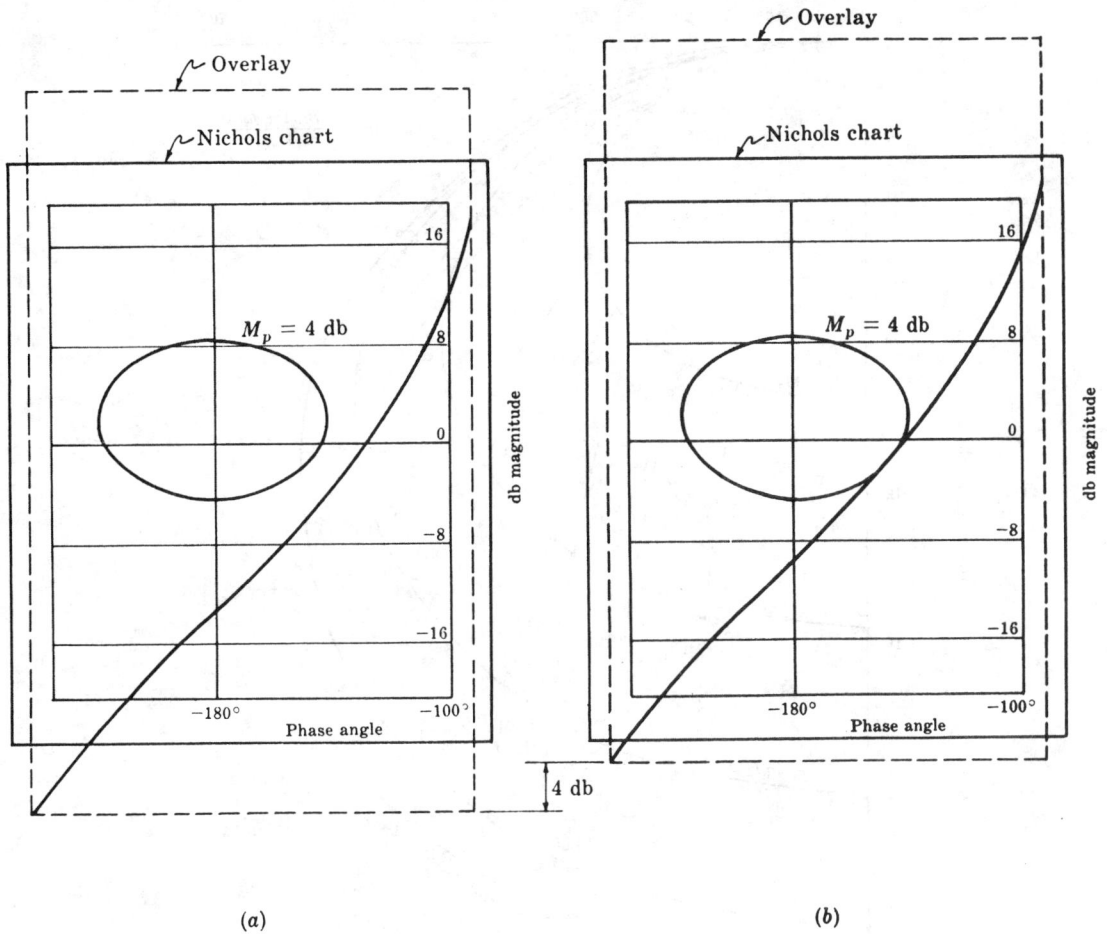

(a)   (b)

Fig. 18-3

18.4 LEAD COMPENSATION

The Bode form of the transfer function for a lead network is

$$P_{\text{Lead}} \; = \; \frac{(a/b)(1 + s/a)}{1 + s/b} \tag{18.1}$$

where $a/b < 1$. The db magnitude-phase angle plots of P_{Lead} for several values of b/a and with the normalized frequency ω/a as the parameter are shown in Fig. 18-4 below.

For some systems in which lead compensation in the forward loop is applicable, appropriate choice of a and b permits an increase in K_B, providing greater accuracy and less sensitivity, without adversely affecting transient performance. Conversely, for a given K_B, the transient performance can be improved. It is also possible to improve both the steady-state and transient responses with lead compensation.

The important properties of a lead network compensator are its phase-lead contribution in the low to medium frequency range (the vicinity of the resonant frequency ω_p) and its negligible attenuation at high frequencies. If a very large phase-lead is required, several lead networks may be cascaded.

Lead compensation generally increases the bandwidth of a system.

Phase angle

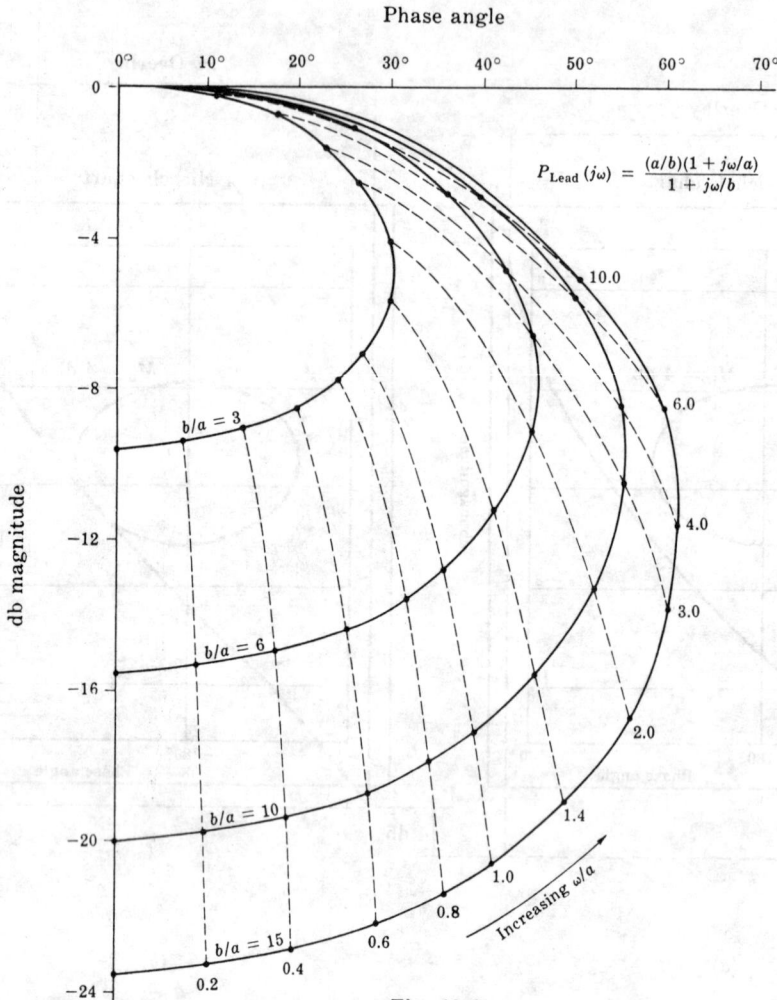

$$P_{\text{Lead}}(j\omega) = \frac{(a/b)(1 + j\omega/a)}{1 + j\omega/b}$$

Fig. 18-4

Example 18.3

The uncompensated unity feedback system whose open-loop transfer function is

$$GH = \frac{2}{s(1 + s)(1 + s/3)}$$

is to be designed to meet the following performance specifications:

(1) When the input is a unit ramp function, the steady-state error in position must be less than 0.25.

(2) $\phi_{\text{PM}} \cong 40°$.

(3) Resonance peak $\cong 1.7$ db.

Note that the Bode gain is equal to the velocity error constant K_v. Therefore the steady-state error for the uncompensated system is $e(\infty) = 1/K_v = 1/2$ [Equation (9.5), page 164]. From the db magnitude-phase angle plot of GH in Fig. 18-5 below, we see that $\phi_{\text{PM}} = 18°$ and $M_p = 5$ db.

The steady-state error is too large by a factor of 2; therefore the Bode gain must be increased by a factor of 2 (6 db). If we increase the Bode gain by 6 db, we obtain the plot labeled GH_1 in Fig. 18-5. The phase margin of GH_1 is about zero and the resonant peak is near infinity. Therefore the system is on the verge of instability.

Phase lead compensation can be used to improve the relative stability of the system. The compensated open-loop transfer function is

$$GH_2 = \frac{K_B(a/b)(1 + s/a)}{s(1 + s)(1 + s/3)(1 + s/b)} = \frac{4(1 + s/a)}{s(1 + s)(1 + s/3)(1 + s/b)}$$

where $K_B = 4(b/a)$ to satisfy the steady-state error.

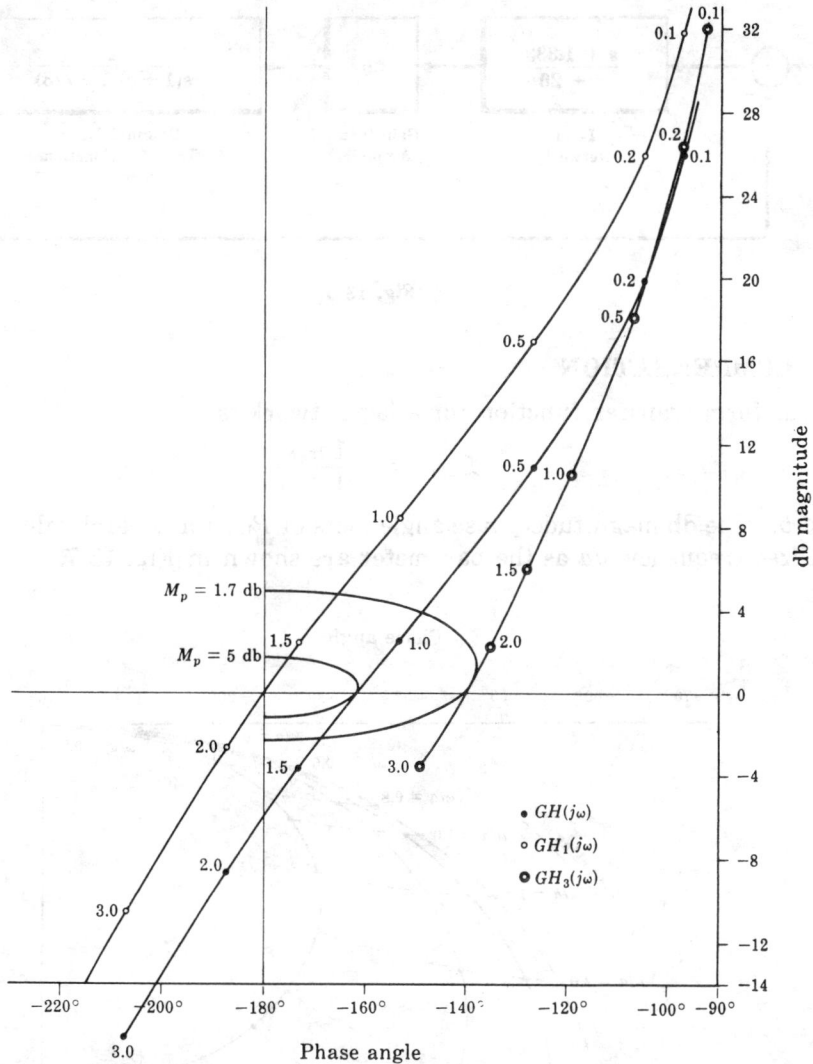

Fig. 18-5

One way of satisfying the requirements on ϕ_{PM} and M_p is to add 40° to 50° of phase lead to the GH_1 curve in the region $1 \leq \omega \leq 2.5$ without substantially changing the db magnitude. We have already chosen $K_B = 4(b/a)$ to compensate for the gain factor a/b in the lead network. Therefore we need concern ourselves only with the effect that the factor $\dfrac{1 + s/a}{1 + s/b}$ has on the GH_1 curve. Referring to Fig. 18-4, we see that in order to provide the necessary phase lead we will require $b/a \cong 10$. We note that the curves of Fig. 18-4 include the effect of the gain factor a/b of the lead network. Since we have already compensated for this gain factor, we must add $20 \log_{10}(b/a)$ to the db magnitudes on the curve. In order to keep the db magnitude contribution of the lead network small in the region $1 \leq \omega \leq 2.5$, we let $b/a = 15$ and choose a so that only the lower portion of the curve ($\omega/a \leq 3.0$) contributes in the region of interest $1 \leq \omega \leq 2.5$. In particular, we let $a = 1.333$. Then the compensated open-loop transfer function is

$$GH_3 = \frac{4(1 + s/1.333)}{s(1 + s)(1 + s/3)(1 + s/20)}$$

The db magnitude-phase angle plot of GH_3 is shown in Fig. 18-5. We see that $\phi_{PM} = 40.5°$ and $M_p = 1.7$ db. Thus the specifications are all met. We note, however, that the resonant frequency ω_p of the compensated system is about 2.25 rad/s. For the uncompensated system defined by GH it is about 1.2 rad/s. Thus the bandwidth has been increased.

A block diagram of the fully compensated system is shown in Fig. 18-6 below.

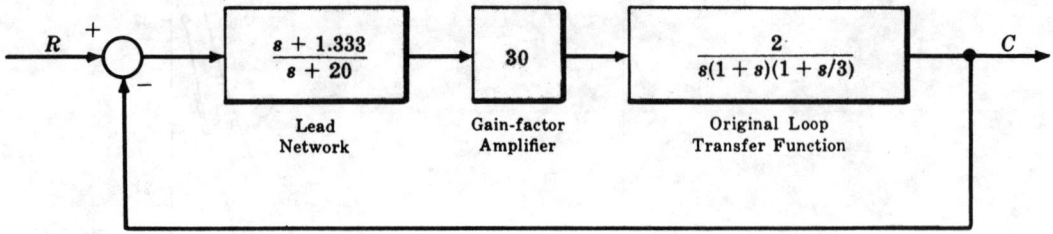

Fig. 18-6

18.5 LAG COMPENSATION

The Bode form transfer function for a lag network is

$$P_{\text{Lag}} \;=\; \frac{1 + s/b}{1 + s/a} \tag{18.2}$$

where $a < b$. The db magnitude-phase angle plots of P_{Lag} for several values of b/a and with the normalized frequency ω/a as the parameter are shown in Fig. 18-7.

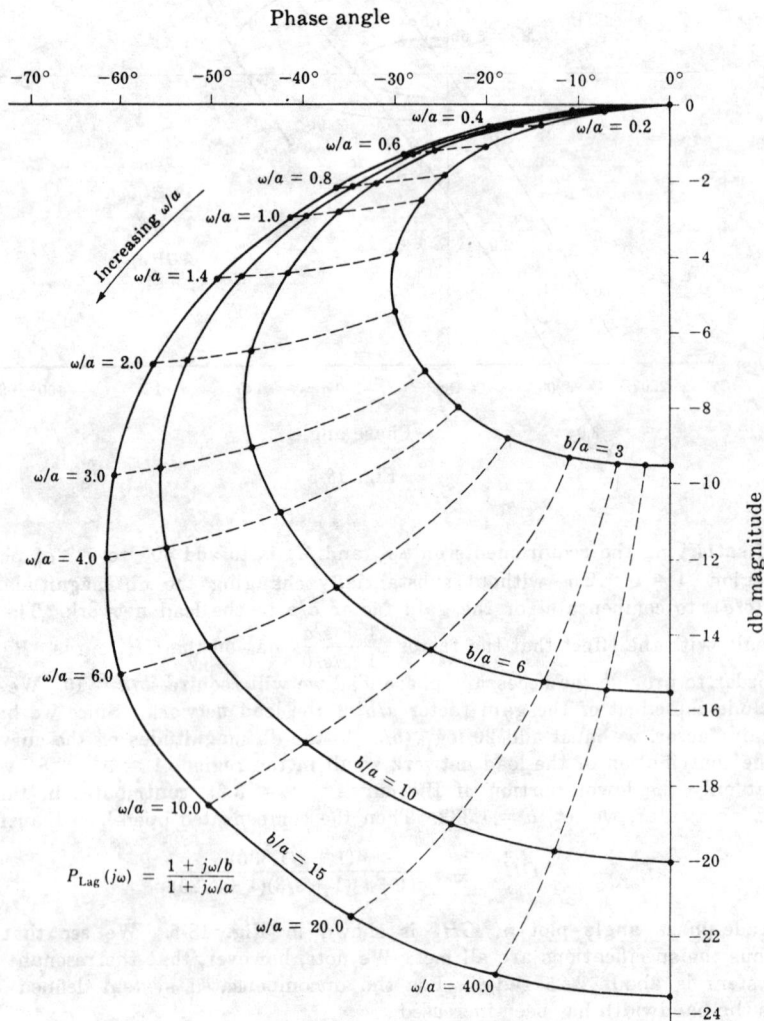

$$P_{\text{Lag}}(j\omega) \;=\; \frac{1 + j\omega/b}{1 + j\omega/a}$$

Fig. 18-7

The lag network provides compensation by attenuating the high frequency portion of the db magnitude-phase angle plot. Higher attenuation is provided by cascading several lag networks.

Several general effects of lag compensation are:

1. The bandwidth of the system is usually decreased.

2. The predominant time constant τ of the system is usually increased, producing a more sluggish system.

3. For a given relative stability, the value of the error constant is increased.

4. For a given error constant, relative stability is improved.

The procedure for using lag compensation is essentially the same as that for lead compensation.

Example 18.4

Let us redesign the system of Example 18.3 using gain factor plus lag compensation. The steady-state specification is again satisfied by GH_1. The db magnitude-phase angle plot of GH_1 is repeated in Fig. 18-8. Since $P_{\text{Lag}}(j0) = 1$, introduction of the lag network after the steady-state specification has been met by gain-factor compensation does not require an additional increase in gain-factor.

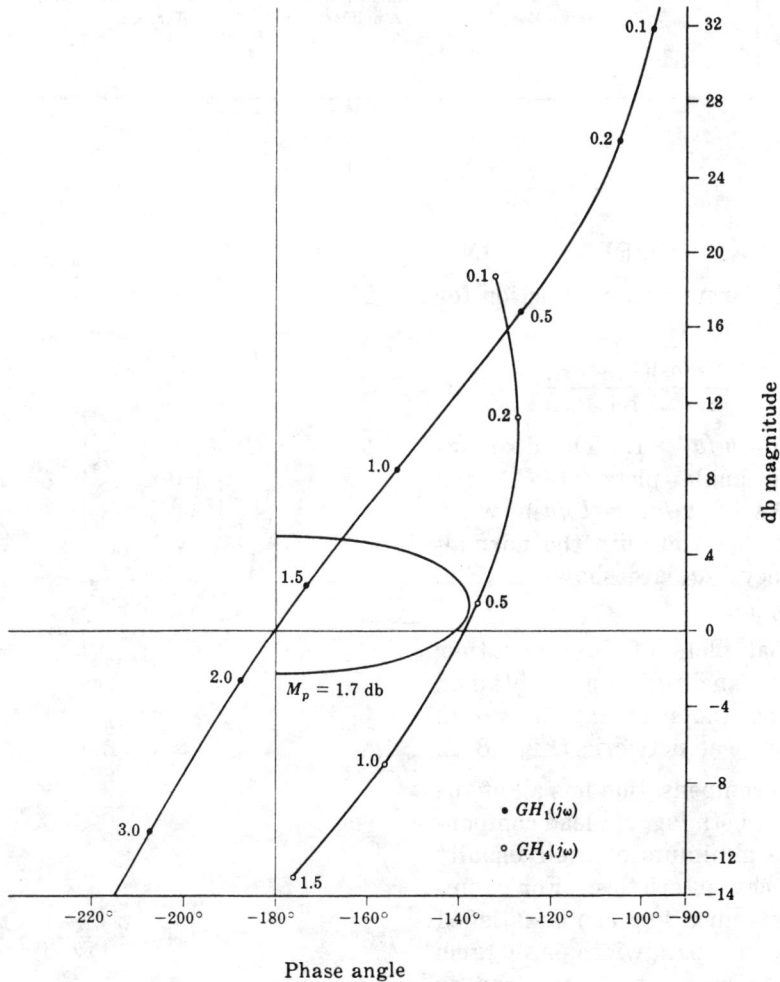

Fig. 18-8

Incorporating the lag network, we get the open-loop transfer function

$$GH_4 = \frac{4(1+s/b)}{s(1+s)(1+s/3)(1+s/a)}$$

One way of satisfying the requirements on ϕ_{PM} and M_p is to choose a and b such that the GH_1 curve is attenuated by about 12 db in the region $0.7 \leq \omega \leq 2.0$ without substantial change in the phase angle. Since the lag network introduces some phase lag, it is necessary to attenuate the curve more than 12 db. Referring to Fig. 18-7, we see that if we choose $b/a = 6$, a maximum of 15.5 db attenuation is possible. If we choose $a = 0.015$, then at a frequency $\omega = 0.5$ ($\omega/a = 33.33$) 15.4 db of attenuation is obtained from the lag network, with a phase lag of $-9°$. GH_4 can now be written as

$$GH_4 = \frac{4(1+s/0.09)}{s(1+s)(1+s/3)(1+s/0.015)}$$

where $b = 6a = 0.09$. The db magnitude-phase angle plot of GH_4 is given in Fig. 18-8. We see that $\phi_{PM} = 41°$ and $M_p \cong 1.7$, which satisfy the specifications. We note that the resonant frequency ω_p of the compensated system is about 0.5 rad/s. For the uncompensated system defined by GH, ω_p is about 1.2 rad/s. A block diagram of the fully compensated system is shown in Fig. 18-9.

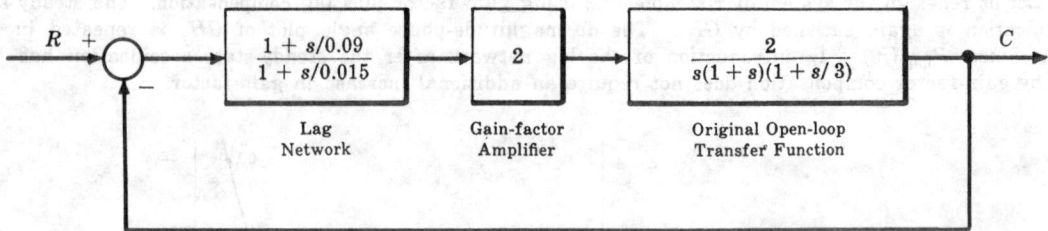

Fig. 18-9

18.6 LAG-LEAD COMPENSATION

The Bode form transfer function for a lag-lead network is

$$P_{LL} = \frac{(1+s/a_1)(1+s/b_2)}{(1+s/b_1)(1+s/a_2)} \quad (18.3)$$

where $b_1/a_1 = b_2/a_2 > 1$. The db magnitude-phase angle plots of P_{LL} for a few values of b_1/a_1 ($= b_2/a_2$), when $a_1/a_2 = 6, 10, 100$, and with the normalized frequency ω/a_2 are shown in Fig. 18-10(a), (b), (c).

Additional plots of P_{LL} for other values of b_1/a_1 and a_1/a_2 can be obtained by combining plots of lag networks (Fig. 18-7) and lead networks (Fig. 18-4).

Lag-lead compensation has all of the advantages of both lag and lead compensation and a minimum of their usually undesirable characteristics. For example, system specifications can be satisfied without excessive bandwidth or sluggish time response caused by phase lead or lag, respectively.

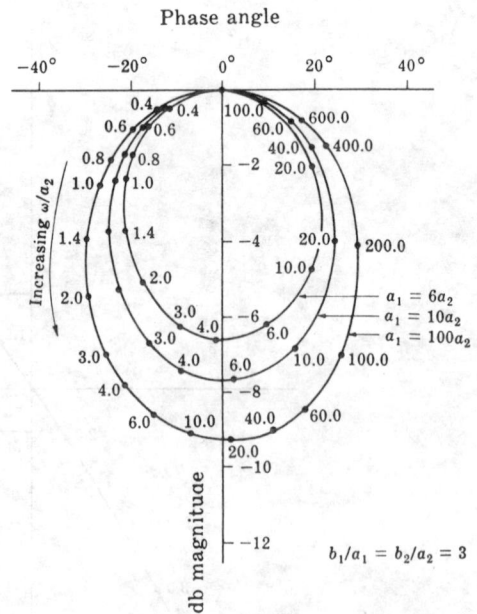

$$P_{LL}(j\omega) = \frac{(1+j\omega/a_1)(1+j\omega/b_2)}{(1+j\omega/b_1)(1+j\omega/a_2)}$$

Fig. 18-10(a)

Phase angle

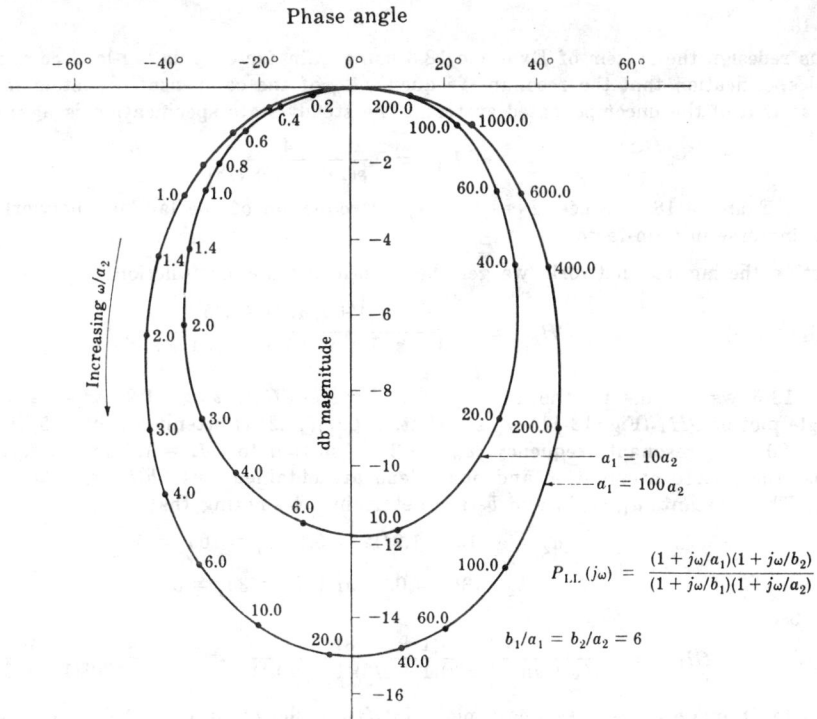

$$P_{\text{I.I.}}(j\omega) = \frac{(1 + j\omega/a_1)(1 + j\omega/b_2)}{(1 + j\omega/b_1)(1 + j\omega/a_2)}$$

$$b_1/a_1 = b_2/a_2 = 6$$

Fig. 18-10(b)

Phase angle

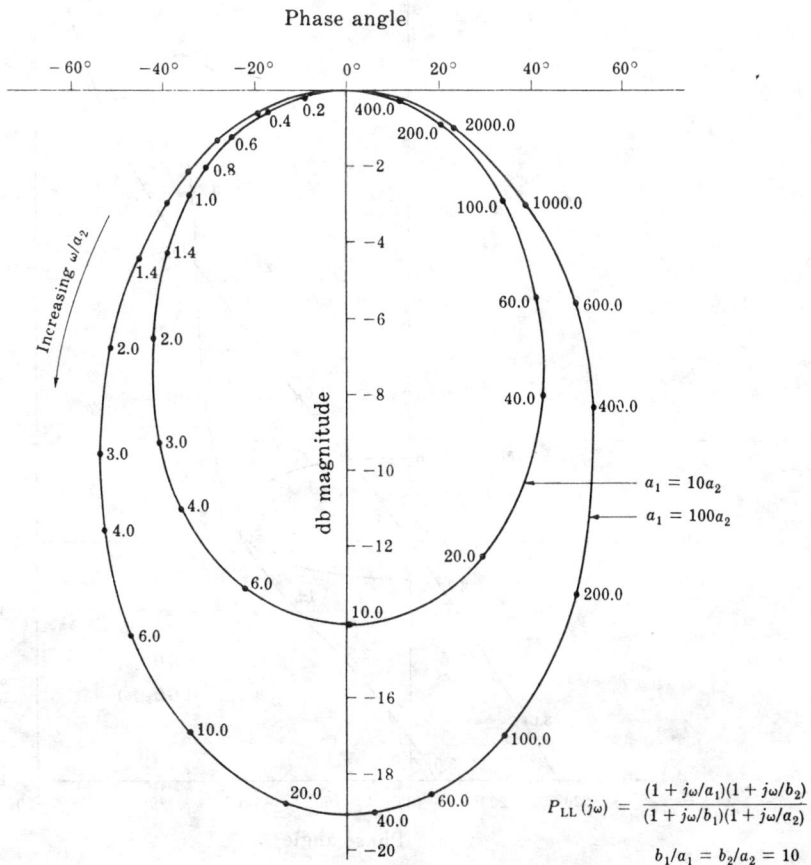

$$P_{\text{LL}}(j\omega) = \frac{(1 + j\omega/a_1)(1 + j\omega/b_2)}{(1 + j\omega/b_1)(1 + j\omega/a_2)}$$

$$b_1/a_1 = b_2/a_2 = 10$$

Fig. 18-10(c)

Example 18.5

Let us redesign the system of Example 18.3 using gain factor plus lag-lead compensation. We add the additional specification that the resonant frequency ω_p of the compensated system must be approximately the same as that of the uncompensated system. The steady-state specification is again satisfied by

$$GH_1 = \frac{4}{s(1+s)(1+s/3)}$$

as shown in Example 18.3. Since $P_{LL}(j0) = 1$, introduction of the lag-lead network does not require an additional increase in gain-factor.

Inserting the lag-lead network, we get the open-loop transfer function

$$GH_5 = \frac{4(1+s/a_1)(1+s/b_2)}{s(1+s)(1+s/3)(1+s/b_1)(1+s/a_2)}$$

From Fig. 18-5, we see that for the uncompensated system GH, $\omega_p = 1.2$ rad/s. From the db magnitude-phase angle plot of GH_1 (Fig. 18-11) we see that, if $GH_1(j1.2)$ is attenuated by 6.5 db and has its phase increased by $20°$, the resonant frequency $\omega_p = 1.2$ is shifted to $M_p = 1.7$ db. Referring to Fig. 18-10(a), we see that the desired attenuation and phase lead are obtained with $b_1/a_1 = b_2/a_2 = 3$, $a_1/a_2 = 10$, and $\omega/a_2 = 12$. The constants a_1, a_2, b_1 and b_2 are determined by noting that

$$a_2 = \omega_p/12 = 1.2/12 = 0.1; \quad a_1 = 10a_2 = 1$$

$$b_2 = 3a_2 = 0.3; \quad \text{and} \quad b_1 = 3a_1 = 3$$

GH_5 then becomes

$$GH_5 = \frac{4(1+s)(1+s/0.3)}{s(1+s)(1+s/3)(1+s/3)(1+s/0.1)} = \frac{4(1+s/0.3)}{s(1+s/3)^2(1+s/0.1)}$$

The complete db magnitude-phase angle plot of GH_5 is shown in Fig. 18-11. We see that $\phi_{PM} = 40.5°$, $M_p = 1.7$ db, and the resonant frequency $\omega_p \cong 1.15$. Thus all specifications have been satisfied.

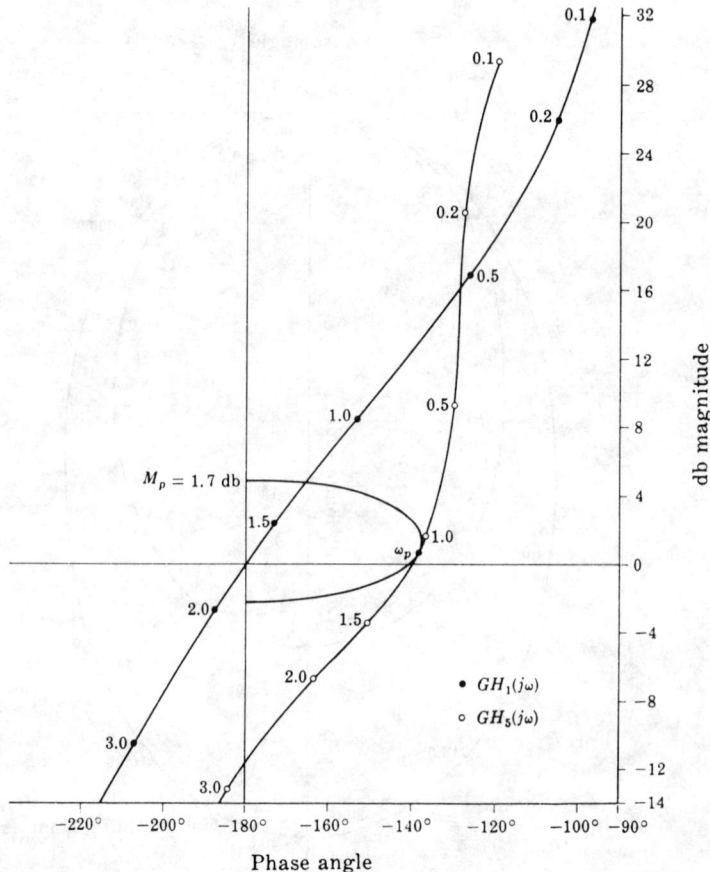

Fig. 18-11

Solved Problems

GAIN FACTOR COMPENSATION

18.1. The db magnitude-phase angle plot of the open-loop frequency response function

$$GH(j\omega) = \frac{K_B[1 - (\omega/2)^2 + j\omega/2]}{j\omega(1 + j\omega/0.5)^2(1 + j\omega/4)}$$

is shown in Fig. 18-12 for $K_B = 1$. The closed-loop system defined by $GH(j\omega)$ is stable for $K_B = 1$. Determine a value of K_B for which the phase margin is 45°.

$\phi_{PM} = 180° + \arg GH(j\omega_1)$, where ω_1 is the gain crossover frequency. For $\phi_{PM} = 45°$, ω_1 must be chosen so that $\arg GH(j\omega_1) = -135°$. If we draw a vertical line with abscissa of $-135°$, it intersects the $GH(j\omega)$ curve at a point $\omega_1' \cong 0.25$ rad/s, where $\arg GH(j\omega_1') = -135°$. The ordinate of this point of intersection is 10.5 db. If we decrease K_B by 10.5 db, the gain crossover frequency becomes ω_1', and $\phi_{PM} = 45°$. A decrease of 10.5 db implies that $20 \log_{10} K_B = -10.5$, or $K_B = 10^{-10.5/20} = 0.3$. Further decrease in K_B increases ϕ_{PM} beyond 45°.

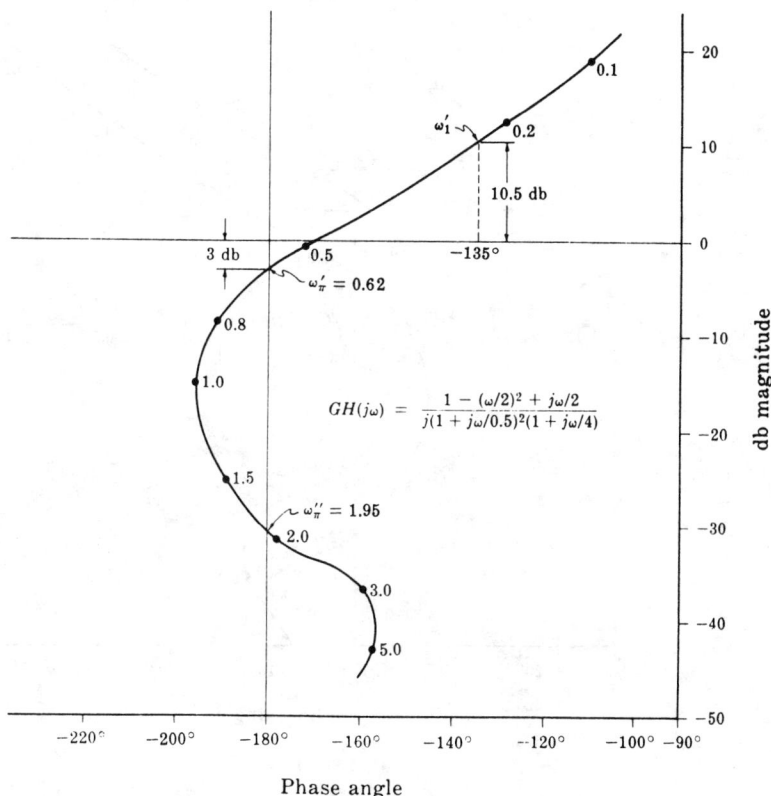

$$GH(j\omega) = \frac{1 - (\omega/2)^2 + j\omega/2}{j(1 + j\omega/0.5)^2(1 + j\omega/4)}$$

Fig. 18-12

18.2. For the system in Problem 18.1, determine the value of K_B for which the system is stable and the gain margin is 10 db.

Gain margin $= -20 \log_{10} |GH(j\omega_\pi)|$ db, where ω_π is the phase crossover frequency. Referring to Fig. 18-12, we see that there are two phase crossover frequencies: $\omega_\pi' \cong 0.62$ rad/s and $\omega_\pi'' \cong 1.95$ rad/s. For $\omega_\pi' = 0.62$, we have $20 \log_{10} |GH(j\omega_\pi')| = -3$ db. Therefore the gain margin is 3 db. It can be increased to 10 db by shifting the $GH(j\omega)$ curve downward by 7 db. The phase crossover frequency ω_π' is the same in the new position, but $20 \log_{10} |GH(j\omega_\pi')| = -10$ db. A gain decrease of 7 db implies that $K_B = 10^{-7/20} = 0.447$. Since the system is stable for $K_B = 1$, it remains stable when the $GH(j\omega)$ curve is shifted downward. Absolute stability is not affected unless the $GH(j\omega)$ curve is shifted upward and across the point defined by 0 db and $-180°$, as would be necessary if $-20 \log_{10} GH(j\omega_\pi'') = 10$ db.

18.3. For the system of Problem 18.1, determine a value for K_B such that: gain margin $\geqq$ 10 db, $\phi_{PM} \geqq 45°$.

In Problem 18.1, it was shown that $\phi_{PM} \geqq 45°$ if $K_B \leqq 0.3$; in Problem 18.2, gain margin $\geqq 10$ db if $K_B \leqq 0.447$. Therefore both requirements can be satisfied by setting $K_B \leqq 0.3$. Note that if we had specified gain margin = 10 db and $\phi_{PM} = 45°$, then the specifications could not be met by gain factor compensation alone.

18.4. Assume that the system of Problem 18.1 is a unity feedback system and determine a value for K_B such that the resonant peak M_p is 2 db.

The db magnitude-phase angle plot of $GH(j\omega)$ for $K_B = 1$ is shown in Fig. 18-13 along with the locus of points for which $\left| \dfrac{C}{R}(j\omega) \right| = 2$ db ($M_p = 2$ db). We see that if K_B is decreased by 8 db, the resulting $GH(j\omega)$ curve is just tangent to the $M_p = 2$ db curve. A decrease of 8 db implies that $K_B = 10^{-8/20} = 0.40$.

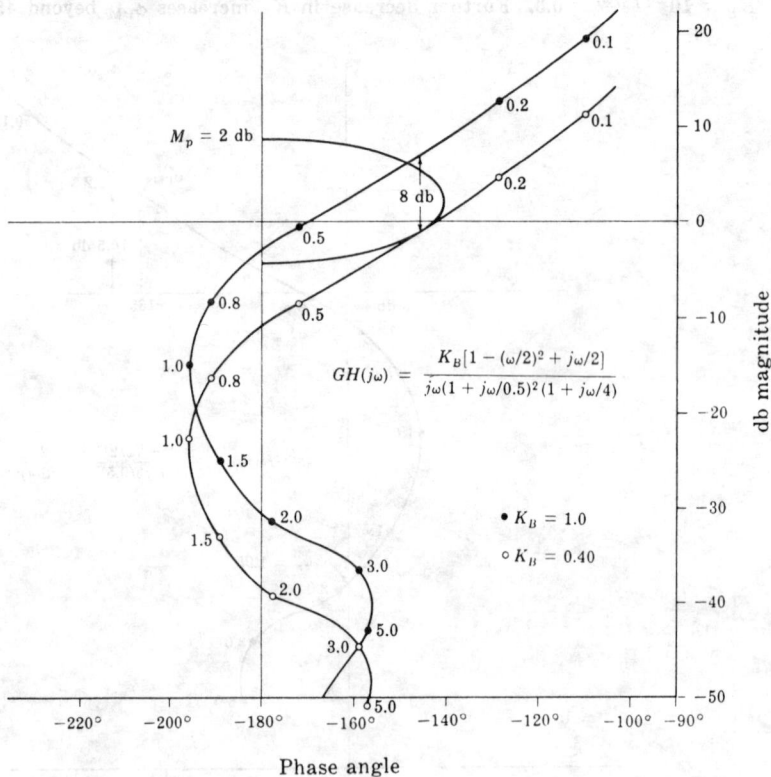

$$GH(j\omega) = \frac{K_B[1 - (\omega/2)^2 + j\omega/2]}{j\omega(1 + j\omega/0.5)^2(1 + j\omega/4)}$$

Fig. 18-13

18.5. The db magnitude-phase angle plot of the open-loop frequency response function

$$GH(j\omega) = \frac{K_B(1 + j\omega/0.5)}{(j\omega)^2[1 - (\omega/2)^2 + j\omega/2]}$$

is given in Fig. 18-14 below for $K_B = 0.5$. The closed-loop system defined by $GH(j\omega)$ is stable for $K_B = 0.5$. Determine the value of K_B which maximizes the phase margin.

$\phi_{PM} = 180° + \arg GH(j\omega_1)$, where ω_1 is the gain crossover frequency. Referring to Fig. 18-14, we see that $\arg GH(j\omega)$ is always negative. Therefore if we maximize $\arg GH(j\omega_1)$, ϕ_{PM} will be maximized. Fig. 18-14 indicates that $\arg GH(j\omega)$ is maximum when $\omega = \omega_1' \cong 0.8$ rad/s and $\arg GH(j\omega_1') = -147°$. The ordinate of the point $GH(j\omega_1')$ is 4.6 db. Therefore if K_B is decreased by 4.6 db, the phase crossover frequency becomes ω_1'; and ϕ_{PM} takes on its maximum value: $\phi_{PM} = 180° + \arg GH(j\omega_1') = 33°$. A decrease of 4.6 db in K_B implies that $20 \log_{10}(K_B/0.5) = -4.6$ db or $K_B/0.5 = 10^{-4.6/20}$. Then $K_B = 0.295$.

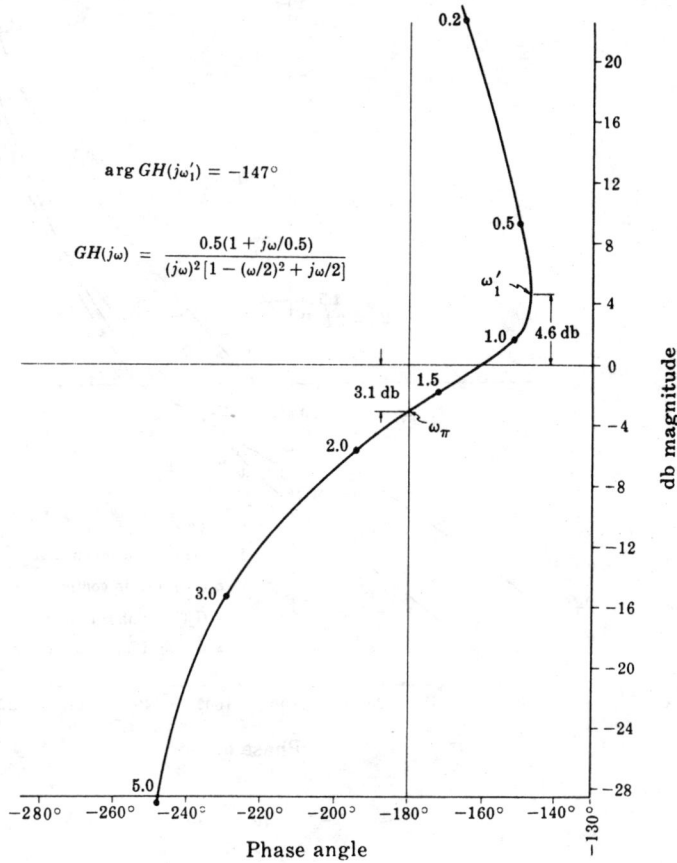

Fig. 18-14

18.6. For the system in Problem 18.5, determine a value of K_B for which the system is stable and the gain margin is 8 db.

Gain margin $= -20 \log_{10} |GH(j\omega_\pi)|$ db. Referring to Fig. 18-14, we see that the gain margin is 3.1 db. This can be increased to 8 db by shifting the curve down by 4.9 db; ω_π remains the same, as it is independent of K_B. A decrease of 4.9 db in K_B implies that $20 \log_{10}(K_B/0.5) = -4.9$ or $K_B = 0.254$.

PHASE COMPENSATION

18.7. The db magnitude-phase angle plot of the open-loop transfer function $G(j\omega)$ for a particular unity feedback system has been determined experimentally as shown in Fig. 18-15. In addition, the steady-state error $e(\infty)$ for a unit ramp function input was measured and found to be $e(\infty) = 0.2$. The open-loop transfer function is known to have a pole at the origin. Determine a combination of phase-lead plus gain compensation such that: $M_p \cong 1.5$ db, $\phi_{PM} \cong 40°$, and the steady state error for a unit ramp input is $e(\infty) = 0.1$.

Since $e(\infty) = 1/K_v = 1/K_B$, the steady-state requirement can be satisfied by doubling K_B. The compensation has the form

$$K'P_{\text{Lead}}(j\omega) = \frac{K'(a/b)(1 + s/a)}{1 + s/b}$$

Hence K_B is doubled by letting $K'(a/b) = 2$, or $K' = 2(b/a)$.

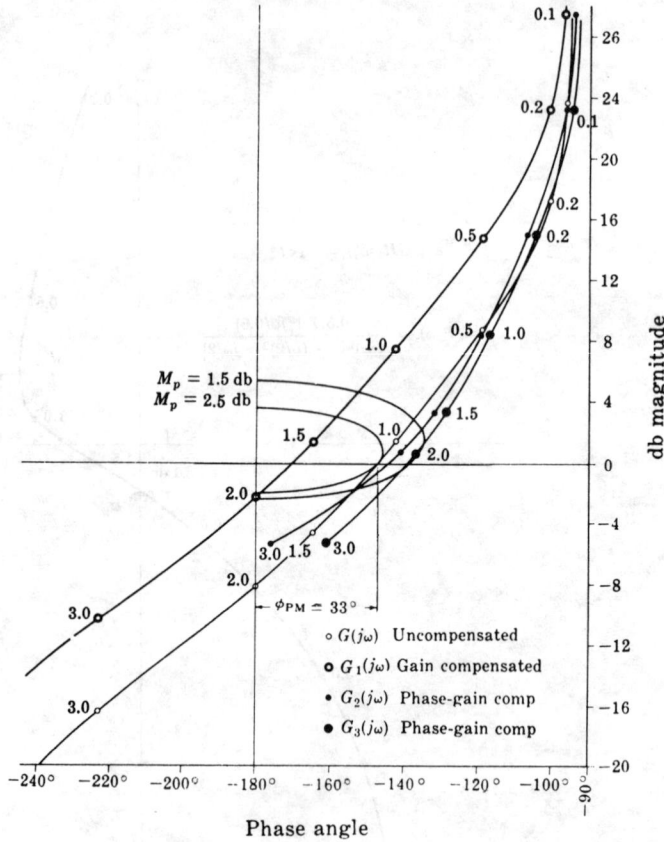

Fig. 18-15

The db magnitude-phase angle plot for the gain compensated open-loop frequency response function

$$G_1(j\omega) = 2G(j\omega)$$

is shown in Fig. 18-15. $G_1(j\omega)$ satisfies the steady-state specification. To satisfy the specifications on M_p and ϕ_{PM}, the $G_1(j\omega)$ curve must be shifted to the right by about 30° to 40° in the region $1.2 \leq \omega \leq 2.5$ without substantially changing the db magnitude. This is done by proper choice of a and b. Referring to Fig. 18-4, we see that, for $b/a = 10$, 30° phase lead is obtained for $\omega/a \cong 0.65$. Since the lead ratio a/b of the lead network is taken into account by designing for the gain factor $K' = 2(b/a) = 20$, we must add $20 \log(b/a) = 20 \log_{10} 10 = 20$ db to all db magnitudes taken from Fig. 18-4.

To obtain 30° or more phase lead in the frequency range of interest, we let $a = 2$. For this choice we have $\omega = (2)(0.65) = 1.3$ and obtain 30° phase lead. Since $b/a = 10$, then $b = 20$. The compensated open-loop frequency response function is

$$G_2(j\omega) = \frac{2(1 + j\omega/2)}{1 + j\omega/20} G(j\omega)$$

The db magnitude-phase angle plot of $G_2(j\omega)$ is shown in Fig. 18-15. We see that $M_p \cong 2.0$ db and $\phi_{PM} = 36°$; therefore the specifications are not satisfied by this compensation. We need to shift $G_2(j\omega)$ 5° to 10° further to the right; hence, additional phase lead is needed. Referring once more to Fig. 18-4, we see that letting $b/a = 15$ increases the phase lead. Again, we let $a = 2$; then $b = 30$. The db magnitude-phase angle plot of

$$G_3(j\omega) = \frac{2(1 + j\omega/2)}{1 + j\omega/30} G(j\omega)$$

is shown in Fig. 18-15. We see that $\phi_{PM} = 41°$ and $M_p \cong 1.5$ db and hence the specifications are met by the compensation

$$30P_{\text{Lead}} = \frac{2(1 + s/2)}{1 + s/30}$$

18.8. **Solve Problem 18.7 using *lag* plus gain compensation.**

In Problem 18.7 we found that the Bode gain K_B must be increased by a factor of 2 to satisfy the steady-state specification. But the Bode gain of a lag network is

$$\lim_{s \to 0} P_{\text{Lag}} = \lim_{s \to 0} \frac{1 + s/b}{1 + s/a} = 1$$

Therefore the compensation required in this problem has the form $\dfrac{2(1 + s/a)}{1 + s/b}$ where the twofold gain factor increase is supplied by an amplifier and a and b for the lag network must be chosen to satisfy the requirements on M_p and ϕ_{PM}. The gain-compensated function is shown as $G_1(j\omega) = 2G(j\omega)$ in Fig. 18-16; $G_1(j\omega)$ must be shifted downward by 7 to 10 db in the region $0.7 \leqq \omega \leqq 2.0$, with no substantial increase in phase lag, to meet the transient specifications.

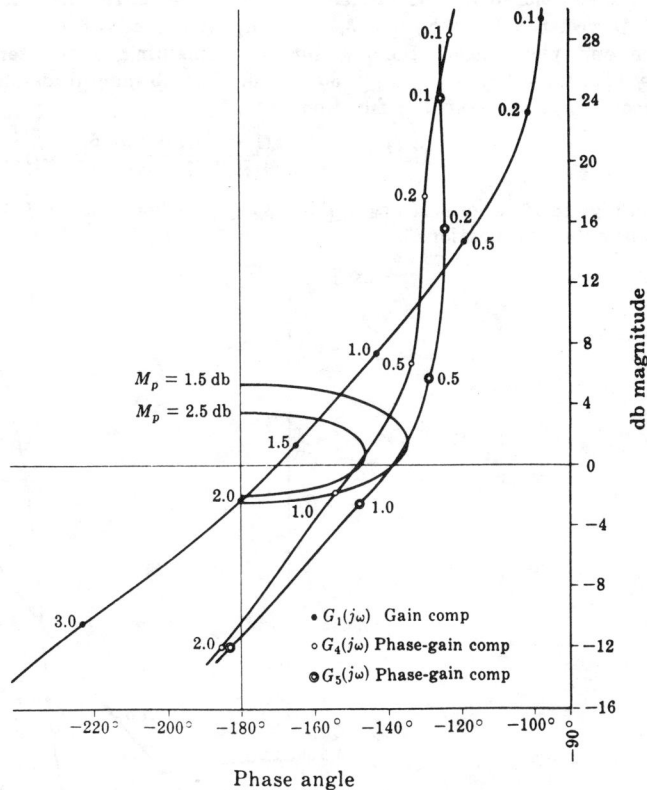

Fig. 18-16

Referring to Fig. 18-7, we see that, for $b/a = 3$, we can obtain a maximum attenuation of 9.5 db. For $a = 0.1$, the phase lag is $-15°$ at $\omega = 0.7$ ($\omega/a = 7$) and $-6°$ at $\omega = 2.0$ ($\omega/a = 20$), i.e. the phase lag is relatively small in the frequency region of interest. The db magnitude-phase angle plot for

$$G_4(j\omega) = \frac{2(1 + j\omega/0.3)}{1 + j\omega/0.1} G(j\omega)$$

is also shown in Fig. 18-16, with $M_p \cong 2.5$ db and $\phi_{\text{PM}} = 32°$; hence this system does not meet the specifications. To decrease the phase lag introduced in the frequency region $0.7 \leqq \omega \leqq 2.0$, we change a to 0.05 and b to 0.15. The phase lag is now $9°$ at $\omega = 0.7$ ($\omega/a = 14$). The db magnitude-phase angle plot for

$$G_5(j\omega) = \frac{2(1 + j\omega/0.15)}{1 + j\omega/0.05} G(j\omega)$$

is shown in Fig. 18-16. We see that $M_p \cong 1.5$ db and $\phi_{\text{PM}} = 41°$. Thus the specifications are satisfied. The desired compensation is given by

$$2P_{\text{Lag}} = \frac{2(1 + s/0.15)}{1 + s/0.05}$$

18.9. Solve Problem 18.7 using *lag-lead* plus gain compensation. In addition to the previous specifications, we require that the resonant frequency ω_p of the compensated system be approximately the same as that for the uncompensated system.

In Problems 18.7 and 18.8 we found that the Bode gain K_B must be increased by a factor of 2 to satisfy the steady-state specification. The frequency response function of the lag-lead plus gain compensation is therefore given by

$$2P_{LL}(j\omega) = \frac{2(1 + j\omega/a_1)(1 + j\omega/b_2)}{(1 + j\omega/b_1)(1 + j\omega/a_2)}$$

We must now choose a_1, b_1, b_2 and a_2 to satisfy the requirements on M_p, ϕ_{PM} and ω_p. Referring to Fig. 18-15, we see that the resonant frequency for the uncompensated system is about 1.1 rad/sec. The db magnitude-phase angle plot of $G_1(j\omega) = 2G(j\omega)$ shown in Fig. 18-17 indicates that, if the $G_1(j\omega)$ curve is attenuated by 6.5 db and 10° of phase lead is added at a frequency of $\omega = 1.0$ rad/s, then the resulting curve will be tangent to the $M_p = 2$ db curve at about 1 rad/s. Referring to Fig. 18-10, if we let $b_1/a_1 = b_2/a_2 = 3$, $a_1 = 6a_2$ and $\omega/a_2 = 6.0$ for $\omega = 1$, we obtain the desired attenuation and phase lead. Solving for the remaining parameters, we get $a_2 = 1/6 = 0.167$, $b_2 = 3a_2 = 0.50$, $a_1 = 6a_2 = 1.0$, $b_1 = 3a_1 = 3.0$. The db magnitude-phase angle plot for the resulting open-loop frequency response function

$$G_6(j\omega) = \frac{2(1 + j\omega)(1 + j\omega/0.5)}{(1 + j\omega/3)(1 + j\omega/0.167)} G(j\omega)$$

is shown in Fig. 18-17, where $M_p \cong 1.5$ db, $\phi_{PM} = 44°$ and $\omega_p \cong 1.0$ rad/s. These values approximately satisfy the specifications.

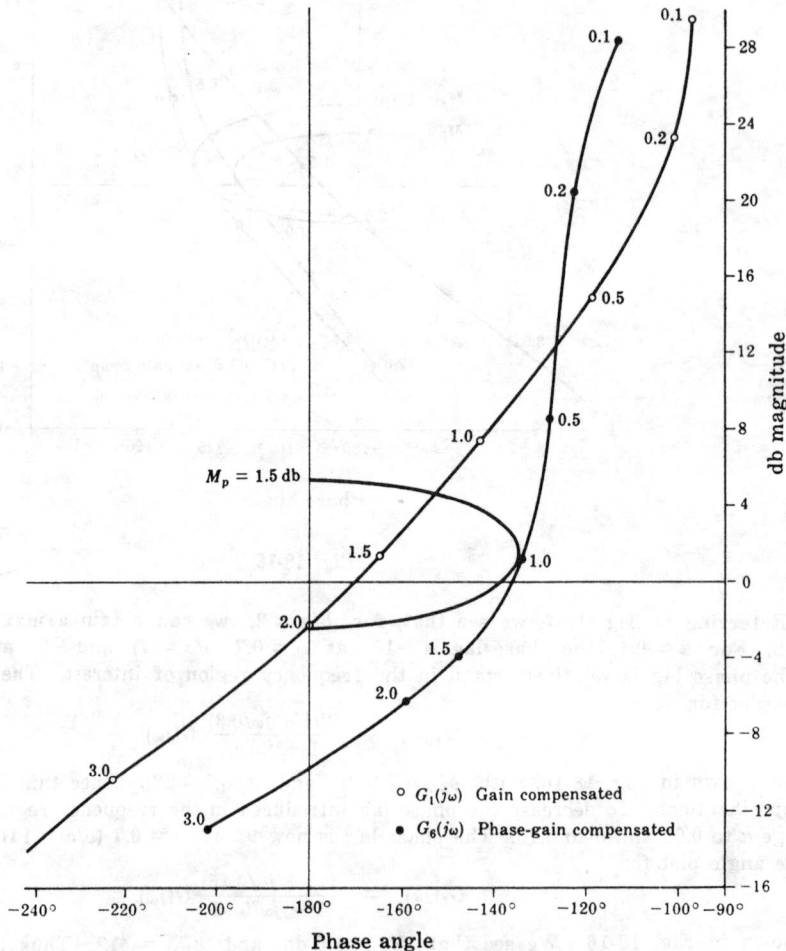

Fig. 18-17

Supplementary Problems

18.10. Find a value of K_B for which the system whose open-loop transfer function is

$$GH = \frac{K_B}{s(1 + s/200)(1 + s/250)}$$

has a resonant peak M_p of 1.4 db. *Ans.* $K_B = 119.4$.

18.11. For the system of Problem 18.10, find gain plus lag compensation such that $M_p \leq 1.7$, $\phi_{PM} \geq 35°$ and $K_v \geq 50$.

18.12. For the system of Problem 18.10, find gain plus lead compensation such that $M_p \leq 1.7$, $\phi_{PM} \geq 50°$ and $K_v \geq 50$.

18.13. For the system of Problem 18.10, find gain plus lag-lead compensation such that $M_p \leq 1.5$, $\phi_{PM} \geq 40°$ and $K_v \geq 100$.

18.14. Find gain plus lag compensation for the system whose open-loop transfer function is

$$GH = \frac{K_B}{s(1 + s/10)(1 + s/5)}$$

such that $K_v = 30$ and $\phi_{PM} \geq 40°$.

18.15. For the system of Problem 18.14, find gain plus lead compensation such that $K_v \geq 30$ and $\phi_{PM} \geq 45°$. *Hint.* Cascade two lead compensation networks.

18.16. Find gain plus lead compensation for the system whose open-loop transfer function is

$$GH = \frac{K_B}{s(1 + s/2)}$$

such that $K_v = 20$ and $\phi_{PM} = 45°$.

Chapter 19

Advanced Topics

19.1 INTRODUCTION

A very general definition of control systems was presented in Chapter 1. In subsequent chapters, however, we confined the discussion to systems describable by linear time-invariant ordinary differential equations excited by deterministic (i.e. not random) Laplace transformable input functions. The techniques we developed for studying these systems and inputs are relatively straightforward and usually lead to simple and practical control system designs. While it is true that no physical system is *exactly* linear and time-invariant and no excitation is *completely* deterministic, such models are often adequate approximations. As a result, the methods treated in this book have broad application. There are many situations, however, for which these approximate representations are invalid or inappropriate. In such cases more sophisticated models are required.

The theories and methods available for analysis and design of complex and advanced control systems constitute a large body of knowledge, beyond the scope of the present work. The purpose of this chapter is to survey some of the prevailing techniques. As a prerequisite to an in-depth study of these more advanced ideas, a thorough knowledge of the material in this book, or a similar book, is strongly recommended.

19.2 NONLINEAR CONTROL SYSTEMS

Linear systems are described in Definition 3.8, page 30. Any system which does not satisfy this definition is nonlinear.

Nonlinear control system problems usually arise because: (1) the structure or fixed elements of the system are inherently nonlinear; and/or (2) nonlinear compensation is introduced into the system for the purpose of improving its behavior.

Example 19.1.

Fig. 19-1(a) is a block diagram of a nonlinear feedback system containing a linear element represented by the transfer function $\frac{1}{D(D+1)}$, where $D \equiv d/dt$ is the *differential* operator defined on page 33, and a nonlinear element N whose transfer characteristic is defined by Fig. 19-1(b). Such nonlinearities are called **saturation** elements. D is used instead of s in the linear transfer function because the Laplace transform and its inverse are generally not suitable for nonlinear analysis.

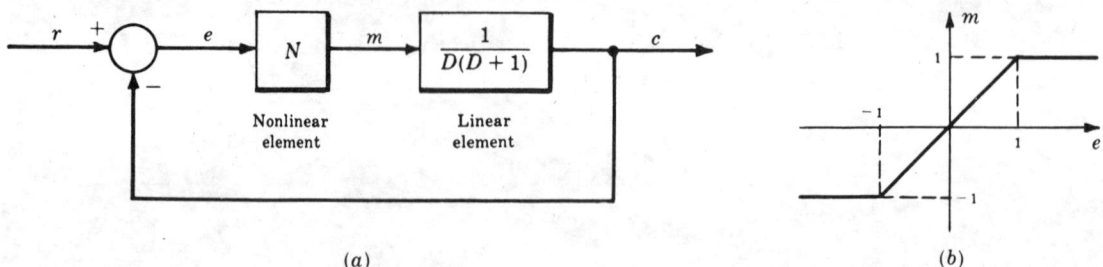

(a)

(b)

Fig. 19-1

Example 19.2.

If the earth is assumed spherical and all external forces other than gravity are negligible, then the motion of an earth satellite lies in a plane called the *orbit plane*. This motion is defined by the following set of nonlinear differential equations (see Problem 3.3, page 41):

$$r\frac{d^2\theta}{dt^2} + 2\frac{dr}{dt}\frac{d\theta}{dt} = 0 \qquad \text{(transverse force equation)}$$

$$\frac{d^2r}{dt^2} - r\left(\frac{d\theta}{dt}\right)^2 = -\frac{k^2}{pr^2} \qquad \text{(radial force equation)}$$

The satellite, together with any controller designed to modify its motion, constitutes a nonlinear control system.

Several popular methods for nonlinear analysis are summarized below.

Linearization

The technique of linearization was discussed in Section 3.8, page 32, and Problems 3.14, page 46, and 3.15, page 47. The methods presented in earlier chapters can be directly applied to time-invariant linearized systems having deterministic inputs.

Describing Functions

The describing function technique is an extension of frequency response methods to nonlinear systems. Classically, it is applicable only to nonlinear elements whose outputs in response to sinusoidal inputs with period T (i.e. of the form $A \sin 2\pi t/T$) are also periodic with period T. The output may then be written as a Fourier series:

$$\sum_{n=1}^{\infty} B_n \sin(n\omega t + \phi_n)$$

The **describing function** is the ratio of the complex Fourier coefficient $B_1 e^{j\phi_1}$ of the fundamental frequency of the output to the amplitude A of the input. That is, the describing function is the complex number $(B_1/A)e^{j\phi_1}$ which is essentially a frequency response function of an approximation of the nonlinear element. In general, B_1 and ϕ_1 are functions of both the input frequency $\omega = 2\pi/T$ and the input amplitude A. Therefore we may write $B_1 = B_1(A, \omega)$ and $\phi_1 = \phi_1(A, \omega)$.

To apply the method, the describing functions must first replace the nonlinearities in the block diagram. Then the frequency domain techniques of Chapters 11, 12, and 15 through 18 can be used to analyze the system, with some modification to account for the dependence of B_1 and ϕ_1 on A.

Example 19.3.

The describing function for the saturation element in Example 19.1 is given by

$$\frac{B_1}{A} e^{j\phi_1} = \frac{2}{\pi}\left[\sin^{-1}\frac{1}{A} + \frac{1}{A}\cos\sin^{-1}\frac{1}{A}\right]$$

This function is computed in Problem 19.2, page 357.

The Phase Plane

A second-order differential equation of the form

$$d^2x/dt^2 = f(x, dx/dt) \tag{19.1}$$

can be rewritten as a pair of first-order differential equations by making the change of variables $x \equiv x_1$ and $dx/dt \equiv x_2$, yielding

$$dx_1/dt = x_2 \tag{19.2}$$

$$dx_2/dt = f(x_1, x_2) \tag{19.3}$$

The two-tuple, or pair, (x_1, x_2) may be considered as a point in the plane of Cartesian coordinates. As the time t increases, $(x_1(t), x_2(t))$ describes a *path* or *trajectory* in the plane. This plane is called the **phase plane**.

If we eliminate time as the independent variable in Equations (19.2) and (19.3), we obtain the first order differential equation

$$\frac{dx_1}{dx_2} = \frac{x_2}{f(x_1, x_2)} \qquad (19.4)$$

Solution of (19.4) for x_1 as a function of x_2 (or vice versa) defines a trajectory in the phase plane. By solving this equation for various initial conditions on x_1 and x_2 and examining the resulting phase plane trajectories, we can determine the behavior of the second-order system.

Example 19.4.

The differential equation $\qquad \frac{d^2x}{dt^2} + \left(\frac{dx}{dt}\right)^2 = 0$

with the initial conditions $x(0) = 0$ and $\left.\frac{dx}{dt}\right|_{t=0} = 1$, can be replaced by the two first-order equations

$$dx_1/dt = x_2 \qquad x_1(0) = 0$$
$$dx_2/dt = -x_2^2 \qquad x_2(0) = 1$$

where $x \equiv x_1$ and $dx/dt \equiv x_2$. Eliminating time as the independent variable, we obtain

$$\frac{dx_1}{dx_2} = -\frac{x_2}{x_2^2} = -\frac{1}{x_2} \qquad \text{or} \qquad dx_1 = -\frac{dx_2}{x_2}$$

Integration of this equation for the given initial conditions yields

$$\int_{x_1(0)=0}^{x_1} dx_1' = x_1 = -\int_{x_2(0)=1}^{x_2} \frac{dx_2'}{x_2'} = -\ln x_2 \qquad \text{or} \qquad x_2 = e^{-x_1}$$

The phase plane trajectory defined by this equation is plotted in Fig. 19-2. The direction of the motion in the phase plane is determined by noting that $dx_2/dt = -x_2^2 < 0$ for all $x_2 \neq 0$; therefore x_2 always decreases and we obtain the motion as shown.

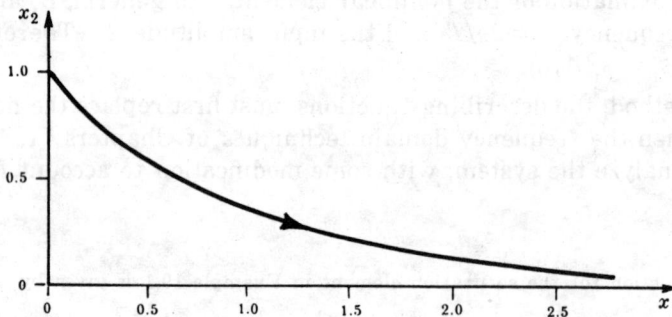

Fig. 19-2

Lyapunov's Stability Criterion

The stability criteria presented in earlier portions of this book cannot be applied to nonlinear systems. A more general method is provided by the Lyapunov theory. It can be used for both linear and nonlinear systems described by sets of simultaneous first-order ordinary differential equations. To simplify the discussion, we consider only the two equation case:

$$\frac{dx_1}{dt} = f_1(x_1, x_2), \qquad \frac{dx_2}{dt} = f_2(x_1, x_2)$$

If we eliminate time as the independent variable, we obtain the single equation

$$\frac{dx_1}{dx_2} = \frac{f_1(x_1, x_2)}{f_2(x_1, x_2)} \qquad (19.5)$$

whose solution describes a trajectory in the phase plane.

A point (x_1, x_2) in the phase plane simultaneously satisfying the two equations $f_1(x_1, x_2) = 0$ and $f_2(x_1, x_2) = 0$ is called a **singular point**. We can let the origin $(0, 0)$ be a singular point. The origin is said to be **stable** if, for any circular region $S(R)$ in the phase plane centered at $(0, 0)$, having radius R, there exists a circular region $S(r)$, centered at the origin with radius $r \leqq R$, in which any trajectory beginning in $S(r)$ remains in $S(R)$ ever after. The origin is **asymptotically stable** if it is stable and all trajectories tend to the origin as time goes to infinity.

Lyapunov's stability criterion states that if the origin is a singular point, then it is stable if a function $V(x_1, x_2)$ can be found such that:

(a) $V(x_1, x_2)$ is positive for all values of x_1 and x_2, except that it may be zero for $x_1 = x_2 = 0$; and

(b) dV/dt is never positive.

If dV/dt is never zero except at the origin, then the origin is *asymptotically stable*.

Example 19.5.

A nonlinear system represented by

$$\frac{d^2x}{dt^2} + \frac{dx}{dt} + \left(\frac{dx}{dt}\right)^3 + x = 0$$

or, equivalently, the pair of equations

$$\frac{dx_1}{dt} = x_2, \qquad \frac{dx_2}{dt} = -x_2 - x_2^3 - x_1$$

where $x_1 \equiv x$, has a singular point at $x_1 = x_2 = 0$. The function $V = x_1^2 + x_2^2$ is positive for all x_1 and x_2, except $x_1 = x_2 = 0$ where $V = 0$. The derivative

$$\frac{dV}{dt} = 2x_1\frac{dx_1}{dt} + 2x_2\frac{dx_2}{dt} = 2x_1x_2 + 2x_2(-x_2 - x_2^3 - x_1) = -2x_2^2 - 2x_2^4$$

is never positive. Therefore the origin is stable.

19.3 DISCRETE TIME SYSTEMS

In our previous discussion we considered only dependent variables defined on the entire interval $0 \leqq t < +\infty$. For some problems, however, it is convenient to choose the domain of the dependent variables as a countable set of points $t_1, t_2, t_3, \ldots$ on this interval. That is, the system inputs and outputs are assumed to be *time sequences* instead of continuous time functions.

Example 19.6.

A function of time $y^*(t)$ is defined as follows on the interval $0 \leqq t < +\infty$:

$$y^*(t) = nT/2, \qquad nT \leqq t < (n+1)T$$

where $n = 0, 1, 2, 3, \ldots$ and T is a positive constant. A graph of $y^*(t)$ is shown in Fig. 19-3.

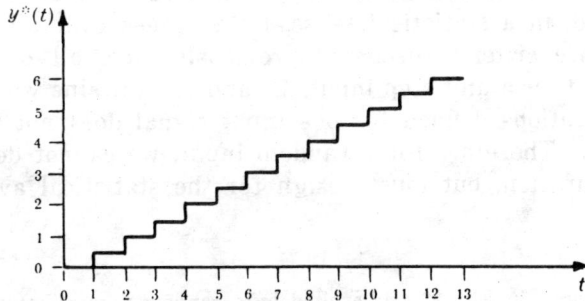

Fig. 19-3

We see that the function $y^*(t)$ is completely determined by the *sequence* of values, $0, T/2, T, 3T/2, \ldots$ which it assumes at the time instants $0, T, 2T, 3T, \ldots$.

Example 19.7.

A digital computer is constructed so that, when performing arithmetic operations, the output changes only at discrete instants of time determined by an internal clock. This output can therefore be defined by a time sequence of values at discrete time instants.

Systems whose inputs and outputs are time sequences are called **discrete time systems** and can be described by *difference equations*. Linear time-invariant lumped parameter discrete time systems may be represented by equations of the form

$$\sum_{j=0}^{n} a_j y[(k+j)T] = x(kT) \qquad (19.6)$$

where the coefficients a_j, $j = 0, 1, \ldots, n$ are constants; $k = 0, 1, 2, 3, \ldots$; T is a positive constant; $y(kT)$ is the output sequence; and $x(kT)$ is the input sequence. Note the similarity between (19.6) and Equation (3.1), page 29.

Since the integral of a sequence is undefined, Laplace transforms of sequences are undefined. Hence the analysis techniques based on Laplace transformations are not applicable to discrete time systems. However, a complete method, analogous to the Laplace transform theory, has been developed to handle discrete time systems; it is called the *z-transform* theory. Its application to discrete time control systems analysis and design problems parallels the use of the Laplace transform in Chapters 4 through 18.

19.4 SYSTEMS WITH RANDOM INPUTS

System stimuli are often random. This means that input functions may sometimes be more appropriately described probabilistically than deterministically. Such excitations are called *random processes*.

A random process can be viewed as a function of two variables, t and η, where t represents time and η a *random event*. The value η assumes is determined by chance.

Example 19.8.

A particular random process is denoted by $x(t, \eta)$. The random event η is the result of tossing an unbiased coin; heads or tails appears with equal probability. We define

$$x(t, \eta) \equiv \begin{cases} \text{a unit step function if } \eta = \text{heads} \\ \text{a unit ramp function if } \eta = \text{tails} \end{cases}$$

Thus $x(t, \eta)$ consists of two simple functions, but is a random process because chance dictates which function occurs.

In practice, random processes consist of an infinity of possible time functions (realizations) and we usually cannot describe them as explicitly as the one in the Example. Instead they must be described, in a statistical sense, by averages over all possible functions of time. The performance criteria discussed previously have all been related to specific inputs (e.g. K_p is defined for a unit step input, M_p and ϕ_{PM} for sine waves). But satisfaction of performance specifications defined for one input signal does not necessarily guarantee satisfaction for others. Therefore for a random input, we cannot design for a *particular* signal, such as step function, but must design for the statistical average of the random input signal.

Example 19.9.

The unity feedback system in Fig. 19-4 is excited by a random process input r having an infinity of possibilities. We want to determine compensation so that the error e is not excessive. There are an infinity of possibilities for r and, therefore, for e. Hence we cannot ask that each possible error satisfy given performance criteria but only that the average errors be small. For instance, we might ask that G_1 be chosen from the set of all causal systems such that, as time goes to infinity, the statistical average of $e^2(t)$ does not exceed some constant, or is minimized.

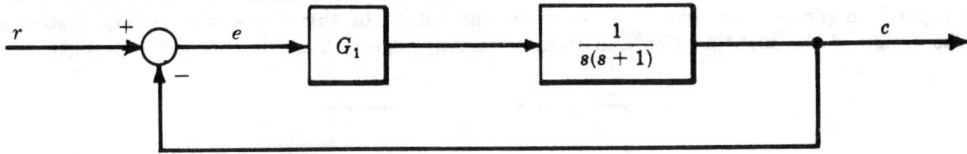

Fig. 19-4

19.5 OPTIMAL CONTROL SYSTEMS

The design problems discussed previously are, in an elementary sense, optimal control problems. The classical measures of system performance such as steady-state error, gain margin and phase margin are essentially criteria of optimality and control system compensators are designed to meet these requirements. In more general optimal control problems the system measure of performance, or *performance index*, is not specified. Instead, compensation is chosen so that the performance index is maximized or minimized. The value of the performance index is unknown until the completion of the optimizing process.

In many problems, the performance index is a measure or function of the error $e(t)$ between the actual and ideal responses. It is formulated in terms of the design parameters chosen, subject to existing physical constraints, to optimize the performance index.

Example 19.10.

For the system in Fig. 19-5 we want to find a nonnegative value of K such that the integral of the square of the error e is minimized when the input is a unit step function. We can formulate this problem as follows: choose $K \geq 0$ such that $\int_0^\infty e^2(t)\, dt$ is minimized, where

$$e(t) = \mathcal{L}^{-1}\left[\frac{s+2}{s^2+2s+K}\right] = \sqrt{\frac{K}{K-1}}\, e^{-t} \sin\left(\sqrt{K-1}\, t + \tan^{-1}\sqrt{K-1}\right)$$

The solution may be obtained using the classical techniques of integral calculus (see Problem 19.9).

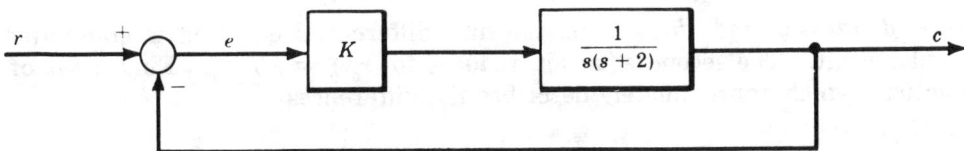

Fig. 19-5

The usual optimal control problems are much more complex than the simple example above and they require sophisticated mathematical techniques for their solution. We can do no more here than mention their existence.

19.6 ADAPTIVE CONTROL SYSTEMS

In some control systems certain parameters are either not constant or vary in an unknown manner. In Chapter 9 we illustrated one way of minimizing the effects of such contingencies by designing for minimum sensitivity. If, however, parameter variations are large or very rapid, it may be desirable to design for the capability of continuously measuring them and changing the compensation so that system performance criteria are always satisfied. This is called *adaptive control* design.

Example 19.11.

Fig. 19-6 is a block diagram of an adaptive control system. The parameters A and B of the plant are known to vary with time. The block labelled "Identification and Parameter Adjustment" continuously measures the input $m(t)$ and output $c(t)$ of the plant to identify the parameters A and B. In this manner, a and b of the lead compensator can be modified by the output of this element to satisfy system specifications. The design of the Identification and Parameter Adjustment block is the major problem of adaptive control.

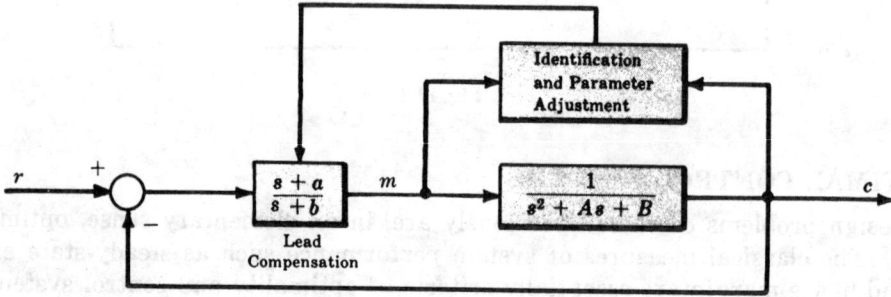

Fig. 19-6

Solved Problems

NONLINEAR SYSTEMS

19.1. The equations describing the motion of an earth satellite in the orbit plane are

$$r\frac{d^2\theta}{dt^2} + 2\frac{dr}{dt}\frac{d\theta}{dt} = 0, \qquad \frac{d^2r}{dt^2} - r\left(\frac{d\theta}{dt}\right)^2 = -\frac{k^2}{pr^2}$$

(See Problem 3.3, page 41, and Example 19.2, page 351, for more details.)   A satellite is in a nearly circular orbit determined by r and $d\theta/dt \equiv \omega$.   An exactly circular orbit is defined by

$$r = r_0 = \text{constant}, \qquad \omega = \omega_0 = \text{constant}$$

Since $dr_0/dt = 0$ and $d\omega_0/dt = 0$, the first differential equation is eliminated for a circular orbit.   The second equation reduces to $r_0\omega_0^2 = k^2/pr_0^2$.   Find a set of linear equations which approximately describes the differences

$$\delta r \equiv r - r_0, \qquad \delta\omega \equiv \omega - \omega_0$$

In the equations of motion we make the substitutions

$$r = r_0 + \delta r, \qquad \omega = \omega_0 + \delta\omega$$

and obtain the equations

$$(r_0 + \delta r)\frac{d(\omega_0 + \delta\omega)}{dt} + 2\frac{d(r_0 + \delta r)}{dt}(\omega_0 + \delta\omega) = 0$$

$$\frac{d^2(r_0 + \delta r)}{dt^2} - (r_0 + \delta r)(\omega_0 + \delta\omega)^2 = -\frac{k}{p(r_0 + \delta r)^2}$$

We note that

$$\frac{d(r_0 + \delta r)}{dt} = \frac{d(\delta r)}{dt}, \qquad \frac{d^2(r_0 + \delta r)}{dt^2} = \frac{d^2(\delta r)}{dt^2}, \qquad \frac{d(\omega_0 + \delta\omega)}{dt} = \frac{d(\delta\omega)}{dt}$$

since both r_0 and ω_0 are constant.   The first differential equation then becomes

$$r_0\frac{d(\delta\omega)}{dt} + (\delta r)\frac{d(\delta\omega)}{dt} + 2\omega_0\frac{d(\delta r)}{dt} + 2\frac{d(\delta r)}{dt}\delta\omega = 0$$

Since the differences δr, $\delta\omega$ and their derivatives are small, the second-order terms $(\delta r)\dfrac{d(\delta\omega)}{dt}$ and $2\dfrac{d(\delta r)}{dt}\delta\omega$ can be assumed negligible and eliminated. The resulting linear equation is

$$r_0\frac{d(\delta\omega)}{dt} + 2\omega_0\frac{d(\delta r)}{dt} = 0$$

which is one of the two desired equations. The second differential equation can be rewritten as

$$\frac{d^2(\delta r)}{dt^2} - r_0\omega_0^2 - 2r_0\omega_0\,\delta\omega - r_0(\delta\omega)^2 - \omega_0^2\,\delta r - 2\omega_0(\delta r)(\delta\omega) - (\delta\omega)^2\,\delta r$$

$$= -\frac{k}{pr_0^2} - \frac{2k\,\delta r}{r_0^3} + \text{higher order terms in } \delta r$$

where the right hand side is the Taylor series expansion of $-k/pr^2$ about r_0. All terms of order two and greater in δr and $\delta\omega$ may again be assumed negligible and eliminated leaving the linear equation

$$\frac{d^2(\delta r)}{dt^2} - r_0\omega_0^2 - 2r_0\omega_0\,\delta\omega - \omega_0^2\,\delta r = -\frac{k}{pr_0^2} - \frac{2k\,\delta r}{pr_0^3}$$

In the problem statement we saw that $r_0\omega_0^2 = k/pr_0^2$. Hence the final equation is

$$\frac{d^2(\delta r)}{dt^2} - 2r_0\omega_0\,\delta\omega - \omega_0^2\,\delta r = -\frac{2k\,\delta r}{pr_0^3}$$

which is the second of the two desired linearized equations.

19.2. Show that the describing function for the saturation element in Example 19.1 is given by

$$\frac{B_1}{A}e^{j\phi_1} = \frac{2}{\pi}\left[\sin^{-1}\frac{1}{A} + \frac{1}{A}\cos\sin^{-1}\frac{1}{A}\right]$$

We see from Fig. 19-1(b) that, when the magnitude of the input is less than 1.0, the output equals the input. When the input exceeds 1.0, then the output equals 1.0. Using the notation of Example 19.1, if

$$e(t) = A\sin\omega t, \quad A > 1$$

then $m(t)$ is as shown in Fig. 19-7 and can be written as

$$m(t) = \begin{cases} A\sin\omega t & \begin{cases} 0 \le t \le t_1 \\ t_2 \le t \le t_3 \\ t_4 \le t \le 2\pi/\omega \end{cases} \\ 1 & t_1 \le t \le t_2 \\ -1 & t_3 \le t \le t_4 \end{cases}$$

The time t_1 is obtained by noting that

$$A\sin\omega t_1 = 1 \quad \text{or} \quad t_1 = \frac{1}{\omega}\sin^{-1}\frac{1}{A}$$

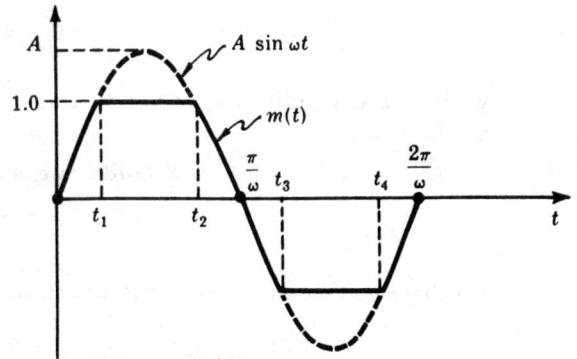

Fig. 19-7

Similarly

$$t_2 = \frac{\pi}{\omega} - \frac{1}{\omega}\sin^{-1}\frac{1}{A}, \quad t_3 = \frac{\pi}{\omega} + \frac{1}{\omega}\sin^{-1}\frac{1}{A}, \quad t_4 = \frac{2\pi}{\omega} - \frac{1}{\omega}\sin^{-1}\frac{1}{A}$$

The magnitude B_1 and phase angle ϕ_1 of the describing function are determined from the expression for the first Fourier coefficient:

$$B_1 = \frac{\omega}{\pi}\int_0^{2\pi/\omega} m(t)\sin\omega t\,dt$$

Since $m(t)$ is an odd function, the phase angle ϕ_1 is zero. The integral defining B_1 can be rewritten as

$$B_1 = \frac{\omega}{\pi}\int_0^{t_1} A\sin^2\omega t\,dt + \frac{\omega}{\pi}\int_{t_1}^{t_2}\sin\omega t\,dt$$

$$+ \frac{\omega}{\pi}\int_{t_2}^{t_3} A\sin^2\omega t\,dt - \frac{\omega}{\pi}\int_{t_3}^{t_4}\sin\omega t\,dt + \frac{\omega}{\pi}\int_{t_4}^{2\pi/\omega} A\sin^2\omega t\,dt$$

But
$$\int_0^{t_1} A\sin^2\omega t\,dt = \int_{t_4}^{2\pi/\omega} A\sin^2\omega t\,dt = \frac{1}{2}\int_{t_2}^{t_3} A\sin^2\omega t\,dt$$

and
$$\int_{t_1}^{t_2}\sin\omega t\,dt = -\int_{t_3}^{t_4}\sin\omega t\,dt = 2\int_{t_1}^{\pi/2\omega}\sin\omega t\,dt$$

We can thus write B_1 as

$$B_1 = \frac{4\omega}{\pi}\int_0^{t_1} A\sin^2\omega t\,dt + \frac{4\omega}{\pi}\int_{t_1}^{\pi/2\omega}\sin\omega t\,dt = \frac{2}{\pi}\left[A\omega t_1 - \frac{A}{2}\sin 2\omega t_1 + 2\cos\omega t_1\right]$$

Substituting $t_1 = \frac{1}{\omega}\sin^{-1}\frac{1}{A}$ and simplifying, we obtain

$$B_1 = \frac{2}{\pi}\left[A\sin^{-1}\frac{1}{A} + \cos\sin^{-1}\frac{1}{A}\right]$$

Finally, the describing function is

$$\frac{B_1}{A} = \frac{2}{\pi}\left[\sin^{-1}\frac{1}{A} + \frac{1}{A}\cos\sin^{-1}\frac{1}{A}\right]$$

19.3. Show that the equation $\dfrac{d^2x}{dt^2} = f\left(x, \dfrac{dx}{dt}\right)$ can be equivalently described by a pair of first-order differential equations.

We define a set of new variables: $x_1 \equiv x$ and $x_2 \equiv dx_1/dt = dx/dt$.

$$d^2x/dt^2 = d^2x_1/dt^2 = dx_2/dt = f(x, dx/dt) = f(x_1, dx_1/dt) = f(x_1, x_2)$$

The two desired equations are therefore $\dfrac{dx_1}{dt} = x_2$, $\dfrac{dx_2}{dt} = f(x_1, x_2)$.

19.4. Show that the phase plane trajectory of the solution of the differential equation

$$\frac{d^2x}{dt^2} + x = 0$$

with initial conditions $x(0) = 0$ and $\dfrac{dx}{dt}\Big|_{t=0} = 1$ is a circle of unit radius centered at the origin.

Letting $x \equiv x_1$ and $x_2 \equiv dx_1/dt$, we obtain the pair of equations

$$dx_1/dt = x_2 \qquad x_1(0) = 0$$
$$dx_2/dt = -x_1 \qquad x_2(0) = 1$$

We eliminate time as the independent variable by writing

$$\frac{dx_1}{dx_2} = -\frac{x_2}{x_1} \qquad \text{or} \qquad x_1\,dx_1 + x_2\,dx_2 = 0$$

Integrating this equation for the given initial conditions, we obtain

$$\int_0^{x_1} x_1'\,dx_1' + \int_1^{x_2} x_2'\,dx_2' = \tfrac{1}{2}x_1^2 + \tfrac{1}{2}x_2^2 - \tfrac{1}{2} = 0 \qquad \text{or} \qquad x_1^2 + x_2^2 = 1$$

which is the equation of a circle of unit radius centered at the origin.

19.5. Determine the equation of the phase plane trajectory of the equation

$$\frac{d^2x}{dt^2} + \frac{dx}{dt} = 0$$

with the initial conditions $x(0) = 0$ and $\dfrac{dx}{dt}\Big|_{t=0} = 1$.

With $x_1 \equiv x$ and $x_2 \equiv dx_1/dt$ we obtain the pair of first order equations

$$dx_1/dt = x_2 \qquad x_1(0) = 0$$
$$dx_2/dt = -x_2 \qquad x_2(0) = 1$$

We eliminate time as the independent variable by writing

$$\frac{dx_1}{dx_2} = -\frac{x_2}{x_2} = -1 \qquad \text{or} \qquad dx_1 + dx_2 = 0$$

Then

$$\int_0^{x_1} dx_1' + \int_1^{x_2} dx_2' = x_1 + x_2 - 1 = 0$$

or $$x_1 + x_2 = 1$$

which is the equation of a straight line, as shown in Fig. 19-8. The direction of the motion in the phase plane is indicated by the arrow and is determined by noting that, initially, $x_2(0) = 1$; therefore $dx_1/dt > 0$ and x_1 is increasing, and $dx_2/dt < 0$ and x_2 is decreasing. The trajectory ends at the point $(x_1, x_2) = (1, 0)$, where $dx_1/dt = dx_2/dt = 0$ and thus motion terminates.

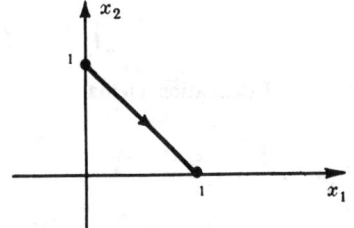

Fig. 19-8

19.6. Find the singular points of the pair of equations

$$\frac{dx_1}{dt} = \sin x_2, \qquad \frac{dx_2}{dt} = x_1 + x_2$$

Singular points are found by setting $\sin x_2 = 0$ and $x_1 + x_2 = 0$. The first equation is satisfied by $x_2 = \pm n\pi$, $n = 0, 1, 2, \ldots$. The second is satisfied by $x_1 = -x_2$. Hence the singular points are defined by

$$x_1 = \mp n\pi, \quad x_2 = \pm n\pi \qquad n = 0, 1, 2, \ldots$$

19.7. The origin is a singular point for the pair of equations

$$\frac{dx_1}{dt} = ax_1 + bx_2, \qquad \frac{dx_2}{dt} = cx_1 + dx_2$$

Using Lyapunov theory, find sufficient conditions on a, b, c and d such that the origin is asymptotically stable.

We choose a function
$$V = x_1^2 + x_2^2$$
which is positive for all x_1, x_2 except $x_1 = x_2 = 0$. The time derivative of V is

$$\frac{dV}{dt} = 2x_1 \frac{dx_1}{dt} + 2x_2 \frac{dx_2}{dt} = 2ax_1^2 + 2bx_1x_2 + 2cx_1x_2 + 2dx_2^2$$

To make dV/dt negative for all x_1, x_2, we might choose $a < 0$, $d < 0$ and $b = -c$. In this case,

$$\frac{dV}{dt} = 2ax_1^2 + 2dx_2^2 < 0$$

except when $x_1 = x_2 = 0$. Hence one set of sufficient conditions for asymptotic stability are $a < 0$, $d < 0$ and $b = -c$. There are other possible solutions to this problem.

DISCRETE TIME SYSTEMS

19.8. Show that a solution of the homogeneous second-order difference equation

$$y(k+2) + 3y(k+1) + 2y(k) = 0$$

is $y(k) = z^k$, where z is a root of the polynomial equation $z^2 + 3z + 2 = 0$.

Substituting $y(k) = z^k$, we get $y(k+1) = z^{k+1}$, $y(k+2) = z^{k+2}$, and

$$z^{k+2} + 3z^{k+1} + 2z^k = (z^2 + 3z + 2)z^k = 0$$

whose two nonzero solutions are the roots of $z^2 + 3z + 2 = 0$, which are $z_1 = -2$ and $z_2 = -1$. Clearly, $y(k) = -2^k$ and $y(k) = -1^k$ satisfy the difference equation.

19.9. Find the minimizing value of K for the optimization problem posed in Example 19.10, page 355.

Using the results of Example 19.10, we obtain

$$\int_0^\infty e^2(t)\, dt = \frac{K}{K-1} \int_0^\infty [e^{-t} \sin(\sqrt{K-1}\, t + \tan^{-1}\sqrt{K-1})]^2\, dt$$

Integration yields

$$\int_0^\infty e^2(t)\, dt = \left(\frac{K}{K-1}\right)\left(\frac{e^{-2t}}{4}\right)\left[-1 - \frac{\cos(2\sqrt{K-1}\, t + 2\tan^{-1}\sqrt{K-1} - \tan^{-1}(-\sqrt{K-1}))}{\sqrt{K}}\right]\Bigg|_0^\infty$$

$$= \frac{K}{4(K-1)}\left[1 + \frac{\cos(2\tan^{-1}\sqrt{K-1} - \tan^{-1}(-\sqrt{K-1}))}{K}\right]$$

But

$$\cos(2\tan^{-1}\sqrt{K-1} - \tan^{-1}(-\sqrt{K-1})) = -\cos 3\sqrt{K-1} = 3\cos\sqrt{K-1} - 4\cos^3\sqrt{K-1}$$

$$= \frac{3K-4}{K\sqrt{K}}$$

Therefore

$$\int_0^\infty e^2(t)\, dt = \frac{K}{4(K-1)}\left(1 + \frac{3K-4}{K^2}\right) = \frac{K}{4(K-1)}\frac{(K-1)(K+4)}{K^2} = \frac{K+4}{4K}$$

The first derivative of $\int_0^\infty e^2(t)\, dt$ with respect to K is given by

$$\frac{d}{dK}\left(\frac{K+4}{4K}\right) = -\frac{1}{K^2}$$

Apparently, $\int_0^\infty e^2(t)\, dt$ decreases monotonically as K increases. Therefore the optimal value of K is $K = \infty$, which is of course unrealizable. For this value of K,

$$\lim_{K\to\infty} \int_0^\infty e^2(t)\, dt = \lim_{K\to\infty}\left(\frac{K+4}{4K}\right) = \frac{1}{4}$$

Note also that the natural frequency ω_n of the optimal system is $\omega_n = \sqrt{K} = \infty$ and the damping ratio $\xi = 1/\omega_n = 0$, making it marginally stable. Therefore only a *suboptimal* (less than optimal) system can be practically realized and its design depends on the specific application.

Supplementary Problems

19.10. Determine the phase plane trajectory of the solution of the differential equation

$$\frac{d^2x}{dt^2} + 2\frac{dx}{dt} + 4x = 0$$

19.11. Using Lyapunov theory, find sufficient conditions on a_1 and a_0 which guarantee that the point $x = 0$, $dx/dt = 0$ is stable for the equation

$$\frac{d^2x}{dt^2} + a_1\frac{dx}{dt} + a_0 x = 0$$

19.12. Find three solutions for the difference equation

$$y(k+3) + 2y(k+2) - y(k+1) - 2y(k) = 0$$

Appendix

SOME LAPLACE TRANSFORM PAIRS USEFUL
FOR CONTROL SYSTEMS ANALYSIS

$F(s)$	$f(t) \qquad t > 0$
1	$\delta(t)$ $\qquad$ unit impulse
e^{-Ts}	$\delta(t - T)$ $\qquad$ delayed impulse
$\dfrac{1}{s + a}$	e^{-at}
$\dfrac{1}{(s + a)^n}$	$\dfrac{1}{(n-1)!}\, t^{n-1}\, e^{-at} \qquad n = 1, 2, 3, \ldots$
$\dfrac{1}{(s + a)(s + b)}$	$\dfrac{1}{b - a}(e^{-at} - e^{-bt})$
$\dfrac{s}{(s + a)(s + b)}$	$\dfrac{1}{a - b}(ae^{-at} - be^{-bt})$
$\dfrac{s + z}{(s + a)(s + b)}$	$\dfrac{1}{b - a}[(z - a)e^{-at} - (z - b)e^{-bt}]$
$\dfrac{1}{(s + a)(s + b)(s + c)}$	$\dfrac{e^{-at}}{(b - a)(c - a)} + \dfrac{e^{-bt}}{(c - b)(a - b)} + \dfrac{e^{-ct}}{(a - c)(b - c)}$
$\dfrac{s + z}{(s + a)(s + b)(s + c)}$	$\dfrac{(z - a)e^{-at}}{(b - a)(c - a)} + \dfrac{(z - b)e^{-bt}}{(c - b)(a - b)} + \dfrac{(z - c)e^{-ct}}{(a - c)(b - c)}$
$\dfrac{\omega}{s^2 + \omega^2}$	$\sin \omega t$
$\dfrac{s}{s^2 + \omega^2}$	$\cos \omega t$
$\dfrac{s + z}{s^2 + \omega^2}$	$\sqrt{\dfrac{z^2 + \omega^2}{\omega^2}} \sin(\omega t + \phi) \qquad \phi \equiv \tan^{-1}(\omega/z)$
$\dfrac{s \sin \phi + \omega \cos \phi}{s^2 + \omega^2}$	$\sin(\omega t + \phi)$
$\dfrac{1}{(s + a)^2 + \omega^2}$	$\dfrac{1}{\omega} e^{-at} \sin \omega t$

$F(s)$	$f(t)$ $\qquad t > 0$
$\dfrac{1}{s^2 + 2\zeta\omega_n s + \omega_n^2}$	$\dfrac{1}{\omega_d} e^{-\zeta\omega_n t} \sin \omega_d t \qquad \omega_d \equiv \omega_n\sqrt{1 - \zeta^2}$
$\dfrac{s + a}{(s + a)^2 + \omega^2}$	$e^{-at}\cos \omega t$
$\dfrac{s + z}{(s + a)^2 + \omega^2}$	$\sqrt{\dfrac{(z - a)^2 + \omega^2}{\omega^2}}\, e^{-at}\sin(\omega t + \phi) \qquad \phi \equiv \tan^{-1}\left(\dfrac{\omega}{z - a}\right)$
$\dfrac{1}{s}$	$u(t)$ or 1 $\qquad$ unit step
$\dfrac{1}{s}e^{-Ts}$	$u(t - T)$ $\qquad$ delayed step
$\dfrac{1}{s}(1 - e^{-Ts})$	$u(t) - u(t - T)$ $\qquad$ rectangular pulse
$\dfrac{1}{s(s + a)}$	$\dfrac{1}{a}(1 - e^{-at})$
$\dfrac{1}{s(s + a)(s + b)}$	$\dfrac{1}{ab}\left(1 - \dfrac{be^{-at}}{b - a} + \dfrac{ae^{-bt}}{b - a}\right)$
$\dfrac{s + z}{s(s + a)(s + b)}$	$\dfrac{1}{ab}\left(z - \dfrac{b(z - a)e^{-at}}{b - a} + \dfrac{a(z - b)e^{-bt}}{b - a}\right)$
$\dfrac{1}{s(s^2 + \omega^2)}$	$\dfrac{1}{\omega^2}(1 - \cos \omega t)$
$\dfrac{s + z}{s(s^2 + \omega^2)}$	$\dfrac{z}{\omega^2} - \sqrt{\dfrac{z^2 + \omega^2}{\omega^4}}\cos(\omega t + \phi) \qquad \phi \equiv \tan^{-1}(\omega/z)$
$\dfrac{1}{s(s^2 + 2\zeta\omega_n s + \omega_n^2)}$	$\dfrac{1}{\omega_n^2} - \dfrac{1}{\omega_n\omega_d} e^{-\zeta\omega_n t}\sin(\omega_d t + \phi)$ $\omega_d \equiv \omega_n\sqrt{1 - \zeta^2} \qquad \phi \equiv \cos^{-1}\zeta$
$\dfrac{1}{s(s + a)^2}$	$\dfrac{1}{a^2}(1 - e^{-at} - ate^{-at})$
$\dfrac{s + z}{s(s + a)^2}$	$\dfrac{1}{a^2}[z - ze^{-at} + a(a - z)te^{-at}]$
$\dfrac{1}{s^2}$	t $\qquad$ unit ramp
$\dfrac{1}{s^2(s + a)}$	$\dfrac{1}{a^2}(at - 1 + e^{-at})$
$\dfrac{1}{s^n} \quad n = 1, 2, 3, \ldots$	$\dfrac{t^{n-1}}{(n-1)!} \qquad 0! = 1$

References

1. Zadeh, L. A. and Desoer, C. A., *Linear System Theory: The State Space Approach*, McGraw-Hill, New York, 1963.

2. Hartline, H. K. and Ratliff, F., "Inhibitory Interaction of Receptor Units in the Eye of the Limulus," *J. Gen. Physiol.*, 40:357, 1957.

3. Bliss, J. C. and Macurdy, W. B., "Linear Models for Contrast Phenomena," *J. Optical Soc. America*, 51:1375, 1961.

4. Reichardt, W. and MacGinitie, "On the Theory of Lateral Inhibition," *Kybernetic* (German), 1:155, 1962.

5. McLachlan, N. W., *Complex Variable Theory and Transform Calculus*, Cambridge University Press, Cambridge, 1955.

6. Churchill, R. V., *Complex Variables and Applications*, Second Edition, McGraw-Hill, New York, 1960.

7. Desoer, C. A., "A General Formulation of the Nyquist Criterion," *IEEE Transactions on Circuit Theory*, Vol. CT-12, No. 2, June 1965.

8. Spiegel, M. R., *College Algebra*, Schaum Publishing Co., New York, 1956.

9. Krall, A. M., "An Extension and Proof of the Root-Locus Method," *Journal of the Society for Industrial and Applied Mathematics*, Vol. 9, No. 4, December 1961, pp. 644-653.

10. Elgerd, O. I. and Stephens, W. C., "Effect of Closed-Loop Transfer Function Pole and Zero Locations on the Transient Response of Linear Control Systems," *AIEE Transactions*, Vol. 78, Part II, May 1959, pp. 121-127.

Bibliography

Aizerman, M. A., *Theory of Automatic Control*, Addison-Wesley, Reading, Massachusetts, 1963.

Bode, H. W., *Network Analysis and Feedback Amplifier Design*, Van Nostrand, Princeton, New Jersey, 1945.

Brown, G. S. and Campbell, D. P., *Principles of Servomechanisms*, John Wiley, New York, 1948.

Bruns, R. A. and Saunders, R. M., *Analysis of Feedback Control Systems*, McGraw-Hill, New York, 1955.

Chestnut H. and Mayer, R. W., *Servomechanisms and Regulating System Design*, Volume I, Second Edition, John Wiley, New York, 1959.

Clark, R. N., *Introduction to Automatic Control Systems*, John Wiley, New York, 1962.

D'Azzo, J. J. and Houpis, C. H., *Feedback Control System Analysis and Synthesis*, Second Edition, McGraw-Hill, New York, 1966.

Del Toro, V. and Parker, S. R., *Principles of Control Systems Engineering*, McGraw-Hill, New York, 1960.

Eckman, D. P., *Automatic Process Control*, John Wiley, New York, 1958.

Evans, W. R., *Control-System Dynamics*, McGraw-Hill, New York, 1954.

Gille, J. C., Pilegrin, M. J. and Decaulne, P., *Feedback Control Systems Analysis, Synthesis, and Design*, McGraw-Hill, New York, 1959.

Horowitz, I. M., *Synthesis of Feedback Systems*, Academic Press, New York, 1963.

James, H. M., Nichols, N. B. and Phillips, R. S., *Theory of Servomechanisms*, McGraw-Hill, New York, 1947.

Kuo, B. C., *Automatic Control Systems*, Prentice-Hall, Englewood Cliffs, New Jersey, 1962.

Lago, G. and Benningfield, L. M., *Control System Theory, Feedback Engineering*, Ronald Press, New York, 1962.

Langill, A. W., *Automatic Control Systems Engineering*, Volume I, Prentice-Hall, Englewood Cliffs, New Jersey, 1965.

Murphy, G. J., *Basic Automatic Control Theory*, Second Edition, Van Nostrand, Princeton, New Jersey, 1966.

Newton, G. C., Gould, L. A. and Kaiser, J. F., *Analytical Design of Linear Feedback Controls*, John Wiley, New York, 1957.

Savant, C. J., *Control System Design*, Second Edition, McGraw-Hill, New York, 1964.

Skinners, S. M., *Control System Design*, John Wiley, New York, 1964.

Smith, O. J. M., *Feedback Control Systems*, McGraw-Hill, New York, 1958.

Takahashi, T., *Mathematics of Automatic Control*, Holt, Rinehart and Winston, New York, 1966.

Thaler, G. J. and Brown, R. G., *Analysis and Design of Feedback Control Systems*, Second Edition, McGraw-Hill, New York, 1960.

INDEX